VOLUME ONE HUNDRED AND THIRTY SEVEN

CURRENT TOPICS IN DEVELOPMENTAL BIOLOGY

Gradients and Tissue Patterning

CURRENT TOPICS IN DEVELOPMENTAL BIOLOGY

"A meeting-ground for critical review and discussion of developmental processes"
A.A. Moscona and Alberto Monroy (Volume 1, 1966)

VOLUME ONE HUNDRED AND THIRTY SEVEN

Current Topics in DEVELOPMENTAL BIOLOGY

Gradients and Tissue Patterning

Edited by

STEPHEN SMALL
Department of Biology, New York University, New York, NY, United States

JAMES BRISCOE
The Francis Crick Institute, London, United Kingdom

ELSEVIER

ACADEMIC PRESS
An imprint of Elsevier

Academic Press is an imprint of Elsevier
50 Hampshire Street, 5th Floor, Cambridge, MA 02139, United States
525 B Street, Suite 1650, San Diego, CA 92101, United States
The Boulevard, Langford Lane, Kidlington, Oxford OX5 1GB, United Kingdom
125 London Wall, London, EC2Y 5AS, United Kingdom

First edition 2020

ISBN: 978-0-12-812790-2
ISSN: 0070-2153

For information on all Academic Press publications
visit our website at https://www.elsevier.com/books-and-journals

Publisher: Zoe Kruze
Editorial Project Manager: Shellie Bryant
Production Project Manager: Denny Mansingh
Cover Designer: Greg Harris

Typeset by SPi Global, India

Contents

Contributors

Prasad U. Bandodkar
Department of Chemical and Biomolecular Engineering, North Carolina State University, Raleigh, NC, United States

Augusto Berrocal
Department of Molecular and Cell Biology, University of California at Berkeley, Berkeley, CA, United States

Kenneth D. Birnbaum
New York University, The Department of Biology, The Center for Genomics and Systems Biology, New York, NY, United States

Elena Camacho-Aguilar
Department of Biosciences, Rice University, Houston, TX, United States

Fernando Casares
CABD (Andalusian Centre for Developmental Biology), GEM-DMC2 Maria de Maeztu Unit, CSIC-Universidad Pablo de Olavide-Junta de Andalucia, Campus UPO, Seville, Spain

Nathalie Dostatni
Institut Curie, PSL Research University, CNRS, Sorbonne Université, Nuclear Dynamics, Paris, France

Andrew D. Economou
Developmental Signalling Laboratory, The Francis Crick Institute, London, United Kingdom

Hernan G. Garcia
Department of Molecular and Cell Biology; Department of Physics; Biophysics Graduate Group; Quantitative Biosciences-QB3, University of California at Berkeley, Berkeley, CA, United States

Diana García-Morales
CABD (Andalusian Centre for Developmental Biology), GEM-DMC2 Maria de Maeztu Unit, CSIC-Universidad Pablo de Olavide-Junta de Andalucia, Campus UPO, Seville, Spain

Nathan W. Goehring
The Francis Crick Institute; Institute for the Physics of Living Systems; MRC Laboratory for Molecular Cell Biology, University College London, London, United Kingdom

Bruno Guillotin
New York University, The Department of Biology, The Center for Genomics and Systems Biology, New York, NY, United States

Anna-Katerina Hadjantonakis
Developmental Biology Program, Sloan Kettering Institute, Memorial Sloan Kettering Cancer Center, New York, NY, United States

Caroline S. Hill
Developmental Signalling Laboratory, The Francis Crick Institute, London, United Kingdom

Kristine Hill
Center for Plant Molecular Biology, University of Tübingen, Tübingen, Germany

Anqi Huang
Mechanobiology Institute, National University of Singapore, Singapore, Singapore

Lars Hubatsch
Max Planck Institute for the Physics of Complex Systems, Dresden, Germany

Johannes Jaeger
Complexity Science Hub (CSH); Department of Molecular Evolution & Development, University of Vienna, Vienna, Austria

Yang Joon Kim
Biophysics Graduate Group, University of California at Berkeley, Berkeley, CA, United States

Simon Klesen
Center for Plant Molecular Biology, University of Tübingen, Tübingen, Germany

Gabriella Martini
Department of Molecular and Cell Biology, University of California at Berkeley, Berkeley, CA, United States

David G. Míguez
Dept. Física de la Materia Condensada, Instituto de Física de la Materia Condensada (IFIMAC), Facultad de Ciencias, and Centro de Biología Molecular Severo Ochoa (CBMSO, CSIC-UAM), Universidad Autónoma de Madrid, Madrid, Spain

Sophie M. Morgani
Developmental Biology Program, Sloan Kettering Institute, Memorial Sloan Kettering Cancer Center, New York, NY, United States; Wellcome Trust-Medical Research Council Cambridge Stem Cell Institute, University of Cambridge, Jeffrey Cheah Biomedical Centre Cambridge Biomedical Campus, Cambridge, United Kingdom

Patrick Müller
Systems Biology of Development Group, Friedrich Miescher Laboratory of the Max Planck Society; Modeling Tumorigenesis Group, Translational Oncology Division, Eberhard Karls University Tübingen, Tübingen, Germany

Gregory T. Reeves
Genetics Program; Department of Chemical and Biomolecular Engineering, North Carolina State University, Raleigh, NC, United States

Katherine W. Rogers
Systems Biology of Development Group, Friedrich Miescher Laboratory of the Max Planck Society, Tübingen, Germany

Timothy E. Saunders
Mechanobiology Institute; Department of Biological Sciences, National University of Singapore; Institute of Molecular and Cell Biology, A*Star, Singapore, Singapore

Allison E. Schloop
Genetics Program, North Carolina State University, Raleigh, NC, United States

Stanislav Y. Shvartsman
The Lewis-Sigler Institute for Integrative Genomics; Department of Molecular Biology; Department of Chemical and Biological Engineering, Princeton University, Princeton, NJ; Center for Computational Biology, Flatiron Institute, Simons Foundation, New York, NY, United States

Celia M. Smits
The Lewis-Sigler Institute for Integrative Genomics; Department of Molecular Biology, Princeton University, Princeton, NJ, United States

Stefan Thor
School of Biomedical Sciences, University of Queensland, Saint Lucia, QLD, Australia

Marja C.P. Timmermans
Center for Plant Molecular Biology, University of Tübingen, Tübingen, Germany

Huy Tran
Institut Curie, PSL Research University, CNRS, Sorbonne Université, Nuclear Dynamics; Ecole Normale Supérieure, PSL Research University, CNRS, Sorbonne Université, Laboratoire de Physique, Paris, France

Berta Verd
Department of Genetics, University of Cambridge, Cambridge, United Kingdom

Aleksandra M. Walczak
Ecole Normale Supérieure, PSL Research University, CNRS, Sorbonne Université, Laboratoire de Physique, Paris, France

Aryeh Warmflash
Department of Biosciences; Department of Bioengineering, Rice University, Houston, TX, United States

Behzad Yaghmaeian Salmani
Department of Cell and Molecular Biology, Karolinska Institute, Stockholm, Sweden

Jiaxi Zhao
Department of Physics, University of California at Berkeley, Berkeley, CA, United States

Preface

At the heart of developmental biology is understanding how multicellular embryos generate the different cell types that comprise their tissues in the correct numbers, and organize them in space and time. Central to this idea has been the morphogen concept, in which diffusible signaling molecules are proposed to coordinate cell fate specification and tissue formation using concentration-dependent mechanisms. This concept has captured the attention of experimental and theoretical biologists for decades, in part, because of its elegance and simplicity. The molecular genetics revolution in the late 1980s and early 1990s began to reveal the identities of several molecules that displayed the hallmarks of morphogens. This led to rigorous molecular and genetic tests of morphogen action, confirming predictions from the past, but also identifying unanticipated complexities. Our knowledge of morphogen activity has continued to grow over the subsequent decades leading to a deeper understanding of how morphogens function and to revisions in the concept itself. Here we have gathered a series of chapters, from the leaders in the field, that brings us up-to-date. The authors discuss molecular details of morphogen patterning in specific tissues, identify outstanding questions and highlight new approaches to tackle these, and reflect on principles and ideas that emerge from the studies.

As with many fields, new technologies and methods have had a major impact on our understanding. The latest techniques for visualizing and inactivating morphogen molecules in cells and embryos in real time are described in Garcia et al. and Rogers and Müeller. These are being used extensively to directly measure rates of morphogen production, movement, and degradation at unprecedented resolution. Other technical advances enable researchers to "count" individual mRNAs produced by target genes that respond to morphogens (Garcia et al.). Together these developments in quantitative imaging have allowed ever more refined observations of how morphogen gradients are formed, and how cells respond to morphogens to differentially regulate target gene expression and cellular responses.

Most of studies reviewed here focus on a number of specific developmental questions: first, what is the molecular nature of the molecules that function in concentration-dependent mechanisms, and how are morphogen gradients reliably matched to the size of a growing tissue or a differentiating cell? Traditionally, the best characterized morphogens were extracellular signaling molecules that pattern fields of adjacent cells and transcription factors and kinases that form gradients in the early syncytial embryo of Drosophila. These morphogens pattern adjacent cells or nuclei, with a range of two to several dozen cell/nucleus diameters and are the focus of several of the chapters here (Economou and Hill; Morgani and Hadjantonakis; Camacho-Aguilar and Warmflash). But the morphogen concept and the principles that have emerged from studying morphogens is not limited to conventional developmental systems. Three chapters probe mechanisms involving other types of molecules that operate at different scales. At the small end of the spectrum (patterning inside a single cell), Hubatsch and Goehring describe protein phosphorylation mechanisms that establish asymmetric distributions of the RNA-binding protein Mex-5 within the one-cell *C. elegans* embryo. A second review identifies microRNAs as asymmetrically localized molecules that are critical for leaf patterning (Klesen et al.). At the other end of the spectrum, the plant hormone auxin, which flows through the vasculature of stems and roots, establishes polarities that affect the activities of families of patterning transcription factors (Guillotin and Birnbaum).

A second question addressed by many of the authors can by summarized as "what is positional information": what are the properties of morphogen gradients that are read by the responding cells and nuclei? Jaeger and Verd propose two ways to define positional information and argue that when both definitions are used to direct experimental and computational studies, the result is a deeper understanding of how *in vivo* tissue patterning is ultimately achieved. A number of reviews in this volume (Schloop et al.; Huang and Saunders; Smits and Shvartsman; Miguez et al.; Economou and Hill; Morgani and Hadjantonakis) describe experimental and computational approaches to determine what genes and quantitative parameters are involved in shaping functional gradients. The details of each system are unique, but it is clear that many mechanisms, including combinatorial interactions between multiple signals and negative feedback, control the composition and function of the positional information that patterns embryos.

Finally, how do cells control differential gene expression to produce proportioned and precise domains target gene expression and specific cell fates in response to morphogen signaling? Salmani and Thor review how differential mitosis rates pattern the mouse central nervous system along the AP axis. At the level of gene expression, Tran et al. outline the challenges in accurately modeling a simple on–off boundary of gene expression in response to a monotonic gradient of the transcription factor Bicoid in the Drosophila embryo. In another attempt to simplify the question of how patterns of gene expression are formed in response to a gradient, Camacho-Aguilar and Warmflash describe recent attempts to use cultured embryonic stem cells to assay gene expression patterns in two-dimensional cultured systems.

Perhaps the most consistent take home message from all the studies reviewed here is that the original morphogen concept as a readout of static concentration thresholds is insufficient to describe or model the complexities of patterning observed in developing embryos. In particular, gradients are not static. They constantly change with time, and it absolutely clear that any quantitative understanding of embryonic patterning is impossible without considering time as a critical parameter. The chapters in this volume show how integrating time into system models deepens our understanding of the complexities and dynamics of morphogen-mediated tissue patterning. These more complex mechanisms argue against the simple ideas of static monotonic gradients and straightforward threshold responses, but reveal a picture of embryonic patterning that is no less elegant and inspiring. Happy reading!

Stephen Small and James Briscoe

CHAPTER ONE

Lighting up the central dogma for predictive developmental biology

Hernan G. Garcia[a,b,c,d,*], Augusto Berrocal[a], Yang Joon Kim[c], Gabriella Martini[a], Jiaxi Zhao[b]

[a]Department of Molecular and Cell Biology, University of California at Berkeley, Berkeley, CA, United States
[b]Department of Physics, University of California at Berkeley, Berkeley, CA, United States
[c]Biophysics Graduate Group, University of California at Berkeley, Berkeley, CA, United States
[d]Quantitative Biosciences-QB3, University of California at Berkeley, Berkeley, CA, United States
[*]Corresponding author: e-mail address: hggarcia@berkeley.edu

Contents

Abstract

Although the last 30years have witnessed the mapping of the wiring diagrams of the gene regulatory networks that dictate cell fate and animal body plans, specific understanding building on such network diagrams that shows how DNA regulatory regions control gene expression lags far behind. These networks have yet to yield the predictive power necessary to, for example, calculate how the concentration dynamics of input transcription factors and DNA regulatory sequence prescribes output patterns of gene expression that, in turn, determine body plans themselves. Here, we argue that reaching a predictive understanding of developmental decision-making calls for an interplay between theory and experiment aimed at revealing how the regulation of the processes of the central dogma dictate network connections and how network topology guides cells toward their ultimate developmental fate. To make this possible, it is crucial to break free from the snapshot-based understanding of embryonic development

Current Topics in Developmental Biology, Volume 137
ISSN 0070-2153
https://doi.org/10.1016/bs.ctdb.2019.10.010

facilitated by fixed-tissue approaches and embrace new technologies that capture the dynamics of developmental decision-making at the single cell level, in living embryos.

1. Introduction

A ubiquitous mystery in nature is how a single cell develops into a multicellular organism. One of the great achievements of genetics and new genome-wide technologies over the last few decades has been the identification of the regulatory molecules that underlie developmental programs. This work has revealed that animal body plans are determined by the action of activators and repressors connected in complex gene regulatory networks. One of the best-studied regulatory networks drives segmentation of the early embryo of the fruit fly *Drosophila melanogaster*. As shown in Fig. 1A, decades of concerted effort have uncovered the identities of the regulatory molecules that determine fly body segments as well as the network connections between these molecules (reviewed in Carroll, Grenier, & Weatherbee, 2001; Davidson, 2006; Peter & Davidson, 2015). However, the amassed data about regulatory proteins and network connections has been mostly descriptive and has not been accompanied by parallel successes in predictively understanding cellular decision-making in developing embryos. To be concrete, is it possible to predict developmental phenotypes from network diagrams? Clearly, if we were to change the names of the genes in the fruit fly segmentation network (Fig. 1B), the network connections—the topology—would be insufficient to predict that this particular network would result in a fly.

The central thesis of this article is that achieving predictive understanding of developmental decision-making requires a two-pronged approach. First, each arrow in these networks must be endowed with molecular and quantitative information that makes it possible to predict how the genome dictates *developmental* input-output functions: the functions relating output protein levels of a gene to the concentrations of its input transcription factors (Fig. 2A). That is, how do the number, placement, and affinity of transcription-factor binding sites within enhancers establish the relation between input transcription-factor concentration and output transcription? How is mRNA processed as it forms a cytoplasmic pattern? How is this mRNA pattern translated into a protein pattern that can be subject to further post-translational regulation and that can ultimately feed back into the

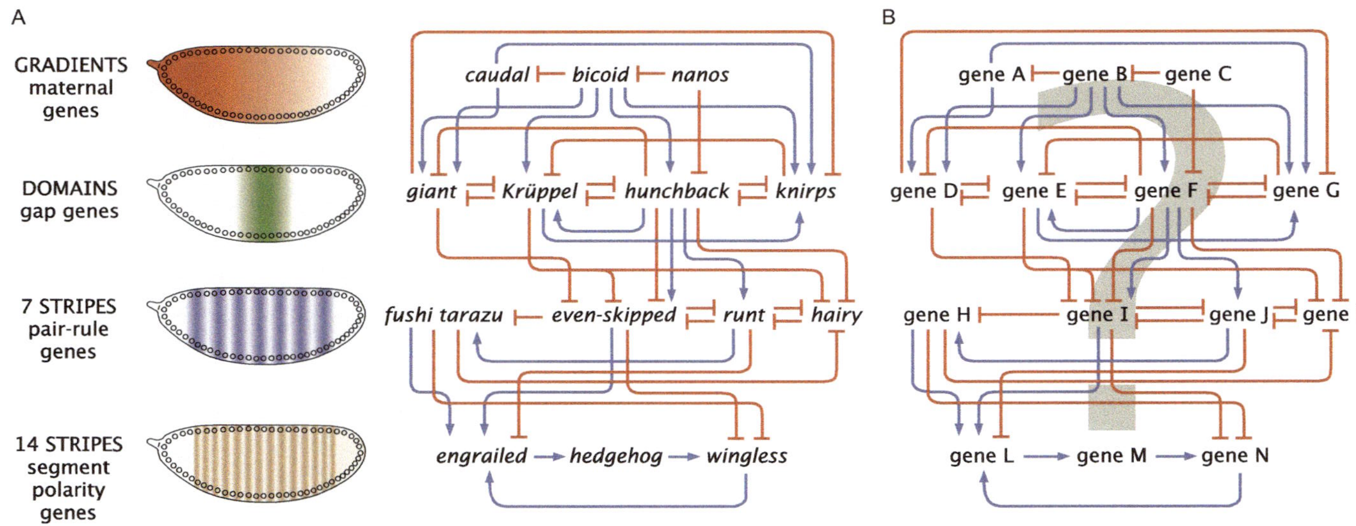

Fig. 1 Gene regulatory networks driving development. (A) Current state of the art in mapping the network that drives segmentation in the fruit fly *Drosophila melanogaster*. (B) Close examination of a network with the same topology, but with different genetic actors, reveals that this description is insufficient to predict that this network will lead to the emergence of a segmented embryo. *Adapted from Carroll, S.B., Grenier, J.K., & Weatherbee, S.D. (2001). From DNA to diversity: Molecular genetics and the evolution of animal design (Malden, Mass: Blackwell Science); Edgar, B.A., Odell, G.M., & Schubiger, G. (1989). A genetic switch, based on negative regulation, sharpens stripes in Drosophila embryos.* Developmental Genetics 10, *124–142; Jaeger, J. (2011). The gap gene network.* Cellular and Molecular Life Sciences 68, *243–274; von Dassow, G., Meir, E., Munro, E.M., & Odell, G.M. (2000). The segment polarity network is a robust developmental module.* Nature, 406, *188–192.*

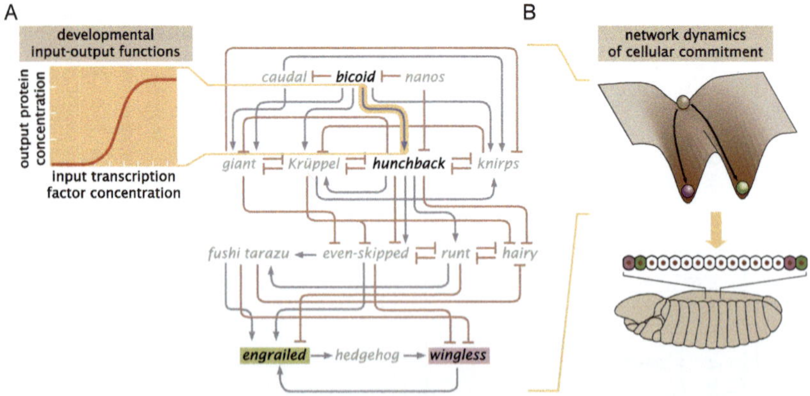

Fig. 2 A two-pronged approach to achieving predictive understanding of developmental decision-making. (A) A quantitative and predictive description of gene regulatory networks in development demands uncovering how the regulation of the processes of the central dogma prescribe developmental input-output functions represented by network arrows. (B) Arrows define the network topology which drives cells within an embryo into discrete and stable developmental states. In this example, the topology defines two distinct developmental trajectories that lead to cells expressing *engrailed or wingless,* and to the adoption of distinct fates that establish body segments in the fly.

network? Second, we must reveal how these network connections conspire together to drive cells into discrete and stable gene expression states that in turn commit these cells to their ultimate developmental fates (Fig. 2B).

In order to achieve this hefty goal of predictive developmental biology, here we argue that developmental biologists urgently require new theoretical models that make precise and falsifiable predictions. Further, we posit that, in order to test the predictions from these models, developmental biology needs to break free from the static view of development shaped by widespread fixed-tissue techniques and establish new technologies that reveal the regulation of the processes of the central dogma at the single-cell level as developmental programs are deployed in real time. Here we focus on the specific case study of the segmentation of the early embryo of the fruit fly. This review is therefore not encyclopedic, and falls short of an exhaustive scholarly survey of the many exciting contributions to quantitative developmental biology in the recent literature. Rather, by focusing on a representative case study, we propose a concrete framework for establishing a quantitative and predictive developmental biology writ large that can be adopted by developmental biologists and biophysicists working on any organism.

2. Turning the fruit fly *Drosophila melanogaster* into a substrate for predictive developmental biology

Drosophila is already a workhorse for developmental biology and genetics research. The fruit fly is also an ideal substrate for realizing predictive developmental biology. The fly network in Fig. 1A shows that many of its fundamental parts (regulatory molecules and connections) have already been identified. The picture that emerges is one of a network that is simple enough that its regulatory interactions could be enumerated, yet intricate enough that it captures the essence of more "complex" developmental processes. As a result, even though some regulatory interactions might remain unmapped, the fly segmentation network offers a unique opportunity to uncover the fundamental and quantitative rules behind developmental decision-making.

What is required to quantitatively predict and control developmental outcomes from regulatory parameters in the fruit fly—and ultimately any organism? The bacterial *lac* operon showcases how to build a predictive understanding of cellular decision-making. Over the last 15 years, theoretical models of this operon, such as those shown in Fig. 3A, have precisely predicted the mean rate of transcription and its cell-to-cell variability as a function of regulatory architecture invoking only a handful of free parameters (Fig. 3B, reviewed in Phillips et al., 2019).

Clearly, flies differ from bacteria; even the architecture of the transcriptional process is different. While bacterial transcription only requires a handful of molecules to be present at the promoter, eukaryotic transcriptional initiation requires the assembly of vast macromolecular complexes such as the preinitiation complex, plus regulatory steps to evict nucleosomes from the DNA to render it accessible to transcription factors (reviewed by Fuda, Ardehali, & Lis, 2009). Similar complexity exists in every step of the central dogma in eukaryotes, from splicing, to transcriptional termination, to translation, to post-translational modifications (Alberts, 2015).

However, the challenges of quantitatively dissecting development go beyond the combinatorial complexity stemming from the numerous molecular machines involved in the processes of the central dogma in eukaryotes. Biology textbooks are dominated by snapshots of static gene-expression patterns. Thus, a large part of our understanding of developmental biology rests on the *assumption* that developmental dynamics can be easily inferred from these static pictures. However, development is a highly dynamic

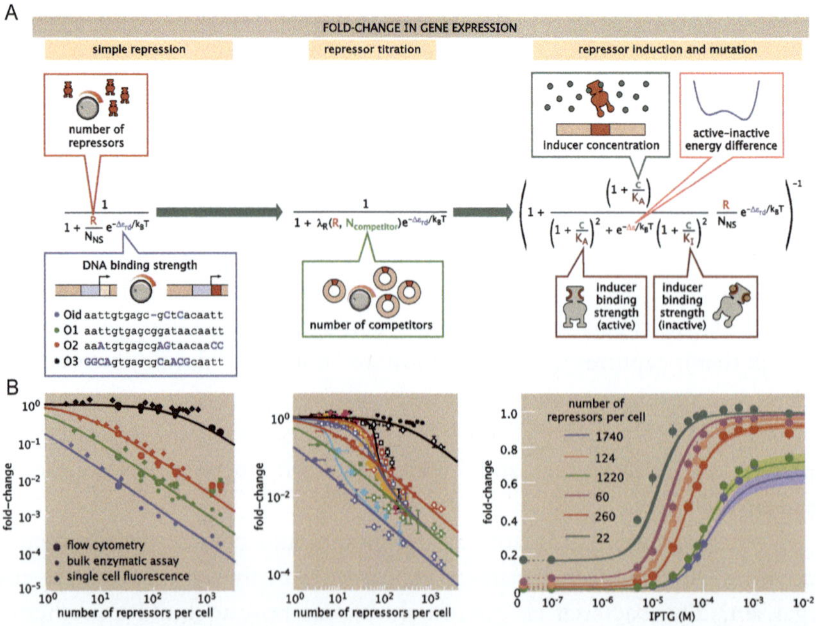

Fig. 3 Unraveling regulation of the *lac* operon through theory and experiment. (A) Examples of theoretical predictions for the fold-change in gene expression of simple repression by Lac repressor (defined as the ratio of gene expression level in the presence of repressor to the gene expression level in in the absence of repressor) for increasingly complex regulatory situations. Each step relies on the parameters learned in the previous iteration. (B) Parameter-free or one-parameter fits to the data demonstrate that simple repression is predictive. *Adapted from* Phillips, R., Belliveau, N.M., Chure, G., Garcia, H.G., Razo-Mejia, M., & Scholes C. (2019). Figure 1 theory meets figure 2 experiments in the study of gene expression, Annual Review of Biophysics, 48, 121–163.

process: choreographed gene expression patterns are rapidly deployed in space and time, and can exist for periods of time as short as 15 min (Bothma et al., 2014). Thus, predicting developmental biology of *Drosophila* not only calls for theoretical frameworks with predictive power, but also demands experimental technologies that reveal how the processes of the central dogma are regulated in real time as single cells commit to their fates and development unfolds.

3. Lighting up the central dogma to assign quantitative and predictive meaning to arrows

Arrows in network diagrams encode *developmental* input-output functions that predict how the concentrations of input transcription factors determine output protein levels (Fig. 4A). These input-output functions, which are the fundamental unit of any quantitative description of developmental programs, arise from the regulation of each step of the central dogma from transcriptional initiation, to mRNA processing, to translation and post-translational modifications (Fig. 4B). There is a specific input-output function for each specific step of the central dogma.

Over the last 40 years, a plethora of theoretical models have sought to predict *transcriptional* input-output functions: how the concentration of input activators and repressors and the arrangement of their binding sites on regulatory DNA dictate the output rate of transcription (Fig. 4C) (Ackers, Johnson, & Shea, 1982; Bintu et al., 2005a, 2005b; Buchler, Gerland, & Hwa, 2003; Estrada, Wong, DePace, & Gunawardena, 2016; Fakhouri et al., 2010; Gregor, Tank, Wieschaus, & Bialek, 2007; Hammar et al., 2014; He, Samee, Blatti, & Sinha, 2010; Kanodia et al., 2012; Li, Cesbron, Oehler, Brunner, & Hofer, 2018; Samee et al., 2015; Sayal, Dresch, Pushel, Taylor, & Arnosti, 2016; Scholes, DePace, & Sanchez, 2017; Segal, Raveh-Sadka, Schroeder, Unnerstall, & Gaul, 2008; Sherman & Cohen, 2012; Vilar & Leibler, 2003; von Hippel, Revzin, Gross, & Wang, 1974). However, despite this wide repertoire of models, only recently did the technology necessary to directly measure transcriptional input-output functions in development become available.

3.1 Lighting up transcriptional dynamics

For years, the state of the art for directly measuring transcriptional activity in developing embryos consisted of fixed-tissue techniques such as *in situ* hybridization, fluorescence *in situ* hybridization (FISH) or single-molecule FISH (Lawrence, Singer, & Marselle, 1989; Raj, Peskin, Tranchina, Vargas, & Tyagi, 2006; Singer & Ward, 1982; Tautz & Pfeifle, 1989). In these techniques, mRNA is labeled such that sites of nascent transcript formation appear as puncta in each nucleus (O'Farrell, Edgar, Lakich, & Lehner, 1989; Zenklusen, Larson, & Singer, 2008). The signal, often fluorescence, emitted by these puncta reports on the number of mRNA molecules being actively transcribed. These technologies have been applied to many biological questions, such as the molecular nature of transcription in development and how mitosis inhibits transcription (Boettiger & Levine, 2013; Bothma, Magliocco, & Levine, 2011; Little, Tikhonov, & Gregor, 2013; Shermoen & O'Farrell, 1991; Zoller, Little, & Gregor, 2018).

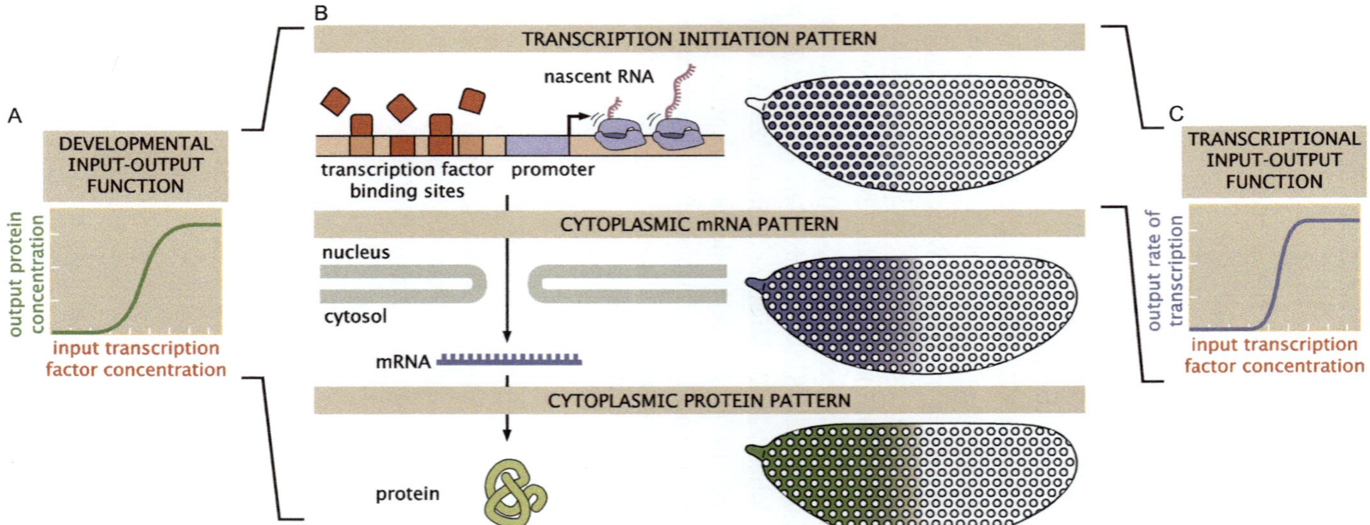

Fig. 4 Regulation along the central dogma. (A) Arrows in gene regulatory networks encode developmental input-output functions that predict output protein concentration as a function of input transcription-factor concentration. (B) Developmental input-output functions are the result of the regulation of multiple steps of the central dogma. (C) Transcriptional input-output functions predict output transcriptional activity as a function of input transcription-factor concentration.

However, the reliance on fixed tissue in these techniques casts doubt on their suitability for measuring dynamical developmental input-output functions; using dead, fixed embryos yields stop-motion "movies" for which each frame requires a new embryo (Dubuis, Samanta, & Gregor, 2013; Poustelnikova, Pisarev, Blagov, Samsonova, & Reinitz, 2004). To measure the output transcription rate of a cell (Fig. 4C), the expression status of a single cell needs to be measured for at least two time points. But since fixed-tissue techniques necessarily only access one time point, they cannot enable the dialog between theory and experiment advocated for in this article. As a result, previous works were constrained to assuming that transcription is in steady-state such that transcriptional dynamics remain largely unaltered during nuclear cycles (Fakhouri et al., 2010; Little et al., 2013; Park et al., 2019; Sayal et al., 2016; Xu, Sepulveda, Figard, Sokac, & Golding, 2015; Xu, Skinner, Sokac, & Golding, 2016; Zoller et al., 2018).

Recently, this critical limitation was overcome by adapting the MS2 system (Bertrand et al., 1998) to measure the instantaneous content of nascent RNA of a specific gene in single cells of a living, developing embryo (Garcia et al., 2013; Lucas et al., 2013). This MS2 system, and its sister PP7 system (Chao, Patskovsky, Almo, & Singer, 2008; Larson, Zenklusen, Wu, Chao, & Singer, 2011), integrate a repetitive DNA sequence into a gene's untranslated region. Upon transcription, the MS2 sequence folds and forms a loop. These mRNA loops are bound by a maternally provided mRNA binding protein fused to GFP (Fig. 5A). As a result, sites of nascent transcript formation become visible as fluorescent puncta due to the localization of active RNA polymerase II molecules to the tagged gene; these puncta are easily visualized using laser scanning two-photon or confocal microscopy, or using light-sheet microscopy (Fig. 5B). Using single-molecule mRNA FISH, the fluorescence value corresponding to each punctum can be converted into an absolute number of polymerase II molecules actively transcribing the gene as a function of time and position along the embryo (Fig. 5C) (Garcia et al., 2013). The result is the first-ever dynamical measurement of transcription in single cells within a living multicellular organism (Bothma & Levine, 2013).

This new ability to interrogate transcriptional activity in real time during development has unveiled new insights into the fundamental and dynamical nature of developmental processes. Here we showcase a few examples:

i. The gene-expression patterns that dictate cellular fate commitment are much more short-lived than previously thought (Bothma et al., 2014; Lammers et al., 2018).

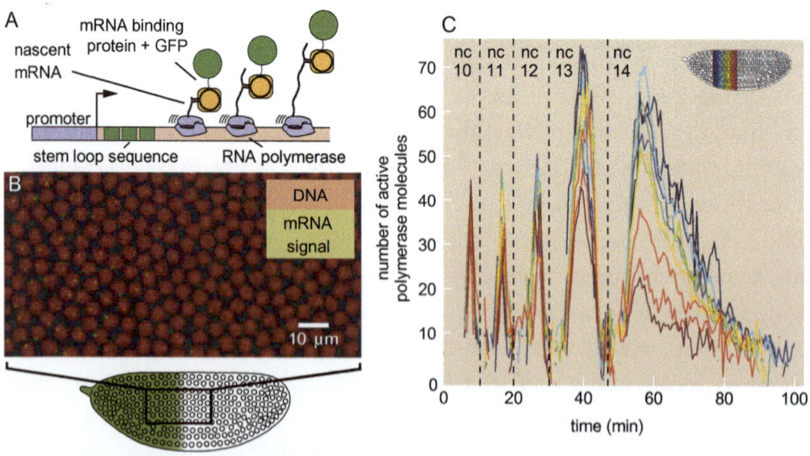

Fig. 5 Accessing transcriptional dynamics in live fly embryos. (A) Repeats of the MS2 sequence are added to a gene that, when transcribed, folds into a stem loop that is recognized by an mRNA binding protein fused to GFP; fluorescence is proportional to transcriptional activity. (B) Typical field of view showing sites of transcription in single nuclei for a reporter of the step-like *hunchback* expression pattern. (C) Number of actively transcribing RNA polymerase II molecules as a function of time for different positions along the embryo's axis. (nc: nuclear cycle.). *Adapted from Garcia, H.G., Tikhonov, M., Lin, A., & Gregor, T. (2013). Quantitative imaging of transcription in living Drosophila embryos links polymerase activity to patterning.* Current Biology, 23, *2140–2145.*

ii. The processes by which enhancers coordinate their activities with each other and with promoters can be directly visualized (Bothma et al., 2015; Chen et al., 2018; El-Sherif & Levine, 2016; Lim, Heist, Levine, & Fukaya, 2018; Scholes, Biette, Harden, & DePace, 2019).

iii. Transcription-factor concentration is read out to generate precise output patterns (Ferraro et al., 2016; Lim, Levine, & Yamazaki, 2017; Tran et al., 2018).

iv. Mitosis and pioneer transcription factors dictate the transcriptional dynamics of embryos (Dufourt et al., 2018; Esposito, Lim, Guessous, Falahati, & Levine, 2016; Yamada et al., 2019).

v. The real-time elongation rate of RNA polymerase II can be captured and quantified (Fukaya, Lim, & Levine, 2017; Garcia et al., 2013).

All of these insights afforded by MS2 will make it possible to rewrite biology textbooks by capturing the processes of cellular commitment in real time and by dramatically overturning or significantly complementing our previous knowledge stemming from fixed-embryos techniques.

Perhaps one of the most captivating outcomes of the tagging of early fly developmental genes with MS2 and PP7 has been the confirmation of the long-suspected presence of transcriptional bursts in development (Little et al., 2013; Pare et al., 2009) *via* their real-time visualization (Bothma et al., 2014). As shown in Fig. 6A and B, the existence of these bursts indicates that the rate of transcriptional initiation is non-zero only during transient, but stochastic, periods of time (Bothma et al., 2014; Chubb, Trcek, Shenoy, & Singer, 2006; Golding, Paulsson, Zawilski, & Cox, 2005; Yunger, Rosenfeld, Garini, & Shav-Tal, 2010). These live-imaging techniques have made it possible to determine the ubiquity of transcriptional bursting in development and to start revealing their molecular control mechanisms (Berrocal, Lammers, Garcia, & Eisen, 2018; Bothma et al., 2014; Desponds et al., 2016; Falo-Sanjuan, Lammers, Garcia, & Bray, 2019; Fukaya, Lim, & Levine, 2016; Lammers et al., 2018; Lim, Fukaya, Heist, & Levine, 2018).

Although MS2 and PP7 made it possible to directly confirm the existence of transcriptional bursts in development, their precise quantitative characterization presents challenges: note that neither MS2 nor PP7 actually report on the rate of transcriptional initiation. An actively transcribing RNA polymerase molecule remains loaded on the DNA, and contributes to the overall fluorescence signal, until transcription terminates (Fig. 6A and B). As a result, the signal from MS2 and PP7 reports on the integrated transcriptional activity over a time window corresponding to the dwell time of RNA polymerase on the gene (given by the time to elongate the mRNA and to terminate transcription). Thus, output fluorescence is not directly related to the instantaneous promoter state and is instead the *convolution* of the promoter activity over a time window (Fig. 6B).

Recently, this fundamental limitation has been circumvented by various strategies. First, by focusing on promoter-enhancer interactions that rarely produce transcriptional bursts, the manual identification and measurement of the properties of these bursts was achieved (Fukaya et al., 2016). Second, by examining the autocorrelation of the output fluorescence signal (Coulon et al., 2014; Coulon & Larson, 2016; Larson et al., 2011), average bursting dynamics (such as the burst size, frequency, and amplitude) were revealed for a wider range of promoter dynamics than those accessible by manual analysis (Desponds et al., 2016). Finally, many computational tools have been recently developed to infer the most likely bursting state of a promoter in a single cell (Bronstein, Zechner, & Koeppl, 2015; Corrigan, Tunnacliffe, Cannon, & Chubb, 2016; Featherstone et al., 2016; Hey et al., 2015; Lammers et al., 2018; Molina et al., 2013; Suter et al., 2011; Zechner, Unger, Pelet, Peter, & Koeppl, 2014; Zoller, Nicolas,

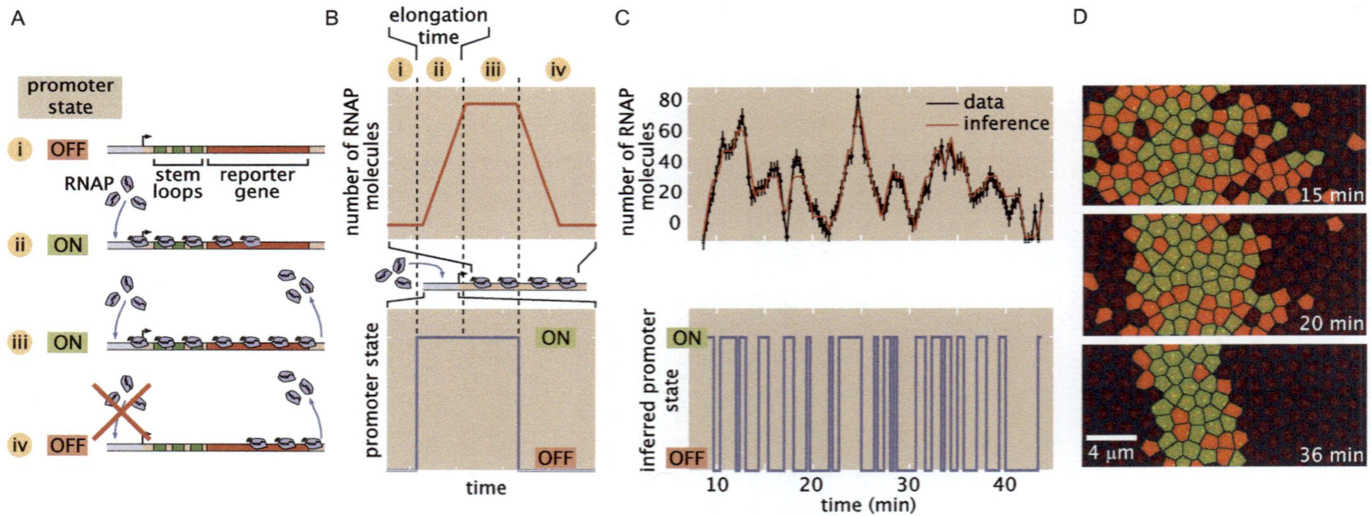

Fig. 6 Relation between MS2 fluorescence and instantaneous transcriptional activity. (A) Loading of RNA polymerase, and transcript elongation and termination as a gene is transiently turned on. (B) These discrete changes in promoter state are convolved with the elongation and termination times, resulting in a trapezoidal-like modulation of the number of RNA polymerase molecules on the reporter, as indicated by MS2 fluorescence. (C) Inference of promoter state from MS2 data using Hidden Markov models. (D) Inference of promoter states for cells expressing a transcriptional reporter of stripe 2 of the *even-skipped* gene. *Panels (C and D) Adapted from Lammers, N.C., Galstyan, V., Reimer, A., Medin, S.A., Wiggins, C.H., & Garcia, H.G. (2018). Multimodal transcriptional control of pattern formation in embryonic development.* bioRxiv, 335919.

Molina, & Naef, 2015). For example, techniques based on Hidden Markov Models enable queries of the instantaneous transcriptional activity of an individual promoter within a single cell as development progresses (Fig. 6C). Thus, novel computational approaches are opening a direct window into the molecular mechanisms of transcription factors by extracting promoter-switching kinetics and correlating these kinetics with the concentrations of input transcription factors (Lammers et al., 2018).

As with any technology to shed light on the inner workings of cells, it is also important to be aware of the potential caveats associated with the implementation of MS2 in development. First, even though single mRNA molecules can be visualized as they are transcribed in bacteria and yeast (Golding et al., 2005; Larson et al., 2011), the signal-to-background present in embryos only allows for the detection of the fluorescence signal of, at the most, three mRNA molecules as they are being transcribed (Garcia et al., 2013). This low signal-to-background stems primarily from the thick optical sections afforded by widespread laser-scanning confocal microscopes which are much wider than the site of transcription and hence capture much of the free mRNA binding protein-GFP fusion in the nucleoplasm. The introduction of new microscopy modalities with higher axial resolution such as lattice light-sheet microscopy (Chen et al., 2014) could enable the single-molecule detection of mRNA molecules as they are being transcribed in an embryo. Further, doubts have been cast on whether the presence of MS2 loops in a transcript affect its stability (Garcia & Parker, 2015; Golding & Cox, 2004; Haimovich et al., 2016; Heinrich, Sidler, Azzalin, & Weis, 2017; Kim, Vieira, Kim, Kesawat, & Park, 2019). While effects on mRNA stability are probably irrelevant if MS2 is used as a reporter of transcriptional activity, these effects could certainly confound downstream measurements of mRNA export and processing, and affect the reliable operation of endogenous genes. New MS2 sequences are actively being developed to circumvent these limitations (Tutucci et al., 2018). In addition, since intronic RNA is rapidly processed during transcription (Coulon et al., 2014), inserting MS2 loops inside introns could prove a reliable strategy to tag endogenous genes without affecting the life cycle of their mRNA molecules. Even if no introns are present in a gene, synthetic introns can be introduced in order to realize this labeling strategy (Bothma, Norstad, Alamos, & Garcia, 2018).

3.2 Lighting up protein dynamics and transcriptional input-output functions

Despite these encouraging breakthroughs in measuring output transcriptional dynamics in real time, biologists have until recently lacked the technology to measure the fast dynamics of translation and degradation of the *input* transcription factors (Fig. 4). Although engineered fluorescent proteins such as GFP have chromophore maturation half-times as low as 6 min *in vitro* or in cultured cells (Nagai et al., 2002), maturation half-times increase to >30 min in embryos of developmental biology workhorses such as frogs, zebrafish, and flies (Little, Tkacik, Kneeland, Wieschaus, & Gregor, 2011). These time scales are much slower than many of the key processes in development. For example, the fruit fly transcription factor Fushi tarazu has a half-life of 8 min (Bothma et al., 2018; Kellerman, Mattson, & Duncan, 1990), and the Hes proteins that drive segmentation in vertebrates have a half-life of ~20 min (Hirata et al., 2004; Schroter et al., 2008). Thus, in many developmental contexts, by the time GFP fusions become fluorescent, the developmental processes these fusions are supposed to report on are already over. This fundamental limitation has prevented developmental biologists from following the central dogma with high spatiotemporal resolution and, more specifically, has made it impossible to measure input transcription-factor concentration dynamics in transcriptional input-output functions.

To enable the real-time measurement of input transcription-factor dynamics over the fast-paced process of development, and to circumvent the confounding effects of fluorescent protein maturation kinetics, the nanobody-based LlamaTag was recently introduced to light up protein concentration dynamics (Bothma et al., 2018). Here, nanobodies, single-chain antibodies raised in llamas against GFP or mCherry variants, are fused to a transcription factor of interest. In parallel, the fluorescent protein is provided maternally such that when development begins, this protein is already mature and uniformly distributed throughout the embryo. Upon translation of the transcription-factor fusion in the cytoplasm, the LlamaTag binds the free fluorescent protein. This complex is translocated into the nucleus *via* the transcription factor's nuclear localization signal, resulting in an enrichment of nuclear fluorescence that directly reports on the nuclear concentration of the complex. Thus, by leveraging localization of fluorescence proteins rather than the (more common) synthesis of new proteins, this technology becomes insensitive to fluorescent-protein maturation (Fig. 7A and B). LlamaTags have already made it possible to correlate bursts in transcriptional activity with bursts in protein concentration, to measure protein degradation, and to reveal the diffusion-mediated coupling between neighboring nuclei that can drive pattern formation in the fly syncytium (Bothma et al., 2018). Excitingly, these tags have also made it possible to quantify

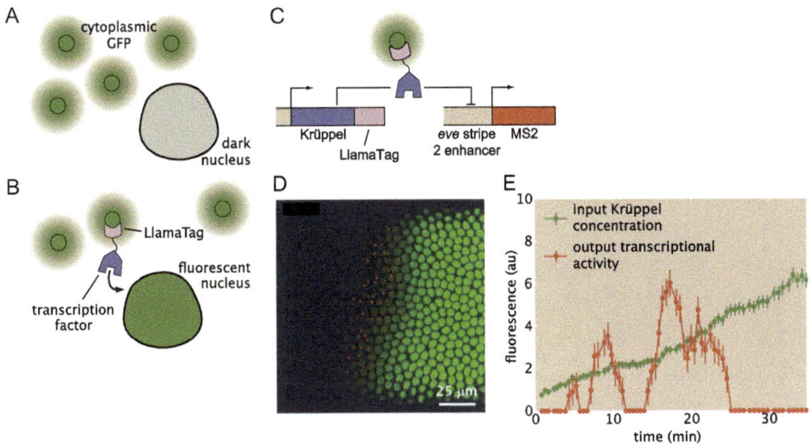

Fig. 7 Beating the fluorescent protein maturation speed limit with LlamaTags. (A) GFP expressed in the cytoplasm is (B) bound by a fusion of a LlamaTag to a transcription factor of interest. The increase in nuclear fluorescence upon translocation of the fusion to the nucleus reports transcription-factor concentration. (C) Combination of LlamaTag and MS2 tagging to simultaneously measure Krüppel repressor concentration and *even-skipped* (*eve*) stripe 2 transcriptional activity. (D) Snapshot of a fly embryo expressing Krüppel-LlamaTag and reporting on *eve* transcriptional activity using MS2 27 min into nuclear cycle 14. (E) Measured input and output dynamics in a nucleus within the stripe. *Panels (D and E) Adapted from Bothma, J.P., Norstad, M.R., Alamos, S., & Garcia, H.G. (2018). LlamaTags: A versatile tool to image transcription factor dynamics in live embryos.* Cell, *173, 1810–1822.*

transcriptional input-output functions at the single-cell level by enabling real-time measurement of instantaneous input transcription factor concentration and output transcriptional activity (Fig. 7C–E).

Just like regular fusions of transcription factors to fluorescent proteins, LlamaTag fusions can affect endogenous protein function. Further, these tags are limited to tagging proteins that undergo translocation after translation, such as transcription factors, and cannot report on the concentration dynamics of proteins that remain in the cytoplasm to perform their function. However, due to the nascent nature of LlamaTags, the full set of potential caveats associated with these tags, and of possible solutions to those caveats, is yet to be revealed as this technique is adopted by developmental biologists.

3.3 Wiring up the synthetic embryo

For the first time, developmental biologists are positioned to directly measure transcriptional input-output functions that capture rapid modulations in the concentration dynamics of input transcription factors and the resulting output transcriptional activity. A crucial next step is to identify regulatory architectures amenable to theoretical modeling that can be attacked with this new arsenal of tools.

Over the last three decades, a great deal of research has focused on the role of transcription factor binding sites in transcriptional input-output functions (Chen, Xu, Mei, Yu, & Small, 2012; Crocker et al., 2015; Crocker, Ilsley, & Stern, 2016; Driever, Thoma, & Nusslein-Volhard, 1989; Fakhouri et al., 2010; Hare, Peterson, Iyer, Meier, & Eisen, 2008; Harrison, Li, Kaplan, Botchan, & Eisen, 2011; Jiang & Levine, 1993; Park et al., 2019; Sayal et al., 2016; Small, Blair, & Levine, 1992; Stathopoulos & Levine, 2005; Swanson, Evans, & Barolo, 2010). Often, complex gene-regulatory regions featuring dozens of binding sites for several transcription factors are dissected *via* systematic deletions of these sites. Such approaches have revolutionized our understanding of the spatial control of developmental patterning, as exemplified by the famed dissection of the regulatory logic of the enhancer that regulates stripe 2 of the *even-skipped* gene, which revealed how activators and repressors work together to create precise gene expression patterns in the fly embryo (Arnosti, Barolo, Levine, & Small, 1996; Small et al., 1992; Small, Kraut, Hoey, Warrior, & Levine, 1991).

Nonetheless, these approaches often face an insurmountable barrier when moving from the qualitative realm to a quantitative understanding that makes it possible to predict transcriptional input-output functions. Complex regulatory architectures, by definition, demand complex theoretical models that in turn are plagued by a plethora of unknown regulatory parameters. Consider the *hunchback* P2 enhancer, perhaps one of the simplest and most studied regulatory architectures in all of development (Driever et al., 1989; Margolis et al., 1995; Park et al., 2019; Perry, Bothma, Luu, & Levine, 2012). The Bicoid activator can bind at least six sites in this enhancer (Driever et al., 1989). Predicting the transcriptional input-output function of *hunchback* P2 activation by Bicoid using even simple models based on equilibrium statistical mechanics demands previous knowledge of at least 27 parameters (Fig. 8; Garcia, Brewster, and Phillips (2016), see also Garcia, Kondev, Orme, Theriot, and Phillips (2007) for an introduction to statistical mechanics for life scientists). This number only grows as assumptions regarding equilibrium are relaxed (Estrada et al., 2016). Inferring these parameters from the measurement of a transcriptional input-output function is both a massive computational and—more critically—conceptual challenge (Garcia et al., 2016).

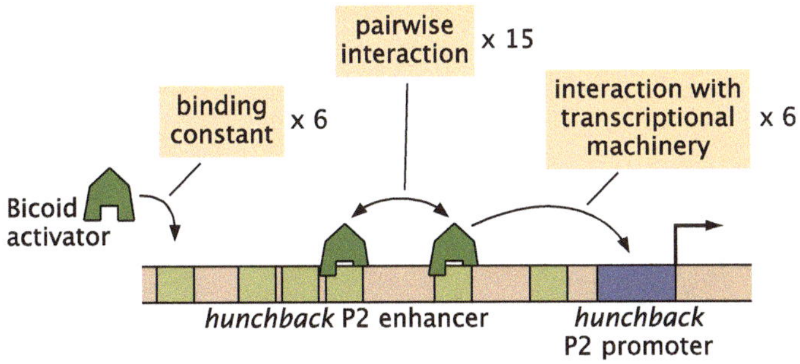

Fig. 8 Combinatorial complexity of endogenous gene regulatory regions. The *hunchback* P2 enhancer is bound by at least six Bicoid activators to regulate *hunchback*. A simple model featuring only binding energies, pairwise interactions between bound activators, and pairwise interactions between each activator and the transcriptional machinery would demand the fitting of 27 unknown parameters.

Synthetic biology could empower our dissection of developmental enhancers. Inspired by work in bacteria, an alternative to fitting complex theoretical models to complex experimental architectures in development is to *bend nature to understand it* (Garcia et al., 2016; Phillips et al., 2019). Specifically, building synthetic enhancers bearing only one binding site for an activator such as Dorsal or Bicoid dramatically reduces regulatory complexity. To be concrete, we consider an activator that is distributed in an exponential gradient along the embryo (Fig. 9A) resulting in a step-like output pattern of gene expression. As shown in Fig. 9B, a thermodynamic model describing this simple regulatory architecture has only two free parameters: the binding affinity of the activator to the DNA (K_d) and a parameter that captures the strength with which a bound activator drives gene expression (r_{AP}) and that depends on the distance between the activator binding site and the promoter. Thus, by measuring the height and position of the activator-driven developmental boundary, these two key parameters can be obtained.

This synthetic approach offers an opportunity to iteratively embrace regulatory complexity. Specifically, consider the case where the complexity of the synthetic enhancer is increased by adding a second activator binding site (Fig. 9C). If we rely on the parameters obtained in the previous iteration (Fig. 9B), and if we assume only pair-wise interactions between bound transcription factors, then only one new unknown parameter emerges. This new parameter, ω_{AA}, characterizes protein-protein interactions that lead to cooperativity and to the sharpening of the boundary. Thus, by harnessing the knowledge obtained in previous iterations, each successive iteration

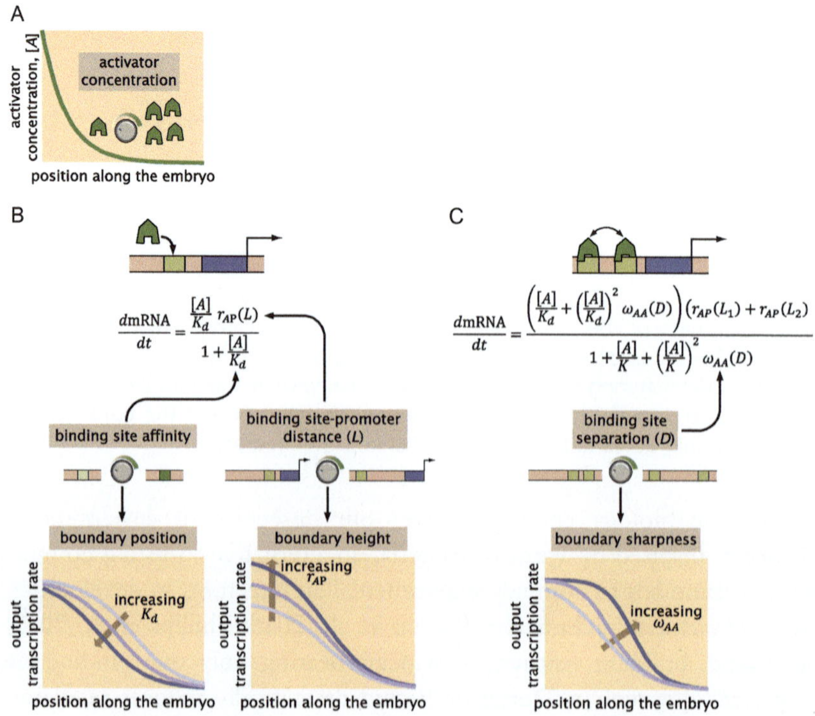

Fig. 9 A synthetic approach to uncovering the governing equations of gene regulatory regions in development. (A) Exponential activator concentration profile along the embryo assumed for this illustrative example. (B, C) Equations, regulatory parameters, and developmental patterns for synthetic enhancers containing (B) one or (C) two activator binding sites. (B) A reporter construct with a single activator binding site drives a step-like pattern whose boundary position is determined only by the binding site affinity (K_d), and whose boundary height is governed by the interaction between the activator and the transcriptional machinery (r_{AP}). (C) Adding a second activator binding site introduces only one new free parameter accounting for activator-activator interactions (ω_{AA}). This parameter controls boundary sharpness. For simplicity, we do not account for the existence of a basal rate of transcription. However, the addition of this parameter to the model would not modify the overall synthetic strategy significantly.

of this synthetic approach only requires the fitting of one or two new parameters. After multiple iterations, the synthetic architectures converge onto endogenous ones—accompanied by increasingly complex, but still predictive, theoretical models.

We speculate that this approach could be used to dissect complex regulatory architectures featuring multiple activators and repressors. Crucially, the key components of synthetic dissection already exist: multiple examples of minimal regulatory architectures featuring binding sites of transcription factors such as Bicoid, Dorsal, Giant, Snail, and Twist have been shown to drive detectable levels of gene expression (Burz & Hanes, 2001; Burz, Rivera-Pomar, Jackle, & Hanes, 1998; Driever et al., 1989; Erceg et al., 2014; Fakhouri et al., 2010; Hanes, Riddihough, Ish-Horowicz, & Brent, 1994; Jiang & Levine, 1993; Lebrecht et al., 2005; Ma, Yuan, Diepold, Scarborough, & Ma, 1996; Park et al., 2019; Ronchi, Treisman, Dostatni, Struhl, & Desplan, 1993; Sayal et al., 2016; Simpson-Brose, Treisman, & Desplan, 1994; Szymanski & Levine, 1995). Recent work has demonstrated the feasibility of this synthetic approach to testing theoretical models of transcriptional regulation in development in the context of activation by Dorsal and Twist, repression by Giant and Snail, and for synthetic transcription factors (Crocker et al., 2016; Fakhouri et al., 2010; Sayal et al., 2016).

3.4 Predicting the central dogma beyond transcription

So far, we have concerned ourselves with the prediction and measurement of transcriptional input-output functions. However, it is important to keep in mind that the information encoded by each arrow in cartoons of gene regulatory networks accounts for multiple steps along the central dogma (Fig. 4B). From chromatin accessibility to alternative splicing to post-translational modifications, seemingly simple arrows capture multiple molecular steps, all of which can be subject to regulation—calling, once again, for an interplay between theory and experiment to uncover the governing equations corresponding to each regulatory step. However, despite huge leaps in genomics (Goodwin, McPherson, & McCombie, 2016; Koboldt, Steinberg, Larson, Wilson, & Mardis, 2013; Shlyueva, Stampfel, & Stark, 2014), technologies to measure chromatin accessibility and modifications, protein binding to the DNA, promoter-enhancer interactions, mRNA processing, translational regulation, and post-translational modifications in single cells within living embryos have lagged behind (Buenrostro, Wu, Chang, & Greenleaf, 2015; Matera & Wang, 2014; Mayer, Landry, & Churchman, 2017; Park, 2009).

First, chromatin must be accessible for transcription factors to bind DNA. However, technology to reveal this accessibility or the epigenetic state of histones in the vicinity of a binding site has been mostly limited to genome-wide or fixed-tissue approaches (Blythe & Wieschaus, 2016; Boettiger et al., 2016; Cusanovich et al., 2018; Haines & Eisen, 2018; Li, Burkhardt, Gross, & Weissman, 2014). The recent development of genetically encoded modification-specific intracellular antibodies (mintbodies) that bind chromatin with specific modifications such as H3K9 acetylation and H4K20 methylation (Sato et al., 2013, 2016), as well as split-luciferase probes to image H3K9 and H3K27 methylation (Sekar, Foygel, Gelovani, & Paulmurugan, 2015), will enable concrete progress in the real-time monitoring of chromatin state in development at the single-cell level.

New imaging technologies and improved fluorescent probes have made it possible to image individual transcription factors as they bind DNA inside living fly embryos (Chen et al., 2014; Mir et al., 2017; Tsai et al., 2017). These measurements have revealed that, while transcription factors appear to spend no more than a few seconds bound to DNA, their binding throughout the nucleus is not uniform: hubs or domains of increased local concentration (and of increased binding frequency) have been found for both Bicoid and Ultrabithorax. While some of these regions of increased binding probability may depend on the pioneer transcription factor Zelda (Mir et al., 2017), their functional role remains unclear. To make progress toward a molecular understanding of how genes read out transcription-factor concentration, it will be necessary to directly correlate this binding with output transcriptional activity—which is only now becoming possible in single cells (Cho et al., 2016; Chong et al., 2018; Donovan et al., 2019; Li et al., 2019), and for which feasibility in live embryos was recently demonstrated (Mir et al., 2018).

Most developmental enhancers do not reside in the vicinity of their target promoter; they are supposed to loop or to translocate over vast distances of DNA in order to carry out their regulatory function (for a recent review on the subject, see Furlong & Levine, 2018). Recently, enhancer position and promoter activity were simultaneously visualized in the early fly embryo in the context of DNA looping (Chen et al., 2018) and transvection (Lim, Heist, et al., 2018). These works demonstrated that bringing enhancers and promoters in close proximity is necessary but not sufficient to activate transcription. These results, plus speculation about larger structures involved in transcriptional regulation (Mir et al., 2017, 2018; Tsai et al., 2017), and reports that stable promoter-enhancer contacts might not be needed for transcriptional activation (Alexander et al., 2019; Benabdallah et al., 2019; Gu et al., 2018) suggest that the classical paradigm of direct contact between enhancers and promoters may have to be revisited.

Regulation does not cease after transcription initiation: the rate of mRNA elongation can be under regulatory control, and fly embryos process mRNA through splicing as well as RNA polymerase II pausing and termination to achieve precise and rapid development (Bentley, 2014;

Core & Adelman, 2019; Larschan et al., 2011; Richard & Manley, 2009). While current genome-wide techniques have been powerful for revealing correlations among large sets of genes, visualizing mRNA processing could shed further light on the role of this processing in development. By combining MS2 and PP7 to label different parts of the same nascent RNA in the human β-globin gene, the life history of an mRNA was revealed as it underwent transcription, splicing, and termination (Coulon et al., 2014). This approach is being adopted in the fly embryo to, for example, determine transcript elongation rates (Fukaya et al., 2017).

The regulation of translation is also widespread in gene regulatory networks. For example, Bicoid represses Caudal translation (Dubnau & Struhl, 1996; Niessing, Blanke, & Jackle, 2002; Rivera-Pomar, Niessing, Schmidt-Ott, Gehring, & Jackle, 1996), while Nanos downregulates Hunchback post-transcriptionally, through either a decrease in translation or an increase in mRNA degradation (Cho et al., 2006; Irish, Lehmann, & Akam, 1989; Murata & Wharton, 1995; Struhl, 1989; Wang & Lehmann, 1991; Wharton & Struhl, 1991). However, we know much less about how translation is regulated at the single mRNA level than we know about the details of transcription. For example, is the translation of specific mRNA molecules downregulated by decreasing the peptide elongation rate of all ribosomes, or by decreasing the fraction of mRNA molecules that are translated? These questions and others can be answered by implementing recently developed reporters for measuring the first round of translation (Halstead et al., 2015), and by examining the translational dynamics of individual mRNA molecules at the single-cell level (Morisaki et al., 2016; Wang, Han, Zhou, & Zhuang, 2016; Wu, Eliscovich, Yoon, & Singer, 2016; Yan, Hoek, Vale, & Tanenbaum, 2016).

Finally, many developmental decisions are mediated by the post-translational modification of proteins. Regulation *via* protein phosphorylation is ubiquitous in development (for a review on this subject see, for example, Peter & Davidson, 2015; Ubersax & Ferrell, 2007). Antibodies cannot always distinguish between phosphorylated and non-phosphorylated protein forms, which hinders our ability to determine how signaling dynamics dictate development. When phosphorylation drives the nuclear localization of a transcription factor, such as for the transcription factor Capicua (Grimm et al., 2012), tracking its nuclear localization using a fusion to a fluorescent protein or a LlamaTag directly reports on the protein's signaling state. New sensors reveal kinase and phosphatase activity without requiring modulation of the cellular localization of their substrates; novel *kinase translocation reporters* can be engineered to become targets of a particular signaling pathway (Kudo et al., 2018; Oldach & Zhang, 2014; Regot, Hughey, Bajar, Carrasco, & Covert, 2014). Upon phosphorylation by the kinase of interest, these sensors change their fluorescence or are translocated to the nucleus, where they report on signaling activity. Implementing these technologies in the embryo could open the door to systematic dissection, at the single-cell level, of the signaling cascades that underlie protein post-translational modifications during development.

Technology is already available to quantify the flow of information along each step of the central dogma in real time and at the single-cell level, as highlighted by the various approaches showcased above. Thus, the main challenge ahead is not one of technology development, but one of implementing these technologies in developing embryos. The new and exciting data generated by these rising technologies must be matched with new theoretical models that draw us closer to a quantitative and predictive understanding of how the regulation of the processes of the central dogma impact cellular decision-making.

4. Developmental programs as dynamical systems

So far, we have focused on new technologies and theoretical approaches that enable the predictive dissection of the input-output functions encoded by each arrow in gene regulatory networks (Fig. 2A). However, predictive understanding of the parts that make a network does not guarantee understanding of how those arrows work together to realize developmental programs. It has been repeatedly hypothesized that the ultimate developmental fate of each cell arises from the trajectory of the gene-expression state of a cell as it traverses the regulatory landscape shaped by the network topology, the patterns of connections between network elements (Fig. 2B, reviewed in Jaeger, Manu, & Reinitz, 2012; Jaeger & Monk, 2014).

By borrowing tools from dynamical systems theory, multiple teams have attempted to describe how network topology prescribes these developmental trajectories. While some of these works have sought to model multiple layers of the network simultaneously, others have focused on isolated network motifs, such as the widespread mutual repression regulatory architecture (Edgar et al., 1989; Gursky et al., 2011; Jaeger, Blagov, et al., 2004; Jaeger, Surkova, et al., 2004; Lopes, Vieira, Holloway, Bisch, & Spirov, 2008; Manu et al., 2009; Papatsenko & Levine, 2011; von Dassow et al., 2000;

Von Dassow & Odell, 2002). Using gene expression data from fixed embryos sorted into temporal classes, these studies have, for instance, revealed how gene expression domains shift along the embryo as development progresses (Jaeger, Surkova, et al., 2004), and how multiple arrows work together to "lock" individual cells into specific developmental fates (Papatsenko & Levine, 2011).

These investigations have been complemented by the realization that the landscape shaped by these arrows is not static. For example, temporal changes in the concentration of transcription factors such as that of the Bicoid activator over development can propagate through the network, effectively modulating the network's topology and impacting cellular, and therefore embryonic, phenotype (Verd et al., 2018; Verd, Crombach, & Jaeger, 2017; Verd, Monk, & Jaeger, 2019). We urgently require theoretical tools to deal with such *non-autonomous dynamical systems*, where network parameters are modulated in time. Further, to test the predictions of these models, it will be necessary to simultaneously visualize the transcriptional activity and protein products of multiple genes in single cells as these networks are deployed. Currently, it is possible to simultaneously image only one input transcription factor and the transcriptional activity of one of its target genes (Bothma et al., 2018). This limitation to multiplexing underscores the need for new fluorescent probes with a large repertoire of spectral ranges, as well as advances in microscopy techniques that make it possible to spectrally resolve these different probes.

5. Toward quantitative and predictive developmental biology

The experimental technologies and theoretical approaches reviewed in this article are the means to the ultimate goal of a predictive understanding of developmental decision-making. Demanding a quantitative and predictive understanding of biological phenomena sharpens our questions and makes our inquiries more sensitive to inconsistencies that may reveal new biological insights that would have remained hidden from qualitative approaches (Cohen, 2004; Garcia, Sanchez, Kuhlman, Kondev, & Phillips, 2010). However, in our opinion, the discovery of new molecular players does not constitute a guiding objective in and of itself (Phillips, 2015). Even in the absence of new discoveries, we would like to define successful physical biology of embryonic development as the demonstration that developmental programs can be predictive, much as it has been shown in the context of gene regulatory programs in bacteria (Garcia et al., 2016; Phillips et al., 2019).

Although such predictive understanding calls for a quantitative view of how all the processes of the central dogma are regulated in development, the topics covered in this article have been vastly biased toward the regulation of transcriptional initiation. We believe that this bias reflects the state of the art in the field, as it is now possible to monitor transcriptional initiation and the concentration dynamics of the transcription factors that direct this initiation in real time during development. However, new technologies, some of which were briefly reviewed here, enable real-time, single-cell, high-precision, *in vivo* measurements of other steps of the central dogma. We therefore envision that, as these technologies are unleashed to unravel development, they will yield the dialog between theory and experiment that has been a defining factor in our understanding of the regulation of transcriptional initiation.

Of course, we must not forget that development transcends regulation of the central dogma! Ultimately, expression patterns arising from gene regulatory networks drive the morphogenic movements that bring about tissue growth and biological shape, and these movements further determine, in turn, gene expression patterns (for reviews on the subject, see Chan, Heisenberg, & Hiiragi, 2017; Gilmour, Rembold, & Leptin, 2017; Mammoto, Mammoto, & Ingber, 2012; Totaro, Panciera, & Piccolo, 2018). The capacity to measure and manipulate actomyosin networks is now making it possible to relate the activity of these regulatory networks to the massive cellular rearrangements that characterize morphogenesis and to control them synthetically (Campas, 2016; Farrell, Weitz, Magnasco, & Zallen, 2017; Guglielmi, Barry, Huber, & De Renzis, 2015; He, Martin, & Wieschaus, 2016; Kale et al., 2018; Martin, Kaschube, & Wieschaus, 2009; Streichan, Lefebvre, Noll, Wieschaus, & Shraiman, 2018). These new measurements and allied theoretical and computational approaches promise to close the gap between our understanding of morphogen gradients and our understanding of morphogenesis.

Finally, although this review limited its scope to the fruit fly, no one species holds all the keys to predictively understanding development. A key challenge will be to demonstrate that the strategies put forth here can also reveal the physical biology of embryos of other developmental biology workhorses such as worms, fish, and mice. Excitingly, the real-time visualization of transcription and mRNA processing was recently achieved in all three of these model organisms (Campbell, Chao, Singer, & Marlow, 2015;

Hadzhiev et al., 2019; Lee, Shin, & Kimble, 2019; Lionnet et al., 2011). Thus, the technologies discussed in this review article are ushering in a new era in developmental biology in which the focus on spatial, almost static, control of developmental programs is being replaced by a dynamical view that embraces the quantitative spatiotemporal control of development (Berrocal et al., 2018; Bothma & Levine, 2013). This new language will empower the discourse between theory and experiment that will revolutionize our ability to predict—and ultimately manipulate—developmental programs at will.

Acknowledgments

The authors are grateful to Jack Bateman, Jacques Bothma, James Briscoe, Leigh Harris, Thomas Lecuit, Mustafa Mir, Rob Phillips, Clarissa Scholes, Stephen Small and Michael Stadler for discussions and/or comments on this manuscript. However, the opinions and point of view expressed here are the authors' and those acknowledged above should not be blamed for these views. HGG was supported by the Burroughs Wellcome Fund Career Award at the Scientific Interface, the Sloan Research Foundation, the Human Frontiers Science Program, the Searle Scholars Program, the Shurl & Kay Curci Foundation, the Hellman Foundation, the NIH Director's New Innovator Award (DP2 OD024541-01), and an NSF CAREER Award (1652236). Y.J.K. was supported by Korea Foundation for Advanced Studies. G.M. was supported by an NIH NRSA Training Grant (T32 HG 00047) and by an NSF Graduate Fellowship. A.B. was supported by the University of California Institute for Mexico and the United States (UC MEXUS). Finally, the references cited throughout the text are meant as a guide to the reader to the literature, and do not attempt to provide a scholarly assessment of the whole field of fly development or of developmental biology writ large.

References

Ackers, G. K., Johnson, A. D., & Shea, M. A. (1982). Quantitative model for gene regulation by lambda phage repressor. *Proceedings of the National Academy of Sciences of the United States of America*, *79*, 1129–1133.

Alberts, B. (2015). *Molecular biology of the cell* (6th ed.). New York, NY: Garland Science, Taylor and Francis Group.

Alexander, J. M., Guan, J., Li, B., Maliskova, L., Song, M., Shen, Y., et al. (2019). Live-cell imaging reveals enhancer-dependent Sox2 transcription in the absence of enhancer proximity. *eLife*, *8*, e41769.

Arnosti, D. N., Barolo, S., Levine, M., & Small, S. (1996). The eve stripe 2 enhancer employs multiple modes of transcriptional synergy. *Development*, *122*, 205–214.

Benabdallah, N. S., Williamson, I., Illingworth, R. S., Kane, L., Boyle, S., Sengupta, D., et al. (2019). Decreased enhancer-promoter proximity accompanying enhancer activation. *Molecular Cell*, *76*(3), 473–484.e477.

Bentley, D. L. (2014). Coupling mRNA processing with transcription in time and space. *Nature Reviews. Genetics*, *15*, 163–175.

Berrocal, A., Lammers, N. C., Garcia, H. G., & Eisen, M. B. (2018). Kinetic sculpting of the seven stripes of the Drosophila even-skipped gene. *bioRxiv*, 335901.

Bertrand, E., Chartrand, P., Schaefer, M., Shenoy, S. M., Singer, R. H., & Long, R. M. (1998). Localization of ASH1 mRNA particles in living yeast. *Molecular Cell*, *2*, 437–445.

Bintu, L., Buchler, N. E., Garcia, H. G., Gerland, U., Hwa, T., Kondev, J., et al. (2005a). Transcriptional regulation by the numbers: Applications. *Current Opinion in Genetics & Development*, *15*, 125–135.

Bintu, L., Buchler, N. E., Garcia, H. G., Gerland, U., Hwa, T., Kondev, J., et al. (2005b). Transcriptional regulation by the numbers: Models. *Current Opinion in Genetics & Development*, *15*, 116–124.

Blythe, S. A., & Wieschaus, E. F. (2016). Establishment and maintenance of heritable chromatin structure during early Drosophila embryogenesis. *eLife*, *5*, e20148.

Boettiger, A. N., Bintu, B., Moffitt, J. R., Wang, S., Beliveau, B. J., Fudenberg, G., et al. (2016). Super-resolution imaging reveals distinct chromatin folding for different epigenetic states. *Nature*, *529*, 418–422.

Boettiger, A. N., & Levine, M. (2013). Rapid transcription Fosters coordinate snail expression in the Drosophila embryo. *Cell Reports*, *3*(1), 8–15.

Bothma, J. P., Garcia, H. G., Esposito, E., Schlissel, G., Gregor, T., & Levine, M. (2014). Dynamic regulation of eve stripe 2 expression reveals transcriptional bursts in living Drosophila embryos. *Proceedings of the National Academy of Sciences of the United States of America*, *111*, 10598–10603.

Bothma, J. P., Garcia, H. G., Ng, S., Perry, M. W., Gregor, T., & Levine, M. (2015). Enhancer additivity and non-additivity are determined by enhancer strength in the Drosophila embryo. *eLife*, *4*, e07956.

Bothma, J., & Levine, M. (2013). Development: Lights, camera, action—The Drosophila embryo goes live! *Current Biology*, *23*, R965.

Bothma, J. P., Magliocco, J., & Levine, M. (2011). The snail repressor inhibits release, not elongation, of paused Pol II in the Drosophila embryo. *Current Biology*, *21*, 1571–1577.

Bothma, J. P., Norstad, M. R., Alamos, S., & Garcia, H. G. (2018). LlamaTags: A versatile tool to image transcription factor dynamics in live embryos. *Cell*, *173*, 1810–1822.

Bronstein, L., Zechner, C., & Koeppl, H. (2015). Bayesian inference of reaction kinetics from single-cell recordings across a heterogeneous cell population. *Methods (San Diego, Calif)*, *85*, 22–35.

Buchler, N. E., Gerland, U., & Hwa, T. (2003). On schemes of combinatorial transcription logic. *Proceedings of the National Academy of Sciences of the United States of America*, *100*, 5136–5141.

Buenrostro, J. D., Wu, B., Chang, H. Y., & Greenleaf, W. J. (2015). ATAC-seq: A method for assaying chromatin accessibility genome-wide. *Current Protocols in Molecular Biology*, *109*, 21–29.

Burz, D. S., & Hanes, S. D. (2001). Isolation of mutations that disrupt cooperative DNA binding by the Drosophila bicoid protein. *Journal of Molecular Biology*, *305*, 219–230.

Burz, B. S., Rivera-Pomar, R., Jackle, H., & Hanes, S. D. (1998). Cooperative DNA-binding by Bicoid provides a mechanism for threshold-dependent gene activation in the Drosophila embryo. *The EMBO Journal*, *17*, 5998–6009.

Campas, O. (2016). A toolbox to explore the mechanics of living embryonic tissues. *Seminars in Cell & Developmental Biology*, *55*, 119–130.

Campbell, P. D., Chao, J. A., Singer, R. H., & Marlow, F. L. (2015). Dynamic visualization of transcription and RNA subcellular localization in zebrafish. *Development*, *142*, 1368–1374.

Carroll, S. B., Grenier, J. K., & Weatherbee, S. D. (2001). *From DNA to diversity: Molecular genetics and the evolution of animal design*. Malden, Mass: Blackwell Science.

Chan, C. J., Heisenberg, C. P., & Hiiragi, T. (2017). Coordination of morphogenesis and cell-fate specification in development. *Current Biology*, *27*, R1024–R1035.

Chao, J. A., Patskovsky, Y., Almo, S. C., & Singer, R. H. (2008). Structural basis for the coevolution of a viral RNA-protein complex. *Nature Structural & Molecular Biology*, *15*(1), 103–105.

Chen, B. C., Legant, W. R., Wang, K., Shao, L., Milkie, D. E., Davidson, M. W., et al. (2014). Lattice light-sheet microscopy: Imaging molecules to embryos at high spatiotemporal resolution. *Science*, *346*, 1257998.

Chen, H., Levo, M., Barinov, L., Fujioka, M., Jaynes, J. B., & Gregor, T. (2018). Dynamic interplay between enhancer-promoter topology and gene activity. *Nature Genetics*, *50*, 1296–1303.

Chen, H., Xu, Z., Mei, C., Yu, D., & Small, S. (2012). A system of repressor gradients spatially organizes the boundaries of bicoid-dependent target genes. *Cell*, *149*, 618–629.

Cho, P. F., Gamberi, C., Cho-Park, Y. A., Cho-Park, I. B., Lasko, P., & Sonenberg, N. (2006). Cap-dependent translational inhibition establishes two opposing morphogen gradients in Drosophila embryos. *Current Biology*, *16*, 2035–2041.

Cho, W. K., Jayanth, N., English, B. P., Inoue, T., Andrews, J. O., Conway, W., et al. (2016). RNA Polymerase II cluster dynamics predict mRNA output in living cells. *eLife*, *5*, e13617.

Chong, S., Dugast-Darzacq, C., Liu, Z., Dong, P., Dailey, G. M., Cattoglio, C., et al. (2018). Imaging dynamic and selective low-complexity domain interactions that control gene transcription. *Science*, *361*(6400), pii: eaar2555.

Chubb, J. R., Trcek, T., Shenoy, S. M., & Singer, R. H. (2006). Transcriptional pulsing of a developmental gene. *Current Biology*, *16*, 1018–1025.

Cohen, J. E. (2004). Mathematics is biology's next microscope, only better; biology is mathematics' next physics, only better. *PLoS Biology*, *2*, e439.

Core, L., & Adelman, K. (2019). Promoter-proximal pausing of RNA polymerase II: A nexus of gene regulation. *Genes & Development*, *33*, 960–982.

Corrigan, A. M., Tunnacliffe, E., Cannon, D., & Chubb, J. R. (2016). A continuum model of transcriptional bursting. *eLife*, *5*, e13051.

Coulon, A., Ferguson, M. L., de Turris, V., Palangat, M., Chow, C. C., & Larson, D. R. (2014). Kinetic competition during the transcription cycle results in stochastic RNA processing. *eLife*, *3*, e03939.

Coulon, A., & Larson, D. R. (2016). Fluctuation analysis: Dissecting transcriptional kinetics with signal theory. *Methods in Enzymology*, *572*, 159–191.

Crocker, J., Abe, N., Rinaldi, L., McGregor, A. P., Frankel, N., Wang, S., et al. (2015). Low affinity binding site clusters confer hox specificity and regulatory robustness. *Cell*, *160*, 191–203.

Crocker, J., Ilsley, G. R., & Stern, D. L. (2016). Quantitatively predictable control of Drosophila transcriptional enhancers in vivo with engineered transcription factors. *Nature Genetics*, *48*, 292–298.

Cusanovich, D. A., Reddington, J. P., Garfield, D. A., Daza, R. M., Aghamirzaie, D., Marco-Ferreres, R., et al. (2018). The cis-regulatory dynamics of embryonic development at single-cell resolution. *Nature*, *555*, 538–542.

Davidson, E. H. (2006). *The regulatory genome: Gene regulatory networks in development and evolution*. Burlington, MA; San Diego: Academic.

Desponds, J., Tran, H., Ferraro, T., Lucas, T., Perez Romero, C., Guillou, A., et al. (2016). Precision of readout at the hunchback gene: Analyzing short transcription time traces in living fly embryos. *PLoS Computational Biology*, *12*, e1005256.

Donovan, B. T., Huynh, A., Ball, D. A., Patel, H. P., Poirier, M. G., Larson, D. R., et al. (2019). Live-cell imaging reveals the interplay between transcription factors, nucleosomes, and bursting. *The EMBO Journal*, *38*, e100809.

Driever, W., Thoma, G., & Nusslein-Volhard, C. (1989). Determination of spatial domains of zygotic gene expression in the Drosophila embryo by the affinity of binding sites for the bicoid morphogen. *Nature*, *340*, 363–367.

Dubnau, J., & Struhl, G. (1996). RNA recognition and translational regulation by a homeodomain protein. *Nature*, *379*, 694–699.

Dubuis, J. O., Samanta, R., & Gregor, T. (2013). Accurate measurements of dynamics and reproducibility in small genetic networks. *Molecular Systems Biology*, *9*, 639, [Electronic Resource].

Dufourt, J., Trullo, A., Hunter, J., Fernandez, C., Lazaro, J., Dejean, M., et al. (2018). Temporal control of gene expression by the pioneer factor Zelda through transient interactions in hubs. *Nature Communications*, *9*, 5194.

Edgar, B. A., Odell, G. M., & Schubiger, G. (1989). A genetic switch, based on negative regulation, sharpens stripes in Drosophila embryos. *Developmental Genetics*, *10*, 124–142.

El-Sherif, E., & Levine, M. (2016). Shadow enhancers mediate dynamic shifts of gap gene expression in the Drosophila embryo. *Current Biology*, *26*, 1164–1169.

Erceg, J., Saunders, T. E., Girardot, C., Devos, D. P., Hufnagel, L., & Furlong, E. E. (2014). Subtle changes in motif positioning cause tissue-specific effects on robustness of an enhancer's activity. *PLoS Genetics*, *10*, e1004060.

Esposito, E., Lim, B., Guessous, G., Falahati, H., & Levine, M. (2016). Mitosis-associated repression in development. *Genes Development*, *30*, 1503–1508.

Estrada, J., Wong, F., DePace, A., & Gunawardena, J. (2016). Information integration and energy expenditure in gene regulation. *Cell*, *166*, 234–244.

Fakhouri, W. D., Ay, A., Sayal, R., Dresch, J., Dayringer, E., & Arnosti, D. N. (2010). Deciphering a transcriptional regulatory code: Modeling short-range repression in the Drosophila embryo. *Molecular Systems Biology*, *6*, 341, [electronic resource].

Falo-Sanjuan, J., Lammers, N. C., Garcia, H. G., & Bray, S. (2019). Enhancer priming enables fast and sustained transcriptional responses to Notch signaling. *Developmental Cell*, *50*, 411–425.

Farrell, D. L., Weitz, O., Magnasco, M. O., & Zallen, J. A. (2017). SEGGA: A toolset for rapid automated analysis of epithelial cell polarity and dynamics. *Development*, *144*, 1725–1734.

Featherstone, K., Hey, K., Momiji, H., McNamara, A. V., Patist, A. L., Woodburn, J., et al. (2016). Spatially coordinated dynamic gene transcription in living pituitary tissue. *eLife*, *5*, e08494.

Ferraro, T., Esposito, E., Mancini, L., Ng, S., Lucas, T., Coppey, M., et al. (2016). Transcriptional memory in the Drosophila embryo. *Current Biology*, *26*, 212–218.

Fuda, N. J., Ardehali, M. B., & Lis, J. T. (2009). Defining mechanisms that regulate RNA polymerase II transcription in vivo. *Nature*, *461*, 186–192.

Fukaya, T., Lim, B., & Levine, M. (2016). Enhancer control of transcriptional bursting. *Cell*, *166*, 358–368.

Fukaya, T., Lim, B., & Levine, M. (2017). Rapid rates of Pol II elongation in the Drosophila embryo. *Current Biology*, *27*, 1387–1391.

Furlong, E. E. M., & Levine, M. (2018). Developmental enhancers and chromosome topology. *Science*, *361*, 1341–1345.

Garcia, H. G., Brewster, R. C., & Phillips, R. (2016). Using synthetic biology to make cells tomorrow's test tubes. *Integrative Biology (Camb)*, *8*, 431–450.

Garcia, H. G., Kondev, J., Orme, N., Theriot, J. A., & Phillips, R. (2007). A first exposure to statistical mechanics for life scientists. *arXiv*. arXiv preprint arXiv: 0708.1899.

Garcia, J. F., & Parker, R. (2015). MS2 coat proteins bound to yeast mRNAs block 5' to 3' degradation and trap mRNA decay products: implications for the localization of mRNAs by MS2-MCP system. *RNA (New York, NY)*, *21*, 1393–1395.

Garcia, H. G., Sanchez, A., Kuhlman, T., Kondev, J., & Phillips, R. (2010). Transcription by the numbers redux: Experiments and calculations that surprise. *Trends in Cell Biology*, *20*, 723–733.

Garcia, H. G., Tikhonov, M., Lin, A., & Gregor, T. (2013). Quantitative imaging of transcription in living Drosophila embryos links polymerase activity to patterning. *Current Biology*, *23*, 2140–2145.

Gilmour, D., Rembold, M., & Leptin, M. (2017). From morphogen to morphogenesis and back. *Nature*, *541*, 311–320.

Golding, I., & Cox, E. C. (2004). RNA dynamics in live Escherichia coli cells. *Proceedings of the National Academy of Sciences of the United States of America*, *101*, 11310–11315.

Golding, I., Paulsson, J., Zawilski, S. M., & Cox, E. C. (2005). Real-time kinetics of gene activity in individual bacteria. *Cell*, *123*, 1025–1036.

Goodwin, S., McPherson, J. D., & McCombie, W. R. (2016). Coming of age: Ten years of next-generation sequencing technologies. *Nature Reviews. Genetics*, *17*, 333–351.

Gregor, T., Tank, D. W., Wieschaus, E. F., & Bialek, W. (2007). Probing the limits to positional information. *Cell*, *130*, 153–164.

Grimm, O., Sanchez Zini, V., Kim, Y., Casanova, J., Shvartsman, S. Y., & Wieschaus, E. (2012). Torso RTK controls Capicua degradation by changing its subcellular localization. *Development*, *139*, 3962–3968.

Gu, B., Swigut, T., Spencley, A., Bauer, M. R., Chung, M., Meyer, T., et al. (2018). Transcription-coupled changes in nuclear mobility of mammalian cis-regulatory elements. *Science*, *359*, 1050–1055.

Guglielmi, G., Barry, J. D., Huber, W., & De Renzis, S. (2015). An optogenetic method to modulate cell contractility during tissue morphogenesis. *Developmental Cell*, *35*, 646–660.

Gursky, V. V., Panok, L., Myasnikova, E. M., Manu, Samsonova, M. G., Reinitz, J., et al. (2011). Mechanisms of gap gene expression canalization in the Drosophila blastoderm. *BMC Systems Biology*, *5*, 118.

Hadzhiev, Y., Qureshi, H. K., Wheatley, L., Cooper, L., Jasiulewicz, A., Van Nguyen, H., et al. (2019). A cell cycle-coordinated polymerase II transcription compartment encompasses gene expression before global genome activation. *Nature Communications*, *10*, 691.

Haimovich, G., Zabezhinsky, D., Haas, B., Slobodin, B., Purushothaman, P., Fan, L., et al. (2016). Use of the MS2 aptamer and coat protein for RNA localization in yeast: A response to "MS2 coat proteins bound to yeast mRNAs block 5' to 3' degradation and trap mRNA decay products: implications for the localization of mRNAs by MS2-MCP system" *RNA (New York, NY)*, *22*, 660–666.

Haines, J. E., & Eisen, M. B. (2018). Patterns of chromatin accessibility along the anterior-posterior axis in the early Drosophila embryo. *PLoS Genetics*, *14*, e1007367.

Halstead, J. M., Lionnet, T., Wilbertz, J. H., Wippich, F., Ephrussi, A., Singer, R. H., et al. (2015). Translation. An RNA biosensor for imaging the first round of translation from single cells to living animals. *Science*, *347*, 1367–1671.

Hammar, P., Wallden, M., Fange, D., Persson, F., Baltekin, O., Ullman, G., et al. (2014). Direct measurement of transcription factor dissociation excludes a simple operator occupancy model for gene regulation. *Nature Genetics*, *46*, 405–408.

Hanes, S. D., Riddihough, G., Ish-Horowicz, D., & Brent, R. (1994). Specific DNA recognition and intersite spacing are critical for action of the bicoid morphogen. *Molecular and Cellular Biology*, *14*, 3364–3375.

Hare, E. E., Peterson, B. K., Iyer, V. N., Meier, R., & Eisen, M. B. (2008). Sepsid even-skipped enhancers are functionally conserved in Drosophila despite lack of sequence conservation. *PLoS Genetics*, *4*, e1000106.

Harrison, M. M., Li, X. Y., Kaplan, T., Botchan, M. R., & Eisen, M. B. (2011). Zelda binding in the early Drosophila melanogaster embryo marks regions subsequently activated at the maternal-to-zygotic transition. *PLoS Genetics*, *7*, e1002266.

He, B., Martin, A., & Wieschaus, E. (2016). Flow-dependent myosin recruitment during Drosophila cellularization requires zygotic dunk activity. *Development*, *143*, 2417–2430.

He, X., Samee, M. A., Blatti, C., & Sinha, S. (2010). Thermodynamics-based models of transcriptional regulation by enhancers: The roles of synergistic activation, cooperative binding and short-range repression. *PLoS Computational Biology*, *6*, 1–15.

Heinrich, S., Sidler, C. L., Azzalin, C. M., & Weis, K. (2017). Stem-loop RNA labeling can affect nuclear and cytoplasmic mRNA processing. *RNA (New York, NY)*, *23*, 134–141.

Hey, K. L., Momiji, H., Featherstone, K., Davis, J. R., White, M. R., Rand, D. A., et al. (2015). A stochastic transcriptional switch model for single cell imaging data. *Biostatistics*, *16*, 655–669.

Hirata, H., Bessho, Y., Kokubu, H., Masamizu, Y., Yamada, S., Lewis, J., et al. (2004). Instability of Hes7 protein is crucial for the somite segmentation clock. *Nature Genetics*, *36*, 750–754.

Irish, V., Lehmann, R., & Akam, M. (1989). The Drosophila posterior-group gene nanos functions by repressing hunchback activity. *Nature*, *338*, 646–648.

Jaeger, J., Blagov, M., Kosman, D., Kozlov, K. N., Manu, Myasnikova, E., et al. (2004). Dynamical analysis of regulatory interactions in the gap gene system of Drosophila melanogaster. *Genetics*, *167*, 1721–1737.

Jaeger, J., Manu, & Reinitz, J. (2012). Drosophila blastoderm patterning. *Current Opinion in Genetics & Development*, *22*, 533–541.

Jaeger, J., & Monk, N. (2014). Bioattractors: Dynamical systems theory and the evolution of regulatory processes. *The Journal of Physiology*, *592*, 2267–2281.

Jaeger, J., Surkova, S., Blagov, M., Janssens, H., Kosman, D., Kozlov, K. N., et al. (2004). Dynamic control of positional information in the early Drosophila embryo. *Nature*, *430*, 368–371.

Jiang, J., & Levine, M. (1993). Binding affinities and cooperative interactions with bHLH activators delimit threshold responses to the dorsal gradient morphogen. *Cell*, *72*, 741–752.

Kale, G. R., Yang, X., Philippe, J. M., Mani, M., Lenne, P. F., & Lecuit, T. (2018). Distinct contributions of tensile and shear stress on E-cadherin levels during morphogenesis. *Nature Communications*, *9*, 5021.

Kanodia, J. S., Liang, H. L., Kim, Y., Lim, B., Zhan, M., Lu, H., et al. (2012). Pattern formation by graded and uniform signals in the early Drosophila embryo. *Biophysical Journal*, *102*, 427–433.

Kellerman, K. A., Mattson, D. M., & Duncan, I. (1990). Mutations affecting the stability of the fushi tarazu protein of Drosophila. *Genes Development*, *4*, 1936–1950.

Kim, S. H., Vieira, M., Kim, H. J., Kesawat, M. S., & Park, H. Y. (2019). MS2 labeling of endogenous Beta-Actin mRNA does not result in stabilization of degradation intermediates. *Molecules and Cells*, *42*, 356–362.

Koboldt, D. C., Steinberg, K. M., Larson, D. E., Wilson, R. K., & Mardis, E. R. (2013). The next-generation sequencing revolution and its impact on genomics. *Cell*, *155*, 27–38.

Kudo, T., Jeknic, S., Macklin, D. N., Akhter, S., Hughey, J. J., Regot, S., et al. (2018). Live-cell measurements of kinase activity in single cells using translocation reporters. *Nature Protocols*, *13*, 155–169.

Lammers, N. C., Galstyan, V., Reimer, A., Medin, S. A., Wiggins, C. H., & Garcia, H. G. (2018). Multimodal transcriptional control of pattern formation in embryonic development. *bioRxiv*, 335919.

Larschan, E., Bishop, E. P., Kharchenko, P. V., Core, L. J., Lis, J. T., Park, P. J., et al. (2011). X chromosome dosage compensation via enhanced transcriptional elongation in Drosophila. *Nature*, *471*, 115–118.

Larson, D. R., Zenklusen, D., Wu, B., Chao, J. A., & Singer, R. H. (2011). Real-time observation of transcription initiation and elongation on an endogenous yeast gene. *Science*, *332*, 475–478.

Lawrence, J. B., Singer, R. H., & Marselle, L. M. (1989). Highly localized tracks of specific transcripts within interphase nuclei visualized by in situ hybridization. *Cell*, *57*, 493–502.
Lebrecht, D., Foehr, M., Smith, E., Lopes, F. J., Vanario-Alonso, C. E., Reinitz, J., et al. (2005). Bicoid cooperative DNA binding is critical for embryonic patterning in Drosophila. *Proceedings of the National Academy of Sciences of the United States of America*, *102*, 13176–13181.
Lee, C., Shin, H., & Kimble, J. (2019). Dynamics of Notch-dependent transcriptional bursting in its native context. *bioRxiv, Developmental Cell*, *50*, 426–435.
Li, G. W., Burkhardt, D., Gross, C., & Weissman, J. S. (2014). Quantifying absolute protein synthesis rates reveals principles underlying allocation of cellular resources. *Cell*, *157*, 624–635.
Li, C., Cesbron, F., Oehler, M., Brunner, M., & Hofer, T. (2018). Frequency modulation of transcriptional bursting enables sensitive and rapid gene regulation. *Cell Systems*, *6*, 409-423 e411.
Li, J., Dong, A., Saydaminova, K., Chang, H., Wang, G., Ochiai, H., et al. (2019). Single-molecule nanoscopy elucidates RNA polymerase II transcription at single genes in live cells. *Cell*, *178*, 491–506.
Lim, B., Fukaya, T., Heist, T., & Levine, M. (2018). Temporal dynamics of pair-rule stripes in living Drosophila embryos. *Proceedings of the National Academy of Sciences of the United States of America*, *115*, 8376–8381.
Lim, B., Heist, T., Levine, M., & Fukaya, T. (2018). Visualization of transvection in living drosophila embryos. *Molecular Cell*, *70*, 287-296 e286.
Lim, B., Levine, M., & Yamazaki, Y. (2017). Transcriptional pre-patterning of Drosophila gastrulation. *Current Biology*, *27*, 286–290.
Lionnet, T., Czaplinski, K., Darzacq, X., Shav-Tal, Y., Wells, A. L., Chao, J. A., et al. (2011). A transgenic mouse for in vivo detection of endogenous labeled mRNA. *Nature Methods*, *8*, 165–170.
Little, S. C., Tikhonov, M., & Gregor, T. (2013). Precise developmental gene expression arises from globally stochastic transcriptional activity. *Cell*, *154*, 789–800.
Little, S. C., Tkacik, G., Kneeland, T. B., Wieschaus, E. F., & Gregor, T. (2011). The formation of the Bicoid morphogen gradient requires protein movement from anteriorly localized mRNA. *PLoS Biology*, *9*, e1000596.
Lopes, F. J., Vieira, F. M., Holloway, D. M., Bisch, P. M., & Spirov, A. V. (2008). Spatial bistability generates hunchback expression sharpness in the Drosophila embryo. *PLoS Computational Biology*, *4*, e1000184.
Lucas, T., Ferraro, T., Roelens, B., De Las Heras Chanes, J., Walczak, A. M., Coppey, M., et al. (2013). Live imaging of bicoid-dependent transcription in Drosophila embryos. *Current Biology*, *23*, 2135–2139.
Ma, X., Yuan, D., Diepold, K., Scarborough, T., & Ma, J. (1996). The Drosophila morphogenetic protein Bicoid binds DNA cooperatively. *Development*, *122*, 1195–1206.
Mammoto, A., Mammoto, T., & Ingber, D. E. (2012). Mechanosensitive mechanisms in transcriptional regulation. *Journal of Cell Science*, *125*, 3061–3073.
Manu, Surkova, S., Spirov, A. V., Gursky, V. V., Janssens, H., Kim, A. R., et al. (2009). Canalization of gene expression and domain shifts in the Drosophila blastoderm by dynamical attractors. *PLoS Computational Biology*, *5*, e1000303.
Margolis, J. S., Borowsky, M. L., Steingrimsson, E., Shim, C. W., Lengyel, J. A., & Posakony, J. W. (1995). Posterior stripe expression of hunchback is driven from two promoters by a common enhancer element. *Development*, *121*, 3067–3077.
Martin, A. C., Kaschube, M., & Wieschaus, E. F. (2009). Pulsed contractions of an actin-myosin network drive apical constriction. *Nature*, *457*, 495–499.
Matera, A. G., & Wang, Z. (2014). A day in the life of the spliceosome. *Nature Reviews. Molecular Cell Biology*, *15*, 108–121.

Mayer, A., Landry, H. M., & Churchman, L. S. (2017). Pause & go: From the discovery of RNA polymerase pausing to its functional implications. *Current Opinion in Cell Biology, 46*, 72–80.
Mir, M., Reimer, A., Haines, J. E., Li, X. Y., Stadler, M., Garcia, H., et al. (2017). Dense Bicoid hubs accentuate binding along the morphogen gradient. *Genes & Development, 31*, 1784–1794.
Mir, M., Stadler, M. R., Ortiz, S. A., Hannon, C. E., Harrison, M. M., Darzacq, X., et al. (2018). Dynamic multifactor hubs interact transiently with sites of active transcription in Drosophila embryos. *eLife*, 7, e40497.
Molina, N., Suter, D. M., Cannavo, R., Zoller, B., Gotic, I., & Naef, F. (2013). Stimulus-induced modulation of transcriptional bursting in a single mammalian gene. *Proceedings of the National Academy of Sciences of the United States of America, 110*, 20563–20568.
Morisaki, T., Lyon, K., DeLuca, K. F., DeLuca, J. G., English, B. P., Zhang, Z., et al. (2016). Real-time quantification of single RNA translation dynamics in living cells. *Science, 352*, 1425–1429.
Murata, Y., & Wharton, R. P. (1995). Binding of pumilio to maternal hunchback mRNA is required for posterior patterning in Drosophila embryos. *Cell, 80*, 747–756.
Nagai, T., Ibata, K., Park, E. S., Kubota, M., Mikoshiba, K., & Miyawaki, A. (2002). A variant of yellow fluorescent protein with fast and efficient maturation for cell-biological applications. *Nature Biotechnology, 20*, 87–90.
Niessing, D., Blanke, S., & Jackle, H. (2002). Bicoid associates with the 5'-cap-bound complex of caudal mRNA and represses translation. *Genes & Development, 16*, 2576–2582.
O'Farrell, P. H., Edgar, B. A., Lakich, D., & Lehner, C. F. (1989). Directing cell division during development. *Science, 246*, 635–640.
Oldach, L., & Zhang, J. (2014). Genetically encoded fluorescent biosensors for live-cell visualization of protein phosphorylation. *Chemistry & Biology, 21*, 186–197.
Papatsenko, D., & Levine, M. (2011). The Drosophila gap gene network is composed of two parallel toggle switches. *PLoS One, 6*, e21145.
Pare, A., Lemons, D., Kosman, D., Beaver, W., Freund, Y., & McGinnis, W. (2009). Visualization of individual Scr mRNAs during Drosophila embryogenesis yields evidence for transcriptional bursting. *Current Biology, 19*, 2037–2042.
Park, P. J. (2009). ChIP-seq: Advantages and challenges of a maturing technology. *Nature Reviews. Genetics, 10*, 669–680.
Park, J., Estrada, J., Johnson, G., Vincent, B. J., Ricci-Tam, C., Bragdon, M. D., Shulgina, Y., Cha, A., Wunderlich, Z., Gunawardena, J., & DePace, A. H. (2019). Dissecting the sharp response of a canonical developmental enhancer reveals multiple sources of cooperativity. *eLife, 8*, e41266.
Perry, M. W., Bothma, J. P., Luu, R. D., & Levine, M. (2012). Precision of hunchback expression in the Drosophila embryo. *Current Biology, 22*, 2247–2252.
Peter, I. S., & Davidson, E. H. (2015). *Genomic control process: Development and evolution*. London, UK; San Diego, CA, USA: Academic Press is an imprint of Elsevier.
Phillips, R. (2015). Theory in biology: Figure 1 or figure 7? *Trends in Cell Biology*.
Phillips, R., Belliveau, N. M., Chure, G., Garcia, H. G., Razo-Mejia, M., & Scholes, C. (2019). Figure 1 theory meets figure 2 experiments in the study of gene expression. *Annual Review of Biophysics, 48*, 121–163.
Poustelnikova, E., Pisarev, A., Blagov, M., Samsonova, M., & Reinitz, J. (2004). A database for management of gene expression data in situ. *Bioinformatics, 20*, 2212–2221.
Raj, A., Peskin, C. S., Tranchina, D., Vargas, D. Y., & Tyagi, S. (2006). Stochastic mRNA synthesis in mammalian cells. *PLoS Biology, 4*, e309.
Regot, S., Hughey, J. J., Bajar, B. T., Carrasco, S., & Covert, M. W. (2014). High-sensitivity measurements of multiple kinase activities in live single cells. *Cell, 157*, 1724–1734.

Richard, P., & Manley, J. L. (2009). Transcription termination by nuclear RNA polymerases. *Genes & Development*, *23*, 1247–1269.

Rivera-Pomar, R., Niessing, D., Schmidt-Ott, U., Gehring, W. J., & Jackle, H. (1996). RNA binding and translational suppression by bicoid. *Nature*, *379*, 746–749.

Ronchi, E., Treisman, J., Dostatni, N., Struhl, G., & Desplan, C. (1993). Down-regulation of the Drosophila morphogen bicoid by the torso receptor-mediated signal transduction cascade. *Cell*, *74*, 347–355.

Samee, M. A., Lim, B., Samper, N., Lu, H., Rushlow, C. A., Jimenez, G., et al. (2015). A systematic ensemble approach to thermodynamic modeling of gene expression from sequence data. *Cell Systems*, *1*, 396–407.

Sato, Y., Kujirai, T., Arai, R., Asakawa, H., Ohtsuki, C., Horikoshi, N., et al. (2016). A genetically encoded probe for live-cell imaging of H4K20 monomethylation. *Journal of Molecular Biology*, *428*, 3885–3902.

Sato, Y., Mukai, M., Ueda, J., Muraki, M., Stasevich, T. J., Horikoshi, N., et al. (2013). Genetically encoded system to track histone modification in vivo. *Scientific Reports*, *3*, 2436.

Sayal, R., Dresch, J. M., Pushel, I., Taylor, B. R., & Arnosti, D. N. (2016). Quantitative perturbation-based analysis of gene expression predicts enhancer activity in early Drosophila embryo. *eLife*, *5*.

Scholes, C., Biette, K. M., Harden, T. T., & DePace, A. H. (2019). Signal integration by shadow enhancers and enhancer duplications varies across the Drosophila embryo. *Cell Reports*, *26*, 2407-2418 e2405.

Scholes, C., DePace, A. H., & Sanchez, A. (2017). Combinatorial gene regulation through kinetic control of the transcription cycle. *Cell Systems*, *4*, 97-108 e109.

Schroter, C., Herrgen, L., Cardona, A., Brouhard, G. J., Feldman, B., & Oates, A. C. (2008). Dynamics of zebrafish somitogenesis. *Developmental Dynamics: An Official Publication of the American Association of the Anatomists*, *237*, 545–553.

Segal, E., Raveh-Sadka, T., Schroeder, M., Unnerstall, U., & Gaul, U. (2008). Predicting expression patterns from regulatory sequence in Drosophila segmentation. *Nature*, *451*, 535–540.

Sekar, T. V., Foygel, K., Gelovani, J. G., & Paulmurugan, R. (2015). Genetically encoded molecular biosensors to image histone methylation in living animals. *Analytical Chemistry*, *87*, 892–899.

Sherman, M. S., & Cohen, B. A. (2012). Thermodynamic state ensemble models of cis-Regulation. *PLoS Computational Biology*, *8*, e1002407.

Shermoen, A. W., & O'Farrell, P. H. (1991). Progression of the cell cycle through mitosis leads to abortion of nascent transcripts. *Cell*, *67*, 303–310.

Shlyueva, D., Stampfel, G., & Stark, A. (2014). Transcriptional enhancers: From properties to genome-wide predictions. *Nature Reviews. Genetics*, *15*, 272–286.

Simpson-Brose, M., Treisman, J., & Desplan, C. (1994). Synergy between the hunchback and bicoid morphogens is required for anterior patterning in Drosophila. *Cell*, *78*, 855–865.

Singer, R. H., & Ward, D. C. (1982). Actin gene expression visualized in chicken muscle tissue culture by using in situ hybridization with a biotinated nucleotide analog. *Proceedings of the National Academy of Sciences of the United States of America*, *79*, 7331–7335.

Small, S., Blair, A., & Levine, M. (1992). Regulation of *even-skipped* stripe 2 in the *Drosophila* embryo. *The EMBO Journal*, *11*, 4047–4057.

Small, S., Kraut, R., Hoey, T., Warrior, R., & Levine, M. (1991). Transcriptional regulation of a pair-rule stripe in Drosophila. *Genes & Development*, *5*, 827–839.

Stathopoulos, A., & Levine, M. (2005). Localized repressors delineate the neurogenic ectoderm in the early Drosophila embryo. *Developmental Biology*, *280*, 482–493.

Streichan, S. J., Lefebvre, M. F., Noll, N., Wieschaus, E. F., & Shraiman, B. I. (2018). Global morphogenetic flow is accurately predicted by the spatial distribution of myosin motors. *eLife*, 7, e27454.

Struhl, G. (1989). Differing strategies for organizing anterior and posterior body pattern in Drosophila embryos. *Nature*, *338*, 741–744.

Suter, D. M., Molina, N., Gatfield, D., Schneider, K., Schibler, U., & Naef, F. (2011). Mammalian genes are transcribed with widely different bursting kinetics. *Science*, *332*, 472–474.

Swanson, C. I., Evans, N. C., & Barolo, S. (2010). Structural rules and complex regulatory circuitry constrain expression of a Notch- and EGFR-regulated eye enhancer. *Developmental Cell*, *18*, 359–370.

Szymanski, P., & Levine, M. (1995). Multiple modes of dorsal-bHLH transcriptional synergy in the Drosophila embryo. *The EMBO Journal*, *14*, 2229–2238.

Tautz, D., & Pfeifle, C. (1989). A non-radioactive in situ hybridization method for the localization of specific RNAs in Drosophila embryos reveals translational control of the segmentation gene hunchback. *Chromosoma*, *98*, 81–85.

Totaro, A., Panciera, T., & Piccolo, S. (2018). YAP/TAZ upstream signals and downstream responses. *Nature Cell Biology*, *20*, 888–899.

Tran, H., Desponds, J., Perez Romero, C. A., Coppey, M., Fradin, C., Dostatni, N., et al. (2018). Precision in a rush: Trade-offs between reproducibility and steepness of the hunchback expression pattern. *PLoS Computational Biology*, *14*, e1006513.

Tsai, A., Muthusamy, A. K., Alves, M. R., Lavis, L. D., Singer, R. H., Stern, D. L., et al. (2017). Nuclear microenvironments modulate transcription from low-affinity enhancers. *eLife*, *6*.

Tutucci, E., Vera, M., Biswas, J., Garcia, J., Parker, R., & Singer, R. H. (2018). An improved MS2 system for accurate reporting of the mRNA life cycle. *Nature Methods*, *15*, 81–89.

Ubersax, J. A., & Ferrell, J. E., Jr. (2007). Mechanisms of specificity in protein phosphorylation. *Nature Reviews. Molecular Cell Biology*, *8*, 530–541.

Verd, B., Clark, E., Wotton, K. R., Janssens, H., Jimenez-Guri, E., Crombach, A., et al. (2018). A damped oscillator imposes temporal order on posterior gap gene expression in Drosophila. *PLoS Biology*, *16*, e2003174.

Verd, B., Crombach, A., & Jaeger, J. (2017). Dynamic maternal gradients control timing and shift-rates for Drosophila gap gene expression. *PLoS Computational Biology*, *13*, e1005285.

Verd, B., Monk, N. A., & Jaeger, J. (2019). Modularity, criticality, and evolvability of a developmental gene regulatory network. *eLife*, *8*.

Vilar, J. M., & Leibler, S. (2003). DNA looping and physical constraints on transcription regulation. *Journal of Molecular Biology*, *331*, 981–989.

von Dassow, G., Meir, E., Munro, E. M., & Odell, G. M. (2000). The segment polarity network is a robust developmental module. *Nature*, *406*, 188–192.

Von Dassow, G., & Odell, G. M. (2002). Design and constraints of the Drosophila segment polarity module: Robust spatial patterning emerges from intertwined cell state switches. *The Journal of Experimental Zoology*, *294*, 179–215.

von Hippel, P. H., Revzin, A., Gross, C. A., & Wang, A. C. (1974). Non-specific DNA binding of genome regulating proteins as a biological control mechanism: I. The lac operon: Equilibrium aspects. *Proceedings of the National Academy of Sciences of the United States of America*, *71*, 4808–4812.

Wang, C., Han, B., Zhou, R., & Zhuang, X. (2016). Real-time imaging of translation on single mRNA transcripts in live cells. *Cell*, *165*, 990–1001.

Wang, C., & Lehmann, R. (1991). Nanos is the localized posterior determinant in Drosophila. *Cell*, *66*, 637–647.

Wharton, R. P., & Struhl, G. (1991). RNA regulatory elements mediate control of Drosophila body pattern by the posterior morphogen nanos. *Cell*, *67*, 955–967.

Wu, B., Eliscovich, C., Yoon, Y. J., & Singer, R. H. (2016). Translation dynamics of single mRNAs in live cells and neurons. *Science*, *352*, 1430–1435.

Xu, H., Sepulveda, L. A., Figard, L., Sokac, A. M., & Golding, I. (2015). Combining protein and mRNA quantification to decipher transcriptional regulation. *Nature Methods*, *12*, 739–742.

Xu, H., Skinner, S. O., Sokac, A. M., & Golding, I. (2016). Stochastic kinetics of Nascent RNA. *Physical Review Letters*, *117*, 128101.

Yamada, S., Whitney, P. H., Huang, S. K., Eck, E. C., Garcia, H. G., & Rushlow, C. A. (2019). The Drosophila pioneer factor Zelda modulates the nuclear microenvironment of a dorsal target enhancer to potentiate transcriptional output. *Current Biology*, *29*, 1387-1393 e1385.

Yan, X., Hoek, T. A., Vale, R. D., & Tanenbaum, M. E. (2016). Dynamics of translation of single mRNA molecules in vivo. *Cell*, *165*, 976–989.

Yunger, S., Rosenfeld, L., Garini, Y., & Shav-Tal, Y. (2010). Single-allele analysis of transcription kinetics in living mammalian cells. *Nature Methods*, *7*, 631–633.

Zechner, C., Unger, M., Pelet, S., Peter, M., & Koeppl, H. (2014). Scalable inference of heterogeneous reaction kinetics from pooled single-cell recordings. *Nature Methods*, *11*, 197–202.

Zenklusen, D., Larson, D. R., & Singer, R. H. (2008). Single-RNA counting reveals alternative modes of gene expression in yeast. *Nature Structural & Molecular Biology*, *15*, 1263–1271.

Zoller, B., Little, S. C., & Gregor, T. (2018). Diverse spatial expression patterns emerge from unified kinetics of transcriptional bursting. *Cell*, *175*, 835-847 e825.

Zoller, B., Nicolas, D., Molina, N., & Naef, F. (2015). Structure of silent transcription intervals and noise characteristics of mammalian genes. *Molecular Systems Biology [Electronic Resource]*, *11*, 823.

CHAPTER TWO

Optogenetic approaches to investigate spatiotemporal signaling during development

Katherine W. Rogers[a], **Patrick Müller**[a,b,*]

[a]Systems Biology of Development Group, Friedrich Miescher Laboratory of the Max Planck Society, Tübingen, Germany

[b]Modeling Tumorigenesis Group, Translational Oncology Division, Eberhard Karls University Tübingen, Tübingen, Germany

[*]Corresponding author: e-mail address: patrick.mueller@tuebingen.mpg.de

Contents

Abstract

Embryogenesis is coordinated by signaling pathways that pattern the developing organism. Many aspects of this process are not fully understood, including how signaling molecules spread through embryonic tissues, how signaling amplitude and dynamics are decoded, and how multiple signaling pathways cooperate to pattern the body plan. Optogenetic approaches can be used to address these questions by providing precise experimental control over a variety of biological processes. Here, we review how these strategies have provided new insights into developmental signaling and discuss how they could contribute to future investigations.

Current Topics in Developmental Biology, Volume 137
ISSN 0070-2153
https://doi.org/10.1016/bs.ctdb.2019.11.009

1. Optogenetic approaches

A major goal of developmental biology is to understand how specific tissues are reliably generated in different regions of the embryo. While decades of work have demonstrated that this process is orchestrated by signaling molecules and the pathways and genes they control, the specific underlying mechanisms remain incompletely characterized. A comprehensive understanding of embryogenesis requires deeper dissection of these mechanisms.

Optogenetic approaches use genetically encoded light-responsive proteins to manipulate biological functions, often with a high degree of spatiotemporal control. These methods can yield insights into embryogenesis by granting unprecedented experimental control over processes related to development, such as signaling and gene expression. Here, we first introduce light-responsive proteins and describe several of their applications and advantages. We then highlight outstanding questions in developmental biology, discuss how optogenetic approaches over the last 10 years have begun to address some of these questions, and suggest how similar strategies could provide novel insights. We end with a brief discussion of some practical considerations when using optogenetic approaches.

1.1 Introduction to light-responsive proteins

The use of optogenetic approaches in neuroscience gained momentum around 2005 following the discovery of channelrhodopsins, algal light-responsive ion channels that can be ectopically expressed in neurons to manipulate neuronal activity (Boyden, Zhang, Bamberg, Nagel, & Deisseroth, 2005; Josselyn, 2018; Li et al., 2005; Nagel et al., 2005, 2003). Other light-responsive proteins from bacteria, fungi, algae, and plants exhibit a range of reversible light-dependent behaviors such as hetero- and homodimerization, oligomerization, unfolding, and reactive oxygen species generation.[a] These functionalities have been exploited to manipulate diverse biological processes, including signaling, gene expression and epigenetic modification, apoptosis, cytoskeletal interactions, cell/nucleus motility, endocytosis, and protein localization, trafficking, and stability. Optogenetic systems using light-responsive proteins have been successfully applied *in vitro* as well as in

[a] More information on light-responsive proteins and their applications can be found at https://www.optobase.org, a searchable optogenetic literature and reagent database (Kolar, Knobloch, Stork, Znidaric, & Weber, 2018)

organisms as diverse as *E. coli*, *Dictyostelium*, yeast, *C. elegans*, sea urchins, *Drosophila*, *Xenopus*, zebrafish, and mice. Typically, these systems feature a light-responsive protein fused to an effector, such as a signal transducer or transcription factor, whose activity, localization, or binding interactions are then altered in response to light.

In this review, we focus mainly on three popular light-responsive proteins: the blue light-responsive LOV (Light-Oxygen-Voltage) domain and cryptochrome CRY2, and the red light-responsive phytochrome PhyB. LOV domains are a subset of the PAS (Period-ARNT-Singleminded) family, and over 6000 LOV variants have been identified in archaea, bacteria, protists, fungi, algae, and plants (Glantz et al., 2016; Losi, Gardner, & Möglich, 2018). The LOV domain has two reversible responses to blue light exposure that are commonly exploited in optogenetic approaches: homodimerization and the unfolding of its C-terminal Jα helix (Fig. 1A). Cryptochromes such as CRY2 have roles in regulating circadian rhythms and plant flowering (Losi et al., 2018; Yu, Liu, Klejnot, & Lin, 2010). In the presence of blue light, CRY2 can either homo-oligomerize (Taslimi et al., 2014) or heterodimerize with its light-insensitive binding partner CIB (cryptochrome-interacting basic helix-loop-helix) (Fig. 1B). Finally, the red light-responsive phytochrome PhyB has multiple roles in plants including shade avoidance, regulation of germination, and thermosensation (Franklin & Quail, 2010; Jung et al., 2016; Schepens, Duek, & Fankhauser, 2004), and it can heterodimerize with its light-insensitive binding partner PIF (phytochrome-interacting factor) (Fig. 1C) (Quail, 2002).

The light-induced responses of LOV, CRY2/CIB, and PhyB/PIF are reversible after light removal, but the timescales of reversion to the dark state differ between variants (Zoltowski, Vaccaro, & Crane, 2009). Many optogenetic systems based on LOV or CRY2/CIB use variants with rapid (seconds to minutes) reversion kinetics. In contrast, the photoactivated state of PhyB/PIF has an hours-long lifetime, but far-red light exposure rapidly induces reversion to the dark state and dissociation (Ni, Tepperman, & Quail, 1999).

Many light-responsive proteins bind extrinsic chromophores that mediate light absorption (Anders & Essen, 2015; Losi et al., 2018), although some exceptions such as the UV light-responsive UVR8 (Crefcoeur, Yin, Ulm, & Halazonetis, 2013; Rizzini et al., 2011) and cyan light-responsive Dronpa variants (Mizuno et al., 2008; Zhou, Chung, Lam, & Lin, 2012; Zhou, Fan, Li, Shen, & Lin, 2017) form intrinsic chromophores similar to GFP. Blue light-responsive proteins such as LOV and CRY2 use flavin

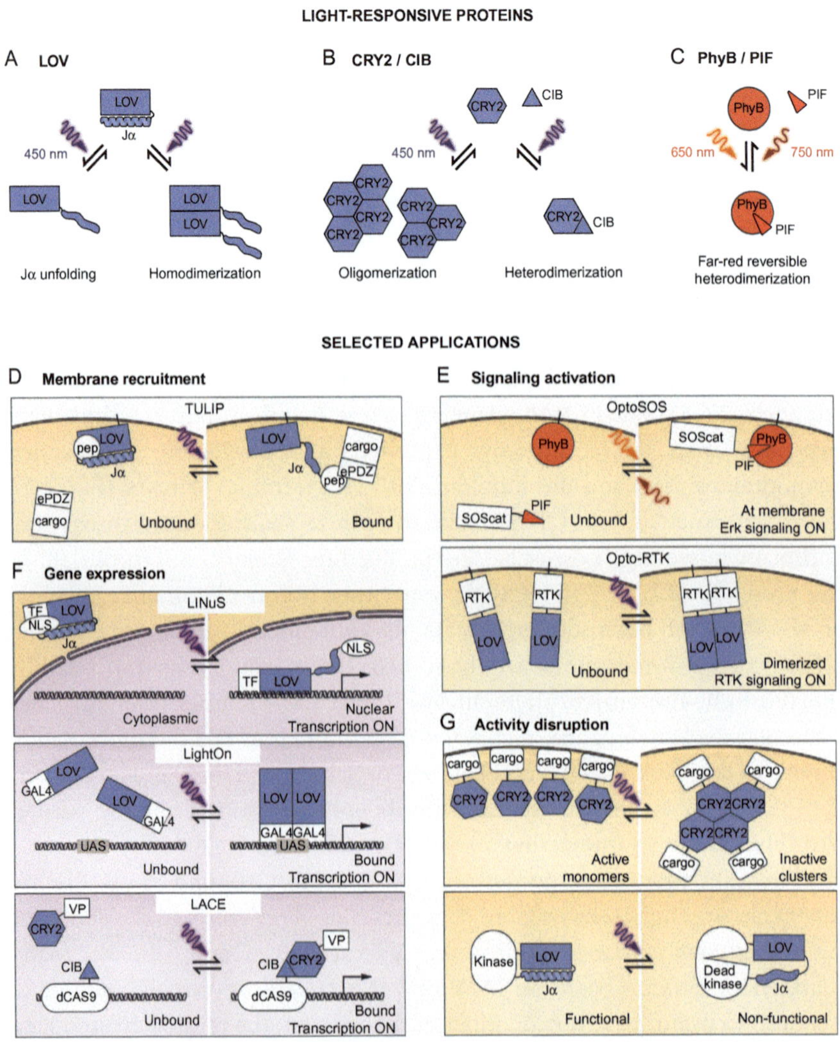

Fig. 1 Selected light-responsive proteins and their applications. (A) Blue light illumination induces unfolding of the LOV Jα helix and homodimerization. (B) Blue light illumination induces CRY2 oligomerization or heterodimerization with its binding partner CIB. (C) Red light illumination induces heterodimerization of PhyB and its binding partner PIF. Dimerization is reversible with far-red illumination. (D–G) Selected applications of light-responsive domains. (D) With TULIP, light-mediated unfolding of the LOV domain Jα helix exposes a peptide (pep), which binds an ePDZ-cargo fusion protein. Cargo can be recruited to distinct cellular compartments (e.g., plasma membrane) by targeting LOVpep to the desired compartment. (E) Signaling activation strategies. (Top) Erk signaling can be reversibly activated by recruiting SOScat to the membrane using membrane-bound PhyB and a SOScat-PIF fusion. (Bottom) Signaling pathways that are activated by the assembly of receptor kinase complexes (such as those

derivatives as their chromophores (flavin mononucleotide and flavin adenine dinucleotide, respectively). Due to the convenient abundance of flavins in cells, exogenous chromophore is not required with LOV- or CRY2-based systems. In contrast, red light-responsive proteins such as PhyB use phycocyanobilin or phytochromobilin (PCB) as their chromophore, which is not naturally produced by many bacterial, fungal, or animal cells and therefore must be added (Kyriakakis et al., 2018). This can be advantageous, as systems requiring exogenous chromophores can be handled in the light prior to PCB addition (Adrian, Nijenhuis, Hoogstraaten, Willems, & Kapitein, 2017; Yang, Jost, Weiner, & Tang, 2013). Alternatively, PCB biosynthesis pathways engineered into *E. coli* (Beyer et al., 2018; Zhang et al., 2009) and mammalian cells (Kyriakakis et al., 2018; Müller, Engesser, Timmer, et al., 2013; Uda et al., 2017) can provide the chromophore for optogenetic systems based on PhyB/PIF.

1.2 Applications of light-responsive proteins

Here, we discuss how light-responsive proteins have been used to manipulate protein localization, signaling pathways, gene expression and chromatin modification, and protein function. We cover a subset of applications relevant for developmental signaling (see Section 2), but many additional techniques are available (reviewed in de Mena, Rizk, & Rincon-Limas, 2018; Guglielmi, Falk, & De Renzis, 2016; Isomura & Kageyama, 2017; Johnson & Toettcher, 2018; Khamo, Krishnamurthy, Sharum, Mondal, & Zhang, 2017; Kowalik & Chen, 2017; Krueger et al., 2019; Losi et al., 2018; Tischer & Weiner, 2014), including approaches for studying tissue morphogenesis (Čapek et al., 2019; Guglielmi, Barry, Huber, & De Renzis, 2015; Izquierdo, Quinkler, & De Renzis, 2018; Krueger,

mediated by RTKs) can be activated optogenetically by fusing LOV domains to receptor kinase domains. (F) Strategies to manipulate gene expression. (Top) LINuS can be used to transport a LOV-transcription factor fusion into the nucleus by revealing a nuclear localization signal (NLS) fused to the Jα helix when illuminated. (Middle) In the LightON system, an obligate dimer transcriptional activator (GAL4) is fused to a LOV domain. Light induces dimerization, promoting expression of genes with upstream activating sequences (UAS). (Bottom) With LACE, nuclease-dead dCas9 can be used to direct CIB to a desired locus. CRY2 fused to a transcriptional effector (VP) can then activate expression in the presence of light. (G) Methods to disrupt protein function. (Top) Light-mediated clustering of CRY2 or CRY2/CIB fusions can affect protein function. (Bottom) Light-mediated conformational changes experienced by LOV domains can be used to affect the function of fusion partners like kinases.

Tardivo, Nguyen, & De Renzis, 2018), signal transduction (Ramachandran et al., 2018), and drug screening (Agus & Janovjak, 2017; Ingles-Prieto et al., 2015; Zhou et al., 2017).

Optogenetic strategies to manipulate protein localization (Levskaya, Weiner, Lim, & Voigt, 2009) have been used in applications ranging from signaling pathway activation (Johnson et al., 2017; Johnson & Toettcher, 2019; Toettcher, Weiner, & Lim, 2013; Wilson, Ravindran, Lim, & Toettcher, 2017) to moving organelles (Adrian et al., 2017) to "knocksideways" approaches in which a protein is functionally inactivated by depleting it from its normal location (Benedetti et al., 2018; Nguyen et al., 2016; Robinson, Sahlender, & Foster, 2010; Wang et al., 2016; Yang et al., 2013). One powerful strategy for relocating proteins is the "tunable light-inducible dimerization tag" (TULIP), which exploits both the light-mediated Jα helix unfolding of the LOV domain and the high-affinity interaction between an engineered ePDZ domain and its target peptide (Fig. 1D) (Strickland et al., 2012). The small target peptide is fused to the LOV Jα helix, which unfolds and exposes the peptide in response to blue light. This "LOVpep" fusion can be localized to different cellular compartments, such that cargo fused to ePDZ can be recruited reversibly to the desired compartment. The "improved light-inducible dimer" (iLID) system (Guntas et al., 2015) is conceptually similar, but utilizes the high affinity binding partners SspB/SsrA from *E. coli* instead of ePDZ/peptide to minimize possible cross-talk with endogenous signaling pathways.

Techniques to manipulate protein localization have been used to activate signaling pathways by recruiting signaling effectors to the membrane (Levskaya et al., 2009). OptoSOS reversibly activates Erk signaling *in vitro* using a two-component system: membrane-tethered PhyB, and PIF fused to SOScat, a truncated protein that activates Ras at the membrane (Fig. 1E, upper panel) (Toettcher et al., 2013; Wilson et al., 2017). An iLID-based OptoSOS system was developed for *in vivo* applications to activate Erk signaling in *Drosophila* embryos (Johnson et al., 2017; Johnson, Shvartsman, & Toettcher, 2019; Johnson & Toettcher, 2019). Another strategy to control signaling pathways exploits the fact that the transduction of many signals—including FGF, EGF, Ret, Nodal, BMP, and insulin—is mediated by the assembly of receptor complexes upon ligand binding (Derynck & Budi, 2019; Lemmon & Schlessinger, 2010). Reversible light-mediated activation of the FGF, EGF, and Ret signaling pathways has been achieved *in vitro* by fusing LOV or CRY2 proteins to their respective receptor tyrosine kinases (RTKs) to generate Opto-RTKs (Fig. 1E, lower panel) (Grusch et al., 2014; Kim et al., 2014). Blue light exposure mimics ligand binding by causing Opto-RTKs to form complexes and activate signaling. Nodal signaling can be optogenetically activated in zebrafish embryos using a similar approach (Sako et al., 2016; Vopalensky, Pralow, & Vastenhouw, 2018). Alternative optogenetic methods have also been developed to control Wnt (Bugaj, Choksi, Mesuda, Kane, & Schaffer, 2013; Čapek et al., 2019), Erk (Dine, Gil, Uribe, Brangwynne, & Toettcher, 2018; Kainrath, Stadler, Reichhart, Distel, & Janovjak, 2017), and other signaling pathways (Morri et al., 2018).

Many optogenetic strategies have been developed to manipulate gene expression and chromatin modifications in cells ranging from yeast to mouse neurons *in vivo* (Fig. 1F) (Bubeck et al., 2018; Crefcoeur et al., 2013; Hughes, Bolger, Tapadia, & Tucker, 2012; Masuda, Nakatani, Ren, & Tanaka, 2013; Müller, Engesser, Metzger, et al., 2013; Ohlendorf, Vidavski, Eldar, Moffat, & Möglich, 2012; Rullan, Benzinger, Schmidt, Milias-Argeitis, & Khammash, 2018; Sokolik et al., 2015). Several systems exploit light-mediated revelation of nuclear localization/export signals to shuttle transcription factors or histone modifiers into or out of the nucleus, including LINuS (Fig. 1F, top panel) (Niopek et al., 2014), LANS (Yumerefendi et al., 2015), LINX (Yumerefendi et al., 2016), LEXY (Niopek, Wehler, Roensch, Eils, & Di Ventura, 2016), and LANSTRAP (Yumerefendi et al., 2018). LightOn (Fig. 1F, middle panel) (Wang, Chen, & Yang, 2012), EL222 (Motta-Mena et al., 2014), and TAEL (Reade et al., 2017) systems contain LOV domains that activate transcriptional effectors in response to blue light, which in turn activate expression of genes modified to harbor the appropriate upstream activating sequence. Alternatively, several unmodified loci have been optogenetically manipulated using customizable approaches involving two components: a light-responsive protein fused to a transcriptional effector or chromatin modifier, and a binding partner fused to a programmable DNA-binding domain such as zinc finger proteins (Polstein & Gersbach, 2012), TALEs (Konermann et al., 2013; Lo, Choudhury, Irudayaraj, & Zhou, 2017), or nuclease-dead dCas9, as used in the LACE approach (Fig. 1F, bottom panel) (Nguyen, He, Martinez-Moczygemba, Huang, & Zhou, 2018; Nihongaki, Yamamoto, Kawano, Suzuki, & Sato, 2015; Polstein & Gersbach, 2015).

Finally, protein function has been manipulated optogenetically using light-regulated clustering, allosteric regulation, and steric inhibition

(Fig. 1G). Techniques involving light-regulated clustering of light-responsive protein fusions have been successful in transiently activating (Bugaj et al., 2013; Dine et al., 2018) or inhibiting (Huang, Amourda, Zhang, Tolwinski, & Saunders, 2017; Kainrath et al., 2017; Lee et al., 2014; Taslimi et al., 2014) protein function. A second strategy to manipulate protein function uses the light-induced conformational changes in LOV domains to allosterically regulate the activity of a fusion partner, such as a kinase or GEF (Dagliyan et al., 2016; Gehrig et al., 2017; Lee et al., 2008). This approach has recently been used to control gene editing by fusing a LOV domain internally into a Cas9-targeting antibody: Cas9 function is inhibited by the antibody in the dark, but activated by light exposure (Bubeck et al., 2018). A third strategy based on steric inhibition created by light-dissociable Dronpa dimers has also been used to manipulate protein function. Dronpa monomers can be fused to the N- and C-termini (or loops) of target proteins including kinases, GEFs, and proteases (Zhou et al., 2012, 2017). Dronpa dimerization inhibits target protein function until exposure to cyan light dissociates the dimer and reveals the active site.

The diversity of natural and engineered light-responsive proteins and their potential applications suggest many exciting experimental possibilities. Even G-protein coupled receptors (GPCRs)—which have complex structures involving seven transmembrane domains—have been successfully engineered to optogenetically activate non-canonical Wnt signaling (Čapek et al., 2019) or neuronal activity (van Wyk, Pielecka-Fortuna, Lowel, & Kleinlogel, 2015), to identify functions of orphan GPCRs (Morri et al., 2018), and to direct immune cells toward tumors (Xu, Hyun, et al., 2014).

1.3 Key advantages of optogenetic approaches

In contrast to many genetic and pharmacological methods, optogenetic approaches can provide tunable, reversible experimental manipulations with a high degree of spatial and temporal control. We highlight these key advantages below, and in Section 2 we describe how they can be used to provide new insights into developmental signaling.

1.3.1 Tunability

Light intensity-dependent "tunability" has been demonstrated in many optogenetic systems (Chen et al., 2019; Chen & Wegner, 2017; Chervyachkova & Wegner, 2018; Guglielmi et al., 2015; Lee et al., 2014; Ohlendorf et al., 2012; Rullan et al., 2018; Tichy, Gerrard,

Legrand, Hobbs, & Janovjak, 2019; Yang et al., 2013). For example, membrane recruitment of SOScat in *Drosophila* (Johnson et al., 2017) using the iLID system (Guntas et al., 2015) and of a cell cycle protein in sea urchin embryos (Uchida & Yajima, 2018) using the TULIP system (Strickland et al., 2012) was stronger at higher light intensities. Optogenetic Erk pathway modifiers based on CRY2 (Aoki et al., 2013; Kim et al., 2014) or CBD (Kainrath et al., 2017) exert light intensity-dependent effects on Erk signaling in mammalian cells, and in *Drosophila* a CRY2-based system that inhibits a developmental signaling pathway is more effective at higher intensities (Huang et al., 2017). Light intensity-dependent gene expression has been demonstrated in vertebrate cells using the LightOn system (Sokolik et al., 2015; Wang et al., 2012) and other gene regulation systems (Crefcoeur et al., 2013; Müller, Engesser, Metzger, et al., 2013; Polstein & Gersbach, 2012). Similarly, optogenetically regulated protein degradation in yeast and mammalian cells is faster at higher intensities (Renicke, Schuster, Usherenko, Essen, & Taxis, 2013; Sun et al., 2017; Usherenko et al., 2014), and optogenetic induction of liquid-like protein droplets was light intensity-dependent (Shin et al., 2017). Phenotypic outputs can also be intensity-dependent: neurons induced from cultured rat cells by an Erk activation system based on CRY2/CIB had longer outgrowths at higher light intensities (Zhang et al., 2014), and cellular tension could be modulated in an intensity-dependent manner in *Drosophila* embryos using a CRY2/CIB-based system to target an actin-depleting protein to the membrane (Guglielmi et al., 2015). In addition, the activity of PhyB/PIF-based systems can be tuned by altering the ratio of red to far-red light exposure (Fig. 1C) (Toettcher, Gong, Lim, & Weiner, 2011; Toettcher et al., 2013).

Optogenetic systems have also been tuned by introducing mutations or chemicals such as imidazole that affect dimerization or photo-response kinetics (Chen & Wegner, 2017; Strickland et al., 2012; Usherenko et al., 2014; Wang et al., 2016; Zoltowski et al., 2009), by expressing different levels of optogenetic constructs (Bubeck et al., 2018; Huang et al., 2017; Shin et al., 2017), or by altering the concentration of PCB chromophore in PhyB/PIF-based systems (Hughes et al., 2012; Müller, Engesser, Metzger, et al., 2013). Further, it has been observed for a variety of systems—across different light-responsive proteins, biological processes, and model organisms—that light pulses produce distinct, often stronger responses than continuous exposure (Aoki et al., 2013; Bugaj et al., 2013; Chen et al., 2019; Imayoshi et al., 2013; Izquierdo et al., 2018; Lee et al., 2014; Toettcher et al., 2013; Wang et al., 2016; Wilson et al., 2017).

For example, in *C. elegans* light-induced neuronal ablation using miniSOG was more effective when light exposure was pulsed rather than continuous (Qi, Garren, Shu, Tsien, & Jin, 2012), and TAEL-mediated reporter gene expression in zebrafish embryos was stronger with pulsed light (Reade et al., 2017). Photodamage (Reade et al., 2017) or adaptation mechanisms such as negative feedback (Heemskerk et al., 2019; Wilson et al., 2017; Zhou et al., 2017) likely have roles in differential responses to pulsed versus continuous light, but in many cases the explanations for this observation remain unexplored.

1.3.2 Spatial control

Optogenetic approaches are particularly well-suited for precise, reversible spatial manipulations that are difficult or impossible with genetic or pharmacological methods (Faden, Mielke, Lange, & Dissmeyer, 2014). For example, manipulations involving tissue-specific promoters are restricted by promoter availability and potentially cumbersome generation of transgenic animals (McGuire, Roman, & Davis, 2004). Small molecule agonists/antagonists and recombinant proteins are highly diffusible and their application is difficult to spatially restrict (Paliwal et al., 2007). Other approaches such as localized electroporation, Cre/Lox clones, and bead or cell transplantations can be challenging to implement and difficult to control and interpret.

In contrast, many optogenetic approaches have achieved cellular or even subcellular spatial resolution both *in vitro* and *in vivo*, especially when paired with digital micromirror devices (DMDs), confocal, lightsheet, or 2-photon illumination (Adrian et al., 2017; Arrenberg, Stainier, Baier, & Huisken, 2010; Beyer et al., 2018; Buckley et al., 2016; Crefcoeur et al., 2013; Deneke et al., 2019; Dine et al., 2018; Grusch et al., 2014; Huang et al., 2017; Johnson et al., 2017; Johnson & Toettcher, 2019; Kaur, Saunders, & Tolwinski, 2017; Lee et al., 2014; Niopek et al., 2016; Ohlendorf et al., 2012; Polstein & Gersbach, 2012; Polstein, Juhas, Hanna, Bursac, & Gersbach, 2017; Reade et al., 2017; Shin et al., 2017; Strickland et al., 2012; Taslimi et al., 2014; Uchida & Yajima, 2018; Wood, Larocque, Clarke, Sarkar, & Royle, 2017; Yang et al., 2013; Zhou et al., 2012). For example, localized lamellipodia induction has been demonstrated in cultured cells using focused illumination and PhyB/PIF to recruit cytoskeleton-modifying RhoGEFs to the membrane (Levskaya et al., 2009), or using CRY2 to induce oligomerization of an FGF receptor, resulting in migration toward localized illuminated regions (Kim et al., 2014). PI3P was specifically depleted from single endosomes in mammalian cells—and not from neighboring endosomes—using LOV-based "Magnets"

(Kawano, Suzuki, Furuya, & Sato, 2015) and a confocal microscope to locally recruit a PI3P phosphatase, achieving organelle-specific resolution *in vitro* (Benedetti et al., 2018). Sub-cellular optogenetic manipulations have been realized *in vivo* using 2-photon laser illumination (Guglielmi et al., 2015; Izquierdo et al., 2018; Krueger et al., 2018): For example, a CRY2/CIB-based system was used to recruit a RhoGEF specifically to the apical membrane of selected cells in live *Drosophila* embryos, activating myosin II and driving region-specific tissue invagination (Izquierdo et al., 2018).

1.3.3 Temporal control

Temporal manipulations using methods such as heat-shock promoters or temperature-sensitive alleles can be limited by long activation/deactivation kinetics and temperature stress (Connors, Tucker, & Mullins, 2006; Faden et al., 2014; Hagos & Dougan, 2007; Haruki, Nishikawa, & Laemmli, 2008; McGuire et al., 2004; Tuazon & Mullins, 2015; Tucker, Mintzer, & Mullins, 2008). Experimental control over the dynamics of pathway activation are technically difficult to achieve using recombinant ligands or small molecule agonists/antagonists, and may require specialized equipment such as microfluidic devices that are complicated to implement in experiments requiring live microscopy (Hansen & O'Shea, 2016; Hao & O'Shea, 2012; Heemskerk et al., 2019; Lucchetta, Lee, Fu, Patel, & Ismagilov, 2005; Paliwal et al., 2007; Purvis et al., 2012; Ryu et al., 2015; Sorre, Warmflash, Brivanlou, & Siggia, 2014; Tay et al., 2010; Zagorski et al., 2017). Further, ligand concentration in media can be difficult to regulate due to degradation and receptor-mediated endocytosis, and long-lived internalized receptor/ligand complexes can prolong signaling after ligand removal (Haruki et al., 2008; Jullien & Gurdon, 2005; Sorre et al., 2014; Tay et al., 2010; Toettcher et al., 2013).

In contrast, optogenetic approaches are ideal to investigate cellular responses to signaling duration and dynamics. The rapid response kinetics of many light-responsive proteins allow reversible activation or inhibition of biological processes, including signaling, at minute or second timescales (Adrian et al., 2017; Arrenberg et al., 2010; Beyer et al., 2018; Buckley et al., 2016; Bugaj et al., 2013; Chen et al., 2019; Dine et al., 2018; Grusch et al., 2014; Guglielmi et al., 2015; Huang et al., 2017; Johnson et al., 2017; Kim et al., 2014; Lee et al., 2014; Levskaya et al., 2009; Niopek et al., 2016; Rullan et al., 2018; Shin et al., 2017; Taslimi et al., 2014; Tischer & Weiner, 2014; Toettcher et al., 2011, 2013; Wang et al., 2016; Wood et al., 2017; Yang et al., 2013; Yumerefendi et al., 2015, 2018; Zhang et al., 2014). Components of optogenetic gene regulatory systems can also be destabilized—for example, by using unstable 3′ UTRs to destabilize mRNA (Isomura, Ogushi, Kori, & Kageyama, 2017; Wang et al., 2012)—in case fast oscillatory gene expression is desired. The rapid kinetics of many optogenetic systems make them ideal tools for experiments requiring temporally precise manipulations.

2. Optogenetic applications in developmental signaling

Here, we highlight how the unique features of optogenetic systems have been beneficial for investigations of spatiotemporal signaling during development and suggest how these approaches could be further exploited. We use the foundational "French flag" model of tissue patterning as a framework to guide our discussion.

In the simplest contemporary conception of the French flag model, different concentration thresholds of a signaling molecule activate distinct sets of target genes and cell fates (Fig. 2A). The graded distribution of a signaling molecule within an embryonic tissue therefore results in position-dependent cell fate specification (Wolpert, 1968, 1969, 1989, 2016). Many studies over the last 50 years have either supported, refined, or challenged this specific model of embryogenesis, but a crucial role for signaling molecules in coordinating patterning has been clearly demonstrated. Below we discuss how optogenetic approaches can provide mechanistic insights into patterning mediated by signaling molecules including: the spreading of signaling molecules through tissues (Fig. 2B), the spatiotemporal requirements for signaling during embryogenesis (Fig. 2C and D), the decoding of signaling amplitude (Fig. 2E) and duration (Fig. 2F), the impact of noise on embryogenesis (Fig. 2G), and finally, the interpretation of combinatorial signaling (Fig. 2H).

2.1 How do signaling molecules spread through tissues?

The French flag model proposes that spatial gradients of signaling molecules coordinate patterning (Fig. 2A). How do signaling molecules spread through embryonic tissues? They may move by extracellular diffusion from a localized source, or by repeated rounds of cellular uptake and release, or they may propagate down long cellular extensions called cytonemes[b]

[b] Other mechanisms, such as convective flow and transport by extracellular vesicles, are also under active investigation (Müller & Schier, 2011)

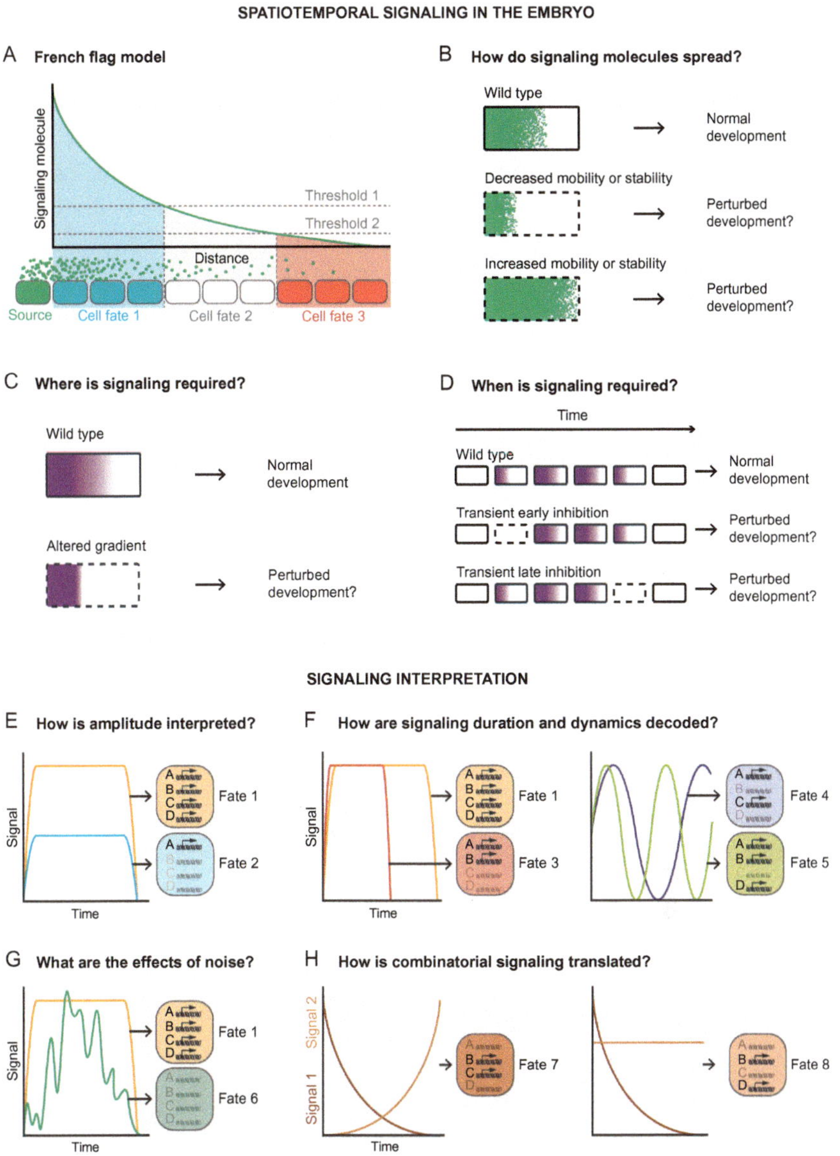

Fig. 2 Optogenetic approaches to investigate spatiotemporal signaling during development. (A) The French flag model states that a spatial gradient of signaling molecules (green dots) patterns tissues by activating concentration-specific cellular responses (blue, white, and red) at different thresholds (gray dotted lines). The gradient may form by diffusion from a localized source (green cell). (B–D) Optogenetic strategies could be used to assess spatiotemporal requirements for signaling during embryogenesis with *in vivo* manipulations (dashed lines) of signaling molecule diffusivity or stability (B), activity gradient shape (C), or timing of signaling (D). (E–H) Optogenetic manipulation of signaling can help determine how signaling amplitude (E), dynamics (F), and noise (G) are converted into gene expression and cell fate decisions. Combining orthogonal optogenetic approaches could help reveal how inputs from multiple signaling pathways are interpreted (H).

(Kornberg & Roy, 2014; Mattes & Scholpp, 2018; Müller, Rogers, Yu, et al., 2013; Müller & Schier, 2011; Rogers & Müller, 2019; Rogers & Schier, 2011; Wartlick, Kicheva, & González-Gaitán, 2009; Yamashita, Inaba, & Buszczak, 2018). Alternatively, some signals may not move far from their sources, but rather be restricted by low mobility to their expression domains (Brankatschk & Dickson, 2006; Dominici et al., 2017; Dubrulle & Pourquié, 2004; Hashimoto-Partyka, Yuge, & Cho, 2003; Ramel & Hill, 2013; van Boxtel et al., 2015; Varadarajan et al., 2017). For example, blocking the movement of a secreted Wnt ligand surprisingly did not abolish its ability to pattern the *Drosophila* wing (Alexandre, Baena-Lopez, & Vincent, 2014). Optogenetic approaches could be used to measure and manipulate the mobility, stability, and trafficking of signaling molecules in their native context in order to understand the mechanisms by which they move through tissues (Fig. 2B).

Extracellular diffusion has been manipulated *in vivo* using different approaches, such as tethering signaling molecule binding partners to cell surfaces (Almuedo-Castillo et al., 2018; Harmansa, Hamaratoglu, Affolter, & Caussinus, 2015; Mörsdorf & Müller, 2019; Pani & Goldstein, 2018). However, these experiments have limited spatial and temporal precision, and coupling such approaches to optogenetics could dramatically expand the possibilities. Extracellular optogenetic interactions have been demonstrated in bacterial and mammalian cells *in vitro* (Baaske et al., 2019; Chen & Wegner, 2017; Ricken, Medda, & Wegner, 2019; Yüz, Rasoulinejad, Mueller, Wegner, & Wegner, 2019; Yüz, Ricken, & Wegner, 2018), and even beads or artificial vesicles decorated with light-responsive proteins have been manipulated with light (Bartelt, Steinkuhler, Dimova, & Wegner, 2018; Chervyachkova & Wegner, 2018). In one example, cell adhesion to glass was controlled optogenetically by coating the glass with a LOV-based light switchable integrin-binding motif (Ricken et al., 2019). Similar strategies to optogenetically expose integrin-binding motifs could be used to immobilize diffusing signaling molecules (or migrating cells) within embryonic tissues. Signaling molecule movement could then be reversibly arrested in specific regions or stages, and the developmental consequences assessed.

The stability of extracellular signaling molecules is thought to affect gradient formation and signaling range, and can be measured *in vivo* (Bläßle & Müller, 2015; Müller et al., 2012; Rogers, Bläßle, Schier, & Müller, 2015). Optogenetic methods have been used to measure intracellular protein half-lives through pulse-chase experiments (Hughes et al., 2012; Sokolik et al.,

2015; Wilson et al., 2017), and could presumably be used in an extracellular context as well. However, manipulating extracellular protein stability is challenging. Optogenetic strategies to destabilize intracellular proteins via proteasome-mediated degradation have been successful (Bonger, Rakhit, Payumo, Chen, & Wandless, 2014; Renicke et al., 2013; Sun et al., 2017; Usherenko et al., 2014), but to our knowledge optogenetic manipulation of extracellular protein stability has not yet been shown. Optogenetically induced intracellular TEV-mediated protein cleavage (Lee, Hyun, Jung, Hannan, & Kwon, 2017; Wang et al., 2017) and light-mediated cleavage of photocleavable proteins (Zhang et al., 2017) have been demonstrated, which could potentially be applied to extracellular signals. Another feasible approach to manipulate extracellular signal stability is to control receptor-mediated endocytosis optogenetically (Guglielmi et al., 2016).

Methods to manipulate endocytosis are also useful for testing signal movement models such as transcytosis, in which signaling molecules move through tissues via repeated rounds of endocytosis and exocytosis (Dierick & Bejsovec, 1998; Entchev, Schwabedissen, & González-Gaitán, 2000; Gallet, Staccini-Lavenant, & Thérond, 2008; González-Gaitán & Jäckle, 1999; Kicheva et al., 2007; Kruse, Pantazis, Bollenbach, Jülicher, & González-Gaitán, 2004). Optogenetic tools that activate and inhibit endocytosis *in vitro* are now available: a TULIP-based method was shown to initiate endocytosis by recruiting clathrin to cell membranes with light (Wood et al., 2017), whereas a hyper-clustering CRY2 variant fused to clathrin reversibly disrupted endocytosis (Taslimi et al., 2014), and a CRY2/CIB-based system was able to block Rab-mediated trafficking at multiple levels, inhibiting either receptor endocytosis or degradation in late endosomes depending on which Rab was targeted (Nguyen et al., 2016). Such strategies could be used to determine how endocytosis affects the movement of signaling molecules within an embryonic tissue *in vivo*. Optogenetic tools could also be used to examine the role of cytonemes in signal transport by, for example, locally inducing (Kim et al., 2014; Levskaya et al., 2009) or inhibiting the formation of filopodia (Du, Sohr, Yan, & Roy, 2018; González-Méndez, Seijo-Barandiarán, & Guerrero, 2017; Mattes et al., 2018; Mattes & Scholpp, 2018; Yamashita et al., 2018).

Finally, signaling molecule distribution can be optogenetically manipulated by either ablating putative signal-producing cells (Makhijani et al., 2017), inhibiting expression of signal-producing genes or activating their inhibitors (Reade et al., 2017), or by blocking signal secretion

(Nguyen et al., 2016). These experiments could help determine whether a dedicated "localized source" of signal-producing cells is required (Fig. 2A), or whether other mechanisms regulate signal distribution (Akiyama & Gibson, 2015; Alexandre et al., 2014; Brankatschk & Dickson, 2006; Dominici et al., 2017; Harmansa et al., 2015; Varadarajan et al., 2017). Alternatively, if signaling molecule production was experimentally slowed rather than completely abolished, the effects of a lower amplitude, shorter-range gradient on developmental patterning could be assessed (Fig. 2C).

So far, optogenetic approaches have not been exploited to probe questions surrounding the spreading of signaling molecules during embryogenesis. However, the existing and feasible optogenetic tools described above could be harnessed to lead to major insights.

2.2 Where is signaling required?

In the context of the French flag model, spatial activity gradients in developing tissues must be highly precise to avoid disastrous patterning consequences (Fig. 2A) (Barkai & Shilo, 2009; Dubuis, Tkacik, Wieschaus, Gregor, & Bialek, 2013; Emberly, 2008; Gregor, Tank, Wieschaus, & Bialek, 2007; He et al., 2008; Houchmandzadeh, Wieschaus, & Leibler, 2002; Porcher et al., 2010; Porcher & Dostatni, 2010; Shilo & Barkai, 2017; Tostevin, ten Wolde, & Howard, 2007; Wieschaus, 2016). However, although the spatial distributions of many signaling molecules or their activity gradients have been quantified *in vivo*, whether these specific distributions are required for patterning is debated. Embryos with perturbed signal distribution or activity gradients can sometimes develop into viable, normal-appearing adults, suggesting alternative patterning mechanisms that either adjust signaling or render spatial precision irrelevant (Alexandre et al., 2014; Kishimoto, Lee, Zon, Hammerschmidt, & Schulte-Merker, 1997; Liu, Morrison, & Gregor, 2013; Lucchetta et al., 2005; Macdonald & Struhl, 1986; Namba, Pazdera, Cerrone, & Minden, 1997; Pei, Williams, Clark, Stemple, & Feldman, 2007; Rogers et al., 2017; Schulte-Merker, Lee, McMahon, & Hammerschmidt, 1997).

Optogenetic techniques can help resolve this question by providing a method to precisely alter spatial signaling gradients and assess the developmental consequences (Fig. 2C). For example, using OptoSOS (Fig. 1E) to reversibly activate Erk signaling with high temporal (1–2 min) and spatial (subcellular) resolution in *Drosophila* embryos, it was shown that ectopic activation of Erk in the middle of the embryo causes lethal patterning defects, indicating that the spatial restriction of Erk signaling to the embryonic termini is paramount for normal gastrulation (Fig. 3A) (Johnson et al., 2017). Strikingly, mutants lacking endogenous terminal Erk activity can be rescued to viability by all-or-none light exposure at the termini, demonstrating that the normal graded distribution of terminal Erk activity is dispensable for patterning (Johnson et al., 2019).

It is possible to deliver light with good spatial precision deep into tissues using standard technologies such as 2-photon illumination, and newer strategies including primed photoconversion (requiring only 1-photon excitation) (Dempsey et al., 2015; Mohr et al., 2017) and multiplexed temporally focused light shaping (Accanto et al., 2018) can also confer high spatial accuracy and penetration. Given the myriad available and improving technologies, precise optogenetic manipulations could become a widely used method to define the spatial requirements for developmental signaling.

2.3 When is signaling required?

In addition to determining *where* signaling is required during development, optogenetic approaches can help determine *when* signaling is required (Fig. 2D). Identifying the developmental windows during which specific processes occur is key to understand how tissues are patterned and how signaling is decoded (Bergmann et al., 2007; Liu et al., 2013; Lucchetta et al., 2005; Sagner & Briscoe, 2017; Tucker et al., 2008; van Boxtel et al., 2015; Zagorski et al., 2017), and is also an important step toward developing therapies for congenital disorders by defining optimal treatment stages (Schneider et al., 2018).

An optogenetic approach was recently used to define developmental windows during which the transcription factor Bicoid is required for patterning in *Drosophila* embryos (Huang et al., 2017). A CRY2-Bicoid fusion was used to transiently block Bicoid-mediated transcription over different durations and developmental stages. Inhibition of Bicoid prior to nuclear cycle 10 did not affect patterning, but inhibition between cycles 11–14 caused lethal patterning defects (Fig. 3B). More generally, genes and morphological features originating at the most anterior portion of the embryo—where Bicoid levels are highest—required longer durations of Bicoid activity, providing insight into how Bicoid signaling is temporally decoded. Interestingly, although light exposure inhibited the transcriptional activity of CRY2-Bicoid, its translational repression activities were unaffected, suggesting an approach to tease apart the contributions of Bicoid-mediated

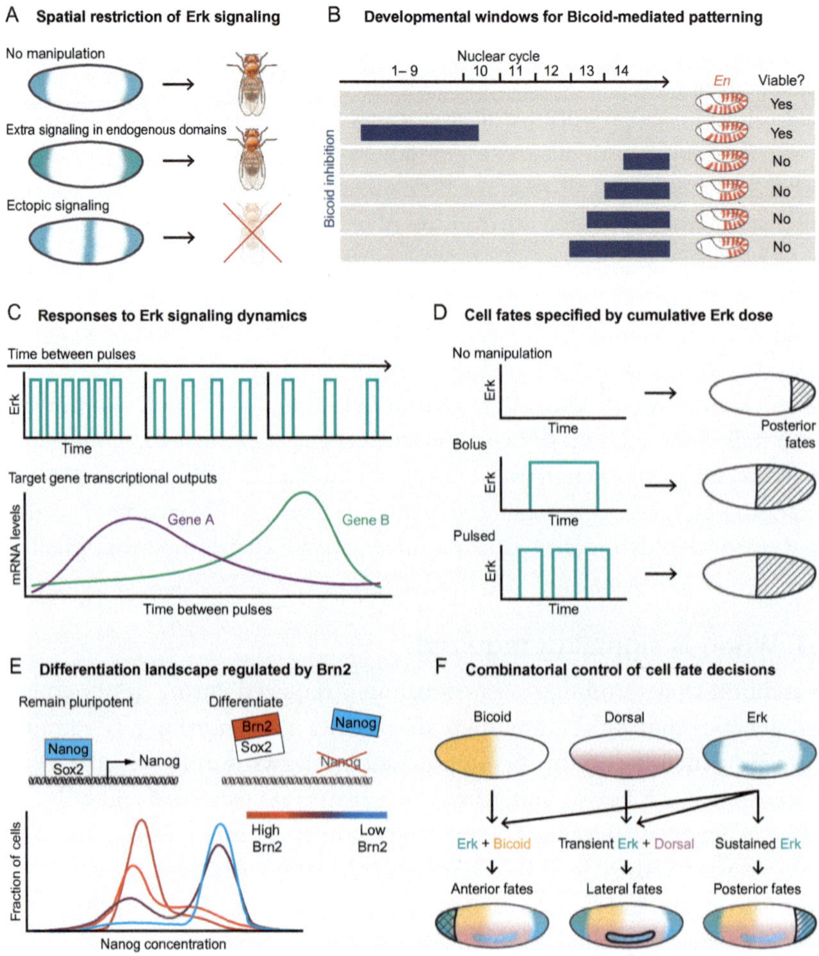

Fig. 3 Insights into developmental signaling using optogenetic approaches. (A) Erk signaling is active at the termini of the *Drosophila* embryo. Increasing Erk signaling levels at the termini is not deleterious, but ectopic activation in the center of the embryo causes lethal defects (Johnson et al., 2017). (B) Inhibition of Bicoid at early stages does not result in defects, but progressively longer inhibition during nuclear cycle 14 causes progressive loss of anterior structures marked by Engrailed (En) (Huang et al., 2017). (C) Erk signaling pulses were delivered to cultured cells with different intervals separating each pulse. Erk target genes had unique frequency-dependent responses (Wilson et al., 2017). (D) *Drosophila* embryos were posteriorized to the same degree when identical doses of Erk signaling were delivered either in one bolus or in pulses, suggesting that posterior fates are determined by the cumulative dose of Erk signaling, rather than a persistence detector (posteriorization shown here inferred from phenotypes) (Johnson & Toettcher, 2019). (E) Nanog supports pluripotency and activates its own expression by binding Sox2. Brn2 promotes differentiation and binds competitively to Sox2, inhibiting Nanog production. Interactions between these transcription factors define an energy landscape that determines the likelihood of differentiation (Sokolik et al., 2015). (F) In *Drosophila* embryos, Erk signaling is activated by Torso at the termini and by EGFR laterally, Bicoid is active in the anterior, and Dorsal is active laterally. Anterior cell fates are defined by a combination of Erk and Bicoid signaling, lateral fates by Dorsal and transient Erk signaling, and posterior fates by sustained Erk signaling (Johnson & Toettcher, 2019).

transcription and translational repression (Rivera-Pomar, Niessing, Schmidt-Ott, Gehring, & Jäckle, 1996) to early embryogenesis.

Optogenetic approaches have also been used to examine developmental windows for Nodal signaling in zebrafish (Reade et al., 2017; Sako et al., 2016; Vopalensky et al., 2018). For example, embryos lose the competence to respond to Nodal ligands by mid-gastrulation. In contrast, Nodal signaling can still be activated at mid-gastrulation using an optogenetic approach that bypasses the requirement for extracellular factors (Vopalensky et al., 2018). A decrease in the levels of a Nodal co-receptor at mid-gastrulation was identified as the factor responsible for competence loss.

Developmental windows for Erk signaling have also been probed optogenetically. Experiments with OptoSOS in *Drosophila* embryos showed that ectopic Erk activity induced after 4h post-fertilization surprisingly resulted in no patterning defects, suggesting that at this stage fate decisions are locked and "safeguarded" from ectopic Erk signaling (Johnson et al., 2017). It was also shown that Erk-dependent posterior cell fates can be induced at different stages, indicating a relatively flexible developmental window for these fates (Johnson & Toettcher, 2019).

The rapid kinetics and reversibility of many optogenetic systems are ideal for defining developmental windows with high temporal resolution. In particular, the development of optogenetic systems that reversibly inhibit signaling pathways (Huang et al., 2017) is a promising avenue for future efforts in this area.

2.4 How do cells respond to different signaling amplitudes?

A key prediction of the French flag model is that target genes should be activated by different signaling level thresholds (Fig. 2A). Accordingly, exposing cells to different signaling amplitudes has revealed level-dependent target gene activation (Dessaud et al., 2007; Dubrulle et al., 2015; Green, New, & Smith, 1992; Green & Smith, 1990; Gurdon, Harger, Mitchell, & Lemaire, 1994; Heemskerk et al., 2019; Paliwal et al., 2007; Tay et al., 2010; Wilson, Lagna, Suzuki, & Hemmati-Brivanlou, 1997; Zagorski et al., 2017). However, whether cells reliably read out their position based on signal concentration is debated (Barkai & Shilo, 2009; Briscoe & Small, 2015; Chen, Xu, Mei, Yu, & Small, 2012; Dubuis et al., 2013; Emberly, 2008; Goldbeter & Wolpert, 1990; Gregor et al., 2007; Gurdon et al., 1999; He et al., 2008; Houchmandzadeh et al., 2002; Ochoa-Espinosa, Yu, Tsirigos, Struffi, & Small, 2009; Ochoa-Espinosa et al., 2005; Petkova, Tkacik,

Bialek, Wieschaus, & Gregor, 2019; Porcher et al., 2010; Porcher & Dostatni, 2010; Shilo & Barkai, 2017; Tostevin et al., 2007; Wieschaus, 2016; Zagorski et al., 2017).

Experimental manipulations of signaling levels, especially *in vivo*, are important to test this idea (Driever & Nüsslein-Volhard, 1988; Harmansa et al., 2015; Liu et al., 2013). For example, when the Bicoid gradient was experimentally flattened and tuned to specific levels in *Drosophila* embryos, target gene expression was altered in ways inconsistent with the French flag model (Ochoa-Espinosa et al., 2009). However, tuning signaling levels can be technically challenging: these experiments relied on a transgene and previously characterized mutations that are not available in all systems. Mutations and overexpression approaches that affect signaling levels can also alter signaling duration, dysregulate other signaling pathways, and disturb earlier developmental processes, making it difficult to unambiguously connect changes in target gene expression to changes in signaling levels (Huang et al., 2017; Johnson & Toettcher, 2019).

The use of tunable optogenetic systems to examine signaling amplitude-dependent responses (Fig. 2E) bypasses some of these challenges, and their utility in such investigations has been recently demonstrated. Experiments with OptoSOS in *Drosophila* showed that increasing Erk signaling levels within its normal domain at the embryonic termini does not affect patterning, indicating that absolute Erk signaling levels are not crucial in this context (Fig. 3A, middle panel) (Johnson et al., 2017). On the other hand, responses to Erk signaling can be amplitude-dependent: optogenetic induction of high, but not low levels of Erk signaling outside of its normal domain led to ectopic specification of posterior cell fates (Johnson & Toettcher, 2019).

Tunable optogenetic systems are ideal tools to assess the relationship between signaling amplitude and cellular responses. Such approaches could be used to manipulate signaling levels—particularly *in vivo*—with unprecedented control, and hold promise to reveal important insights into the decoding of signaling amplitude during development.

2.5 How do signaling dynamics generate diverse responses?

Signaling duration and dynamics such as oscillations are now known to play major roles in signal interpretation (Purvis & Lahav, 2013; Sagner & Briscoe, 2017). Endogenous signaling systems can exhibit complex dynamics that generate distinct cellular responses. For example, mouse neural progenitor cells differentiate into neurons, oligodendrocytes, or astrocytes based on oscillatory input from three transcription factors (Imayoshi et al., 2013). In other contexts, Erk signaling dynamics dictate either proliferation or differentiation (Aoki et al., 2013; Marshall, 1995; Ryu et al., 2015), different NF-κB nuclear dynamics activate either inflammatory or adaptive immune response genes (Werner, Barken, & Hoffmann, 2005), cell cycle arrest or apoptosis is initiated by distinct nuclear p53 dynamics (Purvis et al., 2012), and the yeast transcription factor Msn produces stress-specific gene expression responses dependent on its nuclear dynamics (Hansen & O'Shea, 2016; Hao & O'Shea, 2012). Distinct signaling dynamics can also be instigated by different ligands, such as NGF and EGF in the case of Erk signaling (Lemmon & Schlessinger, 2010; Marshall, 1995), and TGFβ- or BMP-mediated activation of Smad1/5 (Ramachandran et al., 2018). Experiments measuring cellular responses to a given signaling input are crucial to tease out the molecular mechanisms underlying the relationship between signaling dynamics and cellular output (Fig. 2F) (Isomura & Kageyama, 2017; Johnson & Toettcher, 2018; Purvis & Lahav, 2013).

The Ras/Erk MAPK pathway is particularly complex to study because stimulation by different ligands (e.g., FGF, EGF, or NGF) can activate multiple downstream pathways including PI3K/Akt, Src, and Jnk, and these pathways can be activated with distinct kinetics that result in different cellular responses (Albeck, Mills, & Brugge, 2013; Marshall, 1995; Ryu et al., 2015; Toettcher et al., 2013; Wilson et al., 2017). Conveniently, OptoSOS (Fig. 1E) activates only the Ras/Erk signaling branch (Toettcher et al., 2013). Using an elegant system of live-cell reporters to track Erk activity, transcriptional responses, and target protein levels in individual mammalian cells, it was shown that constant Erk activity induced similar transcriptional responses among Erk target genes, but dramatically different responses at the protein level (Wilson et al., 2017). Interestingly, transcriptional responses also adapted—they increased and then decreased in part due to negative feedback from DUSPs, a class of Erk-specific phosphatases. In contrast, transcriptional responses to pulsed signaling varied between target genes, with each gene exhibiting a frequency-dependent maximum induction (Fig. 3C). It was proposed that target genes decode Erk signaling dynamics using unique band-pass filters determined by promoter sensitivities which, together with post-transcriptional regulation, ultimately dictate cellular responses.

Recent work using OptoSOS *in vivo* revealed that the cumulative dose of Erk signaling, not its amplitude or duration alone, controls cell fate decisions in *Drosophila* (Johnson & Toettcher, 2019). In the lateral region of the embryo, Erk signaling is transient and required for specification of lateral cell fates. In contrast, posterior Erk signaling is sustained and necessary for posterior cell fate specification. To clarify whether the duration of Erk signaling determines lateral versus posterior fate decisions, OptoSOS was used to activate ectopic Erk signaling for different durations. Only prolonged activation led to posterior fates, suggesting that either the duration or cumulative dose of signaling specifies this cell type. To distinguish between these possibilities, the same total dose of signaling was optogenetically delivered either as a single bolus, or as several shorter pulses (Fig. 3D). Both sustained and pulsed Erk signaling led the same degree of posteriorization, supporting the idea that the cumulative dose, rather than signaling duration *per se*, is relevant for the lateral/posterior fate decision.

These studies highlight the advantages of optogenetic approaches as tools for understanding how temporal signaling is decoded. The ease of controlling light exposure dynamics combined with the rapid kinetics of many optogenetic systems make them particularly appealing for such investigations. Accordingly, the use of optogenetic strategies in this area is gaining momentum, and will likely continue to be fruitful (Aoki et al., 2013; Guglielmi et al., 2016; Isomura & Kageyama, 2017; Isomura et al., 2017; Johnson & Toettcher, 2019; Khamo et al., 2017; Kim et al., 2014; Sokolik et al., 2015; Toettcher et al., 2013; Wilson et al., 2017).

2.6 How does noise impact development?

Clonal animals or even isogenic cells in identical environments can exhibit diverse gene expression and phenotypes, highlighting the impact of noise in biology (Balazsi, van Oudenaarden, & Collins, 2011; Ji et al., 2013; Paliwal et al., 2007; Raj & van Oudenaarden, 2008; Tay et al., 2010; Wilson et al., 2017). Embryos must deal with noise in gene expression, protein translation, and signaling molecule production and dispersal, among other obstacles (Balazsi et al., 2011; Ji et al., 2013; Sagner & Briscoe, 2017; Urban & Johnston, 2018; Zagorski et al., 2017). Concerns regarding noise have challenged both the French flag model (Fig. 2A) and the classical Waddington landscape view of cellular differentiation (Balazsi et al., 2011; Ferrell, 2012; Ladewig, Koch, & Brüstle, 2013), in which the metaphor of a ball rolling down a hill of bifurcating valleys is used to explain how cell fates are determined or "canalized" (Waddington & Kacser, 1957). Chief among these theoretical concerns is intrinsic noise in gene expression, which might drive cells out of valleys and prevent them from acquiring appropriate fates. How do embryos buffer (or exploit Aoki et al., 2013; Isomura et al., 2017; Turing, 1952; Urban & Johnston, 2018; Wernet et al., 2006) noise to ensure that developmental processes are reproducible and robust (Fig. 2G)?

One optogenetic study has framed the decision between pluripotency and differentiation faced by stem cells as a Waddingtonian energy landscape shaped by interactions among key transcription factors (Sokolik et al., 2015). Pluripotency-supporting Nanog binds to Sox2 and activates its own expression, whereas the pro-neuronal differentiation factor Brn2 binds competitively to Sox2 and blocks Nanog autoregulation (Fig. 3E). Using a LightOn-based system to control Brn2 levels, a switch-like drop in Nanog was found to occur at a specific Brn2 threshold when held for at least 8 h, leading to neuronal differentiation. This Nanog-on/off switching landscape was proposed to be defined by competitive binding between Nanog and Brn2, and the timing of the switch by Nanog stability. Indeed, optogenetic induction of a non-competitive pro-differentiation factor caused gradual, non-switch-like changes in Nanog levels, and an optogenetic pulse-chase experiment determined a Nanog-RFP half-life consistent with the observed switching time. This energy landscape provides a mechanism to buffer noise in Brn2 levels, since the differentiation decision requires consistent, long lasting, high levels of Brn2 activity.

Optogenetic approaches have revealed additional strategies used by biological systems to both buffer and exploit noise (Aoki et al., 2013; Huang et al., 2017; Isomura et al., 2017; Toettcher et al., 2013). Signaling oscillations occur during a variety of patterning processes including somite formation (Pourquié, 2011) and wound healing (Hiratsuka et al., 2015), and Delta/Notch signaling is often involved in oscillatory behaviors. Using a LightOn-based system to drive expression of Delta ligand, it was shown that Delta-pulsing "sender" cells can coordinate Notch signaling oscillations in light-insensitive "receiver" neighbors, and noise is necessary to explain the observed dynamics (Isomura et al., 2017). Optogenetic approaches have also shown that noise in Erk signaling oscillations contributes to the regulation of cell density-dependent proliferation (Aoki et al., 2013). Other optogenetic experiments demonstrated that although responses to Erk vary between cells, individual cells have highly reproducible responses; further, cells suppress responses to "high-frequency" signaling of less than 4 min (Toettcher et al., 2013).

In addition, optogenetic approaches can buffer experimental or intrinsic noise by using real-time feedback to adjust light exposure across populations of cells (Rullan et al., 2018; Toettcher et al., 2011). Together, these studies highlight how optogenetics are a promising avenue for investigations into noise during embryogenesis.

2.7 How are simultaneous inputs from multiple pathways interpreted?

Embryogenesis is controlled by multiple interacting signaling pathways. Different pathways can have spatially and temporally overlapping domains, regulate expression of shared target genes, and enhance or inhibit each other (Briscoe & Small, 2015; Schier & Talbot, 2005). For example, the expression of Bicoid target genes in *Drosophila* is influenced by multiple inputs in addition to the Bicoid gradient (Chen et al., 2012; Liu et al., 2013; Ochoa-Espinosa et al., 2009). Different TGFβ superfamily signaling pathways can be activated by common receptors (Ramachandran et al., 2018) and share target genes (Harvey, Tumpel, Dubrulle, Schier, & Smith, 2010), and interactions between Nodal and BMP are crucial for patterning the vertebrate body plan (Fauny, Thisse, & Thisse, 2009): Entire secondary body axes can be induced in zebrafish embryos by generating opposing ectopic sources of BMP and Nodal (de Olivera-Melo, Xu, Houssin, Thisse, & Thisse, 2018; Thisse & Thisse, 2015; Xu, Houssin, Ferri-Lagneau, Thisse, & Thisse, 2014). In addition, the neural tube is patterned by opposing gradients of BMP and Shh that together create a less noisy patterning system than could be achieved by a single gradient (Tostevin et al., 2007; Zagorski et al., 2017). To understand how multiple signaling pathways interact to coordinate embryogenesis, it would be useful to experimentally manipulate several pathways independently in the same cell or organism (Fig. 2H).

Orthogonal optogenetic systems have been combined to simultaneously manipulate distinct populations with different light wavelengths. For example, mixtures of cells expressing either red light-responsive PhyB or blue light-responsive CRY2 extracellularly were selectively instructed to bind PIF/CIB-coated glass using either red or blue light, respectively (Yüz et al., 2018). Transport of different organelles has also been simultaneously manipulated with orthogonal optogenetic systems: Endosomes containing PhyB were transported to the perinuclear region in the presence of a dynein-PIF fusion and red light, while in the same cell peroxisomes could be driven out of the perinuclear region by promoting their interaction with kinesin using a blue light-activated TULIP system (Adrian et al., 2017).

Reversible self-sorting has also been optogenetically induced using mixtures of beads coated with the orthogonal blue light-activated systems n/pMag and iLID/Nano (Chervyachkova & Wegner, 2018), and multiplexed systems responding to different wavelengths can confer tighter optogenetic control (Redchuk, Kaberniuk, & Verkhusha, 2018; Redchuk, Omelina, Chernov, & Verkhusha, 2017).

Although multiplexed orthogonal optogenetic systems have been demonstrated, to our knowledge no publications have described simultaneous optogenetic manipulation of more than one signaling pathway during development. However, non-multiplexed optogenetic approaches have already provided insights into combinatorial signaling in cellular systems and during development (Johnson et al., 2017; Johnson & Toettcher, 2019; Wilson et al., 2017). In response to continuous Erk signaling induced by OptoSOS in cultured cells, Erk target genes have varying responses at the mRNA and protein levels (Wilson et al., 2017). Combinatorial interactions between Erk and DNA damage-induced signaling were shown to influence these targets, which fall into one of three classes: (1) mRNA and protein are induced by Erk irrespective of DNA damage; (2) mRNA is induced by Erk, DNA damage, or Erk + DNA damage, while protein production requires DNA damage irrespective of Erk; and (3) mRNA is induced by Erk irrespective of DNA damage, but protein production requires Erk and the absence of DNA damage.

Combinatorial interactions between Erk, Bicoid, and Dorsal/NF-κB signaling during *Drosophila* embryogenesis have recently been optogenetically assessed *in vivo* (Johnson & Toettcher, 2019). Using OptoSOS to manipulate Erk signaling, it was demonstrated that anterior cell fates are activated by a combination of Bicoid and Erk signaling, whereas lateral fates are specified by a combination of Dorsal signaling and transient Erk activity, and posterior cell fates are generated by sustained Erk signaling (Fig. 3F). Strikingly, sustained Erk activity can override lateral cell fates and convert them into posterior cells, but this is not true in the anterior, where Bicoid prevents posterior specification even in the presence of sustained Erk signaling.

Optogenetic manipulation of even a single signaling pathway can therefore provide important information about how combinatorial interactions pattern developing tissues. Future approaches utilizing multiplexed orthogonal optogenetic systems to manipulate several pathways simultaneously could provide even deeper insights (Fig. 2H).

3. Practical considerations for optogenetic experiments

Thousands of light-responsive protein variants occur naturally, and many have been engineered to optimize specific features (Glantz et al., 2016; Losi et al., 2018). The choice of a light-responsive protein for a given application should be informed by several factors, including the reversibility, activation/inactivation wavelengths, and chromophore. The on/off kinetics of light-responsive proteins play a major role in an optogenetic system's behavior (Losi et al., 2018; Pudasaini, El-Arab, & Zoltowski, 2015; Taslimi et al., 2014; Usherenko et al., 2014; Wang et al., 2017; Zoltowski et al., 2009). For example, the fast on/off kinetics of LOV-based Magnets can provide high spatial resolution, but at the cost of low levels of steady-state dimerization (Benedetti et al., 2018). Compatibility with fluorescent reporters such as GFP and RFP may also be a factor to consider. Conveniently, near-infrared-responsive proteins do not respond to wavelengths used to detect standard fluorescent reporters (Redchuk et al., 2018, 2017).

Although light-responsive protein variants can be rationally chosen, novel applications typically require extensive optimization. Systems that are well-behaved in one context are sometimes nonfunctional, leaky, or toxic in others. However, these problems can often be resolved by modifying proteins, swapping the positions of fusion partners, using different variants, or by altering linkers (Adrian et al., 2017; Baaske et al., 2019; Bubeck et al., 2018; Čapek et al., 2019; Crefcoeur et al., 2013; Dine et al., 2018; Guntas et al., 2015; Johnson et al., 2017; Kawano et al., 2015; Nguyen et al., 2018; Nihongaki et al., 2015; Niopek et al., 2014; Polstein & Gersbach, 2012; Reade et al., 2017; Renicke et al., 2013; Ricken et al., 2019; Smart et al., 2017; Sun et al., 2017; Taslimi et al., 2014; Wang et al., 2012, 2016; Yang et al., 2013; Yumerefendi et al., 2015). For example, although full-length PhyB is expressed well in cell culture (Levskaya et al., 2009), only a truncated version was expressed and functional in zebrafish (Buckley et al., 2016); efforts to sterically inhibit protein function using light-responsive proteins were sensitive to linker length (Zhou et al., 2017); Opto-RTK (Fig. 1E, lower panel) had a better signal-to-noise ratio with the *Vaucheria frigida* LOV domain compared to other LOV variants (Grusch et al., 2014); and mutagenesis was used to evolve a LOV variant that protected an engineered cleavage site in the dark better than the original variant (Wang et al., 2017). In light of these observations, it is advisable to screen multiple constructs at the start of a novel optogenetic project. In one case, 73 constructs were screened to optimize a system for controlling gene expression and chromatin modification using TALEs, and 12 were ultimately shown to provide a good signal-to-noise ratio (Konermann et al., 2013).

To help overcome the challenges associated with developing novel optogenetic systems, a publicly available optogenetic vector collection containing 11 green, blue, and red light-responsive proteins and 5 of their binding partners was recently created (Tichy et al., 2019). The 29 modular vectors are designed for straightforward generation of optogenetic fusion proteins with flexible fusion sites (NTD or CTD), as well as options for different linkers and incorporation of epitope tags or fluorescent proteins. This resource should facilitate the screening of novel systems by providing a simple method to rapidly generate multiple optogenetic construct variants.

Several efforts have also been made to improve optogenetic systems by decreasing background activity. A good example is the LOVTRAP, developed by engineering the small protein Zdark to bind the Jα helix of the LOV domain in the dark but not when illuminated (Wang et al., 2016). LOVTRAP was used to reduce the leakiness of LOV-based gene expression systems by sequestering LOV-fused transcriptional effectors to cytoplasmic or mitochondrial membranes in the dark (Chen et al., 2019; Yumerefendi et al., 2018). Another sequestration-based approach involves the multiplexing of a blue light-responsive LOV domain with the near-infrared-light-responsive protein BphP and its binding partner, Q-PAS (Redchuk et al., 2018, 2017). A LOV-Q-PAS-NLS fusion was successfully sequestered from the nucleus in the presence of near-infrared light and membrane-bound BphP.

The light sources required for optogenetic experiments depend on the application, and range from simple and inexpensive LED-based systems (Sako et al., 2016; Tichy et al., 2019; Vopalensky et al., 2018) to more costly equipment such as confocal or 2-photon microscopes (Guglielmi et al., 2015; Izquierdo et al., 2018; Krueger et al., 2018) and custom-built systems (Accanto et al., 2018; Dempsey et al., 2015; Mohr et al., 2017). Many light-responsive proteins respond to commonly available light sources and require low, non-toxic light intensities (Grusch et al., 2014; Johnson & Toettcher, 2018; Taslimi et al., 2014; Wang et al., 2012). DMDs are a convenient option that provide flexible spatiotemporal control over light delivery, and can be added to existing microscopes (Arrenberg et al., 2010; Avants, Murphy, Dapello, & Robinson, 2015; Levskaya et al., 2009; Rullan et al., 2018; Scardigli et al., 2018; Toettcher et al., 2011). Inadvertent photoactivation can be avoided by wrapping samples in foil or by covering light sources with low-cost filters that block activating wavelengths, although transgenic animals harboring optogenetic systems may require more involved accommodations (Guglielmi et al., 2015; Izquierdo et al., 2018; Krueger et al., 2018; Qi et al., 2012; Zhou et al., 2017).

4. Conclusions and prospects

Advances in optogenetic methods over the last 10 years have introduced novel opportunities for developmental biology investigations. Optogenetic control over biological processes that are essential to embryogenesis, including signaling and gene expression, has been demonstrated in a range of *in vitro* and *in vivo* models, and insights into several patterning systems have already been gained using these approaches. In particular, the high degree of control offered by some optogenetic systems has the potential to uncover the spatiotemporal requirements for developmental signaling in the near future. Although signaling has been the focus of our discussion, significant progress has also been made using optogenetics to explore tissue morphogenesis and cell cycle control (Čapek et al., 2019; Deneke et al., 2019; Guglielmi et al., 2015, 2016; Izquierdo et al., 2018; Krueger et al., 2018; Uchida & Yajima, 2018), as well as the functions of membrane-less organelles like P-bodies and stress granules (Dine et al., 2018; Shin et al., 2017). More speculative future optogenetic applications include probing mechanical forces during development (Baaske et al., 2019) and expanding tissue engineering efforts (Bartelt et al., 2018; Chervyachkova & Wegner, 2018; Polstein & Gersbach, 2012; Ricken et al., 2019; Yüz et al., 2019, 2018). The rapid recent growth in optogenetic approaches (Kolar et al., 2018) is likely to lead to significant advances in developmental biology in the near future.

Acknowledgments

We are grateful to Harald Janovjak, Jared Toettcher, Daniel Čapek, Mohammad ElGamacy, Amit Landge, David Mörsdorf, Autumn Pomreinke, Hannes Preiß, Timothy Saunders, and Gary Soh, as well as the 2019 EMBO Practical Course on Optogenetics for helpful discussions. We thank Michal Rössler for the illustrations. This work was supported by the Max Planck Society and an HFSP Career Development Award (CDA00031/2013-C).

References

Accanto, N., Molinier, C., Tanese, D., Ronzitti, E., Newman, Z. L., Wyart, C., et al. (2018). Multiplexed temporally focused light shaping for high-resolution multi-cell targeting. *Optica*, *5*, 1478–1491.

Adrian, M., Nijenhuis, W., Hoogstraaten, R. I., Willems, J., & Kapitein, L. C. (2017). A phytochrome-derived photoswitch for intracellular transport. *ACS Synthetic Biology*, *6*, 1248–1256.

Agus, V., & Janovjak, H. (2017). Optogenetic methods in drug screening: Technologies and applications. *Current Opinion in Biotechnology*, *48*, 8–14.

Akiyama, T., & Gibson, M. C. (2015). Decapentaplegic and growth control in the developing *Drosophila* wing. *Nature*, *527*, 375–378.

Albeck, J. G., Mills, G. B., & Brugge, J. S. (2013). Frequency-modulated pulses of ERK activity transmit quantitative proliferation signals. *Molecular Cell*, *49*, 249–261.

Alexandre, C., Baena-Lopez, A., & Vincent, J. P. (2014). Patterning and growth control by membrane-tethered Wingless. *Nature*, *505*, 180–185.

Almuedo-Castillo, M., Bläßle, A., Mörsdorf, D., Marcon, L., Soh, G. H., Rogers, K. W., et al. (2018). Scale-invariant patterning by size-dependent inhibition of Nodal signalling. *Nature Cell Biology*, *20*, 1032–1042.

Anders, K., & Essen, L. O. (2015). The family of phytochrome-like photoreceptors: Diverse, complex and multi-colored, but very useful. *Current Opinion in Structural Biology*, *35*, 7–16.

Aoki, K., Kumagai, Y., Sakurai, A., Komatsu, N., Fujita, Y., Shionyu, C., et al. (2013). Stochastic ERK activation induced by noise and cell-to-cell propagation regulates cell density-dependent proliferation. *Molecular Cell*, *52*, 529–540.

Arrenberg, A. B., Stainier, D. Y., Baier, H., & Huisken, J. (2010). Optogenetic control of cardiac function. *Science*, *330*, 971–974.

Avants, B. W., Murphy, D. B., Dapello, J. A., & Robinson, J. T. (2015). NeuroPG: Open source software for optical pattern generation and data acquisition. *Frontiers in Neuroengineering*, *8*, 1.

Baaske, J., Muhlhauser, W. W. D., Yousefi, O. S., Zanner, S., Radziwill, G., Horner, M., et al. (2019). Optogenetic control of integrin-matrix interaction. *Communications Biology*, *2*, 15.

Balazsi, G., van Oudenaarden, A., & Collins, J. J. (2011). Cellular decision making and biological noise: From microbes to mammals. *Cell*, *144*, 910–925.

Barkai, N., & Shilo, B. Z. (2009). Robust generation and decoding of morphogen gradients. *Cold Spring Harbor Perspectives in Biology*, *1*, a001990.

Bartelt, S. M., Steinkuhler, J., Dimova, R., & Wegner, S. V. (2018). Light-guided motility of a minimal synthetic cell. *Nano Letters*, *18*, 7268–7274.

Benedetti, L., Barentine, A. E. S., Messa, M., Wheeler, H., Bewersdorf, J., & De Camilli, P. (2018). Light-activated protein interaction with high spatial subcellular confinement. *Proceedings of the National Academy of Sciences of the United States of America*, *115*, E2238–E2245.

Bergmann, S., Sandler, O., Sberro, H., Shnider, S., Schejter, E., Shilo, B. Z., et al. (2007). Pre-steady-state decoding of the Bicoid morphogen gradient. *PLoS Biology*, *5*, e46.

Beyer, H. M., Thomas, O. S., Riegel, N., Zurbriggen, M. D., Weber, W., & Horner, M. (2018). Generic and reversible opto-trapping of biomolecules. *Acta Biomaterialia*, *79*, 276–282.

Bläßle, A., & Müller, P. (2015). PyFDAP: automated analysis of fluorescence decay after photoconversion (FDAP) experiments. *Bioinformatics*, *31*, 972–974.

Bonger, K. M., Rakhit, R., Payumo, A. Y., Chen, J. K., & Wandless, T. J. (2014). General method for regulating protein stability with light. *ACS Chemical Biology*, *9*, 111–115.

Boyden, E. S., Zhang, F., Bamberg, E., Nagel, G., & Deisseroth, K. (2005). Millisecond-timescale, genetically targeted optical control of neural activity. *Nature Neuroscience*, *8*, 1263–1268.

Brankatschk, M., & Dickson, B. J. (2006). Netrins guide Drosophila commissural axons at short range. *Nature Neuroscience*, *9*, 188–194.

Briscoe, J., & Small, S. (2015). Morphogen rules: Design principles of gradient-mediated embryo patterning. *Development*, *142*, 3996–4009.

Bubeck, F., Hoffmann, M. D., Harteveld, Z., Aschenbrenner, S., Bietz, A., Waldhauer, M. C., et al. (2018). Engineered anti-CRISPR proteins for optogenetic control of CRISPR-Cas9. *Nature Methods*, *15*, 924–927.

Buckley, C. E., Moore, R. E., Reade, A., Goldberg, A. R., Weiner, O. D., & Clarke, J. D. W. (2016). Reversible optogenetic control of subcellular protein localization in a live vertebrate embryo. *Developmental Cell*, *36*, 117–126.

Bugaj, L. J., Choksi, A. T., Mesuda, C. K., Kane, R. S., & Schaffer, D. V. (2013). Optogenetic protein clustering and signaling activation in mammalian cells. *Nature Methods*, *10*, 249–252.

Čapek, D., Smutny, M., Tichy, A. M., Morri, M., Janovjak, H., & Heisenberg, C. P. (2019). Light-activated Frizzled7 reveals a permissive role of non-canonical wnt signaling in mesendoderm cell migration. *eLife*, *8*, e42093.

Chen, S. Y., Osimiri, L. C., Chevalier, M. W., Bugaj, L. J., Ng, A. H., Stewart-Ornstein, J., et al. (2019). Optogenetic control reveals differential promoter interpretation of transcription factor nuclear translocation dynamics. *bioRxiv*. https://doi.org/10.1101/548255.

Chen, F., & Wegner, S. V. (2017). Blue light switchable bacterial adhesion as a key step toward the design of biofilms. *ACS Synthetic Biology*, *6*, 2170–2174.

Chen, H., Xu, Z., Mei, C., Yu, D., & Small, S. (2012). A system of repressor gradients spatially organizes the boundaries of Bicoid-dependent target genes. *Cell*, *149*, 618–629.

Chervyachkova, E., & Wegner, S. V. (2018). Reversible social self-sorting of colloidal cell-mimics with blue light switchable proteins. *ACS Synthetic Biology*, *7*, 1817–1824.

Connors, S. A., Tucker, J. A., & Mullins, M. C. (2006). Temporal and spatial action of tolloid (mini fin) and chordin to pattern tail tissues. *Developmental Biology*, *293*, 191–202.

Crefcoeur, R. P., Yin, R., Ulm, R., & Halazonetis, T. D. (2013). Ultraviolet-B-mediated induction of protein-protein interactions in mammalian cells. *Nature Communications*, *4*, 1779.

Dagliyan, O., Tarnawski, M., Chu, P. H., Shirvanyants, D., Schlichting, I., Dokholyan, N. V., et al. (2016). Engineering extrinsic disorder to control protein activity in living cells. *Science*, *354*, 1441–1444.

de Mena, L., Rizk, P., & Rincon-Limas, D. E. (2018). Bringing light to transcription: The optogenetics repertoire. *Frontiers in Genetics*, *9*, 518.

de Olivera-Melo, M., Xu, P. F., Houssin, N., Thisse, B., & Thisse, C. (2018). Generation of ectopic morphogen gradients in the zebrafish blastula. In J. Dubrulle (Ed.), *Morphogen gradients. Methods in molecular biology*. New York, NY: Humana Press.

Dempsey, W. P., Georgieva, L., Helbling, P. M., Sonay, A. Y., Truong, T. V., Haffner, M., et al. (2015). In vivo single-cell labeling by confined primed conversion. *Nature Methods*, *12*, 645–648.

Deneke, V. E., Puliafito, A., Krueger, D., Narla, A. V., De Simone, A., Primo, L., et al. (2019). Self-organized nuclear positioning synchronizes the cell cycle in Drosophila embryos. *Cell*, *177*, 925-941.e917.

Derynck, R., & Budi, E. H. (2019). Specificity, versatility, and control of TGF-β family signaling. *Science Signaling*, *12*, eaav5183.

Dessaud, E., Yang, L. L., Hill, K., Cox, B., Ulloa, F., Ribeiro, A., et al. (2007). Interpretation of the sonic hedgehog morphogen gradient by a temporal adaptation mechanism. *Nature, 450*, 717–720.
Dierick, H. A., & Bejsovec, A. (1998). Functional analysis of Wingless reveals a link between intercellular ligand transport and dorsal-cell-specific signaling. *Development, 125*, 4729–4738.
Dine, E., Gil, A. A., Uribe, G., Brangwynne, C. P., & Toettcher, J. E. (2018). Protein phase separation provides long-term memory of transient spatial stimuli. *Cell Systems, 6*, 655-663.e655.
Dominici, C., Moreno-Bravo, J. A., Puiggros, S. R., Rappeneau, Q., Rama, N., Vieugue, P., et al. (2017). Floor-plate-derived netrin-1 is dispensable for commissural axon guidance. *Nature, 545*, 350–354.
Driever, W., & Nüsslein-Volhard, C. (1988). The bicoid protein determines position in the Drosophila embryo in a concentration-dependent manner. *Cell, 54*, 95–104.
Du, L., Sohr, A., Yan, G., & Roy, S. (2018). Feedback regulation of cytoneme-mediated transport shapes a tissue-specific FGF morphogen gradient. *eLife*, 7, e38137.
Dubrulle, J., Jordan, B. M., Akhmetova, L., Farrell, J. A., Kim, S. H., Solnica-Krezel, L., et al. (2015). Response to Nodal morphogen gradient is determined by the kinetics of target gene induction. *eLife, 4*, e05042.
Dubrulle, J., & Pourquié, O. (2004). *fgf8* mRNA decay establishes a gradient that couples axial elongation to patterning in the vertebrate embryo. *Nature, 427*, 419–422.
Dubuis, J. O., Tkacik, G., Wieschaus, E. F., Gregor, T., & Bialek, W. (2013). Positional information, in bits. *Proceedings of the National Academy of Sciences of the United States of America, 110*, 16301–16308.
Emberly, E. (2008). Optimizing the readout of morphogen gradients. *Physical Review. E, Statistical, Nonlinear, and Soft Matter Physics*, 77, 041903.
Entchev, E. V., Schwabedissen, A., & González-Gaitán, M. (2000). Gradient formation of the TGF-β homolog Dpp. *Cell, 103*, 981–991.
Faden, F., Mielke, S., Lange, D., & Dissmeyer, N. (2014). Generic tools for conditionally altering protein abundance and phenotypes on demand. *Biological Chemistry, 395*, 737–762.
Fauny, J. D., Thisse, B., & Thisse, C. (2009). The entire zebrafish blastula-gastrula margin acts as an organizer dependent on the ratio of Nodal to BMP activity. *Development, 136*, 3811–3819.
Ferrell, J. E., Jr. (2012). Bistability, bifurcations, and Waddington's epigenetic landscape. *Current Biology, 22*, R458–R466.
Franklin, K. A., & Quail, P. H. (2010). Phytochrome functions in Arabidopsis development. *Journal of Experimental Botany, 61*, 11–24.
Gallet, A., Staccini-Lavenant, L., & Thérond, P. P. (2008). Cellular trafficking of the glypican Dally-like is required for full-strength Hedgehog signaling and wingless transcytosis. *Developmental Cell, 14*, 712–725.
Gehrig, S., Macpherson, J. A., Driscoll, P. C., Symon, A., Martin, S. R., MacRae, J. I., et al. (2017). An engineered photoswitchable mammalian pyruvate kinase. *The FEBS Journal, 284*, 2955–2980.
Glantz, S. T., Carpenter, E. J., Melkonian, M., Gardner, K. H., Boyden, E. S., Wong, G. K., et al. (2016). Functional and topological diversity of LOV domain photoreceptors. *Proceedings of the National Academy of Sciences of the United States of America, 113*, E1442–E1451.
Goldbeter, A., & Wolpert, L. (1990). Covalent modification of proteins as a threshold mechanism in development. *Journal of Theoretical Biology, 142*, 243–250.

González-Gaitán, M., & Jäckle, H. (1999). The range of spalt-activating Dpp signalling is reduced in endocytosis-defective Drosophila wing discs. *Mechanisms of Development*, *87*, 143–151.
González-Méndez, L., Seijo-Barandiarán, I., & Guerrero, I. (2017). Cytoneme-mediated cell-cell contacts for Hedgehog reception. *eLife*, *6*, e24045.
Green, J. B. A., New, H. V., & Smith, J. C. (1992). Responses of embryonic Xenopus cells to activin and FGF are separated by multiple dose thresholds and correspond to distinct axes of the mesoderm. *Cell*, *71*, 731–739.
Green, J. B., & Smith, J. C. (1990). Graded changes in dose of a Xenopus activin A homologue elicit stepwise transitions in embryonic cell fate. *Nature*, *347*, 391–394.
Gregor, T., Tank, D. W., Wieschaus, E. F., & Bialek, W. (2007). Probing the limits to positional information. *Cell*, *130*, 153–164.
Grusch, M., Schelch, K., Riedler, R., Reichhart, E., Differ, C., Berger, W., et al. (2014). Spatio-temporally precise activation of engineered receptor tyrosine kinases by light. *The EMBO Journal*, *33*, 1713–1726.
Guglielmi, G., Barry, J. D., Huber, W., & De Renzis, S. (2015). An optogenetic method to modulate cell contractility during tissue morphogenesis. *Developmental Cell*, *35*, 646–660.
Guglielmi, G., Falk, H. J., & De Renzis, S. (2016). Optogenetic control of protein function: From intracellular processes to tissue morphogenesis. *Trends in Cell Biology*, *26*, 864–874.
Guntas, G., Hallett, R. A., Zimmerman, S. P., Williams, T., Yumerefendi, H., Bear, J. E., et al. (2015). Engineering an improved light-induced dimer (iLID) for controlling the localization and activity of signaling proteins. *Proceedings of the National Academy of Sciences of the United States of America*, *112*, 112–117.
Gurdon, J. B., Harger, P., Mitchell, A., & Lemaire, P. (1994). Activin signalling and response to a morphogen gradient. *Nature*, *371*, 487–492.
Gurdon, J. B., Standley, H., Dyson, S., Butler, K., Langon, T., Ryan, K., et al. (1999). Single cells can sense their position in a morphogen gradient. *Development*, *126*, 5309–5317.
Hagos, E. G., & Dougan, S. T. (2007). Time-dependent patterning of the mesoderm and endoderm by Nodal signals in zebrafish. *BMC Developmental Biology*, *7*, 22.
Hansen, A. S., & O'Shea, E. K. (2016). Encoding four gene expression programs in the activation dynamics of a single transcription factor. *Current Biology*, *26*, R269–R271.
Hao, N., & O'Shea, E. K. (2012). Signal-dependent dynamics of transcription factor translocation controls gene expression. *Nature Structural & Molecular Biology*, *19*, 31–39.
Harmansa, S., Hamaratoglu, F., Affolter, M., & Caussinus, E. (2015). Dpp spreading is required for medial but not for lateral wing disc growth. *Nature*, *527*, 317–322.
Haruki, H., Nishikawa, J., & Laemmli, U. K. (2008). The anchor-away technique: Rapid, conditional establishment of yeast mutant phenotypes. *Molecular Cell*, *31*, 925–932.
Harvey, S. A., Tumpel, S., Dubrulle, J., Schier, A. F., & Smith, J. C. (2010). No tail integrates two modes of mesoderm induction. *Development*, *137*, 1127–1135.
Hashimoto-Partyka, M. K., Yuge, M., & Cho, K. W. Y. (2003). Nodal signaling in Xenopus gastrulae is cell-autonomous and patterned by β-Catenin. *Developmental Biology*, *253*, 125–138.
He, F., Wen, Y., Deng, J., Lin, X., Lu, L. J., Jiao, R., et al. (2008). Probing intrinsic properties of a robust morphogen gradient in Drosophila. *Developmental Cell*, *15*, 558–567.
Heemskerk, I., Burt, K., Miller, M., Chhabra, S., Guerra, M. C., Liu, L., et al. (2019). Rapid changes in morphogen concentration control self-organized patterning in human embryonic stem cells. *eLife*, *8*, e40526.
Hiratsuka, T., Fujita, Y., Naoki, H., Aoki, K., Kamioka, Y., & Matsuda, M. (2015). Intercellular propagation of extracellular signal-regulated kinase activation revealed by in vivo imaging of mouse skin. *eLife*, *4*, e05178.

Houchmandzadeh, B., Wieschaus, E., & Leibler, S. (2002). Establishment of developmental precision and proportions in the early Drosophila embryo. *Nature*, *415*, 798–802.

Huang, A., Amourda, C., Zhang, S., Tolwinski, N. S., & Saunders, T. E. (2017). Decoding temporal interpretation of the morphogen Bicoid in the early Drosophila embryo. *eLife*, *6*, e26258.

Hughes, R. M., Bolger, S., Tapadia, H., & Tucker, C. L. (2012). Light-mediated control of DNA transcription in yeast. *Methods*, *58*, 385–391.

Imayoshi, I., Isomura, A., Harima, Y., Kawaguchi, K., Kori, H., Miyachi, H., et al. (2013). Oscillatory control of factors determining multipotency and fate in mouse neural progenitors. *Science*, *342*, 1203–1208.

Ingles-Prieto, A., Reichhart, E., Muellner, M. K., Nowak, M., Nijman, S. M., Grusch, M., et al. (2015). Light-assisted small-molecule screening against protein kinases. *Nature Chemical Biology*, *11*, 952–954.

Isomura, A., & Kageyama, R. (2017). Illuminating information transfer in signaling dynamics by optogenetics. *Current Opinion in Cell Biology*, *49*, 9–15.

Isomura, A., Ogushi, F., Kori, H., & Kageyama, R. (2017). Optogenetic perturbation and bioluminescence imaging to analyze cell-to-cell transfer of oscillatory information. *Genes & Development*, *31*, 524–535.

Izquierdo, E., Quinkler, T., & De Renzis, S. (2018). Guided morphogenesis through optogenetic activation of Rho signalling during early Drosophila embryogenesis. *Nature Communications*, *9*, 2366.

Ji, N., Middelkoop, T. C., Mentink, R. A., Betist, M. C., Tonegawa, S., Mooijman, D., et al. (2013). Feedback control of gene expression variability in the *Caenorhabditis elegans* Wnt pathway. *Cell*, *155*, 869–880.

Johnson, H. E., Goyal, Y., Pannucci, N. L., Schupbach, T., Shvartsman, S. Y., & Toettcher, J. E. (2017). The spatiotemporal limits of developmental Erk signaling. *Developmental Cell*, *40*, 185–192.

Johnson, H. E., Shvartsman, S. Y., & Toettcher, J. E. (2019). Optogenetic rescue of a developmental patterning mutant. *bioRxiv*. https://doi.org/10.1101/776120.

Johnson, H. E., & Toettcher, J. E. (2018). Illuminating developmental biology with cellular optogenetics. *Current Opinion in Biotechnology*, *52*, 42–48.

Johnson, H. E., & Toettcher, J. E. (2019). Signaling dynamics control cell fate in the early Drosophila embryo. *Developmental Cell*, *48*, 361-370.e363.

Josselyn, S. A. (2018). The past, present and future of light-gated ion channels and optogenetics. *eLife*, 7, e42367.

Jullien, J., & Gurdon, J. (2005). Morphogen gradient interpretation by a regulated trafficking step during ligand-receptor transduction. *Genes & Development*, *19*, 2682–2694.

Jung, J. H., Domijan, M., Klose, C., Biswas, S., Ezer, D., Gao, M. J., et al. (2016). Phytochromes function as thermosensors in Arabidopsis. *Science*, *354*, 886–889.

Kainrath, S., Stadler, M., Reichhart, E., Distel, M., & Janovjak, H. (2017). Green-light-induced inactivation of receptor signaling using cobalamin-binding domains. *Angewandte Chemie (International Ed. in English)*, *56*, 4608–4611.

Kaur, P., Saunders, T. E., & Tolwinski, N. S. (2017). Coupling optogenetics and light-sheet microscopy, a method to study Wnt signaling during embryogenesis. *Scientific Reports*, 7, 16636.

Kawano, F., Suzuki, H., Furuya, A., & Sato, M. (2015). Engineered pairs of distinct photo-switches for optogenetic control of cellular proteins. *Nature Communications*, *6*, 6256.

Khamo, J. S., Krishnamurthy, V. V., Sharum, S. R., Mondal, P., & Zhang, K. (2017). Applications of optobiology in intact cells and multicellular organisms. *Journal of Molecular Biology*, *429*, 2999–3017.

Kicheva, A., Pantazis, P., Bollenbach, T., Kalaidzidis, Y., Bittig, T., Jülicher, F., et al. (2007). Kinetics of morphogen gradient formation. *Science*, *315*, 521–525.

Kim, N., Kim, J. M., Lee, M., Kim, C. Y., Chang, K. Y., & Heo, W. D. (2014). Spatiotemporal control of fibroblast growth factor receptor signals by blue light. *Chemistry & Biology*, *21*, 903–912.

Kishimoto, Y., Lee, K. H., Zon, L., Hammerschmidt, M., & Schulte-Merker, S. (1997). The molecular nature of zebrafish swirl: BMP2 function is essential during early dorsoventral patterning. *Development*, *124*, 4457–4466.

Kolar, K., Knobloch, C., Stork, H., Znidaric, M., & Weber, W. (2018). OptoBase: A web platform for molecular optogenetics. *ACS Synthetic Biology*, 7, 1825–1828.

Konermann, S., Brigham, M. D., Trevino, A., Hsu, P. D., Heidenreich, M., Cong, L., et al. (2013). Optical control of mammalian endogenous transcription and epigenetic states. *Nature*, *500*, 472–476.

Kornberg, T. B., & Roy, S. (2014). Cytonemes as specialized signaling filopodia. *Development*, *141*, 729–736.

Kowalik, L., & Chen, J. K. (2017). Illuminating developmental biology through photochemistry. *Nature Chemical Biology*, *13*, 587–598.

Krueger, D., Izquierdo, E., Viswanathan, R., Hartmann, J., Pallares Cartes, C., & De Renzis, S. (2019). Principles and applications of optogenetics in developmental biology. *Development*, *146*, dev175067.

Krueger, D., Tardivo, P., Nguyen, C., & De Renzis, S. (2018). Downregulation of basal myosin-II is required for cell shape changes and tissue invagination. *The EMBO Journal*, *37*, e100170.

Kruse, K., Pantazis, P., Bollenbach, T., Jülicher, F., & González-Gaitán, M. (2004). Dpp gradient formation by dynamin-dependent endocytosis: Receptor trafficking and the diffusion model. *Development*, *131*, 4843–4856.

Kyriakakis, P., Catanho, M., Hoffner, N., Thavarajah, W., Hu, V. J., Chao, S. S., et al. (2018). Biosynthesis of orthogonal molecules using ferredoxin and ferredoxin-NADP(+) reductase systems enables genetically encoded PhyB optogenetics. *ACS Synthetic Biology*, 7, 706–717.

Ladewig, J., Koch, P., & Brüstle, O. (2013). Leveling Waddington: The emergence of direct programming and the loss of cell fate hierarchies. *Nature Reviews. Molecular Cell Biology*, *14*, 225–236.

Lee, D., Hyun, J. H., Jung, K., Hannan, P., & Kwon, H. B. (2017). A calcium- and light-gated switch to induce gene expression in activated neurons. *Nature Biotechnology*, *35*, 858–863.

Lee, J., Natarajan, M., Nashine, V. C., Socolich, M., Vo, T., Russ, W. P., et al. (2008). Surface sites for engineering allosteric control in proteins. *Science*, *322*, 438–442.

Lee, S., Park, H., Kyung, T., Kim, N. Y., Kim, S., Kim, J., et al. (2014). Reversible protein inactivation by optogenetic trapping in cells. *Nature Methods*, *11*, 633–636.

Lemmon, M. A., & Schlessinger, J. (2010). Cell signaling by receptor tyrosine kinases. *Cell*, *141*, 1117–1134.

Levskaya, A., Weiner, O. D., Lim, W. A., & Voigt, C. A. (2009). Spatiotemporal control of cell signalling using a light-switchable protein interaction. *Nature*, *461*, 997–1001.

Li, X., Gutierrez, D. V., Hanson, M. G., Han, J., Mark, M. D., Chiel, H., et al. (2005). Fast noninvasive activation and inhibition of neural and network activity by vertebrate rhodopsin and green algae channelrhodopsin. *Proceedings of the National Academy of Sciences of the United States of America*, *102*, 17816–17821.

Liu, F., Morrison, A. H., & Gregor, T. (2013). Dynamic interpretation of maternal inputs by the Drosophila segmentation gene network. *Proceedings of the National Academy of Sciences of the United States of America*, *110*, 6724–6729.

Lo, C. L., Choudhury, S. R., Irudayaraj, J., & Zhou, F. C. (2017). Epigenetic editing of Ascl1 gene in neural stem cells by optogenetics. *Scientific Reports*, 7, 42047.

Losi, A., Gardner, K. H., & Möglich, A. (2018). Blue-light receptors for optogenetics. *Chemical Reviews*, *118*, 10659–10709.

Lucchetta, E. M., Lee, J. H., Fu, L. A., Patel, N. H., & Ismagilov, R. F. (2005). Dynamics of Drosophila embryonic patterning network perturbed in space and time using microfluidics. *Nature*, *434*, 1134–1138.

Macdonald, P. M., & Struhl, G. (1986). A molecular gradient in early Drosophila embryos and its role in specifying the body pattern. *Nature*, *324*, 537–545.

Makhijani, K., To, T. L., Ruiz-González, R., Lafaye, C., Royant, A., & Shu, X. (2017). Precision optogenetic tool for selective single- and multiple-cell ablation in a live animal model system. *Cell Chemical Biology*, *24*, 110–119.

Marshall, C. J. (1995). Specificity of receptor tyrosine kinase signaling: Transient versus sustained extracellular signal-regulated kinase activation. *Cell*, *80*, 179–185.

Masuda, S., Nakatani, Y., Ren, S., & Tanaka, M. (2013). Blue light-mediated manipulation of transcription factor activity in vivo. *ACS Chemical Biology*, *8*, 2649–2653.

Mattes, B., Dang, Y., Greicius, G., Kaufmann, L. T., Prunsche, B., Rosenbauer, J., et al. (2018). Wnt/PCP controls spreading of Wnt/β-catenin signals by cytonemes in vertebrates. *eLife*, *7*, e36953.

Mattes, B., & Scholpp, S. (2018). Emerging role of contact-mediated cell communication in tissue development and diseases. *Histochemistry and Cell Biology*, *150*, 431–442.

McGuire, S. E., Roman, G., & Davis, R. L. (2004). Gene expression systems in Drosophila: A synthesis of time and space. *Trends in Genetics*, *20*, 384–391.

Mizuno, H., Mal, T. K., Walchli, M., Kikuchi, A., Fukano, T., Ando, R., et al. (2008). Light-dependent regulation of structural flexibility in a photochromic fluorescent protein. *Proceedings of the National Academy of Sciences of the United States of America*, *105*, 9227–9232.

Mohr, M. A., Kobitski, A. Y., Sabater, L. R., Nienhaus, K., Obara, C. J., Lippincott-Schwartz, J., et al. (2017). Rational engineering of photoconvertible fluorescent proteins for dual-color fluorescence nanoscopy enabled by a triplet-state mechanism of primed conversion. *Angewandte Chemie (International Ed. in English)*, *56*, 11628–11633.

Morri, M., Sanchez-Romero, I., Tichy, A. M., Kainrath, S., Gerrard, E. J., Hirschfeld, P. P., et al. (2018). Optical functionalization of human Class A orphan G-protein-coupled receptors. *Nature Communications*, *9*, 1950.

Mörsdorf, D., & Müller, P. (2019). Tuning protein diffusivity with membrane tethers. *Biochemistry*, *58*, 177–181.

Motta-Mena, L. B., Reade, A., Mallory, M. J., Glantz, S., Weiner, O. D., Lynch, K. W., et al. (2014). An optogenetic gene expression system with rapid activation and deactivation kinetics. *Nature Chemical Biology*, *10*, 196–202.

Müller, K., Engesser, R., Metzger, S., Schulz, S., Kämpf, M. M., Busacker, M., et al. (2013). A red/far-red light-responsive bi-stable toggle switch to control gene expression in mammalian cells. *Nucleic Acids Research*, *41*, e77.

Müller, K., Engesser, R., Timmer, J., Nagy, F., Zurbriggen, M. D., & Weber, W. (2013). Synthesis of phycocyanobilin in mammalian cells. *Chemical Communications (Cambridge, England)*, *49*, 8970–8972.

Müller, P., Rogers, K. W., Jordan, B. M., Lee, J. S., Robson, D., Ramanathan, S., et al. (2012). Differential diffusivity of Nodal and Lefty underlies a reaction-diffusion patterning system. *Science*, *336*, 721–724.

Müller, P., Rogers, K. W., Yu, S. R., Brand, M., & Schier, A. F. (2013). Morphogen transport. *Development*, *140*, 1621–1638.

Müller, P., & Schier, A. F. (2011). Extracellular movement of signaling molecules. *Developmental Cell*, *21*, 145–158.

Nagel, G., Brauner, M., Liewald, J. F., Adeishvili, N., Bamberg, E., & Gottschalk, A. (2005). Light activation of channelrhodopsin-2 in excitable cells of Caenorhabditis elegans triggers rapid behavioral responses. *Current Biology*, *15*, 2279–2284.

Nagel, G., Szellas, T., Huhn, W., Kateriya, S., Adeishvili, N., Berthold, P., et al. (2003). Channelrhodopsin-2, a directly light-gated cation-selective membrane channel. *Proceedings of the National Academy of Sciences of the United States of America*, *100*, 13940–13945.

Namba, R., Pazdera, T. M., Cerrone, R. L., & Minden, J. S. (1997). Drosophila embryonic pattern repair: How embryos respond to bicoid dosage alteration. *Development*, *124*, 1393–1403.

Nguyen, N. T., He, L., Martinez-Moczygemba, M., Huang, Y., & Zhou, Y. (2018). Rewiring calcium signaling for precise transcriptional reprogramming. *ACS Synthetic Biology*, *7*, 814–821.

Nguyen, M. K., Kim, C. Y., Kim, J. M., Park, B. O., Lee, S., Park, H., et al. (2016). Optogenetic oligomerization of Rab GTPases regulates intracellular membrane trafficking. *Nature Chemical Biology*, *12*, 431–436.

Ni, M., Tepperman, J. M., & Quail, P. H. (1999). Binding of phytochrome B to its nuclear signalling partner PIF3 is reversibly induced by light. *Nature*, *400*, 781–784.

Nihongaki, Y., Yamamoto, S., Kawano, F., Suzuki, H., & Sato, M. (2015). CRISPR-Cas9-based photoactivatable transcription system. *Chemistry & Biology*, *22*, 169–174.

Niopek, D., Benzinger, D., Roensch, J., Draebing, T., Wehler, P., Eils, R., et al. (2014). Engineering light-inducible nuclear localization signals for precise spatiotemporal control of protein dynamics in living cells. *Nature Communications*, *5*, 4404.

Niopek, D., Wehler, P., Roensch, J., Eils, R., & Di Ventura, B. (2016). Optogenetic control of nuclear protein export. *Nature Communications*, *7*, 10624.

Ochoa-Espinosa, A., Yu, D., Tsirigos, A., Struffi, P., & Small, S. (2009). Anterior-posterior positional information in the absence of a strong Bicoid gradient. *Proceedings of the National Academy of Sciences of the United States of America*, *106*, 3823–3828.

Ochoa-Espinosa, A., Yucel, G., Kaplan, L., Pare, A., Pura, N., Oberstein, A., et al. (2005). The role of binding site cluster strength in Bicoid-dependent patterning in Drosophila. *Proceedings of the National Academy of Sciences of the United States of America*, *102*, 4960–4965.

Ohlendorf, R., Vidavski, R. R., Eldar, A., Moffat, K., & Möglich, A. (2012). From dusk till dawn: One-plasmid systems for light-regulated gene expression. *Journal of Molecular Biology*, *416*, 534–542.

Paliwal, S., Iglesias, P. A., Campbell, K., Hilioti, Z., Groisman, A., & Levchenko, A. (2007). MAPK-mediated bimodal gene expression and adaptive gradient sensing in yeast. *Nature*, *446*, 46–51.

Pani, A. M., & Goldstein, B. (2018). Direct visualization of a native Wnt in vivo reveals that a long-range Wnt gradient forms by extracellular dispersal. *eLife*, *7*, e38325.

Pei, W., Williams, P. H., Clark, M. D., Stemple, D. L., & Feldman, B. (2007). Environmental and genetic modifiers of squint penetrance during zebrafish embryogenesis. *Developmental Biology*, *308*, 368–378.

Petkova, M. D., Tkacik, G., Bialek, W., Wieschaus, E. F., & Gregor, T. (2019). Optimal decoding of cellular identities in a genetic network. *Cell*, *176*, 844-855.e815.

Polstein, L. R., & Gersbach, C. A. (2012). Light-inducible spatiotemporal control of gene activation by customizable zinc finger transcription factors. *Journal of the American Chemical Society*, *134*, 16480–16483.

Polstein, L. R., & Gersbach, C. A. (2015). A light-inducible CRISPR-Cas9 system for control of endogenous gene activation. *Nature Chemical Biology*, *11*, 198–200.

Polstein, L. R., Juhas, M., Hanna, G., Bursac, N., & Gersbach, C. A. (2017). An engineered optogenetic switch for spatiotemporal control of gene expression, cell differentiation, and tissue morphogenesis. *ACS Synthetic Biology*, *6*, 2003–2013.

Porcher, A., Abu-Arish, A., Huart, S., Roelens, B., Fradin, C., & Dostatni, N. (2010). The time to measure positional information: maternal hunchback is required for the synchrony of the Bicoid transcriptional response at the onset of zygotic transcription. *Development*, *137*, 2795–2804.

Porcher, A., & Dostatni, N. (2010). The Bicoid morphogen system. *Current Biology*, *20*, R249–R254.

Pourquié, O. (2011). Vertebrate segmentation: From cyclic gene networks to scoliosis. *Cell*, *145*, 650–663.

Pudasaini, A., El-Arab, K. K., & Zoltowski, B. D. (2015). LOV-based optogenetic devices: Light-driven modules to impart photoregulated control of cellular signaling. *Frontiers in Molecular Biosciences*, *2*, 18.

Purvis, J. E., Karhohs, K. W., Mock, C., Batchelor, E., Loewer, A., & Lahav, G. (2012). p53 dynamics control cell fate. *Science*, *336*, 1440–1444.

Purvis, J. E., & Lahav, G. (2013). Encoding and decoding cellular information through signaling dynamics. *Cell*, *152*, 945–956.

Qi, Y. B., Garren, E. J., Shu, X., Tsien, R. Y., & Jin, Y. (2012). Photo-inducible cell ablation in Caenorhabditis elegans using the genetically encoded singlet oxygen generating protein miniSOG. *Proceedings of the National Academy of Sciences of the United States of America*, *109*, 7499–7504.

Quail, P. H. (2002). Phytochrome photosensory signalling networks. *Nature Reviews. Molecular Cell Biology*, *3*, 85–93.

Raj, A., & van Oudenaarden, A. (2008). Nature, nurture, or chance: Stochastic gene expression and its consequences. *Cell*, *135*, 216–226.

Ramachandran, A., Vizan, P., Das, D., Chakravarty, P., Vogt, J., Rogers, K. W., et al. (2018). TGF-β uses a novel mode of receptor activation to phosphorylate SMAD1/5 and induce epithelial-to-mesenchymal transition. *eLife*, 7, e31756.

Ramel, M. C., & Hill, C. S. (2013). The ventral to dorsal BMP activity gradient in the early zebrafish embryo is determined by graded expression of BMP ligands. *Developmental Biology*, *378*, 170–182.

Reade, A., Motta-Mena, L. B., Gardner, K. H., Stainier, D. Y., Weiner, O. D., & Woo, S. (2017). TAEL: A zebrafish-optimized optogenetic gene expression system with fine spatial and temporal control. *Development*, *144*, 345–355.

Redchuk, T. A., Kaberniuk, A. A., & Verkhusha, V. V. (2018). Near-infrared light-controlled systems for gene transcription regulation, protein targeting and spectral multiplexing. *Nature Protocols*, *13*, 1121–1136.

Redchuk, T. A., Omelina, E. S., Chernov, K. G., & Verkhusha, V. V. (2017). Near-infrared optogenetic pair for protein regulation and spectral multiplexing. *Nature Chemical Biology*, *13*, 633–639.

Renicke, C., Schuster, D., Usherenko, S., Essen, L. O., & Taxis, C. (2013). A LOV2 domain-based optogenetic tool to control protein degradation and cellular function. *Chemistry & Biology*, *20*, 619–626.

Ricken, J., Medda, R., & Wegner, S. V. (2019). Photo-ECM: A blue light photoswitchable synthetic extracellular matrix protein for reversible control over cell-matrix adhesion. *Advanced Biosystems*, *3*, 1800302.

Rivera-Pomar, R., Niessing, D., Schmidt-Ott, U., Gehring, W. J., & Jäckle, H. (1996). RNA binding and translational suppression by bicoid. *Nature*, *379*, 746–749.

Rizzini, L., Favory, J. J., Cloix, C., Faggionato, D., O'Hara, A., Kaiserli, E., et al. (2011). Perception of UV-B by the Arabidopsis UVR8 protein. *Science*, *332*, 103–106.

Robinson, M. S., Sahlender, D. A., & Foster, S. D. (2010). Rapid inactivation of proteins by rapamycin-induced rerouting to mitochondria. *Developmental Cell, 18*, 324–331.

Rogers, K. W., Bläßle, A., Schier, A. F., & Müller, P. (2015). Measuring protein stability in living zebrafish embryos using fluorescence decay after photoconversion (FDAP). *Journal of Visualized Experiments*, (95), 52266.

Rogers, K. W., Lord, N. D., Gagnon, J. A., Pauli, A., Zimmerman, S., Aksel, D. C., et al. (2017). Nodal patterning without Lefty inhibitory feedback is functional but fragile. *eLife, 6*, e28785.

Rogers, K. W., & Müller, P. (2019). Nodal and BMP dispersal during early zebrafish development. *Developmental Biology, 447*, 14–23.

Rogers, K. W., & Schier, A. F. (2011). Morphogen gradients: From generation to interpretation. *Annual Review of Cell and Developmental Biology, 27*, 377–407.

Rullan, M., Benzinger, D., Schmidt, G. W., Milias-Argeitis, A., & Khammash, M. (2018). An optogenetic platform for real-time, single-cell interrogation of stochastic transcriptional regulation. *Molecular Cell, 70*, 745-756.e746.

Ryu, H., Chung, M., Dobrzynski, M., Fey, D., Blum, Y., Lee, S. S., et al. (2015). Frequency modulation of ERK activation dynamics rewires cell fate. *Molecular Systems Biology, 11*, 838.

Sagner, A., & Briscoe, J. (2017). Morphogen interpretation: Concentration, time, competence, and signaling dynamics. *Wiley Interdisciplinary Reviews: Developmental Biology, 6*, e271.

Sako, K., Pradhan, S. J., Barone, V., Inglés-Prieto, Á., Müller, P., Ruprecht, V., et al. (2016). Optogenetic control of Nodal signaling reveals a temporal pattern of Nodal signaling regulating cell fate specification during gastrulation. *Cell Reports, 16*, 866–877.

Scardigli, M., Mullenbroich, C., Margoni, E., Cannazzaro, S., Crocini, C., Ferrantini, C., et al. (2018). Real-time optical manipulation of cardiac conduction in intact hearts. *The Journal of Physiology, 596*, 3841–3858.

Schepens, I., Duek, P., & Fankhauser, C. (2004). Phytochrome-mediated light signalling in Arabidopsis. *Current Opinion in Plant Biology, 7*, 564–569.

Schier, A. F., & Talbot, W. S. (2005). Molecular genetics of axis formation in zebrafish. *Annual Review of Genetics, 39*, 561–613.

Schneider, H., Faschingbauer, F., Schuepbach-Mallepell, S., Korber, I., Wohlfart, S., Dick, A., et al. (2018). Prenatal correction of X-Linked hypohidrotic ectodermal dysplasia. *The New England Journal of Medicine, 378*, 1604–1610.

Schulte-Merker, S., Lee, K. J., McMahon, A. P., & Hammerschmidt, M. (1997). The zebrafish organizer requires chordino. *Nature, 387*, 862–863.

Shilo, B. Z., & Barkai, N. (2017). Buffering global variability of morphogen gradients. *Developmental Cell, 40*, 429–438.

Shin, Y., Berry, J., Pannucci, N., Haataja, M. P., Toettcher, J. E., & Brangwynne, C. P. (2017). Spatiotemporal control of intracellular phase transitions using light-activated optoDroplets. *Cell, 168*, 159-171.e114.

Smart, A. D., Pache, R. A., Thomsen, N. D., Kortemme, T., Davis, G. W., & Wells, J. A. (2017). Engineering a light-activated caspase-3 for precise ablation of neurons in vivo. *Proceedings of the National Academy of Sciences of the United States of America, 114*, E8174–E8183.

Sokolik, C., Liu, Y., Bauer, D., McPherson, J., Broeker, M., Heimberg, G., et al. (2015). Transcription factor competition allows embryonic stem cells to distinguish authentic signals from noise. *Cell Systems, 1*, 117–129.

Sorre, B., Warmflash, A., Brivanlou, A. H., & Siggia, E. D. (2014). Encoding of temporal signals by the TGF-β pathway and implications for embryonic patterning. *Developmental Cell, 30*, 334–342.

Strickland, D., Lin, Y., Wagner, E., Hope, C. M., Zayner, J., Antoniou, C., et al. (2012). TULIPs: Tunable, light-controlled interacting protein tags for cell biology. *Nature Methods, 9*, 379–384.

Sun, W., Zhang, W., Zhang, C., Mao, M., Zhao, Y., Chen, X., et al. (2017). Light-induced protein degradation in human-derived cells. *Biochemical and Biophysical Research Communications, 487*, 241–246.

Taslimi, A., Vrana, J. D., Chen, D., Borinskaya, S., Mayer, B. J., Kennedy, M. J., et al. (2014). An optimized optogenetic clustering tool for probing protein interaction and function. *Nature Communications, 5*, 4925.

Tay, S., Hughey, J. J., Lee, T. K., Lipniacki, T., Quake, S. R., & Covert, M. W. (2010). Single-cell NF-kappaB dynamics reveal digital activation and analogue information processing. *Nature, 466*, 267–271.

Thisse, B., & Thisse, C. (2015). Formation of the vertebrate embryo: Moving beyond the Spemann organizer. *Seminars in Cell & Developmental Biology, 42*, 94–102.

Tichy, A. M., Gerrard, E. J., Legrand, J. M. D., Hobbs, R. M., & Janovjak, H. (2019). Engineering strategy and vector library for the rapid generation of modular light-controlled protein-protein interactions. *Journal of Molecular Biology, 431*, 3046–3055.

Tischer, D., & Weiner, O. D. (2014). Illuminating cell signalling with optogenetic tools. *Nature Reviews. Molecular Cell Biology, 15*, 551–558.

Toettcher, J. E., Gong, D., Lim, W. A., & Weiner, O. D. (2011). Light-based feedback for controlling intracellular signaling dynamics. *Nature Methods, 8*, 837–839.

Toettcher, J. E., Weiner, O. D., & Lim, W. A. (2013). Using optogenetics to interrogate the dynamic control of signal transmission by the Ras/Erk module. *Cell, 155*, 1422–1434.

Tostevin, F., ten Wolde, P. R., & Howard, M. (2007). Fundamental limits to position determination by concentration gradients. *PLoS Computational Biology, 3*, e78.

Tuazon, F. B., & Mullins, M. C. (2015). Temporally coordinated signals progressively pattern the anteroposterior and dorsoventral body axes. *Seminars in Cell & Developmental Biology, 42*, 118–133.

Tucker, J. A., Mintzer, K. A., & Mullins, M. C. (2008). The BMP signaling gradient patterns dorsoventral tissues in a temporally progressive manner along the anteroposterior axis. *Developmental Cell, 14*, 108–119.

Turing, A. M. (1952). The chemical basis of morphogenesis. *Philosophical Transactions of the Royal Society B, 237*, 37–72.

Uchida, A., & Yajima, M. (2018). An optogenetic approach to control protein localization during embryogenesis of the sea urchin. *Developmental Biology, 441*, 19–30.

Uda, Y., Goto, Y., Oda, S., Kohchi, T., Matsuda, M., & Aoki, K. (2017). Efficient synthesis of phycocyanobilin in mammalian cells for optogenetic control of cell signaling. *Proceedings of the National Academy of Sciences of the United States of America, 114*, 11962–11967.

Urban, E. A., & Johnston, R. J., Jr. (2018). Buffering and amplifying transcriptional noise during cell fate specification. *Frontiers in Genetics, 9*, 591.

Usherenko, S., Stibbe, H., Musco, M., Essen, L. O., Kostina, E. A., & Taxis, C. (2014). Photo-sensitive degron variants for tuning protein stability by light. *BMC Systems Biology, 8*, 128.

van Boxtel, A. L., Chesebro, J. E., Heliot, C., Ramel, M. C., Stone, R. K., & Hill, C. S. (2015). A temporal window for signal activation dictates the dimensions of a Nodal signaling domain. *Developmental Cell, 35*, 175–185.

van Wyk, M., Pielecka-Fortuna, J., Lowel, S., & Kleinlogel, S. (2015). Restoring the ON switch in blind retinas: Opto-mGluR6, a next-generation, cell-tailored optogenetic tool. *PLoS Biology, 13*, e1002143.

Varadarajan, S. G., Kong, J. H., Phan, K. D., Kao, T. J., Panaitof, S. C., Cardin, J., et al. (2017). Netrin1 produced by neural progenitors, not floor plate cells, is required for axon guidance in the spinal cord. *Neuron*, *94*, 790-799.e793.

Vopalensky, P., Pralow, S., & Vastenhouw, N. L. (2018). Reduced expression of the Nodal co-receptor Oep causes loss of mesendodermal competence in zebrafish. *Development*, *145*, dev.158832.

Waddington, C. H., & Kacser, H. (1957). *The strategy of the genes: A discussion of some aspects of theoretical biology*. London, UK: George Allen & Unwin.

Wang, X., Chen, X., & Yang, Y. (2012). Spatiotemporal control of gene expression by a light-switchable transgene system. *Nature Methods*, *9*, 266–269.

Wang, H., Vilela, M., Winkler, A., Tarnawski, M., Schlichting, I., Yumerefendi, H., et al. (2016). LOVTRAP: An optogenetic system for photoinduced protein dissociation. *Nature Methods*, *13*, 755–758.

Wang, W., Wildes, C. P., Pattarabanjird, T., Sanchez, M. I., Glober, G. F., Matthews, G. A., et al. (2017). A light- and calcium-gated transcription factor for imaging and manipulating activated neurons. *Nature Biotechnology*, *35*, 864–871.

Wartlick, O., Kicheva, A., & González-Gaitán, M. (2009). Morphogen gradient formation. *Cold Spring Harbor Perspectives in Biology*, *1*, a001255.

Werner, S. L., Barken, D., & Hoffmann, A. (2005). Stimulus specificity of gene expression programs determined by temporal control of IKK activity. *Science*, *309*, 1857–1861.

Wernet, M. F., Mazzoni, E. O., Celik, A., Duncan, D. M., Duncan, I., & Desplan, C. (2006). Stochastic spineless expression creates the retinal mosaic for colour vision. *Nature*, *440*, 174–180.

Wieschaus, E. (2016). Positional information and cell fate determination in the early *Drosophila* embryo. *Current Topics in Developmental Biology*, *117*, 567–579.

Wilson, P. A., Lagna, G., Suzuki, A., & Hemmati-Brivanlou, A. (1997). Concentration-dependent patterning of the Xenopus ectoderm by BMP4 and its signal transducer Smad1. *Development*, *124*, 3177–3184.

Wilson, M. Z., Ravindran, P. T., Lim, W. A., & Toettcher, J. E. (2017). Tracing information flow from Erk to target gene induction reveals mechanisms of dynamic and combinatorial control. *Molecular Cell*, *67*, 757-769.e755.

Wolpert, L. (1968). The French Flag Problem: A contribution to the discussion on pattern development and regulation. In C. H. Waddington (Ed.), *Towards a theoretical biology* (pp. 125–133). Edinburgh: Edinburgh University Press.

Wolpert, L. (1969). Positional information and the spatial pattern of cellular differentiation. *Journal of Theoretical Biology*, *25*, 1–47.

Wolpert, L. (1989). Positional information revisited. *Development*, *107*(Suppl), 3–12.

Wolpert, L. (2016). Positional information and pattern formation. *Current Topics in Developmental Biology*, *117*, 597–608.

Wood, L. A., Larocque, G., Clarke, N. I., Sarkar, S., & Royle, S. J. (2017). New tools for "hot-wiring" clathrin-mediated endocytosis with temporal and spatial precision. *The Journal of Cell Biology*, *216*, 4351–4365.

Xu, P. F., Houssin, N., Ferri-Lagneau, K. F., Thisse, B., & Thisse, C. (2014). Construction of a vertebrate embryo from two opposing morphogen gradients. *Science*, *344*, 87–89.

Xu, Y., Hyun, Y. M., Lim, K., Lee, H., Cummings, R. J., Gerber, S. A., et al. (2014). Optogenetic control of chemokine receptor signal and T-cell migration. *Proceedings of the National Academy of Sciences of the United States of America*, *111*, 6371–6376.

Yamashita, Y. M., Inaba, M., & Buszczak, M. (2018). Specialized intercellular communications via cytonemes and nanotubes. *Annual Review of Cell and Developmental Biology*, *34*, 59–84.

Yang, X., Jost, A. P., Weiner, O. D., & Tang, C. (2013). A light-inducible organelle-targeting system for dynamically activating and inactivating signaling in budding yeast. *Molecular Biology of the Cell*, *24*, 2419–2430.

Yu, X., Liu, H., Klejnot, J., & Lin, C. (2010). The cryptochrome blue light receptors. *Arabidopsis Book*, *8*, e0135.

Yumerefendi, H., Dickinson, D. J., Wang, H., Zimmerman, S. P., Bear, J. E., Goldstein, B., et al. (2015). Control of protein activity and cell fate specification via light-mediated nuclear translocation. *PLoS One*, *10*, e0128443.

Yumerefendi, H., Lerner, A. M., Zimmerman, S. P., Hahn, K., Bear, J. E., Strahl, B. D., et al. (2016). Light-induced nuclear export reveals rapid dynamics of epigenetic modifications. *Nature Chemical Biology*, *12*, 399–401.

Yumerefendi, H., Wang, H., Dickinson, D. J., Lerner, A. M., Malkus, P., Goldstein, B., et al. (2018). Light-dependent cytoplasmic recruitment enhances the dynamic range of a nuclear import photoswitch. *Chembiochem*, *19*, 1319–1325.

Yüz, S. G., Rasoulinejad, S., Mueller, M., Wegner, A. E., & Wegner, S. V. (2019). Blue light switchable cell-cell interactions provide reversible and spatiotemporal control towards bottom-up tissue engineering. *Advanced Biosystems*, *3*, 1800310.

Yüz, S. G., Ricken, J., & Wegner, S. V. (2018). Independent control over multiple cell types in space and time using orthogonal blue and red light switchable cell interactions. *Advanced Science (Weinheim, Baden-Württemberg, Germany)*, *5*, 1800446.

Zagorski, M., Tabata, Y., Brandenberg, N., Lutolf, M. P., Tkacik, G., Bollenbach, T., et al. (2017). Decoding of position in the developing neural tube from antiparallel morphogen gradients. *Science*, *356*, 1379–1383.

Zhang, K., Duan, L., Ong, Q., Lin, Z., Varman, P. M., Sung, K., et al. (2014). Light-mediated kinetic control reveals the temporal effect of the Raf/MEK/ERK pathway in PC12 cell neurite outgrowth. *PLoS One*, *9*, e92917.

Zhang, W., Guan, X., Yang, Y., Ge, B., Chen, H., Li, F., et al. (2009). Biosynthesis of fluorescent allophycocyanin alpha-subunits by autocatalysis in Escherichia coli. *Biotechnology and Applied Biochemistry*, *52*, 135–140.

Zhang, W., Lohman, A. W., Zhuravlova, Y., Lu, X., Wiens, M. D., Hoi, H., et al. (2017). Optogenetic control with a photocleavable protein, PhoCl. *Nature Methods*, *14*, 391–394.

Zhou, X. X., Chung, H. K., Lam, A. J., & Lin, M. Z. (2012). Optical control of protein activity by fluorescent protein domains. *Science*, *338*, 810–814.

Zhou, X. X., Fan, L. Z., Li, P., Shen, K., & Lin, M. Z. (2017). Optical control of cell signaling by single-chain photoswitchable kinases. *Science*, *355*, 836–842.

Zoltowski, B. D., Vaccaro, B., & Crane, B. R. (2009). Mechanism-based tuning of a LOV domain photoreceptor. *Nature Chemical Biology*, *5*, 827–834.

CHAPTER THREE

A matter of time: Formation and interpretation of the Bicoid morphogen gradient

Anqi Huang[a,†], Timothy E. Saunders[a,b,c,*]
[a]Mechanobiology Institute, National University of Singapore, Singapore, Singapore
[b]Department of Biological Sciences, National University of Singapore, Singapore, Singapore
[c]Institute of Molecular and Cell Biology, A*Star, Singapore, Singapore
[*]Corresponding author: e-mail address: dbsste@nus.edu.sg

Contents

[†] Present address: The Francis Crick Institute, London, United Kingdom

Current Topics in Developmental Biology, Volume 137
ISSN 0070-2153
https://doi.org/10.1016/bs.ctdb.2019.11.016

Abstract

Spatially distributed signaling molecules, known as morphogens, provide spatial information during development. A host of different morphogens have now been identified, from subcellular gradients through to morphogens that act across a whole embryo. These gradients form over a wide-range of timescales, from seconds to hours, and their time windows for interpretation are also highly variable; the processes of morphogen gradient formation and interpretation are highly dynamic.

The morphogen Bicoid (Bcd), present in the early *Drosophila* embryo, is essential for setting up the future *Drosophila* body segments. Due to its accessibility for both genetic perturbations and imaging, this system has provided key insights into how precise patterning can occur within a highly dynamic system. Here, we review the temporal scales of Bcd gradient formation and interpretation. In particular, we discuss the quantitative evidence for different models of Bcd gradient formation, outline the time windows for Bcd interpretation, and describe how Bcd temporally adapts its own ability to be interpreted. The utilization of temporal information in morphogen readout may provide crucial inputs to ensure precise spatial patterning, particularly in rapidly developing systems.

1. Introduction

Morphogen gradients are a central component of spatial positioning during development. The spatial information encoded within morphogens (Kicheva, Bollenbach, Wartlick, Jülicher, & Gonzalez-Gaitan, 2012; Kicheva, Cohen, & Briscoe, 2012) can be interpreted into precise boundaries, often with single cell precision (Gregor, Tank, Wieschaus, & Bialek, 2007). Yet, morphogen formation and interpretation are both inherently dynamic, and the tissues in which morphogens act are continually changing through growth, cell rearrangements, division and apoptosis (Wartlick, Julicher, & Gonzalez-Gaitan, 2014). Therefore, the question of *when*, not just *where*, is crucial in understanding how gene expression boundaries are precisely defined (Sagner & Briscoe, 2017).

There has been significant debate over the timing of gradient formation and when spatial information from morphogens is deciphered (Bergmann et al., 2007; Bergmann, Tamari, Schejter, Shilo, & Barkai, 2008; Bialek, Gregor, Tank, & Wieschaus, 2008; de Lachapelle & Bergmann, 2010a, 2010b; Gregor, Tank, et al., 2007; Jaeger, 2010; Lander, Nie, & Wan, 2002; Porcher et al., 2010; Wolpert, 2016). Here, we focus on two elements related to temporal interpretation of morphogen gradients. First, how do the underlying biophysical processes driving morphogen gradient formation determine the time for gradients to form? Second, when are morphogen gradients interpreted and how do time constraints affect the mechanisms of interpretation? Importantly, morphogens encode *temporal* information, and recent experimental and theoretical work has started to reveal how systems take advantage of such information to generate precisely defined domains in both space and time.

To elucidate principles underlying the above two questions, we focus primarily on the morphogen Bicoid (Bcd) in the early *Drosophila* embryo (Ephrussi & Johnston, 2004; Porcher & Dostatni, 2010; Surkova, Golubkova, Mamon, & Samsonova, 2018; Wieschaus, 2016), and its role in specifying the downstream gap genes. This system has proven to be an excellent model for quantitative testing of models of morphogen formation and interpretation (Gregor, Tank, et al., 2007; Gregor, Wieschaus, McGregor, Bialek, & Tank, 2007; Grimm, Coppey, & Wieschaus, 2010; Jaeger et al., 2004; Manu et al., 2009a). The Bcd gradient forms in the early *Drosophila* embryo during the blastoderm stage (Driever, Ma, Nüsslein-Volhard, & Ptashne, 1989; Driever & Nüsslein-Volhard, 1988a; Frohnhöfer & Nüsslein-Volhard, 1986), Fig. 1A. As a transcription factor, Bcd plays a critical role in specifying the domains of gap gene expression which provide the framework for downstream patterning events (Driever & Nüsslein-Volhard, 1988b, 1989; Petkova, Tkačik, Bialek, Wieschaus, & Gregor, 2019). Extensive reviews of Bcd are available that provide detail on its function and spatial extent (Grimm et al., 2010; Porcher & Dostatni, 2010; Wieschaus, 2016); here, we focus on the role of time in Bcd gradient formation and interpretation.

We begin this review focusing on the dynamics of Bcd morphogen gradient formation. We outline a general theoretical framework through which experimental results can be encapsulated. In the second half of the review, we discuss recent results on the temporal interpretation of the Bcd morphogen gradient. In particular, Bcd appears to modulate the ability of nuclei to further interpret Bcd—i.e., there is dynamic feedback between morphogen formation and interpretation.

2. Dynamics of morphogen gradient formation

Models of morphogen gradients, such as the French Flag model, Fig. 1B, typically focus on the spatial information encoded in the gradient. However, a morphogen gradient can be defined more generally by its ability to impart spatial information to a system via its spatial and temporal

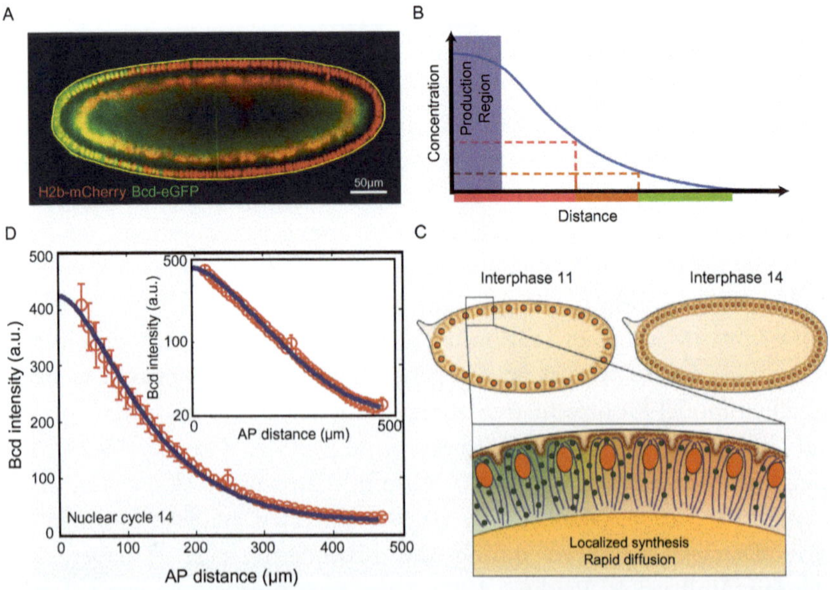

Fig. 1 The Bicoid morphogen gradient. (A) Drosophila embryo expressing Bcd::eGFP and H2b::mCherry, imaged in early n.c. 14 on a light-sheet microscope. (B) Morphogen gradient and readout. Distinct cell fates occur at different concentrations. (C) The local environment changes with time, with nuclear density increasing and the cytoskeletal environment becoming denser (microtubules in blue and actin caps in red). (D) Measured Bcd concentration profile. Inset shows same on logarithmic scale. Blue line is a fit to the SDD model, accounting for eGFP folding time.

inhomogeneity. These inhomogeneities form through asymmetries in morphogen transport, removal (e.g., degradation), and production. How do these different mechanisms alter the *timing* of morphogen gradient formation?

2.1 Morphogen production

Models of morphogen formation typically localize morphogen production to a specific region (Crick, 1970; Wolpert, 1969). This can either be through localization of mRNA, as proposed for *bcd* (Berleth et al., 1988; St Johnston, Driever, Berleth, Richstein, & Nüsslein-Volhard, 1989), or by a specific region of cells acting as source cells, producing the morphogen (Kicheva et al., 2007). More complicated source regions may exist, such as moving sources in growing domains (Wartlick et al., 2014; Zartman et al., 2011), and these have effects on the spatial and temporal range of morphogen gradients (Dalessi, Neves, & Bergmann, 2012). Recent work using in-bred lines of *Drosophila* that lay larger or smaller eggs, has found that production of Bcd is increased in larger embryos (He et al., 2015). Morphogen production is a dynamic process, with potentially spatial and temporal variation during development.

2.2 Morphogen spatial distribution

Protein distribution can be either passive or active. Passive (diffusive) transport requires no specific energy input beyond thermal fluctuations (Berg, 1993). With active transport, molecules are moved in specific directions, e.g., along extended actin fibers, with external energy input (typically via ATP). Both passive and active transport processes have been proposed to form morphogen gradients (Muller, Rogers, Yu, Brand, & Schier, 2013; Wartlick, Kicheva, & Gonzalez-Gaitan, 2009).

A simple way to form a morphogen concentration gradient is to have a localized source of morphogen production and then have morphogen diffuse unhindered (with coefficient D) (Wartlick et al., 2009), known as *Fickian* diffusion (Berg, 1993). Within biologically relevant contexts, D is often an *effective* diffusion coefficient as diffusion is hindered due to, for example, the dense extracellular space around cells (Muller et al., 2013) or around the nuclei in the *Drosophila* blastoderm, Fig. 1C. Consequently, the apparent protein diffusion is often considerably less than that measured for free GFP, which can vary from around $25\,\mu m^2 s^{-1}$ (Swaminathan, Hoang, & Verkman, 1997) in the cytoplasm of eukaryotic cells to $85\,\mu m^2 s^{-1}$ in the extracellular space of zebrafish embryos (Wang, Wang, Wohland, & Sampath, 2016). The assumptions of Fickian diffusion can break down, for example due to chaperone binding (Nybakken & Perrimon, 2002; Tabata, 2004), leading to anomalous diffusion (Hornung, Berkowitz, & Barkai, 2005), though distinguishing such different regimes is challenging (Fradin, 2017). There can also be tissue or fluid flow that alters protein movement via convection (Hecht, Rappel, & Levine, 2009). To estimate the timescale for diffusive transport processes, we consider the average root-mean-square distance, x_{rms} traveled by an ensemble of particles after time t, $x_{rms} \approx \sqrt{2dDt}$, where d is the dimensionality of the system. To generate a gradient over $100\,\mu m$ in 1 h, requires $D \sim 1\,\mu m^2 s^{-1}$. Therefore, diffusion is a plausible mechanism to produce a long-ranged gradient in the early *Drosophila* embryo.

Morphogen gradients can also form by directed morphogen transport, such as polarized auxin transport in plants (Adamowski & Friml, 2015). Can such mechanisms work at the intercellular scales of morphogens in animals?

Filopodia protrusions, known as cytonemes, can set up long-ranged, directed transport between cells (Kornberg, 2014; Stanganello & Scholpp, 2016; Yamashita, Inaba, & Buszczak, 2018). Cytoneme-mediated morphogen transport has been reported for Wnts (Roy, Hsiung, & Kornberg, 2011; Stanganello et al., 2015), TGFβ (Hsiung, Ramirez-Weber, David Iwaki, & Kornberg, 2005; Roy, Huang, Liu, & Kornberg, 2014) and Hedgehog (Sanders, Llagostera, & Barna, 2013). In effect, the system has transferred the positional information from the morphogen to the underlying cytoskeletal network (Muller et al., 2013). The timescale for morphogen transport is rapid once the network is generated. It has been proposed that in the early *Drosophila* embryo, a cytoskeletal network distributes *bcd* mRNA through the embryo, enabling localized—and rapid—production of Bcd protein across the embryo (Ali-Murthy & Kornberg, 2016; Spirov et al., 2009). In Section 3, we detail the experimental evidence for the different hypotheses for Bcd transport.

2.3 Morphogen removal processes

The mechanism of morphogen removal shapes the morphogen profile (Eldar, Rosin, Shilo, & Barkai, 2003; Saunders & Howard, 2009). Morphogen can be removed by, for example, local morphogen sequestration through receptor binding or cellular internalization (Bollenbach et al., 2008; Yu et al., 2009), resulting in effective morphogen degradation. Linear degradation (rate μ) coupled with diffusion (D) leads to the synthesis, diffusion, degradation (SDD) model:

$$\partial_t \rho(\boldsymbol{x}, t) = D\nabla^2 \rho(\boldsymbol{x}, t) - \mu \rho(\boldsymbol{x}, t) \tag{1}$$

where ∇^2 is the n-dimensional Laplacian in space (for example, in two-dimensions, $\nabla^2 = \frac{\partial^2}{\partial x^2} + \frac{\partial^2}{\partial y^2}$). In steady-state ($\partial_t \rho(\boldsymbol{x}, t) = 0$) and a localized source of protein (J) at $x=0$, $\rho(x) = \frac{J}{\sqrt{D\mu}} e^{-x/\lambda}$, where $\lambda = \sqrt{D/\mu}$, Fig. 1D (provided system length $\gg \lambda$). In Section 3.1 we discuss more realistic forms for J—which can change in both space and time—in relation to the Bcd gradient. More complicated (non-linear) morphogen degradation mechanisms are also plausible, and these can give rise to power-law like profiles of morphogen gradients (Eldar et al., 2003), and have been observed for Bcd in *nejire* mutants (He et al., 2010). Measuring the dynamic parameters remains challenging *in vivo* (Muller et al., 2013). In Box 1 we discuss two methods—fluorescence correlation spectroscopy (FCS) and fluorescence recovery after photobleaching (FRAP)—that have been used to quantify the dynamic parameters for Bcd *in vivo*.

BOX 1 FCS and FRAP measurements of morphogen dynamics. FCS measures local intensity variations within a small imaging region to infer particle dynamics; how the signal varies provides information about the dynamic processes (Krieger et al., 2015). Slow moving particles stay within the observation volume longer, resulting in long-time traces in the signal, Fig. B1A. Fast moving particles rapidly leave the imaging volume, resulting in low temporal correlation. These correlations are encapsulated within the autocorrelation function $G(\tau) = \frac{\langle \delta I(t+\tau) \cdot \delta I(t) \rangle}{\langle I \rangle^2}$, where $\delta I(t)$ represents the fluctuations of the signal about the time-average mean intensity $\langle I \rangle$. The fluctuation magnitude scales with the square root of particle number in the observation volume and hence FCS measurements are best performed at low particle numbers. Photo-switchable fluorophores can be used to more precisely control the fluorescent molecule number in the imaging volume and hence improve FCS readout (White et al., 2016). The value of $G(0)$ can be used to estimate particle number. FCS typically measures timescales of events from milliseconds to second scales. FCS is sensitive to different biophysical processes. For example, if there is binding within the imaging volume, then the autocorrelation function has an extended tail at long times. Estimates for both diffusion and binding kinetics are possible from FCS data (Krieger et al., 2015), and this has been performed for Bcd (Abu-Arish et al., 2010).

FRAP measures how rapidly a region depleted of fluorescence recovers, and has been applied extensively to morphogens (Gregor, Wieschaus, et al., 2007; Kicheva et al., 2007; Muller et al., 2012). FRAP typically measures longer timescales (from seconds to minutes scales) than FCS, and measures the effective diffusion, which-combines both diffusion and binding kinetics (De Los Santos, Chang, Mycek, & Cardullo, 2015; Sigaut et al., 2014; Sprague & McNally, 2005), Fig. B1B. However, care must be taken to correct for photobleaching, which can occur on similar timescales (Abu-Arish et al., 2010), and the high laser powers involved can heat the local environment (De Los Santos et al., 2015). Estimates for dynamic parameters including the diffusion coefficient can be inferred from FRAP experiments, but FRAP generally measure the timescales of longer processes, such as binding, and can often miss more rapid diffusive motion (Sprague & McNally, 2005).

Without binding kinetics (or other processes inhibiting diffusion), FCS and FRAP return similar estimates of diffusion coefficients (Sigaut et al., 2014). However, in most biological systems this is rarely the case. It has been argued that since FRAP reports on longer timescales, it is more biologically relevant (Grimm et al., 2010). However, the short time kinetics of the morphogen are important, for both dispersal and signal interpretation (Sigaut et al., 2014). Measurements of the lifetime and decay length of the Bcd gradient find values around 30 min (Durrieu et al., 2018) and 85 μm (Little et al., 2011; Liu et al., 2013) respectively. This gives an effective diffusion constant of around $3\,\mu m^2 s^{-1}$. Both FCS ($D_{Bcd} \sim 5$–$10\,\mu m^2 s^{-1}$) and FRAP ($D_{Bcd} \sim 0.3$–$1\,\mu m^2 s^{-1}$) provide useful information but neither, by itself, is able to definitively describe Bcd dynamics as D is an amalgamation of numerous processes, including diffusion, nuclear trapping, binding, and potential active transport through the cytoskeleton.

Continued

BOX 1 FCS and FRAP measurements of morphogen dynamics.—cont'd

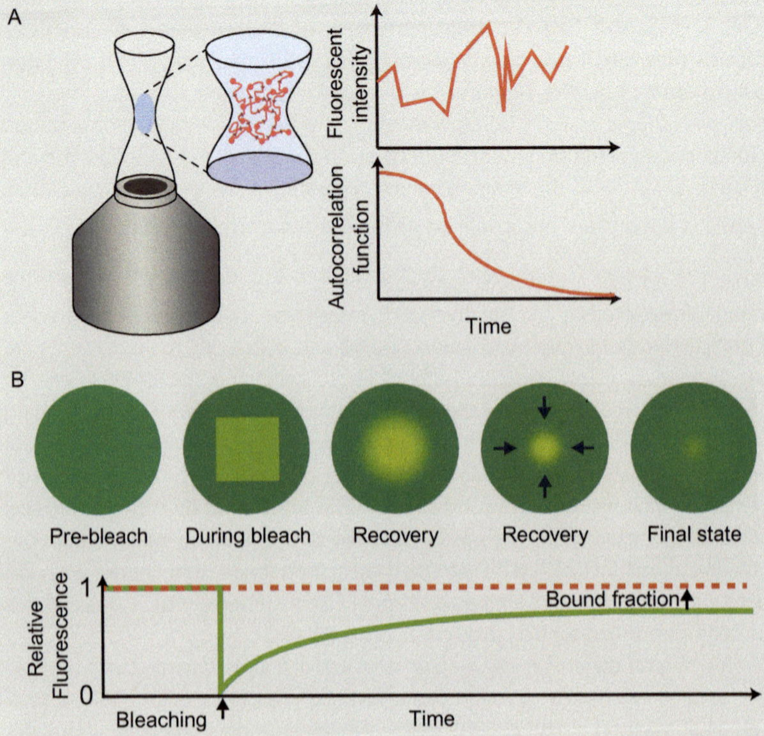

Fig. B1 (A) Principle of fluorescence correlation spectroscopy. Fluorescence intensity is measured within a small focal volume at μs frame rate. Intensity traces (top right) are converted into the autocorrelation function (bottom right), which describes the protein dynamics inside the focal volume. (B) Principle of fluorescence recovery after photobleaching. A region is bleached using a short, high energy laser pulse, and then the subsequent recovery of the signal is recorded (bottom). The rate of recovery and the final intensity level correspond to the diffusivity and bound fraction of protein respectively.

2.4 Timescales of morphogen formation

How long does a morphogen gradient take to form? A simple but powerful framework (Berezhkovskii, Sample, & Shvartsman, 2010, 2011; Gordon, Sample, Berezhkovskii, Muratov, & Shvartsman, 2011) has been introduced for calculating the local accumulation time $\tau(x)$, defined as the time to reach steady-state at a specific position. For the SDD model,

$$\tau(x) = \frac{1}{2\mu}\left(1 + \frac{x}{\lambda}\right) \tag{2}$$

where μ and λ are defined as above. Regions close to the source equilibrate rapidly, in around 30 min for Bcd, determined primarily by degradation. Equilibration further from the source depends also on diffusion, and can take over 2 h using dynamic parameters appropriate for Bcd. Modifying degradation can substantially alter the equilibration time (Kolomeisky, 2011; Teimouri & Kolomeisky, 2014). $\tau(x)$ can be found for more complex models incorporating morphogen binding (Berezhkovskii & Shvartsman, 2013) and intracellular shuttling (Berezhkovskii & Shvartsman, 2015). In systems with global morphogen production, the time to equilibrate is much shorter since proteins are produced near their final position. However, this ignores the timescale for forming the distributed production region; for example, we are not aware of quantifications of the timescale for cytoneme network formation. Developmental time is also temperature sensitive; going from 18 °C to 25 °C halves *Drosophila* embryonic developmental time (Chong, Amourda, & Saunders, 2018; Kuntz & Eisen, 2014). The production rate of Bcd protein is temperature sensitive, with higher production at low (18 °C) and high (29 °C) temperatures compared with 25 °C (Cheung & Ma, 2015). According to the Einstein-Stokes relation, free diffusion scales linearly with the temperature in kelvin, and therefore diffusion coefficients should only be minorly affected by physiological temperature changes. How gradient formation adjusts to dramatic changes in developmental speed under temperature variation remains an intriguing problem, for which there is currently a paucity of high-quality experimental data.

Methods have also recently been developed to quantify the age of morphogen proteins *in vivo* (Donà et al., 2014; Durrieu et al., 2018; Khmelinskii et al., 2012) and this is discussed in detail in Box 2. Applying this approach to Bcd below, we will see how measuring protein age provides a powerful tool for distinguishing models of gradient formation.

BOX 2 Protein age as a measure of morphogen dynamics.

Knowing protein age provides critical information for understanding protein dynamics. To measure age experimentally a tandem reporter system has been developed (Khmelinskii et al., 2012). This involves tagging a protein of interest with two fluorophores with different folding rates. For a young protein, only the fast folding (say green) fluorescent protein will have folded. For older protein, the slower folding (say red) fluorescent protein will have also folded. The ratio between the two fluorescent signals gives information about the protein age, Fig. B2A and B. An advantage of this approach is that dynamic information can be inferred from static, high-quality images as each image contains information about morphogen profile *and* morphogen age.

Continued

BOX 2 Protein age as a measure of morphogen dynamics.—cont'd

For proteins governed by SDD dynamics, the average age, T, of a particle a distance x from the source in steady-state is given by $T(x) = \frac{1}{2\mu}\left(1 + \frac{x}{\lambda}\right)$ (Durrieu et al., 2018) (λ and μ are as defined for Eq. 1) which, interestingly, has the same functional form as Eq. (2). Unless the source term is rapidly changing, the measured protein age is independent of the production rate (Khmelinskii et al., 2012). For proteins produced locally by an mRNA gradient, the protein lifetime $T(x) \sim 1/\mu$; spatial variation is unimportant. The SDD and mRNA gradient models make different predictions for protein age as a function of distance, even though both produce similar protein gradients, Fig. 2D.

To highlight the potential power of this approach we use the following toy system. These (zero spatial dimensional) models have identical temporal behavior for the protein concentration, ρ:

Model 1: $\frac{d\rho_1(t)}{dt} = J - \mu\rho_1(t)$, with solution $\rho_1(t) = \frac{J}{\mu}(1 - e^{-\mu t})$

Model 2: $\frac{d\rho_2(t)}{dt} = Je^{-\mu t}$, with solution $\rho_2(t) = \frac{J}{\mu}(1 - e^{-\mu t})$

Model 1 assumes constant production of protein (rate J) and linear degradation of protein (rate μ). Model 2 has time-dependent production. Initially the production rate is equal to J, but this decreases to zero at long times, also at rate μ. Measurement of protein concentration itself cannot distinguish the two models as $\rho_1(t) = \rho_2(t)$, Fig. B2C. However, the molecular age in each model is distinct. In the first model, the average protein age $T \sim \mu^{-1}$ (at long times), independent of system age, t. In the second model, $T \sim t$ as there is no protein degradation. Measuring the protein age can distinguish the models despite having identical concentration behavior, Fig. B2D.

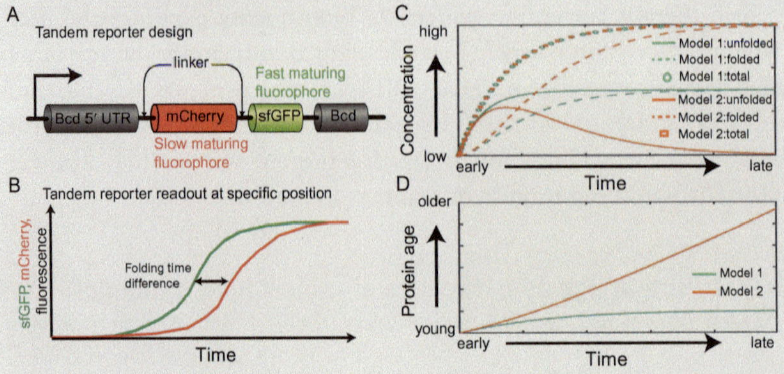

Fig. B2 (A) Design of fluorescent lifetime reporter. The linkers are critical in ensuring appropriate degradation of the construct (Khmelinskii et al., 2016). (B) Fluorescent reporter folding steps. (C) Concentration profiles for the toy models outlined above. We assume the protein described by the models is tagged with a fluorescent protein that has similar folding time to the protein degradation rate μ. Therefore, at a particular time some protein has a folded fluorophore (dashed) and some has unfolded protein (solid line). However, the total amount of protein (circles) is identical in both models. (D) From the equations for Model 1 and Model 2, we can deduce the average protein age at a particular time. Details in Durrieu et al. (2018).

3. Dynamics of Bcd morphogen formation

Having outlined mechanisms of gradient formation and the corresponding timescales, we now discuss the dynamics underlying Bcd gradient formation. We end this section by presenting the most likely formation process based on current experimental data and highlighting open questions.

3.1 Bcd production

Upon identification of the Bcd gradient, *bcd* mRNA was observed localized to the embryo anterior (Driever & Nüsslein-Volhard, 1988a). This observation led to the proposal that Bcd is synthesized in the anterior pole, with long-ranged diffusion setting up the gradient (SDD model). Later analysis revealed that *bcd* mRNA extends into the embryo a significant distance (Ali-Murthy & Kornberg, 2016; Little, Tkačik, Kneeland, Wieschaus, & Gregor, 2011; Spirov et al., 2009). A gradient of *bcd* mRNA is present even in the oocyte (Ali-Murthy & Kornberg, 2016). These observations have led to alternative models for Bcd gradient formation whereby the spatial information is encoded within the *bcd* mRNA distribution (Ali-Murthy & Kornberg, 2016; Spirov et al., 2009). If the *bcd* mRNA profile is commensurate with the Bcd gradient then long-ranged gradients can be formed without long-ranged diffusion, Fig. 2A. However, these models face a number of problems. (1) The observed *bcd* mRNA gradient is significantly shorter than the Bcd protein distribution when carefully quantified (Little et al., 2011). (2) For the Bcd gradient to maintain a similar profile to that of the *bcd* mRNA requires Bcd diffusion to be very small (Dilão & Muraro, 2010), which conflicts with FCS (Abu-Arish, Porcher, Czerwonka, Dostatni, & Fradin, 2010) and Bcd::eGFP live imaging (Mir et al., 2017). (3) The Bcd gradient varies by 10% between embryos (Gregor, Wieschaus, et al., 2007), which can be readily explained within an SDD model (Gregor, Wieschaus, et al., 2007), but similar mechanisms for forming a precise gradient of *bcd* mRNA are unknown. (4) Models with early production of Bcd (prior to n.c. 9) require very long-lived Bcd protein—i.e., protein produced in n.c. 4 needs to last until n.c. 12, where peak Bcd::eGFP signal is observed (Little et al., 2011). But, this conflicts with the measured short Bcd lifetimes, as discussed below. It is worth noting that incorporating a short-ranged *bcd* mRNA gradient into the SDD model does improve fitting to experimental profiles (Little et al., 2011).

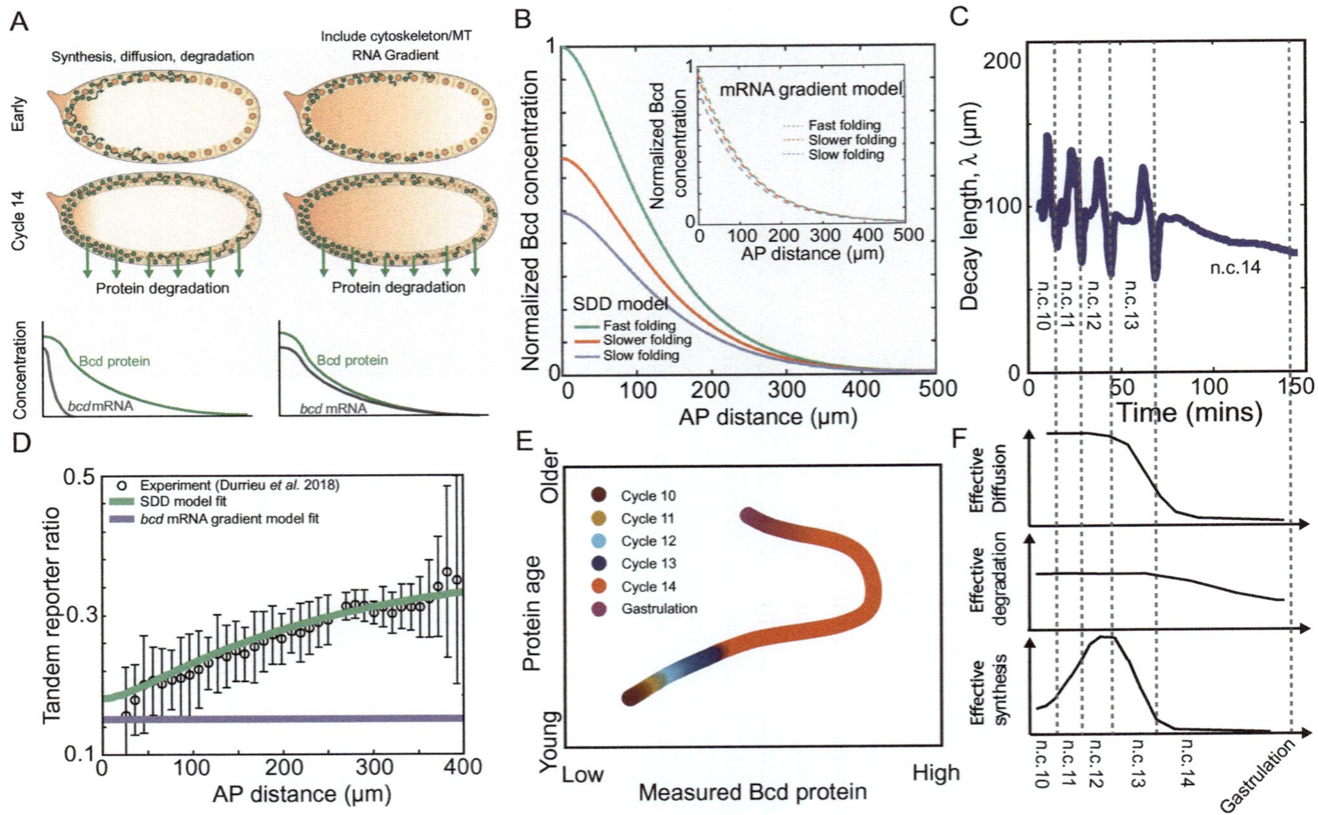

Fig. 2 See legend on opposite page.

Timing delays caused by the underlying biochemical processes can also be used to test models. The measured fluorescence intensity in live imaging does not correspond to the actual concentration of protein at a specific time due to time delay in protein folding (Berezhkovskii & Shvartsman, 2014; Little et al., 2011; Liu, Morrison, & Gregor, 2013). For example, eGFP takes around 45 min to fold in the early *Drosophila* embryo. If Bcd is produced via a long-ranged mRNA *bcd* gradient, then changes in fluorophore folding rates do not change the shape of the measured profile, only the amplitude, Fig. 2B. However, the profiles for Bcd::Venus, Bcd::eGFP, and Bcd::CFP are different (Little et al., 2011), particularly in the anterior, which further supports models whereby the significant majority of Bcd protein is produced in the anterior pole. From n.c. 13 onward, the Bcd gradient is observed to decrease in magnitude (Durrieu et al., 2018; Little et al., 2011). This decrease correlates with an observed decrease in *bcd* mRNA intensity (Little et al., 2011), suggesting that the decrease is due, at least in part, to decreased production of Bcd protein in n.c. 13 and 14. However, Bcd protein is still present in the embryo, even at gastrulation (Durrieu et al., 2018).

Fig. 2 Dynamics of Bicoid gradient formation. (A) Two models of Bcd gradient formation. Left: SDD model, where the Bcd gradient forms through localized production, transport, and removal. Right: RNA gradient model, where Bcd morphogen is produced in a graded fashion across the embryo. Both models can produce similar morphogen profiles. Green dots (arrows) represent Bcd protein (movement). (B) Model demonstrating how the fluorophore folding time can alter the observed gradient for the SDD model and (inset) RNA gradient model. Parameters: Bcd lifetime = 45 min, protein folding times = 10, 20, 30 min for fast, slower, and slow folding fluorophores. SDD model, $D = 2.5\,\mu m^2 s^{-1}$; RNA gradient model, production domain described by $e^{-x/115\mu m}$. (C) Measured decay length of the Bcd morphogen profile across different nuclear cycles. (D) Bcd protein age, as measured by tandem reporter (Box 2), across the embryo. (E) Cartoon of Bcd protein age and abundance from nuclear cycle 10 to gastrulation. Initially, Bcd protein levels are low and protein age is relatively young (bottom left of cartoon). As development progresses, both total Bcd amount and average Bcd protein age increase. In early n.c. 14, protein levels begin to decrease, but protein age continues to increase. This supports cessation of Bcd production. (F) Schematics of dynamic changes in parameters for the SDD model. After n.c. 13 Bcd diffusion may decrease markedly, perhaps due to cellularization. The timer reporter suggests that the effective rate of Bcd degradation decreases around n.c. 13. *From Little et al. (2011), the effective Bcd synthesis rate appears to peak around n.c. 12. Data shown in (C) and (D) collected as described in Durrieu et al. (2018).*

3.2 Bcd spatial distribution

Bcd dynamics have been inferred from imaging of fluorophore-tagged Bcd. Using FRAP with Bcd::eGFP (Gregor, Wieschaus, et al., 2007), Fig. 1A, the Bcd diffusion constant was estimated as $D \approx 0.3\ \mu m^2 s^{-1}$. FCS on Bcd::eGFP (Abu-Arish et al., 2010) or single molecule tracking using either Bcd:eGFP (Mir et al., 2017) or photoswitchable fluorophores (Drocco, Grimm, Tank, & Wieschaus, 2011; Mir et al., 2018) record similar timescales for Bicoid dynamics with an inferred diffusion constant of around $D \approx 5\ \mu m^2 s^{-1}$. The latter value of D is sufficient to form a 500 μm gradient in around 3 h. This difference in measured values for D has led to significant confusion (Wolpert, 2016). As explained in Box 1, these values are not contradictory as the two approaches are measuring different dynamic properties of Bcd; see Sigaut, Pearson, Colman-Lerner, & Ponce Dawson (2014) for detailed discussion. Direct live imaging of small Bcd hubs in the embryo posterior has revealed rapid, random, Bcd dynamics (Mir et al., 2017); i.e., posterior Bcd::eGFP is highly mobile, with diffusive-like behavior.

Bcd transport is relatively slow (morphogens in zebrafish have recorded $D > 30\ \mu m^2 s^{-1}$), suggesting that the cytoplasmic domain in the blastoderm is densely packed, with significant actin and microtubule aggregation (Daniels, Rikhy, & Renz, 2012; Fahmy et al., 2014; Thukral et al., 2019). This implies Bcd undergoes restricted diffusive transport, due to nuclear trapping and/or impediment by the cytoskeletal architecture, Fig. 1C. There has been detailed theoretical analysis of how such impeded transport mechanisms can perturb Bcd gradient formation (Fedotov & Falconer, 2014; Fradin, 2017; Hernández, Varea, & Barrio, 2009; Yuste, Abad, & Lindenberg, 2010) but to date there has not been a systematic quantitative study relating Bcd protein dynamics with the underlying cytoskeletal structure at different stages of development. Bcd transport is limited to the cytoplasmic region between the yolk (without movement through the yolk) and the cortex during the period of signal interpretation (Cai, Akber, Spirov, & Baumgartner, 2017; Gregor, Wieschaus, et al., 2007). Therefore, Bcd is compartmentalized in the cytoplasmic region between the yolk and cortex, and so (restricted) diffusion is a viable mechanism of gradient formation.

3.3 Bcd degradation

Measurements of Bcd degradation using photoconversion of Bcd:Dronpa (Drocco et al., 2011) have estimated the degradation rate at different times in development. Up to n.c. 13, degradation is generally slow, with Bcd half-life around 40 min, but in n.c. 14, Bcd half-life shortens. Western blotting of *Drosophila* S2 cells (Liu, He, & Ma, 2011; Liu & Ma, 2011), found a half-life around 20 min in n.c. 12, but conversely they saw an increase in half-life later in n.c. 14 to around 40 min. Though there is some disagreement between these values, they support the conclusion that Bcd degradation occurs prior to n.c. 14, contrary to models where the Bcd concentration is diluted through nuclear division (Kavousanakis, Kanodia, Kim, Kevrekidis, & Shvartsman, 2010; Shvartsman, Coppey, & Berezhkovskii, 2008). Inhibiting Bcd nuclear localization (Grimm & Wieschaus, 2010) and quantification of Bcd gradient in unfertilized embryos (Drocco, Wieschaus, & Tank, 2012) further support Bcd degradation being relevant in gradient shaping. Finally, Bcd is degraded rapidly during late n.c. 14, as can be seen by the rapidly decreasing decay length λ, Fig. 2C, though this signal is still relevant for signaling (Huang, Amourda, Zhang, Tolwinski, & Saunders, 2017) (Section 5).

3.4 Temporal dynamics of Bcd gradient formation

Information about protein age can distinguish different models of gradient formation with similar concentration profiles, Box 2. To measure protein age, a functional Bcd tandem-fluorescent reporter was generated (Durrieu et al., 2018) that rescues bcd^{E1} mutant embryos. Bcd protein age increases from anterior to posterior, Fig. 2D and E. The SDD model generally fit the data best, though with time-varying kinetic parameters. In particular, Bcd production has to effectively cease around the beginning of n.c. 13 (in qualitative agreement with Bcd::eGFP imaging (Little et al., 2011)) and the degradation rate appears to increase during n.c. 14 (consistent with *in vitro* measurements of Bcd life time (Liu & Ma, 2011)). As there is time dependence in the Bcd dynamic parameters, care must be taken in determining whether a steady-state profile of Bcd occurs. Around n.c. 12, the spatial profile of the Bcd gradient is relatively stable (Little et al., 2011) for about 30 min, but this represents a period of less than 20% of the total developmental time of the blastoderm. Analysis of readout of the Bcd gradient often assumes that the gradient itself is static. If readout is sufficiently rapid (<3 min) and between n.c. 11 and early n.c. 13 this assumption is reasonable. However, as we discuss below, Bcd readout occurs at other times during blastoderm development and during these periods the Bcd gradient can be changing substantially during the period of readout.

3.5 Summary of Bcd dynamics

Bringing together the above results, we highlight key conclusions that are most strongly supported by current experimental quantification and theoretical reasoning:

(1) *Bcd production:* The majority of Bcd protein is produced in the anterior pole region of the embryo. This does not discount some local production across the embryo, but the significant contribution is in the anterior. Bcd production rates vary in time, with a significant decrease around the onset of n.c. 13.

(2) *Bcd transport:* Bcd displays dynamics consistent with restricted diffusion. FRAP and FCS measurements find a range of values for the diffusion coefficient D. However, these measurements are not inconsistent when transport processes are correctly accounted for (Sigaut et al., 2014). The *effective* diffusion of Bcd is sufficiently fast to set up a gradient across the whole embryo from a localized source. Diffusion of Bcd is within the cytoplasmic region, not through the yolk (in contrast to some formulations of the SDD model). Alternative mechanisms, such as active transport, could aid formation of the gradient, but the cumulative effect of such processes results in a system well-described by an *effective* diffusion process.

(3) *Bcd degradation:* Bcd is degraded in the blastoderm, via potentially different mechanisms prior to, and during, n.c. 14. The average Bcd half-life is in the range of 20–40 min.

The SDD model, with time-dependent kinetic parameters Fig. 2F, is the only current model consistent with the quantitative data for Bcd dynamics.

Of course, there remain important questions:

(1) What are the primary contributions to the slow effective Bcd diffusion? Is it cytoplasmic density or nuclear trapping? Relatedly, is there a role for active transport in the cytoplasmic region?

(2) How is Bcd production regulated? There appears to be an abrupt change in Bcd production at the beginning of n.c. 13, yet what underlies this is unknown.

(3) What is regulating Bcd degradation? Inhibiting the proteasome alters the Bcd concentration profile (Durrieu et al., 2018; Liu & Ma, 2011). *fates shifted* has been proposed to mediate Bcd degradation (Liu & Ma, 2011), yet the Bcd tandem reporter showed little change in Bcd age in *fates shifted* mutants.

(4) How do dynamic parameters change in time and with temperature?

(5) Do Bcd isoforms (Baumgartner, 2018) have distinct behaviors in either transport or signaling?

We have reviewed the current understanding of how the Bcd gradient forms over within the first hours of *Drosophila* embryo development. However, as a morphogen, its role is to impart spatial information to downstream targets. We next focus on signaling and how the Bcd gradient can be precisely interpreted in such a short time window.

4. Dynamics of morphogen gradient interpretation

How gradients are interpreted is critical to their function, and recent work has highlighted how temporal changes can play an important role in Bcd readout. The French Flag model provides a simple paradigm for understanding morphogen readout (Wolpert, 1969), Fig. 1B. A specific threshold concentration of morphogen corresponds to a specific gene expression boundary within the tissue (Shimizu & Gurdon, 1999). However, such a model ignores potential *temporal* information encoded within the morphogen, Fig. 3A and fails to account for growth and other morphological variation during development (Jaeger, Irons, & Monk, 2008). We outline general principles of morphogen temporal information readout, and then discuss in the next section Bcd interpretation.

4.1 Interpreting temporal changes in morphogen signal

Few morphogens obtain a steady-state prior to interpretation, so the local morphogen concentration being measured by the cells is continually changing (Bergmann et al., 2007). Evidence for cells utilizing such varying concentration information came from interpretation of the Sonic Hedgehog (Shh) gradient in mouse digit specification (Ahn & Joyner, 2004). Cells in the limb do not respond to specific levels of Shh, but instead to changes in Shh signal during specification. Using cell culture and *Xenopus* extracts, it has been shown that TGFβ signaling is dependent on the rate of change of the morphogen (Sorre, Warmflash, Brivanlou, & Siggia, 2014; Warmflash et al., 2012), not the concentration itself. Sharp changes in TGFβ concentration resulted in changes in Smad4 nuclear localization (Warmflash et al., 2012). Recent work with human embryonic stem cells has demonstrated that cells can simultaneously respond to both absolute concentrations (interpretation of BMP4) and change in concentration (interpretation of Nodal signal) (Heemskerk et al., 2019). Theoretical modeling has aided in understanding the temporal interpretation of morphogen gradients. An *adaptive* signaling pathway—one that effectively interprets the rate of change of morphogen signal, $\frac{\partial \rho(x,t)}{\partial t} = \dot{\rho}$—can determine spatial boundaries even in the

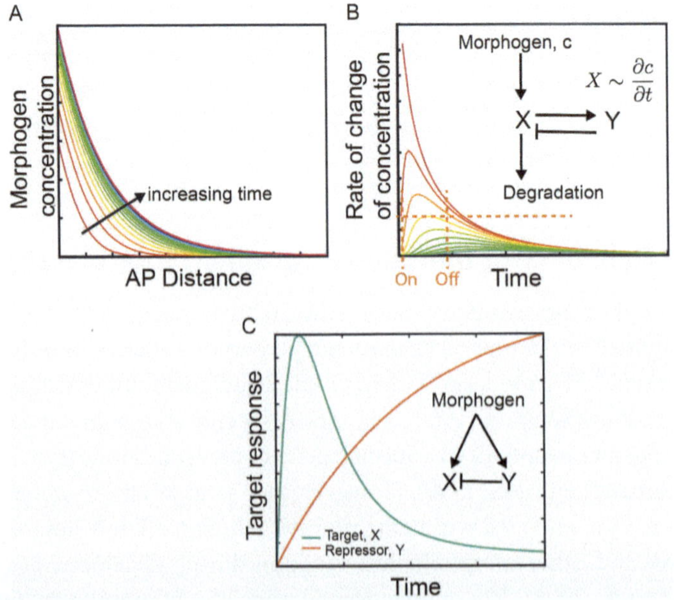

Fig. 3 Temporal information and its interpretation from morphogen gradients. (A) Temporal change in the SDD model, parameters as Fig. 2B. Arrow denotes increasing time (red early, magenta late). (B) The rate of change of morphogen concentration from (A). Red curves denote positions near the source, and green curves denote positions further away. If the cell can measure a specific rate of change (a network that can interpret the rate of change is shown in the inset), then a switch-like response can be defined, as highlighted by the orange dashed lines. (C) A specific time window of system response to a morphogen signal can be defined by an incoherent feedforward loop, as shown in the inset.

absence of a spatial gradient, and can produce robust spatial boundaries rapidly (Sorre et al., 2014). $\dot{\rho}$ is spatially dependent (Richards & Saunders, 2015) and if cells can interpret $\dot{\rho}$, then a switch-like response can be defined, Fig. 3B. By coupling $\dot{\rho}$ interpretation with growth, dynamic tissue scaling can also be achieved (Wartlick et al., 2011). Interpreting the temporal change in signals provides useful information to cells and tissues that can be interpreted through relatively straightforward network architectures.

4.2 Interpreting morphogen signal duration

Cells may respond to the duration of morphogen signal rather than specific concentrations (Alexandre, Baena-Lopez, & Vincent, 2014; Dessaud et al., 2007; Harfe et al., 2004). Interpretation of Shh in the developing vertebrate neural tube is dependent on such temporal decoding

(Sagner & Briscoe, 2017). The signal integration is performed by the downstream gene regulatory network, through a series of mutually repressive gene interactions (Balaskas et al., 2012; Cohen, Page, Perez-Carrasco, Barnes, & Briscoe, 2014). The structure of the network for specification of cell fate in the vertebrate neural tube and the gap gene network in the early *Drosophila* embryo (below) is remarkably similar (Briscoe & Small, 2015), though interpretation of the former takes place on hour timescales whereas the latter is interpreted in minutes.

A system that responds to a cumulative, rather than instantaneous, signal needs to have a form of memory; i.e., the cell needs to record the history of its morphogen exposure. This can be achieved through epigenetic memory, such as histone modification and positive feedback reinforcement (Perez-Carrasco, Guerrero, Briscoe, & Page, 2016; Sagner & Briscoe, 2017; Shvartsman & Baker, 2012). Information can be encoded within the downstream gene regulatory network, stored within the downstream protein concentrations (Sagner & Briscoe, 2017). This appears to be relevant in both the vertebrate neural tube (Balaskas et al., 2012) and the *Drosophila* gap gene network (Huang et al., 2017; Verd et al., 2018). Determining the timescales over which the information is encoded and then stored remains an open problem and is likely highly system specific, depending on factors such as tissue growth, cell rearrangements, and the required speed of cell fate determination.

4.3 Time window for interpretation

When is the information from a morphogen gradient interpreted? The precise timing of morphogen readout can have an effect on both boundary positioning and the spatial precision of such boundaries (Bergmann et al., 2007; de Lachapelle & Bergmann, 2010a). In the developing zebrafish embryo, the time window for Nodal interpretation is determined by the differential production of Nodal and its antagonist Lefty (van Boxtel et al., 2015) and by cell-cell contact duration (Barone et al., 2017). The precise timing of Lefty production is controlled by repression from a micro-RNA (van Boxtel et al., 2015). Micro-RNAs may play a role in temporal coordination by negatively regulating signal interpretation, effectively switching off morphogen integration, Fig. 3C, hence fine-tuning interpretation windows. In tissues with dynamic cell rearrangements, such as the zebrafish somite, the time window for morphogen interpretation can also depend on the timing of cell migration (Yin, Lee, Ono, Ingham, & Saunders, 2018).

A major challenge in determining the timing of morphogen interpretation has been the lack of tools to perform precise temporal manipulations of morphogen activity. Optogenetic approaches are now providing a powerful toolset to perform precise spatial and temporal manipulations to developing systems (Huang et al., 2017; Johnson et al., 2017; Johnson & Toettcher, 2018, 2019; Kaur, Saunders, & Tolwinski, 2017; Sako et al., 2016), Box 3. Optogenetic control of Nodal signaling in zebrafish has shown that extended signaling results in prechordal plate specification and suppresses endoderm formation (Sako et al., 2016). In the early *Drosophila* embryo, by optogenetically controlling pulses of Erk activation, different cell fates can be induced. In particular, it was found that the cell fates are sensitive to the cumulative Erk signal, but not the dynamics of individual pulses of Erk activity (Johnson & Toettcher, 2019).

There is still the question of when downstream networks begin to interpret the input signal in a meaningful manner? Positive feedback can drive a

BOX 3 Optogenetic exploration of temporal signaling integration.

We can easily think of two optogenetic strategies to manipulate the timing of signaling activities *in vivo*. The first is *temporal activation*, whereby the system activates signaling in response to light, and is usually required to be placed under conditions where endogenous signaling is depleted. By providing an activated pathway to the developing organisms with precisely defined stages and durations, and observing the rescuing effects, we can decipher how the same signal input results in differential cellular changes. This strategy is especially useful when studying positional information encoded by extracellular morphogen ligands. Activation can be achieved by either light-induced clustering of transmembrane receptors (Sako et al., 2016) or recruiting activators to the proximity of the membrane (Johnson et al., 2017), both resulting in the transduction of downstream signals.

The second strategy does the exact opposite, i.e., *temporal deprivation*. Here, the target proteins (the morphogen molecules) tagged with the optogenetic cassettes are designed to preserve their native activity in the absence of light exposure. Upon light induction, however, the engineered fusion proteins are converted into inactive forms as a consequence of conformational hindrance. By depriving the developing embryos of the morphogen-encoded information with temporally structured illumination, a temporal atlas of morphogen function can be depicted (Huang et al., 2017; McDaniel et al., 2019).

BOX 3 Optogenetic exploration of temporal signaling integration.—cont'd

Although serving similar purposes, very different considerations should be taken for these two strategies. The advantages of *temporal activation* lie in that the molecular processes underlying light-induced signaling activation are better understood and the signaling dynamics are easily monitored by live imaging, Fig. B3A. Further, as the level of receptor clustering and protein recruitment is proportional to light intensity, it is feasible to fine-tune the signaling activity applied or even examine the effects of over-activation. The difficulty, on the other hand, is to fully recapitulate the endogenous signaling pattern with such engineered systems, especially when the native signaling pattern is spatially graded and temporally evolving. This can be nicely complemented by the second strategy, as the expression of the fusion protein can be placed under the control of the endogenous regulatory sequences, Fig. B3B. The disadvantage of *temporal deprivation* is that the molecular mechanisms of conformational hindrance are less well understood, and caution needs to be taken in assessing the extent and specificity of the light-induced inhibitory effects

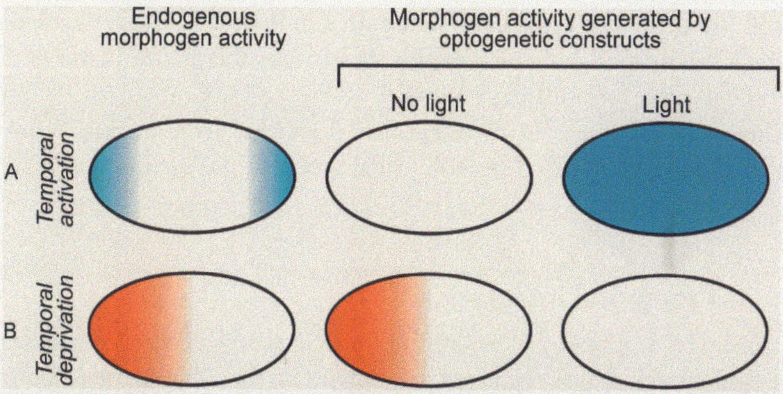

Fig. B3 (A) Temporal activation strategy. In the endogenous null background and in the absence of light, no signaling activity is transduced. In contrast, the signaling pathway is activated wherever and whenever there is suitable light activation. In Johnson et al. (2017), the Torso pathway is activated, which is present at both poles in wild-type conditions. (B) Temporal deprivation strategy. The optogenetic constructs recapitulate endogenous activity in the absence of light and illumination abolishes this activity and deprives the embryo of signaling inputs. Schematics recapitulate the methods employed in Johnson et al. (2017) (A) and Huang et al. (2017) (B), respectively.

switch-like response to time-varying morphogen inputs (Jutras-Dubé, Henry, & François, 2018; Shvartsman & Baker, 2012). Such feedback can take place at the gene regulatory level, or, as discussed below, at the level of the chromatin structure. Recent work on interpretation of the Bcd gradient has begun to reveal how and when the timing of gradient interpretation occurs, and we discuss this next.

5. Bcd temporal interpretation and developmental precision

The central problem of how developmental precision is achieved—in the case of Bcd, with single cell precision—brings us back to why time matters. Given the limited developmental duration, the biophysical processes that we have discussed above not only determine what the morphogen gradient profile should look like at a defined stage, but also constrain how reproducible this gradient can be across different individuals (Bergmann et al., 2007; Gregor, Tank, et al., 2007; Saunders & Howard, 2009). As a consequence, the reproducibility of the gradient influences the accuracy of cell fate decisions. It has been shown that in the case of Bcd, the readout of the local morphogen concentration approaches the physical limit of precision expected for diffusing molecules (Gregor, Tank, et al., 2007; Little, Tikhonov, & Gregor, 2013). Recent experiments have begun to unveil the temporal steps of Bcd decoding that facilitate the system operating at such a limit.

5.1 Bcd temporal integration

Stochastic fluctuations in all biochemical reactions pose challenges to the precise readout of local morphogen concentrations by cells (Berg & Purcell, 1977). The random encounter between receptor (of size a) and target (with local concentration c and diffusion coefficient, D) within a time period T, results in measurement noise, δc, which can be approximated as $\delta c/c \sim 1/\sqrt{DacT}$ (the Berg-Purcell limit). This defines the fundamental limit of biological precision for diffusive processes (Bialek & Setayeshgar, 2005; Kaizu et al., 2014). By simultaneously measuring the absolute Bcd nuclear concentration and the relative expression of one of its target genes, *hunchback (hb)*, the noise in this input/output relation is less than 10% (Gregor, Tank, et al., 2007). According to the Berg-Purcell limit, and given the measured Bcd concentration and dynamics, a time window T from 25 min (Porcher et al., 2010) to around 2 h (Gregor, Tank, et al., 2007) is required, depending on the Bcd diffusion coefficient, to reach the observed precision of 10%. This corresponds to the duration of the entire blastoderm stage, hence complementary mechanisms must exist to reduce the time needed to interpret Bcd.

One mechanism for reducing the required signal integration time is spatial averaging. In effect, if a nucleus can average the signal over its neighbors, then precise readout can be achieved on the order of minutes, rather than hours (Erdmann, Howard, & ten Wolde, 2009; Gregor, Tank, et al., 2007). Using fluorescence *in situ* hybridization, it has been demonstrated that *hb* transcripts can be exported out of nuclei, and subsequently mix in the cytoplasm, thereby potentially spatially averaging the signal (Little et al., 2013). To unravel the underlying dynamics, the MS2-MCP reporter, which highlights transcription of a specific gene (Garcia, Tikhonov, Lin, & Gregor, 2013; Lucas et al., 2013), has been used to quantitatively analyze the Hb response to Bcd. A period of 3 min within interphase was sufficient for subtle (Bcd-dependent) differences in *hb* expression to be apparent between nuclei (Lucas et al., 2018). Theoretical work suggests that a minute could be sufficient for *hb* promoters to interpret the local Bcd concentration if decisions are made with dynamically changing time windows rather than during a fixed time (Desponds, Vergassola, & Walczak, 2019), though this remains to be tested *in vivo*. Theoretical modeling showed that equilibrium cooperative binding of Bcd was insufficient to drive this rapid and spatially controlled response (Lucas et al., 2018). Additional elements may be playing an important role in facilitating rapid and precise readout of Bcd, such as extra information from other gradient signals and possible non-equilibrium binding dynamics (Tran et al., 2018). Analysis of *hb* enhancers suggests that the cumulative response of the different enhancers behaves either additively or sub-additively (i.e., saturated) at low or high Bcd concentrations respectively (Bothma et al., 2015). Such spatially distinct readout modes may be critical in ensuring precise and rapid readout of the Bcd gradient. It is noticeable that these timescales (~minutes) are much shorter than the apparent interpretation timescales in vertebrate systems (~hours). The embryo has only a few hours to go from a single nucleus to >6000 nuclei combined with determining cell fate along both the anterior-posterior and dorsal-ventral axes. Therefore, cell fate in the *Drosophila* blastoderm, determined by interpretation of morphogen gradients such as Bcd, needs to be as rapid as possible.

5.2 Time window for Bcd decoding

The optogenetic manipulation adopting the *temporal deprivation* design (Box 3) has enabled direct testing of when, and for how long, Bcd is temporally integrated *in vivo* (Huang et al., 2017). A fusion protein of Bcd and an N-terminally tagged optogenetic cassette CRY2-mCh was expressed under the native Bcd regulatory sequences. Replacing endogenous Bcd with the engineered optogenetic Bcd, recapitulated the wild-type dynamics and, importantly, cells underwent proper differentiation trajectories. In the dark, the Bcd optogenetic construct was able to rescue embryos lacking endogenous Bcd. Exposure to 488 nm light inhibits Bcd transcriptional activity due to a conformational change in CRY2-mCherry-Bcd, rapidly inhibits information transfer from the morphogen to the nuclei. This was demonstrated through live imaging of *hb* transcripts in the Bcd optogenetic background using the MS2 reporter system. Upon light activation, *hb* transcription was immediately abolished, and such transcriptional inhibition was reversible with fast kinetics upon dark recovery. These results suggested that the optogenetic construct inhibited Bcd-mediated downstream gene transcription (Huang et al., 2017).

By subjecting the embryos to illumination at precisely defined stages, distinct time windows of Bcd activity were defined. Temporal integration is indeed a necessary process in Bcd decoding and the integration time window covers as long as 1.5 h, from n.c. 10 until the onset of gastrulation, Fig. 4A. This is a surprising result, as Bcd profile is highly dynamic during this period, with the amplitude of the gradient first increasing and then decreasing (Little et al., 2011). Meanwhile the nuclei undergo rapid division cycles. How do these two processes couple with each other and give rise to precise cell fate decisions? The answer may lie in the fact that the integration duration is not uniform along the antero-posterior axis, where the Bcd concentration varies. Closer examinations of the embryonic patterning subjected to different illumination time windows unveiled that the anterior nuclei are more susceptible to interruption of Bcd transcriptional activity. In other words, cells exposed to the highest local Bcd concentration integrate Bcd for the longest duration. Comparatively, cells receiving lower Bcd dosage require Bcd inputs for shorter time in order to arrive at the correct cell fates, Fig. 4A. Therefore, this sequential cell fate determination from posterior to anterior may facilitate the precise boundary formation, as nuclei that have reached their final fates may no longer be influenced by fluctuations in the local concentration.

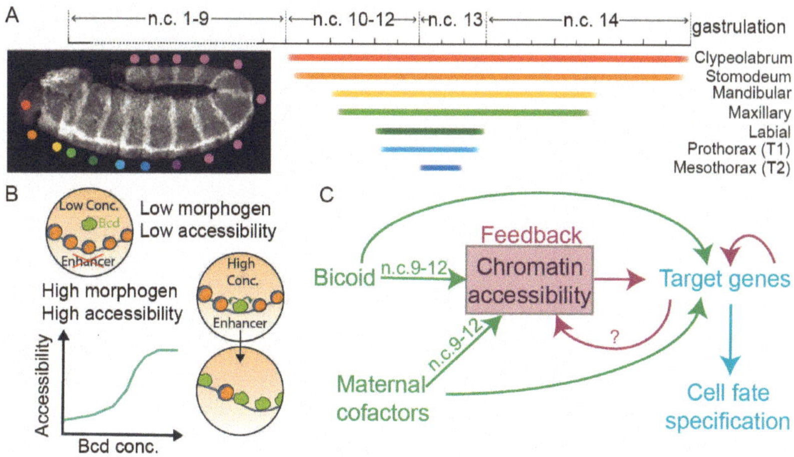

Fig. 4 Temporal interpretation of the Bcd gradient. (A) Schematic representation of time windows for Bcd interpretation, from Huang et al. (2017). More anterior nuclei require Bcd signal earlier in development. Anterior nuclei also require Bcd signal for a longer duration. (B) Morphogen binding to DNA can alter the accessibility of the DNA for further binding, therefore modulating cell response (Hannon, Blythe, & Wieschaus, 2017). (C) Bcd and other inputs (green) drives downstream gene expression of target genes and alters chromatin accessibility (in concert with other facts such as Zelda), enabling spatial and temporal modulation of its readout. Changes in chromatin accessibility to Bcd binding appears to be necessary during n.c. 9–12 for precise readout. Given the gap genes are DNA binding proteins there may be additional feedback to the chromatin structure, but this is currently unknown.

Another key result from the experiments controlling the time window of Bcd signal interpretation is that it reveals when Bcd signaling is *not* required for precise gene boundary specification. While there has been significant work focusing on the readout precision of Bcd in n.c. 11 and earlier (Ali-Murthy & Kornberg, 2016; Bergmann et al., 2007; de Lachapelle & Bergmann, 2010a; Porcher et al., 2010), the optogenetic-derived time windows suggest that such early protein expression plays little role in the final patterning of the gap genes. This is consistent with the careful quantifications of boundary specification (Little et al., 2013). Due to increased nuclear number at later cycles and rapid protein turnover, the number of gap gene proteins produced during n.c. 13 and early n.c. 14 are likely far more numerous than those produced in n.c. 11 and earlier. Essentially, the presence of a signaling response to a morphogen does not imply that this response is important for precise boundary specification.

But, one can then ask why the embryo displays such rapid and precise readout of Bcd in n.c. 11 (Lucas et al., 2018)? One issue in determining when gap genes are required for boundary specification has been the very fast turnover of gap genes, making live imaging very challenging. Recently, Llama-tag has been developed (Bothma, Norstad, Alamos, & Garcia, 2018), which uses genetically encoded antibodies, enabling live imaging of Hunchback protein despite its rapid turnover. It would be very interesting to use photo-convertible fluorophores and/or CRY2-mCherry-Bcd in concert with Llama-tag, to explore how protein dynamics change at different cycles, to quantitatively test where and when Bcd signaling is necessary. Lastly, we note that possible compensatory mechanisms may exist that correct for perturbed Bcd activity in the early embryo. However, given the speed of signal interpretation in the blastoderm is already operating near its theoretical limit, it seems unlikely that compensatory mechanisms can operate sufficiently fast to adjust for optogenetic perturbations.

Interestingly, the critical time windows unveiled by the optogenetic experiments are consistent with the classic studies using temperature sensitive hypomorphic *bcd* alleles (Frohnhöfer & Nüsslein-Volhard, 1987). Most hypomorphic *bcd* alleles show stronger anterior defects at 29 °C compared to 18 °C. By shifting bcd^{E3} embryos from 18 °C to 29 °C, they explored the critical time windows for Bcd activity and found that the embryos were sensitive to such shifts from stages 3 to 5, equivalent to n.c. 9 to 14. It will be interesting to return to these observations in light of the above optogenetic results. In particular, in bcd^{E3} embryos how is the gradient of Bcd affected and its downstream targets at different temperatures? We also note that the phenotypes of the optogenetically perturbed embryos (Huang et al., 2017) do not precisely phenocopy the range of *bcd* alleles (Frohnhöfer & Nüsslein-Volhard, 1986, 1987). This may be due to the presence of a normal Caudal gradient in the optogenetically perturbed embryos, though further analysis of Caudal in hypomorphic *bcd* alleles is required.

A microfluidic tool has also been used to temporally perturb the Bcd gradient by generating embryos with different temperatures at different times along the anterior-posterior axis (Lucchetta, Lee, Fu, Patel, & Ismagilov, 2005; Lucchetta, Vincent, & Ismagilov, 2008). They found the system was particularly sensitive to temperature perturbations around n.c. 7–10. This corresponds to the time period during which nuclei migrate to the cortex, and it is likely that such defects are more likely due to morphological, rather than gene readout, perturbations. Rigorous, quantitative analysis of temperature sensitivity is needed to assess the interplay between temperature and timing on patterning in the early embryo. However, an issue with the microfluidic approach is that temperature changes to the embryo affect more than just Bcd; therefore, disentangling thermal effects from genetic ones remains challenging.

5.3 Bcd's role as a genome organizer

Key questions remain as to what are the molecular processes underlying the spatial and temporal non-linear decoding of Bcd. An important aspect of Bcd-mediated patterning is that Bcd's role goes beyond transcribing a limited number of canonical target genes. Multiple lines of evidence indicate that Bcd influences the transcriptional landscape of cells in a concentration-dependent manner. ChIP-chip experiments have estimated between 500 and 1000 Bcd binding regions throughout the genome (Hannon et al., 2017; Li et al., 2008). Microarray assays identified hundreds of genes differentially regulated in response to varying concentration of Bcd, expressed in distinct anterior-posterior domains (Ochoa-Espinosa, Yu, Tsirigos, Struffi, & Small, 2009). A more profound impact of Bcd activity on cell fate determination is its role in modifying chromatin accessibility. ATAC-seq experiments have shown that in the early *Drosophila* embryo, the regulatory regions of early zygotically transcribed genes gain chromatin accessibility in a sequential manner: first the enhancers become accessible at n.c. 11, followed by the corresponding promoter regions at n.c. 12 and 13 (Blythe & Wieschaus, 2016). Intriguingly, high local Bcd concentration is essential in driving open chromatin states, particularly at the enhancers of high-concentration targets (Hannon et al., 2017), Fig. 4B.

Bcd cofactors, including Hunchback (Simpson-Brose, Treisman, & Desplan, 1994) and Dampened (Liu & Ma, 2013), play an important role in determining the sensitivity of nuclei to Bcd signal. Caudal extends from the posterior and its spatial profile along the anterior-posterior axis is dependent on interactions with Bcd (Ochoa-Espinosa et al., 2009, 2005). Zelda binding (Liang et al., 2008), a putative pioneer factor, is critical in regulating Bcd readout (Hannon et al., 2017; Xu et al., 2014). For enhancers with low affinity Bcd binding sites, there are often high affinity Zelda motifs, which may be acting to maintain an open local chromatin structure and allow other transcription factors to bind and induce downstream patterning. Dynamic imaging of Bcd binding shows that the Bcd binding time is on the order of seconds throughout the embryo (Mir et al., 2017). However, the local accumulation of Bcd is concentration dependent, an effect particularly apparent in the posterior. This increased Bcd accumulation was closely correlated with Zelda binding. Finally, an optogenetic strategy similar to Huang et al. (2017) has been used to explore the time window for when Zelda activity is required (McDaniel et al., 2019). They demonstrate that Zelda activity is required from n.c. 10 onward and that Zelda has to rapidly rebind to its binding motifs after nuclear division to ensure robust patterning. In summary, Bcd modulates its own ability to be interpreted through effective feedforward and feedback loops—sometimes in concert with other factors (Park et al., 2019)—and this provides a mechanism for temporally regulating its readout at different positions in the embryo, Fig. 4C.

5.4 Precise information transfer as a general paradigm in morphogen interpretation?

Thus far, we have focused on the temporal readout of Bcd. However, as is well established for a number of systems (Briscoe & Small, 2015; Sagner & Briscoe, 2017), integration of morphogenetic signal by the downstream gene regulatory network is key in robust signal integration. The gap gene network downstream of Bcd, consists of multiple feedback loops that spatially segregate gene expression. The spatial positions of these boundaries are not static, and they adjust during n.c. 13 and 14 (Little et al., 2013; Liu et al., 2013; Manu et al., 2009b). It has been shown through precise spatio-temporal quantifications, that the initial boundaries of gap genes are tightly correlated with Bcd expression levels. However, the final locations depend on interactions between gap genes and also scaling with embryo size (Liu et al., 2013; Wu, Manu, Jiao, & Ma, 2015). Further, there is positional information beyond Bcd; for example, repressors of Bcd activity, Capicua (Lim et al., 2015; Löhr, Chung, Beller, & Jäckle, 2009) and Runt (Chen, Xu, Mei, Yu, & Small, 2012), are expressed in specific spatial domains along the anterior-posterior axis. Refinement of the spatial information provided by the Bcd gradient through the gap gene and segment polarity networks has been a paradigm model for understanding patterning in the early *Drosophila* embryo.

However, recent work has begun to reveal that a remarkable level of spatial information can be extracted from the Bcd gradient alone (Petkova et al., 2019). Using concepts from information theory, the authors developed an *optimal decoder*, whereby positional information stored in the Bcd gradient is decoded by the gap gene network into single-cell-precise spatial patterns of Even-Skipped and other *Drosophila* patterning genes. By examining the positioning of gene boundaries in *bcd*-only germline clones (where other maternal patterning genes, such as *nanos* and *hb* are excluded), they demonstrate that the Bcd gradient contains sufficient spatial information to pattern most gene expression boundaries, even in the posterior of the embryo (up to the sixth abdomen segment). This work encapsulates how both quantitative experimental data and theoretical reasoning are essential for understanding how complex networks operate during development.

6. Conclusions

Bcd is a transcription factor, whereas most other morphogens are receptor binding ligands. So, what general lessons can we learn from studying Bcd dynamics and its interpretation?

Regarding dynamics: (1) Bcd provides an excellent testbed for quantitative techniques, such as the tandem reporter. Such techniques may help resolve longstanding questions in other systems, particularly with regard to diffusive transport (Kornberg, 2012; Zhou et al., 2012). (2) Detailed analysis of Bcd has clarified how FCS and FRAP measurements can be interpreted and the information they provide. (3) The time dependency in the dynamic parameters is likely relevant to other systems.

With respect to signal interpretation: (1) Dynamic chromatin modification plays a critical role in robust boundary specification and such a mechanism is plausible for other systems (Paliou et al., 2019). (2) Relatedly, there is direct feedback between the morphogen concentration and its ability to be interpreted. The upregulation of receptors (e.g., Patched by Shh) is qualitatively similar, whereby a cell that responds to a morphogen further increases its sensitivity to that morphogen. (3) The detailed optogenetic study discussed above also demonstrates the importance of quantitative analysis of temporal readout. Previously unknown windows of Bcd activity have been identified, such as the late role in n.c. 14 for Bcd in determining the most-anterior body parts. Such long-time temporal dependence on the morphogen input may be particularly important for gene expression near domain boundaries, and this may be relevant to other systems. (4) The ability to spatially and temporally tune morphogen readout by modulating chromatin accessibility is likely necessary to ensure precise transferal of positional information. (5) Finally, for extracellular morphogens such as BMP and

Shh, the signals are ultimately relayed to the nucleus, resulting in spatially distinct transcription factor concentrations. Therefore, understanding how a graded transcription factor (i.e., Bcd) imparts positional information is likely relevant to many systems.

It is worth emphasizing that the timescale of Bcd interpretation—on the order of minutes—is significantly faster than many other patterning systems—on the order of hours, in the case of the vertebrate neural tube. Therefore, care must be taken in drawing analogies between Bcd and other systems. In a sense, patterning of the early *Drosophila* embryo demonstrates what can be achieved by a noisy biological system under extreme time constraints. The time available to incorporate regulatory feedback is severely limited. But, most other patterning systems likely use a more sequential approach, enabling integration and refinement of information over many hours. Systems that integrate signals over longer periods can take advantage of extensive feedback regulation, but they must also be able to compensate for environmental (e.g., temperature shifts) or morphological (e.g., growth (Kicheva et al., 2014)) changes that occur during cell fate specification.

The Bcd morphogen gradient remains a paradigm system for understanding morphogen gradient formation and interpretation. The ability to perform quantitative assays with subcellular spatial resolution and <1 min temporal resolution provides a framework for detailed testing of different models. The processes of gradient formation and interpretation are not independent, with dynamic feedback between them, Fig. 4C. Understanding how spatial and temporal information is interpreted requires a truly systems approach, involving quantitative experiments and modeling, that accounts for genetic interactions, spatial gradients, and biologically relevant sources of noise. Detailed quantitative analysis of these processes in Bcd will hopefully illuminate general principles into how morphogens are precisely interpreted in both space and time. To finish with, it is apt to recall Hans Meinhardt (Meinhardt, 1986): "As a rule, models do not have a good reputation in the field of developmental biology." The use of quantitative models in understanding how Bcd forms within and patterns the developing *Drosophila* embryo hopefully demonstrate conclusively the power of such approaches and their use in deepening our understanding of developmental biology.

Acknowledgments

We thank Olivier Hamant for critical comments on the manuscript. We are grateful to Eric Wieschaus and Trudi Schüpbach for stimulating discussion about Bcd and its role in patterning and to Nathalie Dostatni and Huy Tran for discussion about the timing of

Hunchback readout. This work was supported by the Mechanobiology Institute, National University of Singapore and a Singapore Ministry of Education Academic Research Fund Tier 3 grant (MOE2016-T3-1-005).

References

Abu-Arish, A., Porcher, A., Czerwonka, A., Dostatni, N., & Fradin, C. (2010). High mobility of bicoid captured by fluorescence correlation spectroscopy: Implication for the rapid establishment of its gradient. *Biophysical Journal, 99*, L33–L35.

Adamowski, M., & Friml, J. (2015). PIN-dependent auxin transport: Action, regulation, and evolution. *Plant Cell, 27*, 20–32.

Ahn, S., & Joyner, A. L. (2004). Dynamic changes in the response of cells to positive hedgehog signaling during mouse limb patterning. *Cell, 118*, 505–516.

Alexandre, C., Baena-Lopez, A., & Vincent, J.-P. (2014). Patterning and growth control by membrane-tethered Wingless. *Nature, 505*, 180–185.

Ali-Murthy, Z., & Kornberg, T. B. (2016). Bicoid gradient formation and function in the Drosophila pre-syncytial blastoderm. *eLife, 5*, e13222.

Balaskas, N., Ribeiro, A., Panovska, J., Dessaud, E., Sasai, N., Page, K. M., et al. (2012). Gene regulatory logic for reading the Sonic Hedgehog signaling gradient in the vertebrate neural tube. *Cell, 148*, 273–284.

Barone, V., Lang, M., Krens, S. F. G., Pradhan, S. J., Shamipour, S., Sako, K., et al. (2017). An effective feedback loop between cell-cell contact duration and morphogen signaling determines cell fate. *Developmental Cell, 43*, 198–211.

Baumgartner, S. (2018). Seeing is believing: The Bicoid protein reveals its path. *Hereditas, 155*, 28.

Berezhkovskii, A. M., Sample, C., & Shvartsman, S. Y. (2010). How long does it take to establish a morphogen gradient? *Biophysical Journal, 99*, L59–L61.

Berezhkovskii, A. M., Sample, C., & Shvartsman, S. Y. (2011). Formation of morphogen gradients: Local accumulation time. *Physical Review E, 83*, 051906.

Berezhkovskii, A. M., & Shvartsman, S. Y. (2013). Kinetics of receptor occupancy during morphogen gradient formation. *The Journal of Chemical Physics, 138*, 244105.

Berezhkovskii, A. M., & Shvartsman, S. Y. (2014). On the GFP-based analysis of dynamic concentration profiles. *Biophysical Journal, 106*, L13–L15.

Berezhkovskii, A. M., & Shvartsman, S. Y. (2015). Dynamics of gradient formation by intracellular shuttling. *The Journal of Chemical Physics, 143*, 074116.

Berg, H. C. (1993). *Random walks in biology*. Princeton University Press.

Berg, H. C., & Purcell, E. M. (1977). Physics of chemoreception. *Biophysical Journal, 20*, 193–219. https://doi.org/10.1016/S0006-3495(77)85544-6.

Bergmann, S., Sandler, O., Sberro, H., Shnider, S., Schejter, E., Shilo, B.-Z., et al. (2007). Pre-steady-state decoding of the Bicoid morphogen gradient. *PLoS Biology, 5*, e46.

Bergmann, S., Tamari, Z., Schejter, E., Shilo, B.-Z., & Barkai, N. (2008). Re-examining the stability of the Bicoid morphogen gradient. *Cell, 132*, 15–17.

Berleth, T., Burri, M., Thoma, G., Bopp, D., Richstein, S., Frigerio, G., et al. (1988). The role of localization of bicoid RNA in organizing the anterior pattern of the Drosophila embryo. *The EMBO Journal, 7*, 1749–1756.

Bialek, W., Gregor, T., Tank, D. W., & Wieschaus, E. F. (2008). Response: Can we fit all of the data? *Cell, 132*, 17–18.

Bialek, W., & Setayeshgar, S. (2005). Physical limits to biochemical signaling. *Proceedings of the National Academy of Sciences of the United States of America, 102*, 10040–10045.

Blythe, S. A., & Wieschaus, E. F. (2016). Establishment and maintenance of heritable chromatin structure during early Drosophila embryogenesis. *eLife, 5*, e20148.

Bollenbach, T., Pantazis, P., Kicheva, A., Bokel, C., Gonzalez-Gaitan, M., & Julicher, F. (2008). Precision of the Dpp gradient. *Development*, *135*, 1137–1146.

Bothma, J. P., Garcia, H. G., Ng, S., Perry, M. W., Gregor, T., & Levine, M. (2015). Enhancer additivity and non-additivity are determined by enhancer strength in the Drosophila embryo. *eLife*, *4*, e07956.

Bothma, J. P., Norstad, M. R., Alamos, S., & Garcia, H. G. (2018). LlamaTags: A versatile tool to image transcription factor dynamics in live embryos. *Cell*, *14*, 1810–1822.

Briscoe, J., & Small, S. (2015). Morphogen rules: Design principles of gradient-mediated embryo patterning. *Development*, *142*, 3996–4009.

Cai, X., Akber, M., Spirov, A., & Baumgartner, S. (2017). Cortical movement of Bicoid in early Drosophila embryos is actin- and microtubule-dependent and disagrees with the SDD diffusion model. *PLoS One*, *12*, e0185443.

Chen, H., Xu, Z., Mei, C., Yu, D., & Small, S. (2012). A system of repressor gradients spatially organizes the boundaries of Bicoid-dependent target genes. *Cell*, *149*, 618–629.

Cheung, D., & Ma, J. (2015). Probing the impact of temperature on molecular events in a developmental system. *Scientific Reports*, *5*, 13124.

Chong, J., Amourda, C., & Saunders, T. E. (2018). Temporal development of Drosophila embryos is highly robust across a wide temperature range. *Journal of the Royal Society, Interface*, *15*, 20180304.

Cohen, M., Page, K. M., Perez-Carrasco, R., Barnes, C. P., & Briscoe, J. (2014). A theoretical framework for the regulation of Shh morphogen-controlled gene expression. *Development*, *141*, 3868–3878.

Crick, F. (1970). Diffusion in embryogenesis. *Nature*, *225*, 420–422.

Dalessi, S., Neves, A., & Bergmann, S. (2012). Modeling morphogen gradient formation from arbitrary realistically shaped sources. *Journal of Theoretical Biology*, *294*, 130–138.

Daniels, B. R., Rikhy, R., & Renz, M. (2012). Multiscale diffusion in the mitotic Drosophila melanogaster syncytial blastoderm. *Proceedings of the National Academy of Sciences of the United States of America*, *109*, 8588–8593.

de Lachapelle, A. M., & Bergmann, S. (2010a). Precision and scaling in morphogen gradient read-out. *Molecular Systems Biology*, *6*, 351.

de Lachapelle, A. M., & Bergmann, S. (2010b). Pre-steady and stable morphogen gradients: Can they coexist? *Molecular Systems Biology*, *6*, 428.

De Los Santos, C., Chang, C.-W., Mycek, M.-A., & Cardullo, R. A. (2015). FRAP, FLIM, and FRET: Detection and analysis of cellular dynamics on a molecular scale using fluorescence microscopy. *Molecular Reproduction and Development*, *82*, 587–604.

Desponds, J., Vergassola, M., & Walczak, A. M. (2019). *Hunchback* promoters can readout morphogenetic positional information in less than a minute. *Biorxiv*. https://doi.org/10.1101/676684.

Dessaud, E., Yang, L. L., Hill, K., Cox, B., Ulloa, F., Ribeiro, A., et al. (2007). Interpretation of the sonic hedgehog morphogen gradient by a temporal adaptation mechanism. *Nature*, *450*, 717–720.

Dilão, R., & Muraro, D. (2010). mRNA diffusion explains protein gradients in Drosophila early development. *Journal of Theoretical Biology*, *264*, 847–853.

Donà, E., Barry, J. D., Valentin, G., Quirin, C., Khmelinskii, A., Kunze, A., et al. (2014). Directional tissue migration through a self-generated chemokine gradient. *Nature*, *503*, 285–289.

Driever, W., Ma, J., Nüsslein-Volhard, C., & Ptashne, M. (1989). Rescue of bicoid mutant Drosophila embryos by bicoid fusion proteins containing heterologous activating sequences. *Nature*, *342*, 149–154.

Driever, W., & Nüsslein-Volhard, C. (1988a). A gradient of bicoid protein in Drosophila embryos. *Cell*, *54*, 83–93.

Driever, W., & Nüsslein-Volhard, C. (1988b). The bicoid protein determines position in the Drosophila embryo in a concentration-dependent manner. *Cell, 54*, 95–104.
Driever, W., & Nüsslein-Volhard, C. (1989). The bicoid protein is a positive regulator of hunchback transcription in the early Drosophila embryo. *Nature, 337*, 138–143.
Drocco, J. A., Grimm, O., Tank, D. W., & Wieschaus, E. (2011). Measurement and perturbation of morphogen lifetime: Effects on gradient shape. *Biophysical Journal, 101*, 1807–1815.
Drocco, J. A., Wieschaus, E. F., & Tank, D. W. (2012). The synthesis–diffusion–degradation model explains Bicoid gradient formation in unfertilized eggs. *Physical Biology, 9*, 055004.
Durrieu, L., Kirrmaier, D., Schneidt, T., Kats, I., Raghavan, S., Hufnagel, L., et al. (2018). Bicoid gradient formation mechanism and dynamics revealed by protein lifetime analysis. *Molecular Systems Biology, 14*, e8355.
Eldar, A., Rosin, D., Shilo, B. Z., & Barkai, N. (2003). Self-enhanced ligand degradation underlies robustness of morphogen gradients. *Developmental Cell, 5*, 635–646.
Ephrussi, A., & Johnston, D. S. (2004). Seeing is believing: The bicoid morphogen gradient matures. *Cell, 116*, 143–152.
Erdmann, T., Howard, M., & ten Wolde, P. R. (2009). Role of spatial averaging in the precision of gene expression patterns. *Physical Review Letters, 103*, 258101.
Fahmy, K., Akber, M., Cai, X., Koul, A., Hayder, A., & Baumgartner, S. (2014). αTubulin 67C and Ncd are essential for establishing a cortical microtubular network and formation of the Bicoid mRNA gradient in Drosophila. *PLoS One, 9*, e112053.
Fedotov, S., & Falconer, S. (2014). Nonlinear degradation-enhanced transport of morphogens performing subdiffusion. *Physical Review E, 89*, 012107.
Fradin, C. (2017). On the importance of protein diffusion in biological systems: The example of the Bicoid morphogen gradient. *Biochimica et Biophysica Acta. Proteins and Proteomics, 1865*, 1676–1686.
Frohnhöfer, H. G., & Nüsslein-Volhard, C. (1986). Organization of anterior pattern in the Drosophila embryo by the maternal gene bicoid. *Nature, 324*, 120–125.
Frohnhöfer, H. G., & Nüsslein-Volhard, C. (1987). Maternal genes required for the anterior localization of bicoid activity in the embryo of Drosophila. *Genes & Development, 1*, 880–890.
Garcia, H. G., Tikhonov, M., Lin, A., & Gregor, T. (2013). Quantitative imaging of transcription in living Drosophila embryos links polymerase activity to patterning. *Current Biology, 23*, 2140–2145.
Gordon, P. V., Sample, C., Berezhkovskii, A. M., Muratov, C. B., & Shvartsman, S. Y. (2011). Local kinetics of morphogen gradients. *Proceedings of the National Academy of Sciences of the United States of America, 108*, 6157–6162.
Gregor, T., Tank, D. W., Wieschaus, E. F., & Bialek, W. (2007). Probing the limits to positional information. *Cell, 130*, 153–164.
Gregor, T., Wieschaus, E. F., McGregor, A. P., Bialek, W., & Tank, D. W. (2007). Stability and nuclear dynamics of the Bicoid morphogen gradient. *Cell, 130*, 141–152.
Grimm, O., Coppey, M., & Wieschaus, E. (2010). Modelling the Bicoid gradient. *Development, 137*, 2253–2264.
Grimm, O., & Wieschaus, E. (2010). The Bicoid gradient is shaped independently of nuclei. *Development, 137*, 2857–2862.
Hannon, C. E., Blythe, S. A., & Wieschaus, E. F. (2017). Concentration dependent chromatin states induced by the bicoid morphogen gradient. *eLife, 6*, 3165.
Harfe, B. D., Scherz, P. J., Nissim, S., Tian, H., McMahon, A. P., & Tabin, C. J. (2004). Evidence for an expansion-based temporal Shh gradient in specifying vertebrate digit identities. *Cell, 118*, 517–528.

He, F., Saunders, T. E., Wen, Y., Cheung, D., Jiao, R., ten Wolde, P. R., et al. (2010). Shaping a morphogen gradient for positional precision. *Biophysical Journal, 99*, 697–707.

He, F., Wei, C., Wu, H., Cheung, D., Jiao, R., & Ma, J. (2015). Fundamental origins and limits for scaling a maternal morphogen gradient. *Nature Communications, 6*, 6679.

Hecht, I., Rappel, W.-J., & Levine, H. (2009). Determining the scale of the Bicoid morphogen gradient. *Proceedings of the National Academy of Sciences of the United States of America, 106*, 1710–1715.

Heemskerk, I., Burt, K., Miller, M., Chhabra, S., Guerra, M. C., Liu, L., et al. (2019). Rapid changes in morphogen concentration control self-organized patterning in human embryonic stem cells. *eLife, 8*, e40526.

Hernández, D., Varea, C., & Barrio, R. A. (2009). Dynamics of reaction-diffusion systems in a subdiffusive regime. *Physical Review E, 79*, 297.

Hornung, G., Berkowitz, B., & Barkai, N. (2005). Morphogen gradient formation in a complex environment: An anomalous diffusion model. *Physical Review E, 72*, 041916.

Hsiung, F., Ramirez-Weber, F.-A., David Iwaki, D., & Kornberg, T. B. (2005). Dependence of Drosophila wing imaginal disc cytonemes on Decapentaplegic. *Nature, 437*, 560–563.

Huang, A., Amourda, C., Zhang, S., Tolwinski, N. S., & Saunders, T. E. (2017). Decoding temporal interpretation of the morphogen Bicoid in the early Drosophila embryo. *eLife, 6*, 197.

Jaeger, J. (2010). A matter of timing and precision. *Molecular Systems Biology, 6*, 427.

Jaeger, J., Irons, D., & Monk, N. (2008). Regulative feedback in pattern formation: Towards a general relativistic theory of positional information. *Development, 135*, 3175–3183.

Jaeger, J., Surkova, S., Blagov, M., Janssens, H., Kosman, D., Kozlov, K. N., et al. (2004). Dynamic control of positional information in the early Drosophila embryo. *Nature, 430*, 368–371.

Johnson, H. E., Goyal, Y., Pannucci, N. L., Schüpbach, T., Shvartsman, S. Y., & Toettcher, J. E. (2017). The spatiotemporal limits of developmental Erk signaling. *Developmental Cell, 40*, 185–192.

Johnson, H. E., & Toettcher, J. E. (2018). Illuminating developmental biology with cellular optogenetics. *Current Opinion in Biotechnology, 52*, 42–48.

Johnson, H. E., & Toettcher, J. E. (2019). Signaling dynamics control cell fate in the early Drosophila embryo. *Developmental Cell, 48*, 361–370.e3.

Jutras-Dubé, L., Henry, A., & François, P. (2018). Modelling time-dependent acquisition of positional information. In R. Ireton, K. Montgomery, R. Bumgarner, R. Samudrala, & J. McDermott (Eds.), *Methods in molecular biology, methods and protocols* (pp. 281–301). New York, NY: Springer New York.

Kaizu, K., de Ronde, W., Paijmans, J., Takahashi, K., Tostevin, F., & ten Wolde, P. R. (2014). The Berg-Purcell limit revisited. *Biophysical Journal, 106*, 976–985.

Kaur, P., Saunders, T. E., & Tolwinski, N. S. (2017). Coupling optogenetics and light-sheet microscopy, a method to study Wnt signaling during embryogenesis. *Scientific Reports, 7*, 16636.

Kavousanakis, M. E., Kanodia, J. S., Kim, Y., Kevrekidis, I. G., & Shvartsman, S. Y. (2010). A compartmental model for the bicoid gradient. *Developmental Biology, 345*, 12–17.

Khmelinskii, A., Keller, P. J., Bartosik, A., Meurer, M., Barry, J. D., Mardin, B. R., et al. (2012). Tandem fluorescent protein timers for in vivo analysis of protein dynamics. *Nature Biotechnology, 30*, 708–714.

Khmelinskii, A., Meurer, M., Ho, C.-T., Besenbeck, B., Füller, J., Lemberg, M. K., et al. (2016). Incomplete proteasomal degradation of green fluorescent proteins in the context of tandem fluorescent protein timers. *Molecular Biology of the Cell, 27*, 360–370.

Kicheva, A., Bollenbach, T., Ribeiro, A., Valle, H. P., Lovell-Badge, R., Episkopou, V., et al. (2014). Coordination of progenitor specification and growth in mouse and chick spinal cord. *Science, 345*, 1254927.

Kicheva, A., Bollenbach, T., Wartlick, O., Jülicher, F., & Gonzalez-Gaitan, M. (2012). Investigating the principles of morphogen gradient formation: From tissues to cells. *Current Opinion in Genetics & Development, 22*, 527–532.
Kicheva, A., Cohen, M., & Briscoe, J. (2012). Developmental pattern formation: Insights from physics and biology. *Science, 338*, 210–212.
Kicheva, A., Pantazis, P., Bollenbach, T., Kalaidzidis, Y., Bittig, T., Julicher, F., et al. (2007). Kinetics of morphogen gradient formation. *Science, 315*, 521–525.
Kolomeisky, A. B. (2011). Formation of a morphogen gradient: Acceleration by degradation. *Journal of Physical Chemistry Letters, 2*, 1502–1505.
Kornberg, T. B. (2012). The imperatives of context and contour for morphogen dispersion. *Biophysical Journal, 103*, 2252–2256.
Kornberg, T. B. (2014). Cytonemes and the dispersion of morphogens. *WIREs Developmental Biology, 3*, 445–463.
Krieger, J. W., Singh, A. P., Bag, N., Garbe, C. S., Saunders, T. E., Langowski, J. O. R., et al. (2015). Imaging fluorescence (cross-) correlation spectroscopy in live cells and organisms. *Nature Protocols, 10*, 1948–1974.
Kuntz, S. G., & Eisen, M. B. (2014). Drosophila embryogenesis scales uniformly across temperature in developmentally diverse species. *PLoS Genetics, 10*, e1004293. https://doi.org/10.1371/journal.pgen.1004293.
Lander, A. D., Nie, Q., & Wan, F. Y. M. (2002). Do morphogen gradients arise by diffusion? *Developmental Cell, 2*, 785–796.
Li, X.-Y., MacArthur, S., Bourgon, R., Nix, D., Pollard, D. A., Iyer, V. N., et al. (2008). Transcription factors bind thousands of active and inactive regions in the Drosophila blastoderm. *PLoS Biology, 6*, e27.
Liang, H.-L., Nien, C.-Y., Liu, H.-Y., Metzstein, M. M., Kirov, N., & Rushlow, C. (2008). The zinc-finger protein Zelda is a key activator of the early zygotic genome in Drosophila. *Nature, 456*, 400–403.
Lim, B., Dsilva, C. J., Levario, T. J., Lu, H., Schüpbach, T., Kevrekidis, I. G., et al. (2015). Dynamics of inductive ERK signaling in the Drosophila embryo. *Current Biology, 25*, 1–8.
Little, S. C., Tikhonov, M., & Gregor, T. (2013). Precise developmental gene expression arises from globally stochastic transcriptional activity. *Cell, 154*, 789–800. https://doi.org/10.1016/j.cell.2013.07.025.
Little, S. C., Tkačik, G., Kneeland, T. B., Wieschaus, E. F., & Gregor, T. (2011). The formation of the Bicoid morphogen gradient requires protein movement from anteriorly localized mRNA. *PLoS Biology, 9*, e1000596.
Liu, J., He, F., & Ma, J. (2011). Morphogen gradient formation and action. *Fly, 5*, 242–246.
Liu, J., & Ma, J. (2011). Fates-shifted is an F-box protein that targets Bicoid for degradation and regulates developmental fate determination in Drosophila embryos. *Nature Cell Biology, 13*, 22–29.
Liu, J., & Ma, J. (2013). Dampened regulates the activating potency of Bicoid and the embryonic patterning outcome in Drosophila. *Nature Communications, 4*, 1–11.
Liu, F., Morrison, A. H., & Gregor, T. (2013). Dynamic interpretation of maternal inputs by the Drosophila segmentation gene network. *Proceedings of the National Academy of Sciences of the United States of America, 110*, 6724–6729.
Löhr, U., Chung, H.-R., Beller, M., & Jäckle, H. (2009). Antagonistic action of Bicoid and the repressor Capicua determines the spatial limits of Drosophila head gene expression domains. *Proceedings of the National Academy of Sciences of the United States of America, 106*, 21695–21700.
Lucas, T., Ferraro, T., Roelens, B., De Las Heras Chanes, J., Walczak, A. M., Coppey, M., et al. (2013). Live imaging of bicoid-dependent transcription in Drosophila embryos. *Current Biology, 23*, 2135–2139.

Lucas, T., Tran, H., Perez Romero, C. A., Guillou, A., Fradin, C., Coppey, M., et al. (2018). 3 minutes to precisely measure morphogen concentration. *PLoS Genetics, 14*, e1007676.

Lucchetta, E. M., Lee, J. H., Fu, L. A., Patel, N. H., & Ismagilov, R. F. (2005). Dynamics of Drosophila embryonic patterning network perturbed in space and time using microfluidics. *Nature, 434*, 1134–1138.

Lucchetta, E. M., Vincent, M. E., & Ismagilov, R. F. (2008). A precise Bicoid gradient is nonessential during cycles 11–13 for precise patterning in the Drosophila blastoderm. *PLoS One, 3*, e3651.

Manu, Surkova, S., Spirov, A. V., Gursky, V. V., Janssens, H., Kim, A.-R., et al. (2009a). Canalization of gene expression in the Drosophila blastoderm by gap gene cross regulation. *PLoS Biology*, 7, e1000049.

Manu, Surkova, S., Spirov, A. V., Gursky, V. V., Janssens, H., Kim, A.-R., et al. (2009b). Canalization of gene expression and domain shifts in the Drosophila blastoderm by dynamical attractors. *PLoS Computational Biology, 5*, e1000303.

McDaniel, S. L., Gibson, T. J., Schulz, K. N., Garcia, M. F., Nevil, M., Jain, S. U., et al. (2019). Continued activity of the pioneer factor Zelda is required to drive zygotic genome activation. *Molecular Cell, 74*, 185–195.e4.

Meinhardt, H. (1986). Hierarchical inductions of cell states: A model for segmentation in Drosophila. *Journal of Cell Science. Supplement, 4*, 357–381.

Mir, M., Reimer, A., Haines, J. E., Li, X.-Y., Stadler, M., Garcia, H., et al. (2017). Dense Bicoid hubs accentuate binding along the morphogen gradient. *Genes & Development, 31*, 1784–1794. https://doi.org/10.1101/gad.305078.117.

Mir, M., Stadler, M. R., Ortiz, S. A., Hannon, C. E., Harrison, M. M., Darzacq, X., et al. (2018). Dynamic multifactor hubs interact transiently with sites of active transcription in Drosophila embryos. *eLife*, 7, e40497.

Muller, P., Rogers, K. W., Jordan, B. M., Lee, J. S., Robson, D., Ramanathan, S., et al. (2012). Differential diffusivity of Nodal and Lefty underlies a reaction-diffusion patterning system. *Science, 336*, 721–724.

Muller, P., Rogers, K. W., Yu, S. R., Brand, M., & Schier, A. F. (2013). Morphogen transport. *Development, 140*, 1621–1638.

Nybakken, K., & Perrimon, N. (2002). Heparan sulfate proteoglycan modulation of developmental signaling in Drosophila. *Biochimica et Biophysica Acta, 1573*, 280–291.

Ochoa-Espinosa, A., Yu, D., Tsirigos, A., Struffi, P., & Small, S. (2009). Anterior-posterior positional information in the absence of a strong Bicoid gradient. *Proceedings of the National Academy of Sciences of the United States of America, 106*, 3823–3828.

Ochoa-Espinosa, A., Yucel, G., Kaplan, L., Pare, A., Pura, N., Oberstein, A., et al. (2005). The role of binding site cluster strength in Bicoid-dependent patterning in Drosophila. *Proceedings of the National Academy of Sciences of the United States of America, 102*, 4960–4965.

Paliou, C., Guckelberger, P., Schöpflin, R., Heinrich, V., Esposito, A., Chiariello, A. M., et al. (2019). Preformed chromatin topology assists transcriptional robustness of *Shh* during limb development. In *Proceedings of the National Academy of Sciences of the United States of America, 116*. (pp. 12390–12399).

Park, J., Estrada, J., Johnson, G., Vincent, B. J., Ricci-Tam, C., Bragdon, M. D. J., et al. (2019). Dissecting the sharp response of a canonical developmental enhancer reveals multiple sources of cooperativity. *ELife, 8*, e41266.

Perez-Carrasco, R., Guerrero, P., Briscoe, J., & Page, K. M. (2016). Intrinsic noise profoundly alters the dynamics and steady state of morphogen-controlled bistable genetic switches. *PLoS Computational Biology, 12,* e1005154.

Petkova, M. D., Tkačik, G., Bialek, W., Wieschaus, E. F., & Gregor, T. (2019). Optimal decoding of cellular identities in a genetic network. *Cell, 176*, 1–28.

Porcher, A., Abu-Arish, A., Huart, S., Roelens, B., Fradin, C., & Dostatni, N. (2010). The time to measure positional information: Maternal hunchback is required for the

synchrony of the Bicoid transcriptional response at the onset of zygotic transcription. *Development, 137*, 2795–2804.
Porcher, A., & Dostatni, N. (2010). The Bicoid morphogen system. *Current Biology, 20*, R249–R254.
Richards, D. M., & Saunders, T. E. (2015). Spatiotemporal analysis of different mechanisms for interpreting morphogen gradients. *Biophysical Journal, 108*, 2061–2073.
Roy, S., Hsiung, F., & Kornberg, T. B. (2011). Specificity of Drosophila cytonemes for distinct signaling pathways. *Science, 332*, 354–358.
Roy, S., Huang, H., Liu, S., & Kornberg, T. B. (2014). Cytoneme-mediated contact-dependent transport of the Drosophila decapentaplegic signaling protein. *Science, 343*, 1244624.
Sagner, A., & Briscoe, J. (2017). Morphogen interpretation: Concentration, time, competence, and signaling dynamics. *WIREs Developmental Biology, 6*, e271.
Sako, K., Pradhan, S. J., Barone, V., Inglés-Prieto, Á., Müller, P., Ruprecht, V., et al. (2016). Optogenetic control of nodal signaling reveals a temporal pattern of nodal signaling regulating cell fate specification during gastrulation. *Cell Reports, 16*, 866–877.
Sanders, T. A., Llagostera, E., & Barna, M. (2013). Specialized filopodia direct long-range transport of SHH during vertebrate tissue patterning. *Nature, 497*, 628–632.
Saunders, T., & Howard, M. (2009). Morphogen profiles can be optimized to buffer against noise. *Physical Review E, 80*, 041902.
Shimizu, K., & Gurdon, J. B. (1999). A quantitative analysis of signal transduction from activin receptor to nucleus and its relevance to morphogen gradient interpretation. *Proceedings of the National Academy of Sciences of the United States of America, 96*, 6791–6796.
Shvartsman, S. Y., & Baker, R. E. (2012). Mathematical models of morphogen gradients and their effects on gene expression. *WIREs Developmental Biology, 1*, 715–730.
Shvartsman, S. Y., Coppey, M., & Berezhkovskii, A. M. (2008). Dynamics of maternal morphogen gradients in Drosophila. *Current Opinion in Genetics & Development, 18*, 342–347.
Sigaut, L., Pearson, J. E., Colman-Lerner, A., & Ponce Dawson, S. (2014). Messages do diffuse faster than messengers: Reconciling disparate estimates of the morphogen bicoid diffusion coefficient. *PLoS Computational Biology, 10*, e1003629.
Simpson-Brose, M., Treisman, J., & Desplan, C. (1994). Synergy between the hunchback and bicoid morphogens is required for anterior patterning in Drosophila. *Cell, 78*, 855–865.
Sorre, B., Warmflash, A., Brivanlou, A. H., & Siggia, E. D. (2014). Encoding of temporal signals by the TGF-β pathway and implications for embryonic patterning. *Developmental Cell, 30*, 334–342.
Spirov, A., Fahmy, K., Schneider, M., Frei, E., Noll, M., & Baumgartner, S. (2009). Formation of the bicoid morphogen gradient: An mRNA gradient dictates the protein gradient. *Development, 136*, 605–614.
Sprague, B., & McNally, J. (2005). FRAP analysis of binding: Proper and fitting. *Trends in Cell Biology, 15*, 84–91.
Stanganello, E., Hagemann, A. I. H., Mattes, B., Sinner, C., Meyen, D., Weber, S., et al. (2015). Filopodia-based Wnt transport during vertebrate tissue patterning. *Nature Communications, 6*, 5846.
Stanganello, E., & Scholpp, S. (2016). Role of cytonemes in Wnt transport. *Journal of Cell Science, 129*, 665–672.
St Johnston, D., Driever, W., Berleth, T., Richstein, S., & Nüsslein-Volhard, C. (1989). Multiple steps in the localization of bicoid RNA to the anterior pole of the Drosophila oocyte. *Development, 107*, 13–19.
Surkova, S., Golubkova, E., Mamon, L., & Samsonova, M. (2018). Dynamic maternal gradients and morphogenetic networks in Drosophila early embryo. *Biosystems, 173*, 207–213.
Swaminathan, R., Hoang, C. P., & Verkman, A. S. (1997). Photobleaching recovery and anisotropy decay of green fluorescent protein GFP-S65T in solution and cells:

Cytoplasmic viscosity probed by green fluorescent protein translational and rotational diffusion. *Biophysical Journal, 72*, 1900–1907.

Tabata, T. (2004). Morphogens, their identification and regulation. *Development, 131*, 703–712.

Teimouri, H., & Kolomeisky, A. B. (2014). Development of morphogen gradient: The role of dimension and discreteness. *The Journal of Chemical Physics, 140,* 085102.

Thukral, S., Kaity, B., Dey, B., Sharma, S., Nandi, A., Mitra, M., et al. (2019). Cytoarchitecture constrains a photoactivation induced tubulin gradient in the syncytial Drosophila embryo. *bioRxiv*. https://www.biorxiv.org/content/10.1101/520031v1.

Tran, H., Desponds, J., Perez Romero, C. A., Coppey, M., Fradin, C., Dostatni, N., et al. (2018). Precision in a rush: Trade-offs between reproducibility and steepness of the hunchback expression pattern. *PLoS Computational Biology, 14,* e1006513.

van Boxtel, A. L., Chesebro, J. E., Heliot, C., Ramel, M.-C., Stone, R. K., & Hill, C. S. (2015). A temporal window for signal activation dictates the dimensions of a nodal signaling domain. *Developmental Cell, 35*, 175–185.

Verd, B., Clark, E., Wotton, K. R., Janssens, H., Jiménez-Guri, E., Crombach, A., et al. (2018). A damped oscillator imposes temporal order on posterior gap gene expression in Drosophila. *PLoS Biology, 16,* e2003174.

Wang, Y., Wang, X., Wohland, T., & Sampath, K. (2016). Extracellular interactions and ligand degradation shape the nodal morphogen gradient. *eLife, 5*, 91.

Warmflash, A., Zhang, Q., Sorre, B., Vonica, A., Siggia, E. D., & Brivanlou, A. H. (2012). Dynamics of TGF-β signaling reveal adaptive and pulsatile behaviors reflected in the nuclear localization of transcription factor Smad4. *Proceedings of the National Academy of Sciences of the United States of America, 109*, E1947–E1956.

Wartlick, O., Julicher, F., & Gonzalez-Gaitan, M. (2014). Growth control by a moving morphogen gradient during Drosophila eye development. *Development, 141*, 1884–1893.

Wartlick, O., Kicheva, A., & Gonzalez-Gaitan, M. (2009). Morphogen gradient formation. *Cold Spring Harbor Perspectives in Biology, 1*, a001255.

Wartlick, O., Mumcu, P., Kicheva, A., Bittig, T., Seum, C., Julicher, F., et al. (2011). Dynamics of Dpp signaling and proliferation control. *Science, 331*, 1154–1159.

White, M. D., Angiolini, J. F., Alvarez, Y. D., Kaur, G., Zhao, Z. W., Mocskos, E., et al. (2016). Long-lived binding of Sox2 to DNA predicts cell fate in the four-cell mouse embryo. *Cell, 165*, 75–87.

Wieschaus, E. (2016). Positional information and cell fate determination in the early Drosophila embryo. *Current Topics in Developmental Biology, 117*, 567–579.

Wolpert, L. (1969). Positional information and the spatial pattern of cellular differentiation. *Journal of Theoretical Biology, 25*, 1–47.

Wolpert, L. (2016). Positional information and pattern formation. *Current Topics in Developmental Biology, 117*, 597–608.

Wu, H., Manu, Jiao, R., & Ma, J. (2015). Temporal and spatial dynamics of scaling-specific features of a gene regulatory network in Drosophila. *Nature Communications, 6*, 1–13.

Xu, Z., Chen, H., Ling, J., Yu, D., Struffi, P., & Small, S. (2014). Impacts of the ubiquitous factor Zelda on Bicoid-dependent DNA binding and transcription in Drosophila. *Genes & Development, 28*, 608–621.

Yamashita, Y. M., Inaba, M., & Buszczak, M. (2018). Specialized intercellular communications via cytonemes and nanotubes. *Annual Review of Cell and Developmental Biology, 34*, 59–84.

Yin, J., Lee, R., Ono, Y., Ingham, P. W., & Saunders, T. E. (2018). Spatiotemporal coordination of FGF and Shh signaling underlies the specification of myoblasts in the zebrafish embryo. *Developmental Cell, 46*, 735–750.

Yu, S. R., Burkhardt, M., Nowak, M., Ries, J., Petrášek, Z., Scholpp, S., et al. (2009). Fgf8 morphogen gradient forms by a source-sink mechanism with freely diffusing molecules. *Nature*, *461*, 533–536.

Yuste, S. B., Abad, E., & Lindenberg, K. (2010). Reaction-subdiffusion model of morphogen gradient formation. *Physical Review. E*, *82*, 061123.

Zartman, J. J., Cheung, L. S., Niepielko, M. G., Bonini, C., Haley, B., Yakoby, N., et al. (2011). Pattern formation by a moving morphogen source. *Physical Biology*, *8*, 045003.

Zhou, S., Lo, W.-C., Suhalim, J. L., Digman, M. A., Gratton, E., Nie, Q., et al. (2012). Free extracellular diffusion creates the Dpp morphogen gradient of the Drosophila wing disc. *Current Biology*, *22*, 668–675.

CHAPTER FOUR

Constraints and limitations on the transcriptional response downstream of the Bicoid morphogen gradient

Huy Tran[a,b], Aleksandra M. Walczak[b,*], Nathalie Dostatni[a,*]
[a]Institut Curie, PSL Research University, CNRS, Sorbonne Université, Nuclear Dynamics, Paris, France
[b]Ecole Normale Supérieure, PSL Research University, CNRS, Sorbonne Université, Laboratoire de Physique, Paris, France
*Corresponding authors: e-mail address: awalczak@lpt.ens.fr; nathalie.dostatni@curie.fr

Contents

Abstract

The regulation of the *hunchback* promoter expression by the maternal Bicoid gradient has been studied as a model system in development for many years. Yet, at the level of quantitative agreement between data and theoretical models, even the first step of this regulation, transcription, continues to be challenging. This situation is slowly progressing, thanks to quantitative live-imaging techniques coupled to advanced statistical data analysis and modeling. Here, we outline the current state of our knowledge of this apparently "simple" step, highlighting the newly appreciated role of bursty transcription dynamics and its regulation.

Current Topics in Developmental Biology, Volume 137
ISSN 0070-2153
https://doi.org/10.1016/bs.ctdb.2019.12.002

1. Introduction

The development of properly proportioned individuals relies on cells precisely reading out positional information and committing to a specific cell fate. In a developing organisms, non-uniformly distributed morphogens gradients provide such positional information to each cell by turning on the expression of specific target genes in a concentration-dependent manner (Karaiskos et al., 2017; Liu et al., 2009; Petkova, Bialek, Wieschaus, & Gregor, 2019). Although these gradients are known to be essential since decades in many developmental systems, in most cases we do not know how they are established or how they activate target gene expression in a dose-dependent manner. Here, we discuss recent developments in describing how a noisy transcriptional process can be tuned very rapidly into a reproducible functional developmental pattern by focusing on one well-studied morphogen-target gene pair—the Bicoid (Bcd) morphogen controlling the *hunchback* (*hb*) gap gene in early fly development.

The *hb* gap gene is involved in antero-posterior (AP) patterning of young fruit fly embryos. *hb* expression is first detected at the onset of zygotic transcription (nc 8), around 1 h after fertilization. No more than 30 min later (nc 11), *hb* transcription occurs in an anterior domain with most nuclei highly expressing the *hb* gene. The border separating this expressing anterior domain from the posterior domain with mostly *hb*-silent nuclei is extremely sharp (Little, Tikhonov, & Gregor, 2013; Porcher et al., 2010) (Fig. 1A). This type of expression pattern defines a step-like pattern. Following the *hb* transcription step-like pattern, the Hb protein expression quantified at nc14 is also step-like (Houchmandzadeh, Wieschaus, & Leibler, 2002) and exhibits very low variability in concentration between nuclei at the same position along the AP axis (Gregor, Tank, Wieschaus, & Bialek, 2007). This precise Hb pattern, combined with other gap gene patterns, was proposed to contain enough positional information for nuclei to predict their position in the embryo with ~99% accuracy (Dubuis, Tkacik, Wieschaus, Gregor, & Bialek, 2013; Petkova et al., 2019).

In young embryos, the major regulator of *hb* transcription is the homeodomain-containing Bcd (Driever & Nusslein-Volhard, 1988). Translated from maternal mRNAs anchored at the anterior pole, Bcd proteins form an AP exponential gradient and activate *hb* transcription in the anterior half of the embryo. Increasing or decreasing the amount of

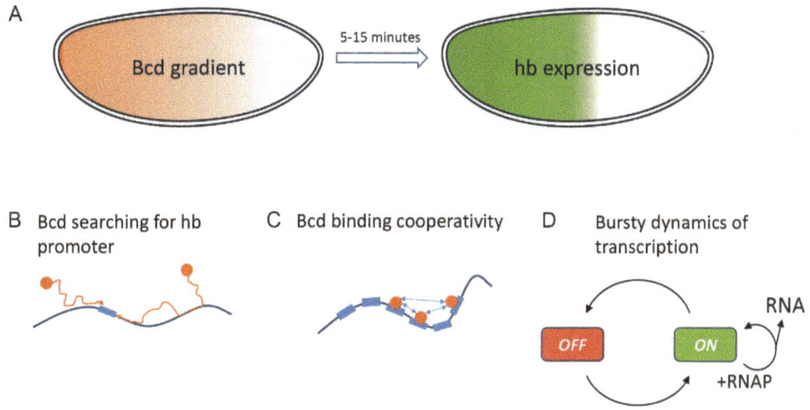

Fig. 1 (A) How is positional information contained in the shallow Bcd concentration gradient transformed into a robust step-like *hb* expression pattern in the 5–15 min of the nuclear cycle 11–13? (B–D) The rate-limiting steps of *hb* expression in response to the Bcd gradient: (B) The Bcd searches for the *hb* promoter, which can involve either 3D diffusion inside the nuclear space or 1D diffusion along the DNA. (C) Bcd binds cooperatively to *hb* promoter, separating the embryo into a rich Bcd-bound anterior region and a no Bcd-bound posterior region. This process can involve exclusively Bcd molecules, other DNA bound transcription factors or factors facilitating chromatin accessibility. (D) The production of *hb* RNA was shown to be bursty (Desponds et al., 2016), with relatively infrequent switching of the *hb* promoter between the ON and OFF expression states.

Bicoid in the embryo induces a posterior or anterior shift, respectively, of the *hb* step-like pattern, arguing that the expression of *hb* is dependent on Bcd concentration (Driever & Nusslein-Volhard, 1988). The discovery of Bcd's role in *hb* transcription was exciting as it provided the first example illustrating the elegant idea of the French Flag model for morphogenesis. However, 30 years after its discovery, the mechanisms responsible for the sharpness of the *hb* step-like pattern and its reproducibility are not yet fully understood. The first quantitative studies of the Bcd gradient and Hb protein pattern reproducibility (Gregor, Wieschaus, McGregor, Bialek, & Tank, 2007; Houchmandzadeh et al., 2002) raised doubts about the pattern emerging solely from diffusive biochemical interactions between transcription factors and the gene promoter region. Gregor, Tank, et al. (2007) used the Berg–Purcell scheme (Berg & Purcell, 1977), in which the concentration of a diffusing ligand is estimated by counting the number of ligand-receptor binding events within a specific time. In this simplest scheme of concentration sensing, the time required for a Bcd binding site of the *hb* promoter to read the Bcd concentration with 10% accuracy was estimated to be of the order of 2h (Gregor, Tank, et al., 2007), much longer than the interphase duration of nuclear cycles. These early attempts to explain the establishment of the Hb step-like pattern marked a shift in experimental studies toward more quantitative approaches. The advent of new measurement methods, based on live imaging especially of transcription dynamics (Bertrand et al., 1998), allowed for a deeper view of the processes of positional readout in young fruit fly embryos. Specifically it allowed one to focus on the outcome of the first steps of regulation, transcription at the *hb* locus, to determine how this stage of early development is controlled. Thanks to advances in imaging discussed below, the pattern of transcription from the *hb* locus is observed to be established sharply in about 3min at nc11, despite being generated by noisy gene expression from a bursty promoter. In this review, we discuss recent work and challenges for advancing our understanding of positional information propagation from the Bcd gradient to the *hb* expression pattern (Fig. 1). We made the choice to only focus on the first step, the transcription process, which in itself contains several steps (Fig. 1B–D), and highlight why this system remains challenging 30 years on. More details about the establishment of the Bcd gradient can be found in the chapter by Huang and Saunders in the same issue (Huang & Saunders, 2020).

2. The Bcd gradient

2.1 Positional information

The Bcd concentration gradient arises from maternal mRNA anchored at the anterior pole of the syncytial embryo (Grimm, Coppey, & Wieschaus, 2010). Large scale analysis of the Bcd gradient, through immunofluorescent staining of the endogenous protein or analysis of the fluorescent-fusion Bcd-eGFP protein (Gregor, Wieschaus, et al., 2007) (Fig. 2A), indicated that the Bcd gradient is exponential with a concentration decay length around 80–120 μm (between one-fifth and one-fourth of embryo length) (Abu-Arish, Porcher, Czerwonka, Dostatni, & Fradin, 2010; Gregor, Bialek, de Ruyter van Steveninck, Tank, & Wieschaus, 2005; Gregor, Wieschaus, et al., 2007). The concentration of this gradient is a source of positional information for each nuclei along the AP axis. All the experiments performed to measure the absolute concentration of the Bcd gradient along the AP axis used the Bcd-eGFP expressed from a transgene in the embryo which rescues the viability of the bcd^{E1} null allele. The total concentration of Bcd at the anterior pole, measured *via* fluorescent Bcd-eGFP ranges from 90 nM, measured

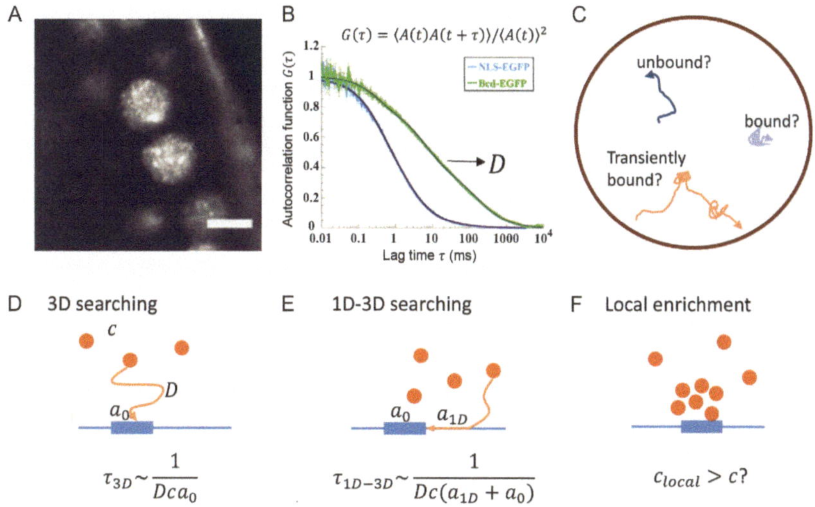

Fig. 2 Studying Bcd motility in living fruit fly embryos. (A) Fluorescent-tagged Bcd molecules were shown to be distributed in foci in the nuclear space of fixed embryos (Xu, Sepúlveda, Figard, Sokac, & Golding, 2015) and hubs of fluorescently-tagged Bcd (Mir et al., 2017; Mir, Stadler, Harrison, Darzacq, & Eisen, 2018). (B and C) Different motilities of Bcd molecules were inferred from Fluorescent Correlation Spectroscopy (FCS) analysis of the Bcd-eGFP signal in nuclei (Porcher et al., 2010), with a Bcd population with a high (free) diffusion coefficient (B), and (C), by single-molecule tracking approaches revealing bound, unbound and transiently bound molecules (Mir et al., 2018). (D–F) various scenarios of Bcd searching for its binding sites in the *hb* promoter. (D) Bcd molecules can diffuse in 3D nuclear space to search for Bcd binding sites on the target promoter. The time, τ_{3D}, for the binding site to be found is inversely proportional to the product of *D* (coefficient of free diffusion), *c* (concentration) and a_0 (size of the binding site); (E) Bcd molecules can "hop" to non-specific locations on the DNA and slide along DNA segments in 1D to search for specific binding sites. The average sliding distance (or footprint) a_{1D} can be much bigger than the size of a Bcd binding site and therefore reduces the search time from the 3D case; (F) Local enrichment of transcription factor (TF) concentration ($c_{local} > c$) at the promoter region can also reduce the promoter search time. *Panel (A) figure reused from Mir, M., Reimer, A., Haines, J. E., Li, X. Y., Stadler, M., Garcia, H., et al. (2017). Dense bicoid hubs accentuate binding along the morphogen gradient. Genes and Development, 31(17), 1784–1794 and panel (B) figure reused from Porcher, A., Abu-Arish, A., Huart, S., Roelens, B., Fradin, C., & Dostatni, N. (2010). The time to measure positional information: Maternal hunchback is required for the synchrony of the Bicoid transcriptional response at the onset of zygotic transcription.* Development, 137*(16), 2795–2804.*

by comparing the fluorescence inside nuclei with the fluorescence of an eGFP solution at a given concentration (Gregor, Garcia, & Little, 2014), to 140 nM (Abu-Arish et al., 2010), equivalent to, respectively, 4.4 (estimated to be ~700 Bcd molecules per nuclei at nc14; Gregor, Wieschaus, et al., 2007) to

7 molecules/μm^3 where the *hb* pattern boundary is established. Even though these measurements are consistent, they were obtained with the same fluorescent Bcd-eGFP and are likely to be underestimates of the real concentration because a proportion of the eGFP might not be fluorescent. Further analysis using a Bcd fusion carrying two fluorescent domains (Durrieu et al., 2018) might help resolve the issue of absolute Bcd concentration measurements in the embryo. Similarly, taking into account the maturation time of the eGFP points to a slight overestimation of the Bcd gradient decay length of about 15% (Durrieu et al., 2018; Liu, Morrison, & Gregor, 2013). After corrections, the length constant of the Bcd gradient is estimated as low as one-sixth of embryo length (16.5 ± 0.7% EL).

2.2 The motility of Bcd molecules

The *hb* locus extracts positional information from the local Bcd concentration *via* interactions with Bcd molecules. Given the short time window for positional readout in each interphase, the Bcd search time for the *hb* promoter τ_{search} is critical in determining the limit of positional readout precision. Therefore, several studies analyzed Bcd motility, using FRAP (Gregor, Tank, et al., 2007) or FCS (Abu-Arish et al., 2010) on fluorescent Bcd-eGFP or single-molecule tracking (Fig. 2A) (Drocco, Grimm, Tank, & Wieschaus, 2011; Mir et al., 2018). In the initial FRAP experiments, Bcd motility in the cytoplasm turned out to be quite slow (~0.3 μm^2/s) (Gregor, Wieschaus, et al., 2007). FCS experiments performed both in the cytoplasm and the interphase nuclei revealed the existence of Bcd molecules with different motilities: best fitting of the data to the two-species diffusion model indicated that (i) in the cytoplasm, 18% of the Bcd molecules are slow-moving while 82% of the Bcd molecules are fast-moving with an average diffusion coefficient of ~7.4 μm^2/s (Abu-Arish et al., 2010) and (ii) in the nucleus, 43% of the Bcd molecuels are slow-moving (~0.22 μm^2/s) and 57% fast-moving (~7.7 μm^2/s) (Porcher et al., 2010). Fast moving Bcd molecules (~4 μm^2/s) were also observed using a photoactivable Dronpa-Bcd (Drocco et al., 2011) and the existence of at least two populations of Bcd molecules was further confirmed by high resolution single molecule imaging suggesting that in nuclei, Bcd molecules spend the same amount of time on nuclear exploration (searching for a binding target) and on binding to chromatin with surprisingly high unbinding rates, distributed with long tails (Mir et al., 2018).

2.3 The time it takes for Bcd to find its target sites on the *hb* promoter

Transcription factors (TFs) explore, usually by passive diffusion, the nuclear space in "search" of their DNA binding sites on the regulatory sequences of their target genes. In the case of *Drosophila* embryogenesis, the time it takes for the Bcd protein to find its target sites on the *hb* promoter is critical to explain how *hb* expression in response to Bcd can be so fast yet precise despite the limited Bcd concentration in the mid-embryo region. Therefore, many studies have used modeling to estimate the time it takes Bcd to find its target sites and proposed several strategies to shorten this search time (Abu-Arish et al., 2010; Gregor, Tank, et al., 2007; Mir et al., 2017, 2018; Mirny et al., 2009; Normanno, Dahan, & Darzacq, 2012).

In an initial attempt to estimate the Bcd search time for the *hb* promoter, Gregor and colleagues proposed that Bcd molecules can diffuse in 3D inside the nuclear space to come in contact with the specific Bcd binding sites on the *hb* promoter (Fig. 2D). They proposed that the lapse time between each contacts was inversely proportional to the diffusion coefficient of Bcd D, the local Bcd concentration c and the size of the Bcd binding site a.

$$\tau_{search} = \tau_{3D} \sim 1/Dca \qquad (1)$$

The binding site for TFs, including Bcd, are commonly 10 bp long which corresponds to ~3 nm. Given this value and assuming that the searching Bcd molecules are diffusing freely (i.e., are fast-moving molecules) and that Bcd binding is diffusion limited (each collision between a Bcd molecule and a binding site results in a successful binding event), the binding site search time τ_{3D} is estimated to be on the order of 10 s. However, this is an optimistic estimate given that a is likely 10-fold too large since displacement by a single bp may lead to an entirely different DNA sequence that is not recognizable by the protein (Slutsky & Mirny, 2004). In this later case, a would be ~0.3 nm and the search time for Bcd would be 100 s. In addition, activation by Bcd requires in general not just the binding of one Bcd molecule to the *hb* promoter but several (Burz, Rivera-Pomar, Jäckle, & Hanes, 1998; Driever, Thoma, & Nüsslein-Volhard, 1989; Ronchi, Treisman, Dostatni, Struhl, & Desplan, 1993). Thus, clustering of binding sites in the regulatory sequences of the target genes is another factor that may lengthen the time it takes to bind enough Bcd molecules together to get a transcriptional response.

The rapid establishment of the *hb* pattern, despite a long search time in 3D, leads to hypotheses that Bcd molecules may use a combination of 1D and 3D diffusion to search for their target sites (Mirny et al., 2009; Slutsky & Mirny, 2004). This mode of searching was first observed in *Escherichia coli* (Elf, Li, & Xie, 2007; Hammar et al., 2012; Riggs, Bourgeois, & Cohn, 1970) for the *lacI* repressor where a combination of a 1D and 3D search was found to reduce the search time by up to 100-fold compared to a pure 3D search. In this scheme, proteins bind non-specifically to DNA and, while doing so, slide along the DNA segment in search for the specific target site (Fig. 2E). Therefore, the effective size of the target site is increased by the TF's sliding footprint along the DNA a_{1D} (up to a100 bp, compared to the size of a binding site $a = 10$ bp). Another hypothesis is that the target locus could be located in a micro-environments with enhanced TF concentration ($c_{local} > c$) (Fig. 2F), thus speeding up the search, as proposed in the case of the Ultrabithorax protein (Tsai et al., 2017). The recent observation of Bcd concentration in dense hubs (Mir et al., 2017, 2018) opens up the possibility that micro-environments that enhance local Bcd concentration could contribute to reduce the Bcd search time for the *hb* promoter. However, one should note that these mechanisms can also introduce non-linearities into the position sensing process: Bcd hubs were found to persist even in the posterior region of low Bcd concentration, leading to a much flatter Bcd concentration profile in hubs than in the cytoplasm. In addition, Hammar et al. observed that bound lacI molecules may interfere with the 1D-sliding molecules when the distance between the target sites is shorter than the sliding footprint (Hammar et al., 2012). If Bcd employs this mode of searching, the very short distances between Bcd binding sites on the *hb* promoter (as short as 12 bp) may introduce negative feedback to Bcd binding, instead of the positive feedback normally linked to a sharp *hb* pattern (Ma, Yuan, Diepold, Scarborough, & Ma, 1996). It is experimentally challenging to directly observe the scheme of the Bcd target site search. Currently, it is not possible using a confocal microscope to directly identify which one of the Bcd subpopulations with characterized motilities are directly searching for target sites. Observing interactions between Bcd and its target sites is still limited to epigenomics approaches such as ChIP on fixed samples (Bradley et al., 2010; Li et al., 2011) performed on population of nuclei that have very different transcription features. Recent advances in sample preparation (Combs & Eisen, 2013; Karaiskos et al., 2017) may allow us to observe how these interactions correlate with the varying degree of inhomogeneity in Bcd motility along the embryo AP axis and help identify the target site search mode.

2.4 Activation of transcription by Bcd

The Bcd protein is able, on its own, to activate transcription when bound to a promoter containing as few as three of its DNA binding sites (Crauk & Dostatni, 2005; Ronchi et al., 1993). However, how this is achieved remains largely unknown. Structure-function analyses of the Bcd transcription factor indicated that it contains many redundant functional domains (Schaeffer, Janody, Loss, Desplan, & Wimmer, 1999). Besides its homeodomain which allows binding to DNA (Hanes & Brent, 1989; Trelsman, Gönczy, Vashishtha, Harris, & Desplan, 1989), the Bcd protein contains several independent activation domains which can activate transcription on their own when multimerized and fused to a Gal4 DNA binding domain *in vitro* (Sauer, Hansen, & Tjian, 1995) or in the early embryo (Janody, Sturny, Schaeffer, Azou, & Dostatni, 2001). These include a Glutamine-rich domain, a ST-rich domain and a C-terminal acidic domain. The Bicoid protein also contains independent inhibitory domains which reduce its activation potential (Janody et al., 2001; Zhao et al., 2002). Finally, activation by Bcd is enhanced by other TFs binding to the promoter. These include the maternal contribution of the Hunchback protein itself (Porcher et al., 2010; Simpson-Brose, Treisman, & Desplan, 1994) or Zelda (Mir et al., 2017; Xu et al., 2014). Yet, the mechanisms underlying these essential synergistic effects are poorly understood.

3. *hb* transcription dynamics

3.1 Visualizing *hb* transcription dynamics

RNA-FISH on fixed embryos allowed for the monitoring of *hb* nascent transcript accumulation at their site of synthesis inside each nucleus of the embryo, making it an initial marker to study ongoing transcription and promoter dynamics at a given locus. This allowed subsequently for the detection of single mature mRNAs in the cytoplasm and in the nucleus (Little et al., 2013; Perry, Bothma, Luu, & Levine, 2012; Zoller, Little, & Gregor, 2018). Distributions of nascent and cytoplasmic *hb* RNA along the AP axis demonstrate sharp step-like patterns at the transcription level, with an expression boundary width varying from 10% egg length to 8% egg length at nc11 and nc13, respectively. The observed data suggested, despite low heterogeneity at the protein level (Gregor, Tank, et al., 2007), a very noisy transcription process occurring with periods of promoter activity and inactivity (Xu et al., 2015).

RNA FISH requires fixation of the sample and can only provide a snap shot view of the transcription process at a given time (the time of fixation) during nuclear interphase. Following the pioneering work of R. Singer (Bertrand et al., 1998), the MS2 fluorescent RNA-tagging system has been implemented to monitor transcription dynamics in living early *Drosophila* embryo development (Garcia, Tikhonov, Lin, & Gregor, 2013; Lucas et al., 2013). The system takes advantage of strong interactions between the MCP coating proteins and its RNA stem loops from the MS2 bacteriophage. As nascent RNA containing stem loops are being transcribed, they are bound by fusion proteins MCP-GFP, making ongoing transcription at the loci visible as bright fluorescent spots under the confocal microscope (Ferraro, Lucas, et al., 2016).

Recently, the MS2 system allowed for the direct visualization in real-time of position-dependent activation of the proximal *hb* P2 promoter (700 bp) (Lucas et al., 2018): at each nuclear interphase (from nc11 to nc13), *hb* expression (detected *via* MCP-GFP foci) first occurs in the most anterior then proceeds progressively though very rapidly from the anterior to the boundary region (at ~45% of embryo length from the anterior pole). Of note, this progressive appearance of first *hb* foci after mitosis is measured considering as an origin of time the onset of mitosis of each nucleus and is thus not a consequence of the mitotic wave. This dynamics from the anterior pole to the center of the embryo occurs even within the anterior region where Bcd is presumably saturated. These findings suggest that Bcd concentration is not only rate-limiting in the boundary region, as observed from the pattern dynamics at the steady-state, but also within the anterior region. In this region, differences in Bcd concentrations leads to differences in *hb* activation times (up to 3 min), which can take significant portion of the early nuclear interphase and consequently affects the total amount of transcripts produced. In addition, analysis of the MCP-GFP time trace of each individual nucleus shows variability in the transcription process even in the anterior region with high Bcd concentration (Desponds et al., 2016), indicating that Bcd is not the sole factor contributing to the noise in *hb* transcription.

3.2 Characterizing *hb* transcription dynamics

The fluorescent time traces acquired with the MS2 system provide an indirect observation of transcription dynamics. The signal is noisy, convoluting both experimental and intrinsic noise with the properties of the MS2 probe. To obtain a specific fluorescent signal sufficiently strong to overcome background fluorescence due to unbound MCP-GFP molecules, a long probe of 24 RNA stem loops was used (Garcia et al., 2013; Lucas et al., 2013). As the signal is only detected while the probe is being transcribed (Fig. 3A and B), this introduces a significant delay (buffering time) between each instant the promoter is ON and the corresponding fluorescent detection. This buffering time of ~1 min (Coulon et al., 2014; Fukaya, Lim, & Levine, 2017; Garcia et al., 2013) and the short length of the traces (5–15 min) prevent traditional analysis based on OFF time distributions or autocorrelation functions to quantify the statistics of the activation and inactivation times.

Desponds et al. developed a tailored autocorrelation analysis of the fluorescent time traces to overcome these limitations (Desponds et al., 2016). Combining this analysis with models of transcription initiation (Fig. 3C) and estimates of the precision of the transcriptional readout, provided evidence for bursty transcription initiation in nuclear cycles 12–13 (Desponds et al., 2016). Namely, they find the dynamics in agreement with a telegraph model, in which the promoter switches between the ON and OFF states. Only during the ON state can RNA polymerase arrive and initiate transcription successively (Fig. 3C). The switching between the ON and OFF promoter states can be driven by polymerase pausing which was proposed to prevent new initiation between transcription bursts (Shao & Zeitlinger, 2017), the binding/unbinding of other TFs required for activation, or DNA looping for distal enhancer-promoter contacts (Fukaya, Lim, & Levine, 2016). The best-fit switching period (for a full ON-OFF cycle) is in the order of ~30 s, with the probability to be in the ON state of ~50% at the anterior and of ~10% at the boundary.

It should be noted that the autocorrelation function analysis alone is not able to distinguish reliably between different models for promoter activation and requires complementary information about the precision of the transcriptional readout to conclude that transcription is most likely bursty (Fig. 3D). Recently, an inference method based on hidden Markov model and maximum likelihood has been developed (Corrigan, Tunnacliffe, Cannon, & Jonathan, 2016) and tailored for the MS2 system in *Drosophila* (Lammers et al., 2019). This discrete-time model employs a hidden compound state, which records the previous promoter states during the elongation time (Fig. 3B). This compound state is used to map RNA Polymerase (RNAP)'s position on the reporter gene segment and calculate the active loci intensity at a given time. The rates of switching between the promoter states in each time step are fitted based on maximum likelihood.

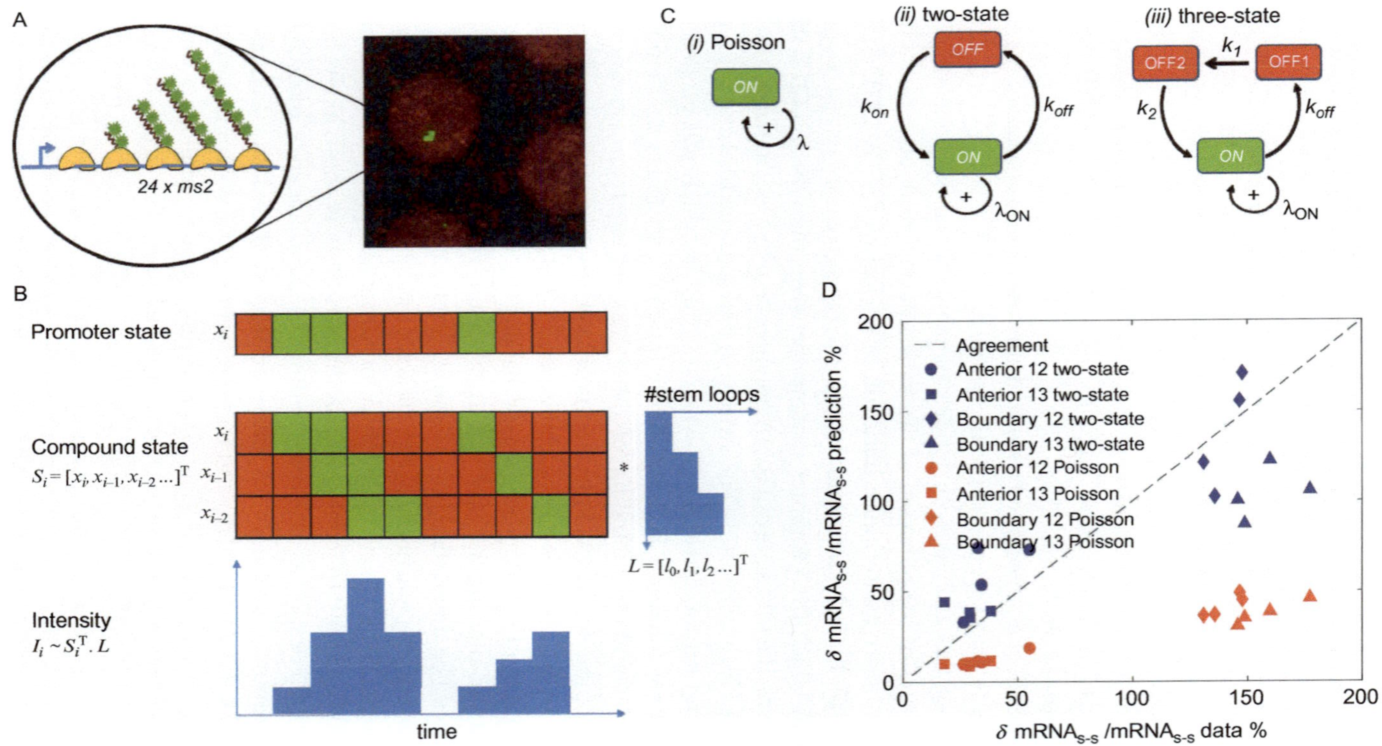

Fig. 3 See legend on next page

While computationally expensive, the method allows for direct model selection and shows that transcription bursts are prevalent in stripe gene expression in later stages of fly development (Berrocal, Lammers, Garcia, & Eisen, 2018; Bothma et al., 2014; Lammers et al., 2019).

3.3 Transcription regulation of *hb* gene by Bcd proteins

Data obtained from the MS2 system provided insights not only at the molecular level about the kinetics of the promoter behavior but also at the cellular level when considering individual nuclei along the AP axis and individual loci in each of these nuclei. In particular, it was possible to analyze the transcription dynamics of each *hb-MS2* locus at the scale of the whole embryo. This analysis indicated that depending on its position along the AP axis, each locus was able to either turn ON when positioned in the anterior or remain silent when positioned in the posterior. Surprisingly, the steep border forms in under 3 min at each nuclear interphase 11–13 (Lucas et al., 2018). This indicates that the system is able to measure extremely rapidly very subtle differences of Bcd concentration and produce a complete sharp border. This rapid responsiveness is fascinating because it is almost 10 times faster than predicted by previous theoretical models assuming that the Bcd gradient is the only driver for the *hb* transcription process.

Fig. 3 Inferring promoter dynamics from MS2 loci intensity: (A) Visualization of active transcription loci: as RNAs containing MS2 stem loops are transcribed by RNA Polymerases (RNAP, in yellow), they are bound by fluorescent MCP-GFP molecules (in green). The succession of several RNAPs transcribing the gene allows for the accumulation of several fluorescently tagged MS2-containing RNA at the same location which become visible under the confocal microscope as green bright spots (right panel). (B) Transformation from the promoter state dynamics to MS2 spot intensity dynamics: promoter state in discrete time $\mathbf{x}_i$ indicates whether the promoter is ON (green) or OFF (red). This state is encoded in the compound state vector $\mathbf{S}_i = [\mathbf{x}_i, \mathbf{x}_{i-1}, \mathbf{x}_{i-2} \ldots]$, which also maps the position of RNAP on the MS2 cassette. RNAP arriving at time *i* will be transcribing a nascent RNA containing $\mathbf{L}_j$ MS2 stem loops at time $\mathbf{i+j}$. $\mathbf{L}$ depends on the length and on the arrangement of the MS2 stem loops on the reporter gene. The active loci intensity $\mathbf{I}_i$ at time $\mathbf{i}$ is given by the product of $\mathbf{S}_i$ and $\mathbf{L}$. (C) Different models of promoter dynamics: (i) Poisson model: random RNAP arrival and initiation of transcription; (ii) two-state and (iii) three-state models, where promoters switch successively between ON and OFF states. During the ON state, RNAPs arrive and initiate transcription in bursts, with maximum rates. (D) Comparison of readout noise (**δmRNA/mRNA**) at steady state generated from Poisson (red) and bursty two-state models (blue) and data (dashed). *Figure reused from Desponds, J., Tran, H., Ferraro, T., Lucas, T., Perez Romero, C., Guillou, A., et al. (2016). Precision of readout at the hunchback gene: Analyzing short transcription time traces in living fly embryos. PLoS Computational Biology, 12(12), e1005256.*

The steep Bcd-dependent *hb* pattern, given the smooth Bcd gradient, demonstrates a strong nonlinear regulation of the *hb* gene by Bcd. The presence of multiple Bcd binding sites on the *hb* promoter (Driever & Nusslein-Volhard, 1988) suggests that such strong nonlinearities can be achieved by high cooperativity of Bcd binding to the *hb* promoter site. Cooperative binding of Bcd to multimerized binding sites was observed *in vitro* (Burz et al., 1998; Ma et al., 1996) but remains too weak to account for the extremely steep Bcd-dependent *hb* pattern observed *in vivo*. Synthetic reporters with only Bcd binding sites are weakly expressed in very anterior domains which harbor, however, remarkably steep posterior boundaries (Crauk & Dostatni, 2005; Ronchi et al., 1993). This suggests that Bcd and Bcd binding sites are sufficient to generate a steep posterior border and models of regulation by Bcd binding/unbinding can help understand how this is achieved.

Using powerful optogenetics to manipulate Bcd activity in real time, Huang and colleagues demonstrated that the full functional Bcd is critical in the early nuclear cycles (nc11 to 13) for the development of the embryo's mesothorax, where the *hb* boundary is established (Huang, Amourda, Zhang, Tolwinski, & Saunders, 2017). During this time window, light-induced conformational changes of the Bcd molecules, which leads to their functional inactivation, significantly reduces transcription at the *hb* promoter, as shown using the MCP-GFP system. This study links the transcriptional activity of TF bound to their target sites with the temporal dynamics of transcription and provide an unprecedented time resolution of the TF-dependent transcription process. A general model of transcription regulation *via* binding/unbinding of TF to the binding sites on the target promoter (Fig. 4A), demonstrates that the pattern's degree of steepness, conventionally characterized by a Hill coefficient H of the fitted Hill function, is limited by the number of TF binding sites N (Estrada, Wong, DePace, Gunawardena, 2016). In this model, the promoter states ν_i, each associated with a transcription rate, are updated with every TF binding and unbinding event. When the model satisfies detailed balance, the forward and reversed transitions between any two states ν_i and ν_j are equilibrated:

$$P(\nu_i).k_{ij} = P(\nu_j).k_{ji}, \tag{2}$$

with k_{ij} and k_{ji} the forward and reverse rate constants, $P(\nu_i)$ and $P(\nu_j)$ the probability of the promoter states. Thus, the model can be collapsed into the linearized model as in Fig. 4B, where only the number of occupied

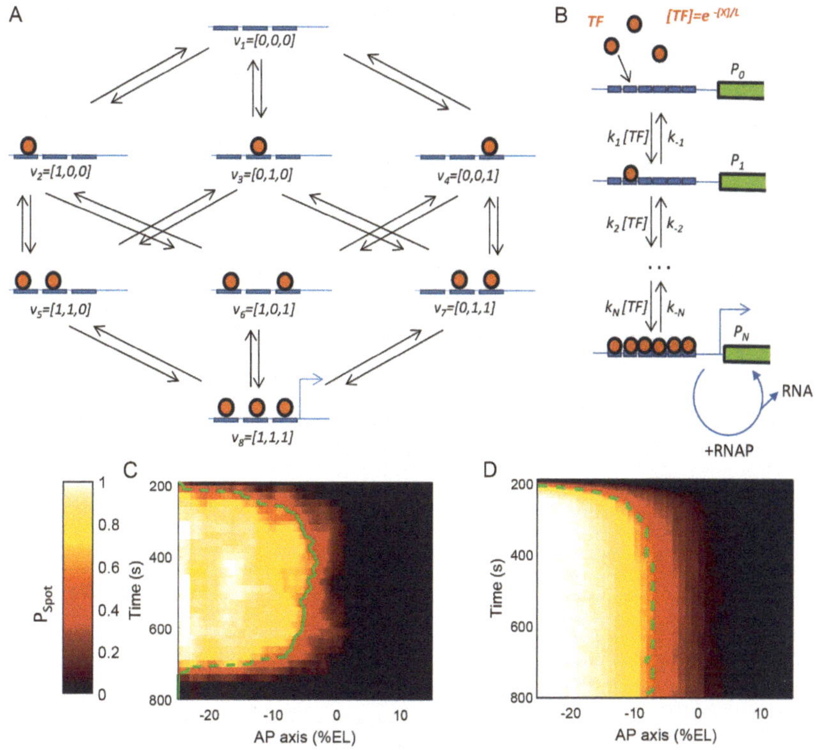

Fig. 4 Modeling transcriptional regulation by the Bcd transcription factor through interactions with binding sites on *hb* promoter. (A) A general model of transcription factors (TF) (orange) binding/unbinding to *N* binding sites of target promoter ($N=3$). Each node *v* corresponds to a unique ordered TF-bound promoter state. (B) A simplified promoter binding model assuming detailed balance to account for energy expenditure in the unbinding process (Estrada et al., 2016), coupled with transcription initiation. When many Bcd binding sites on the *hb* promoter are occupied, RNAP can randomly bind to the promoter and initiate transcription. (C and D) The probability of an active transcription locus (P_{Spot}, color bar) for the *hb* locus as a function of time in the nuclear cycle and position along the AP axis obtained from the MS2-MCP data (C, nc13) and from the model in B (D, with $N=6$, not accounting for mitosis at the end of interphase). *Panel (A) figure adapted from Estrada, J., Wong, F., DePace, A., & Gunawardena, J. (2016). Information integration and energy expenditure in gene regulation. Cell, 166(1), 234–244 and panels (B–D) figure are reused from Lucas, T., Tran, H., Romero, C. A. P., Guillou, A., Fradin, C., Coppey, M., et al. (2018). 3 minutes to precisely measure morphogen concentration. PLoS Genetics, 14(10), e1007676.*

Bcd binding sites and not their identity matters. In this case, *H* cannot be above *N*. Experimental *H* values obtained when observing the protein level (Gregor, Tank, et al., 2007) and several features of the *hb* transcription dynamics (e.g., total amount of RNA produced; Garcia et al., 2013;

Lucas et al., 2018) range from ~5 to ~7, roughly equal to the number of known Bcd binding sites on *hb* promoter. This leads to assumptions that the six binding sites, with an unstable first Bcd-bound state (large k_{-1} in Fig. 4B) and a stable fully bound state (small k_{-N} in Fig. 4B), are sufficient to explain the observed pattern steepness in static measurements.

However, Estrada et al. did not consider the search time issue (Estrada et al., 2016) and their theory cannot explain how the *hb* pattern can be established in such a short time of 3 min following mitosis, as observed in live imaging data (Lucas et al., 2018) (Fig. 4C). Considering a model which accounts for the Bcd search time for the *hb* promoter, Tran et al. (2018) found that, at the mid-boundary position, very high pattern steepness ($H \approx N$) requires a very slow promoter switching time (called τ_{active}) between Bcd-bound states allowing transcription and Bcd-free states prohibiting transcription. Therefore, according to the model, it should take a very long time for the pattern to be established and this is in contradiction with the experimental data from the MS2 system (Lucas et al., 2018). In addition, slow promoter dynamics results in high nuclei-to-nuclei variability in the amount of total RNA produced in each nuclear cycle. Thus, it would require even more averaging time to achieve the robust protein pattern (10% variability) observed in nuclear cycle 14 (Gregor, Tank, et al., 2007).

The failure of the simple model to explain both the observed high pattern steepness and fast establishment time begs for the reconsideration of the model's assumptions. Most obvious candidates are either the underestimation of the number of Bicoid binding sites N or overestimation of the time it takes for Bcd to find the promoter τ_{search}. Estrada et al. suggested that energy expenditure, which removes the detailed balance assumption from the binding and unbinding process, can expand the model's limit on pattern steepness H beyond the binding site number N, allowing both high steepness and fast formation time at the same time (Estrada et al., 2016). However, including non-equilibrium binding does not resolve the problem of obtaining a steep yet precise boundary in a short time. Alternatively, Desponds, Vergassola, and Walczak (2019) suggested that instead of probing the concentration for a fixed amount of time and then making the decision about the positioning of the nucleus, constantly updating the odds of being in an anterior *vs* posterior position always results in much faster decisions for a fixed accuracy. Assuming a promoter with 6 Bcd binding sites, they showed that the decision time can be reduced by an order of magnitude compared to the classical Berg–Purcell scheme, possibly below the 3 min limit. Unlike the classical time averaging (Berg–Purcell) scheme, which theoretically can reach 100% accuracy given enough sensing time, the target gene "commits" to expression or silence when the odds favoring the anterior or posterior reach an acceptable confidence, even before the end of the exposure window. Possible molecular mechanisms for how this decision scheme could be implemented in fly embryos are still needed.

3.4 Dissecting noise in *hb* transcription

Noise in transcription dictates the variability of transcript and protein readouts after each interphase and might play a role in determining nuclei identity in downstream processes (Holloway et al., 2011). However, beyond its characterization from observed data, we still lack the mechanistic understanding of processes responsible for this noise.

As transcription bursts are prevalent across the embryo in the very short early nuclear cycles, *hb* transcription dynamics is well-fitted by a two-state model, in which the switching rates between the ON and OFF states are modulated by the nuclei's position or Bcd concentration (Desponds et al., 2016; Xu et al., 2015; Xu, Skinner, Sokac, & Golding, 2016; Zoller et al., 2018). It should be noted that these ON and OFF states do not correspond to Bcd-free and Bcd-bound states of the promoter as in (Estrada et al., 2016; Tran et al., 2018): in the Bcd-saturating anterior region, the *hb* promoter is constantly active (i.e., bound by Bcd molecules) but transcription still occurs in bursts with the switching time between ON and OFF states ~50 s (Desponds et al., 2016). This suggests that promoter bursting may be an inherent property of transcription in this phase of development (Bothma et al., 2014; Zoller et al., 2018). The early transcription of *hb* is also regulated by other TFs such as maternal Hb (Lopes, Spirov, & Bisch, 2011; Porcher et al., 2010; Simpson-Brose et al., 1994) or Zelda (Harrison, Li, Kaplan, Botchan, & Eisen, 2011; Lucas et al., 2018; Nien et al., 2011; Xu et al., 2014). Optogenetics was also used to inactivate the transcription factor Zelda in early embryos and reveals that this pioneer factor continuously regulates zygotic gene expression from nc10 to nc14 (McDaniel et al., 2019). Though these factors other than Bcd may not act as a source of positional information, their concentration may be rate-limiting and therefore responsible for bursts.

In the *hb* boundary region, where cell fate decision is critical, *hb* transcription readout is more variable than in the anterior region (Desponds et al., 2016; Lucas et al., 2018). This was initially thought to be due to extrinsic noise from Bcd variability (Gregor, Wieschaus, et al., 2007) being amplified in this region. However, given the very high steepness observed from the *hb* pattern (Lucas et al., 2018; Xu et al., 2015), the switching time between Bcd-dependent active and inactive states of the promoter (τ_{active}) is expected to be at least one order of magnitude greater than the Bcd search time τ_{search} (Tran et al., 2018). If the search is done *via* 3D diffusion ($\tau_{search} \sim 10s$), τ_{active} at *hb* boundary is at least of a similar time scale as the switching time between ON and OFF states at the anterior (~50 s). In the context of very rapid embryo development (interphase duration of 5–15 min in nc11–nc13), Bcd-dependent promoter switching becomes a non-negligible source of intrinsic noise that contributes substantially to the higher readout variability observed.

4. Perspectives

Despite several decades since the identification of the Bcd gradient, we still lack a quantitative description of the process allowing for the transcription of its main target gene *hb* in a step-like pattern. The short timescales of early development and its remarkable precision are questioning how fundamental limits coming from stochastic processes such as diffusion or bursty regulation influence the molecular encoding of regulation. To pursue these issues, recent experimental advances are allowing us to rigorously test theoretical ideas, and call for the creation of new models.

This simple example of developmental biology, is turning out to also be an ideal *in vivo* testbed for advances in single molecule techniques to study protein motility (Drocco et al., 2011; Mir et al., 2018) and promises to bring a more definitive view on how TF can find their target promoter and activate transcription. The different motilities of TF (Abu-Arish et al., 2010; Mir et al., 2018) and their inhomogeneous distribution in those so-called hubs (Mir et al., 2017) as seen with the Bcd case support the idea that the search time can be improved by 1D-diffusion along the DNA or via micro-environments of enriched TF concentration. However, with fluorescent-tagged molecules alone, it remains difficult to identify *in vivo* the location of the binding molecule along the chromosome, with enough resolution to detect 1D-diffusion or local enrichment. Conversely, such epigenomics tools as ChIP-seq give information on TF binding along the chromosome, but lack the temporal dynamics of these interactions. With advances in sample preparation and super-resolution microscopy, it might be possible to combine these two types of experiments to understand how TF searches for their specific sites. Another important question is the mechanisms of TF binding "cooperativity" after the search. In the Bcd case, binding cooperativity has been suggested in theoretical work to explain *hb*'s high pattern steepness at steady-state (Estrada et al., 2016; Gregor, Tank, et al., 2007; Lopes et al., 2011; Tran et al., 2018). While high steepness is observed experimentally (corresponding to a Hill coefficient ~7 to 8) (Lucas et al., 2018), *in vivo* measurements of Bcd binding cooperativity of such high order at the gene loci remains a challenge (Mir et al., 2018; Xu et al., 2015). *In vitro* studies showed that Bcd binding cooperativity is of low order (Hill coefficient ~2 to 3) and robust to different target sites' spacing (Ma et al., 1996), suggesting that it may originate from protein-protein interactions rather than from transcription factor-induced binding site conformational changes (Bray & Duke, 2004). More direct experimental evidence is required to fully elucidate high-order cooperativity observed in the expression pattern, possibly coming from different transcription factors.

Advances in live imaging of transcription offered a direct view on *hb* transcription pattern establishment in early nuclear cycles: the pattern is established in a considerable portion of the very short interphase and, in contrast to the low variability observed at the protein level, transcription is highly noisy, with prevalent bursty behavior across the embryo. This marks a shift of our approaches toward the understanding of transcription regulation during development from previously static and usually steady-state views to more theoretical and quantitative studies focusing on the out-of-steady-state dynamics of gene expression (Petkova et al., 2019; Tran et al., 2018). The question of TF searching for its targets remains of great interests to explain the rapidity of the *hb* pattern establishment despite a noisy transcription process. Transcription activation following TF binding, even though intensively studied for decades, also remains a mystery. In the context of development, it will be important to understand how the measurement process of positional information manages to combine the rapidity of the response with accuracy (Tran et al., 2018). These observations lead to suggestions that transcription activation may be non-reversible or involve transcriptional memory or mitotic bookmarking (Desponds et al., 2019; Ferraro, Esposito et al., 2016; Zhao, Nakamura, Fu, Lazar, & Spector, 2011). The development of powerful optogenetic tools with high spatial and temporal resolution might shed light on this mechanism, as it is now possible to manipulate TF properties and observe the effects of these manipulations in real time under the microscope (Huang et al., 2017; McDaniel et al., 2019).

Finally, *hb* protein expression in nc14 is still remarkably robust, suggesting that temporal and spatial averaging downstream of noisy transcription is at work. Recent works have shown that positional information can be accurately decoded at the level of the gap genes (Petkova et al., 2019). However, decoding as well as encoding mechanisms in the earlier cell cycles remain unknown and the current experimental and theoretical methods are ready to tackle these questions. At last, it is likely that Bcd may no longer play a major role in maintaining *hb* pattern during the very long nuclear cycle 14 (Durrieu et al., 2018; Perry, Boettiger, & Levine, 2011; Perry et al., 2012). Given this, an intriguing aspect of this system is why it has been selected to provide such fast step-like pattern dynamics in the earlier nuclear cycles.

Acknowledgments

We thank M. Andrieu, M. Coppey, G. Fernandes, C. Fradin, C. Perez-Romero, A. Ramaekers for stimulating discussions. Work in the Walczak and Dostatni labs is supported by PSL IDEX REFLEX Grant for Mesoscopic Biology (AMW & ND), ANR-11-BSV2-0024 Axomorph (AMW & ND), ARC PJA20151203341 (ND) and ANR-11-LABX-0044 DEEP Labex (ND). H.T. was supported by the PSL IDEX REFLEX Grant, the Institut Curie and the DEEP Labex.

References

Abu-Arish, A., Porcher, A., Czerwonka, A., Dostatni, N., & Fradin, C. (2010). High mobility of bicoid captured by fluorescence correlation spectroscopy: Implication for the rapid establishment of its gradient. *Biophysical Journal, 99*(4), 33–35.

Berg, H. C., & Purcell, E. M. (1977). Physics of chemoreception. *Biophysical Journal, 20*(2), 193–219.

Berrocal, A., Lammers, N. C., Garcia, H. G., & Eisen, M. B. (2018). Kinetic sculpting of the seven stripes of the Drosophila even-skipped gene. *BioRxiv*, 335901.

Bertrand, E., Chartrand, P., Schaefer, M., Shenoy, S. M., Singer, R. H., & Long, R. M. (1998). Localization of ASH1 mRNA particles in living yeast. *Molecular Cell, 2*(4), 437–445.

Bothma, J. P., Garcia, H. G., Esposito, E., Schlissel, G., Gregor, T., & Levine, M. (2014). Dynamic regulation of eve stripe 2 expression reveals transcriptional bursts in living Drosophila embryos. *Proceedings of the National Academy of Sciences of the United States of America, 111*(29), 10598–10603.

Bradley, R. K., Li, X. Y., Trapnell, C., Davidson, S., Pachter, L., Chu, H. C., et al. (2010). Binding site turnover produces pervasive quantitative changes in transcription factor binding between closely related drosophila species. *PLoS Biology, 8*(3), e1000343.

Bray, D., & Duke, T. (2004). Conformational spread: The propagation of allosteric states in large multiprotein complexes. *Annual Review of Biophysics and Biomolecular Structure, 33*, 53–73.

Burz, D. S., Rivera-Pomar, R., Jäckle, H., & Hanes, S. D. (1998). Cooperative DNA-binding by Bicoid provides a mechanism for threshold-dependent gene activation in the Drosophila embryo. *EMBO Journal, 17*(20), 5998–6009.

Combs, P. A., & Eisen, M. B. (2013). Sequencing mRNA from Cryo-sliced Drosophila embryos to determine genome-wide spatial patterns of gene expression. *PLoS One, 8*(8), 2–8.

Corrigan, A. M., Tunnacliffe, E., Cannon, D., & Jonathan, R. C. (2016). A continuum model of transcriptional bursting. *eLife*, *5*, e13051.
Coulon, A., Ferguson, M. L., de Turris, V., Palangat, M., Chow, C. C., & Larson, D. R. (2014). Kinetic competition during the transcription cycle results in stochastic RNA processing. *eLife*, *3*, 1–22.
Crauk, O., & Dostatni, N. (2005). Bicoid determines sharp and precise target gene expression in the Drosophila embryo. *Current Biology*, *15*(21), 1888–1898.
Desponds, J., Tran, H., Ferraro, T., Lucas, T., Perez Romero, C., Guillou, A., et al. (2016). Precision of readout at the hunchback gene: Analyzing short transcription time traces in living fly embryos. *PLoS Computational Biology*, *12*(12), e1005256.
Desponds, J., Vergassola, M., & Walczak, A. M. (2019). hunchback promoters can readout morphogenetic positional information in less than a minute. *BioRxiv*, 676684.
Driever, W., & Nusslein-Volhard, C. (1988). The Bicoid protein determines position in the Drosophila embryo in a concentration-dependent manner. *Cell*, *54*(1), 95–104.
Driever, W., Thoma, G., & Nüsslein-Volhard, C. (1989). Determination of spatial domains of zygotic gene expression in the Drosophila embryo by the affinity of binding sites for the bicoid morphogen. *Nature*, *340*(6232), 363–367.
Drocco, J. A., Grimm, O., Tank, D. W., & Wieschaus, E. (2011). Measurement and perturbation of morphogen lifetime: Effects on gradient shape. *Biophysical Journal*, *101*(8), 1807–1815.
Dubuis, J. O., Tkacik, G., Wieschaus, E. F., Gregor, T., & Bialek, W. (2013). Positional information, in bits. *Proceedings of the National Academy of Sciences of the United States of America*, *110*(41), 16301–16308.
Durrieu, L., Kirrmaier, D., Schneidt, T., Kats, I., Raghavan, S., Hufnagel, L., et al. (2018). Bicoid gradient formation mechanism and dynamics revealed by protein lifetime analysis. *Molecular Systems Biology*, *14*(9), e8355.
Elf, J., Li, G.-W., & Xie, X. S. (2007). Probing transcription factor dynamics at the single-molecule level in a living cell. *Science*, *316*(5828), 1191–1194.
Estrada, J., Wong, F., DePace, A., & Gunawardena, J. (2016). Information integration and energy expenditure in gene regulation. *Cell*, *166*(1), 234–244.
Ferraro, T., Esposito, E., Mancini, L., Ng, S., Lucas, T., Coppey, M., et al. (2016). Transcriptional memory in the Drosophila embryo. *Current Biology*, *26*, 1–7.
Ferraro, T., Lucas, T., Clémot, M., De Las Heras Chanes, J., Desponds, J., Coppey, M., et al. (2016). New methods to image transcription in living fly embryos: The insights so far, and the prospects. *Wiley Interdisciplinary Reviews: Developmental Biology*, *5*(3), 296–310.
Fukaya, T., Lim, B., & Levine, M. (2016). Enhancer control of transcriptional bursting. *Cell*, *166*, 1–11.
Fukaya, T., Lim, B., & Levine, M. (2017). Rapid rates of pol II elongation in the Drosophila embryo. *Current Biology*, *27*(9), 1387–1391.
Garcia, H. G., Tikhonov, M., Lin, A., & Gregor, T. (2013). Quantitative imaging of transcription in living Drosophila embryos links polymerase activity to patterning. *Current Biology*, *23*(21), 2140–2145.
Gregor, T., Bialek, W., de Ruyter van Steveninck, R. R., Tank, D. W., & Wieschaus, E. (2005). Diffusion and scaling during early embryonic pattern formation. *Proceedings of the National Academy of Sciences of the United States of America*, *102*(51), 18403–18407.
Gregor, T., Garcia, H. G., & Little, S. C. (2014). The embryo as a laboratory: Quantifying transcription in Drosophila. *Trends in Genetics*, *30*(8), 1–12.
Gregor, T., Tank, D. W., Wieschaus, E., & Bialek, W. (2007). Probing the limits to positional information. *Cell*, *130*(1), 153–164.
Gregor, T., Wieschaus, E., McGregor, A. P., Bialek, W., & Tank, D. W. (2007). Stability and nuclear dynamics of the Bicoid morphogen gradient. *Cell*, *130*(1), 141–152.

Grimm, O., Coppey, M., & Wieschaus, E. (2010). Modelling the Bicoid gradient. *Development, 137*(14), 2253–2264.

Hammar, P., Leroy, P., Mahmutovic, A., Marklund, E. G., Berg, O. G., & Elf, J. (2012). The lac repressor displays facilitated diffusion in living cells. *Science, 336*(6088), 1595–1598.

Hanes, S. D., & Brent, R. (1989). DNA specificity of the bicoid activator protein is determined by homeodomain recognition helix residue 9. *Cell, 57*(7), 1275–1283.

Harrison, M. M., Li, X. Y., Kaplan, T., Botchan, M. R., & Eisen, M. B. (2011). Zelda binding in the early Drosophila melanogaster embryo marks regions subsequently activated at the maternal-to-zygotic transition. *PLoS Genetics*, 7(10), e1002266.

Holloway, D. M., Lopes, F. J. P., da Fontoura Costa, L., Travençolo, B. A. N., Golyandina, N., Usevich, K., et al. (2011). Gene expression noise in spatial patterning: Hunchback promoter structure affects noise amplitude and distribution in Drosophila segmentation. *PLoS Computational Biology*, 7(2), e1001069.

Houchmandzadeh, B., Wieschaus, E., & Leibler, S. (2002). Establishment of developmental precision and proportions in the early Drosophila embryo. *Nature, 415*(6873), 798–802.

Huang, A., Amourda, C., Zhang, S., Tolwinski, N. S., & Saunders, T. E. (2017). Decoding temporal interpretation of the morphogen Bicoid in the early Drosophila embryo. *eLife, 6*, 1–21.

Huang, A., & Saunders, T. E. (2020). A matter of time: Formation and interpretation of the Bicoid morphogen gradient. *Current Topics in Developmental Biology, 137*, 79–117.

Janody, F., Sturny, R., Schaeffer, V., Azou, Y., & Dostatni, N. (2001). Two distinct domains of Bicoid mediate its transcriptional downregulation by the torso pathway. *Development, 128*(12), 2281–2290.

Karaiskos, N., Wahle, P., Alles, J., Boltengagen, A., Ayoub, S., Kipar, C., et al. (2017). The Drosophila embryo at single-cell transcriptome resolution. *Science, 358*(6360), 194–199.

Lammers, N. C., Galstyan, V., Reimer, A., Medin, S. A., Wiggins, C. H., & Garcia, H. G. (2019). Multimodal transcriptional control of pattern formation in embryonic development. *BioRxiv*, 335919.

Li, X. Y., Thomas, S., Sabo, P. J., Eisen, M. B., Stamatoyannopoulos, J. A., & Biggin, M. D. (2011). The role of chromatin accessibility in directing the widespread, overlapping patterns of Drosophila transcription factor binding. *Genome Biology, 12*(4), R34.

Little, S. C., Tikhonov, M., & Gregor, T. (2013). Precise developmental gene expression arises from globally stochastic transcriptional activity. *Cell, 154*(4), 789–800.

Liu, X., Long, F., Peng, H., Aerni, S. J., Jiang, M., Sánchez-Blanco, A., et al. (2009). Analysis of cell fate from single-cell gene expression profiles in C. elegans. *Cell, 139*(3), 623–633.

Liu, F., Morrison, A. H., & Gregor, T. (2013). Dynamic interpretation of maternal inputs by the Drosophila segmentation gene network. *Proceedings of the National Academy of Sciences of the United States of America, 110*(17), 6724–6729.

Lopes, F. J., Spirov, A. V., & Bisch, P. M. (2011). The role of Bicoid cooperative binding in the patterning of sharp borders in Drosophila melanogaster. *Developmental Biology*, 72(2), 181–204.

Lucas, T., Ferraro, T., Roelens, B., Chanes, J. D. L. H., Walczak, A. M., Coppey, M., et al. (2013). Live imaging of Bicoid-dependent transcription in Drosophila embryos. *Current Biology, 23*(21), 2135–2139.

Lucas, T., Tran, H., Romero, C. A. P., Guillou, A., Fradin, C., Coppey, M., et al. (2018). 3 minutes to precisely measure morphogen concentration. *PLoS Genetics, 14*(10), e1007676.

Ma, X., Yuan, D., Diepold, K., Scarborough, T., & Ma, J. (1996). The Drosophila morphogenetic protein Bicoid binds DNA cooperatively. *Development (Cambridge, England), 122*(4), 1195–1206.

McDaniel, S. L., Gibson, T. J., Schulz, K. N., Fernandez Garcia, M., Nevil, M., Jain, S. U., et al. (2019). Continued activity of the pioneer factor Zelda is required to drive zygotic genome activation. *Molecular Cell, 74*(1), 185–195.e4.

Mir, M., Reimer, A., Haines, J. E., Li, X. Y., Stadler, M., Garcia, H., et al. (2017). Dense bicoid hubs accentuate binding along the morphogen gradient. *Genes and Development, 31*(17), 1784–1794.

Mir, M., Stadler, M. R., Harrison, M. M., Darzacq, X., & Eisen, M. B. (2018). Dynamic multifactor hubs interact transiently with sites of active transcription in Drosophila embryos. *eLife*, 7, e40497.

Mirny, L., Slutsky, M., Wunderlich, Z., Tafvizi, A., Leith, J., & Kosmrlj, A. (2009). How a protein searches for its site on DNA: The mechanism of facilitated diffusion. *Journal of Physics A: Mathematical and Theoretical, 42*(43), 434013.

Nien, C. Y., Liang, H. L., Butcher, S., Sun, Y., Fu, S., Gocha, T., et al. (2011). Temporal coordination of gene networks by Zelda in the early Drosophila embryo. *PLoS Genetics*, 7(10), e1002339.

Normanno, D., Dahan, M., & Darzacq, X. (2012). Intra-nuclear mobility and target search mechanisms of transcription factors: A single-molecule perspective on gene expression. *Biochimica et Biophysica Acta - Gene Regulatory Mechanisms, 1819*(6), 482–493.

Perry, M. W., Boettiger, A. N., & Levine, M. (2011). Multiple enhancers ensure precision of gap gene-expression patterns in the Drosophila embryo. *Proceedings of the National Academy of Sciences of the United States of America, 108*(33), 1–12.

Perry, M. W., Bothma, J. P., Luu, R. D., & Levine, M. (2012). Precision of hunchback expression in the Drosophila embryo. *Current Biology, 22*(23), 2247–2252.

Petkova, M. D., Bialek, W., Wieschaus, E. F., & Gregor, T. (2019). Optimal decoding of cellular identities in a genetic network. *Cell, 176*(4), 844–855.

Porcher, A., Abu-Arish, A., Huart, S., Roelens, B., Fradin, C., & Dostatni, N. (2010). The time to measure positional information: Maternal hunchback is required for the synchrony of the Bicoid transcriptional response at the onset of zygotic transcription. *Development, 137*(16), 2795–2804.

Riggs, A. D., Bourgeois, S., & Cohn, M. (1970). The lac repressor-operator interaction. 3. Kinetic studies. *Journal of Molecular Biology, 53*(3), 401–417.

Ronchi, E., Treisman, J., Dostatni, N., Struhl, G., & Desplan, C. (1993). Down-regulation of the Drosophila morphogen bicoid by the torso receptor-mediated signal transduction cascade. *Cell, 74*(2), 347–355.

Sauer, F., Hansen, S. K., & Tjian, R. (1995). DNA template and activator-coactivator requirements for transcriptional synergism by Drosophila Bicoid. *Science, 270*(5243), 1825–1828.

Schaeffer, V., Janody, F., Loss, C., Desplan, C., & Wimmer, E. A. (1999). Bicoid functions without its TATA-binding protein-associated factor interaction domains. *Proceedings of the National Academy of Sciences of the United States of America, 96*(8), 4461–4466.

Shao, W., & Zeitlinger, J. (2017). Paused RNA polymerase II inhibits new transcriptional initiation. *Nature Genetics, 49*(7), 1045–1051.

Simpson-Brose, M., Treisman, J., & Desplan, C. (1994). Synergy between the hunchback and bicoid morphogens is required for anterior patterning in Drosophila. *Cell, 78*(5), 855–865.

Slutsky, M., & Mirny, L. A. (2004). Kinetics of protein-DNA interaction: Facilitated target location in sequence-dependent potential. *Biophysical Journal, 87*(6), 4021–4035.

Tran, H., Desponds, J., Romero, C. A. P., Coppey, M., Fradin, C., Dostatni, N., et al. (2018). Precision in a rush: Trade-offs between reproducibility and steepness of the hunchback expression pattern. *PLoS Computational Biology, 14*(10), e1006513.

Trelsman, J., Gönczy, P., Vashishtha, M., Harris, E., & Desplan, C. (1989). A single amino acid can determine the DNA binding specificity of homeodomain proteins. *Cell*, *59*(3), 553–562.

Tsai, A., Muthusamy, A. K., Alves, M. R. P., Lavis, L. D., Singer, R. H., Stern, D. L., et al. (2017). Nuclear microenvironments modulate transcription from low-affinity enhancers. *eLife*, *6*, 1–18.

Xu, Z., Chen, H., Ling, J., Yu, D., Struffi, P., & Small, S. (2014). Impacts of the ubiquitous factor Zelda on Bicoid-dependent DNA binding and transcription in Drosophila. *Genes and Development*, *28*(6), 608–621.

Xu, H., Sepúlveda, L. A., Figard, L., Sokac, A. M., & Golding, I. (2015). Combining protein and mRNA quantification to decipher transcriptional regulation. *Nature Methods*, *12*(8), 739–742.

Xu, H., Skinner, S. O., Sokac, A. M., & Golding, I. (2016). Stochastic kinetics of nascent RNA. *Physical Review Letters*, *117*(12), 1–6.

Zhao, R., Nakamura, T., Fu, Y., Lazar, Z., & Spector, D. L. (2011). Gene bookmarking accelerates the kinetics of post-mitotic transcriptional re-activation. *Nature Cell Biology*, *13*(11), 1295–1304.

Zhao, C., York, A., Yang, F., Forsthoefel, D. J., Dave, V., Fu, D., et al. (2002). The activity of the Drosophila morphogenetic protein Bicoid is inhibited by a domain located outside its homeodomain. *Development*, *129*(7), 1669–1680.

Zoller, B., Little, S. C., & Gregor, T. (2018). Diverse spatial expression patterns emerge from unified kinetics of transcriptional bursting. *Cell*, *175*(3), 835–847.e25.

Further reading

Mirny, L. A. (2010). Nucleosome-mediated cooperativity between transcription factors. *Proceedings of the National Academy of Sciences of the United States of America*, *107*(52), 22534–22539.

CHAPTER FIVE

Formation, interpretation, and regulation of the *Drosophila* Dorsal/NF-κB gradient

Allison E. Schloop[a,†], Prasad U. Bandodkar[b,†], Gregory T. Reeves[a,b,*]
[a]Genetics Program, North Carolina State University, Raleigh, NC, United States
[b]Department of Chemical and Biomolecular Engineering, North Carolina State University, Raleigh, NC, United States
[*]Corresponding author: e-mail address: gtreeves@ncsu.edu

Contents

[†] Both authors are contributed equally.

Current Topics in Developmental Biology, Volume 137
ISSN 0070-2153
https://doi.org/10.1016/bs.ctdb.2019.11.007

Abstract

The morphogen gradient of the transcription factor Dorsal in the early *Drosophila* embryo has become one of the most widely studied tissue patterning systems. Dorsal is a *Drosophila* homolog of mammalian NF-κB and patterns the dorsal-ventral axis of the blastoderm embryo into several tissue types by spatially regulating upwards of 100 zygotic genes. Recent studies using fluorescence microscopy and live imaging have quantified the Dorsal gradient and its target genes, which has paved the way for mechanistic modeling of the gradient. In this review, we describe the mechanisms behind the initiation of the Dorsal gradient and its regulation of target genes. The main focus of the review is a discussion of quantitative and computational studies of the Dl gradient system, including regulation of the Dl gradient. We conclude with a discussion of potential future directions.

1. Introduction

The Dorsal network has been one of the best-studied tissue patterning systems since its discovery four decades ago (Anderson, Jürgens, & Nüsslein-Volhard, 1985; Anderson & Nüsslein-Volhard, 1984; Nusslein-Volhard, 1979; Nüsslein-Volhard, Lohs-Schardin, Sander, & Cremer, 1980; Roth, Stein, & Nüsslein-Volhard, 1989; Rushlow, Han, Manley, & Levine, 1989; Steward, 1989; Steward, Ambrosio, & Schedl, 1985; Steward, McNally, & Schedl, 1984; Steward, Zusman, Huang, & Schedl, 1988). It is a textbook example of a morphogen system, and is the subject of periodic reviews (see, for example, Anderson, 1987; Belvin & Anderson, 1996; Morisato & Anderson, 1995; Moussian & Roth, 2005; Reeves & Stathopoulos, 2009; Roth, 2003; Rushlow & Shvartsman, 2012; Sandler & Stathopoulos, 2016a; Stathopoulos & Levine, 2005; Stathopoulos, Van Drenth, Erives, Markstein, & Levine, 2002; Stein & Stevens, 2014; Steward & Govind, 1993). Dorsal (Dl) is one of three *Drosophila* homologs of mammalian NF-κB (Ghosh et al., 1990; Kieran et al., 1990; Sen & Baltimore, 1986; Steward, 1987), and a large fraction of the pathway structure and components are also well conserved, including the inhibitor Cactus, the receptor Toll, the adaptor protein Myd88, and the kinase Pelle (Govind, 1999; Hetru & Hoffmann, 2009).

Mammalian NF-κB functions in multiple cellular responses, including cancer, inflammation, apoptosis, proliferation, and innate immunity (Dev, Iyer, Razani, & Cheng, 2010; Hoesel & Schmid, 2013; Liu, Zhang, Joo, & Sun, 2017; Tornatore, Thotakura, Bennett, Moretti, & Franzoso, 2012; Xia, Shen, & Verma, 2014). In particular, the role of Dl—and other NF-κB homologs Dif and Relish (Dushay, Asling, & Hultmark, 1996; Ip et al., 1993)—in the *Drosophila* innate immune system is functionally mirrored in mammals (Govind, 1999; Hetru & Hoffmann, 2009; Lemaitre et al., 1995). The conservation of the Dl/NF-κB network enables Dl gradient studies to impact developmental biology, innate immunity, and myriad other cellular responses in a tractable model system that allows for repeatable, simultaneous observation of multiple levels of NF-κB activity in vivo (Belvin & Anderson, 1996; Drier & Steward, 1997; Govind, 1999; Hetru & Hoffmann, 2009; Minakhina & Steward, 2006).

The Dl nuclear gradient patterns the dorsal-ventral (DV) axis of the *Drosophila* blastoderm-stage embryo (roughly 1.5–3 h after fertilization; Fig. 1A). During this time frame, four nuclear divisions take place (nuclear cycles 10–13), and after the 13th round of divisions, nuclear cycle (nc) 14 interphase lasts roughly 45 min (Foe & Alberts, 1983). Dl protein is ubiquitously expressed and is translated in the early embryo from maternally deposited *dl* mRNA, which is loaded into the egg during oogenesis (Roth et al., 1989; Steward et al., 1985, 1988). Dl is retained in the cytoplasm through binding to the *Drosophila* IκB homolog Cactus (Cact), and thus cannot enter the nuclei to regulate its target genes (Belvin, Jin, & Anderson, 1995; Bergmann et al., 1996; Kidd, 1992; Reach et al., 1996; Roth, Hiromi, Godt, & Nüsslein-Volhard, 1991; Whalen & Steward, 1993). Signaling through the Toll receptor, the *Drosophila* homolog of the interleukin 1 receptor (Schneider, Hudson, Lin, & Anderson, 1991; Whitham et al., 1994), results in the dissociation of the Dl/Cact complex, freeing Dl to enter the nuclei (Fig. 1B; Hashimoto, Gerttula, & Anderson, 1991; Schneider et al., 1991; Stein, Roth, Vogelsang, & Nüsslein-Volhard, 1991; Whalen & Steward, 1993). Because Toll signaling is spatially asymmetric, Dl is present in high nuclear concentrations in the ventral regions, has a graded nuclear localization along the lateral regions of the embryo, and reaches non-zero basal levels on the dorsal side (see Fig. 1C; Roth et al., 1989; Rushlow et al., 1989; Steward et al., 1988).

The Dl gradient regulates upwards of 100 genes (Chopra & Levine, 2009; Reeves & Stathopoulos, 2009; Stathopoulos et al., 2002). The number of identified targets of Dl has increased exponentially over time as more sophisticated technologies have been put into use

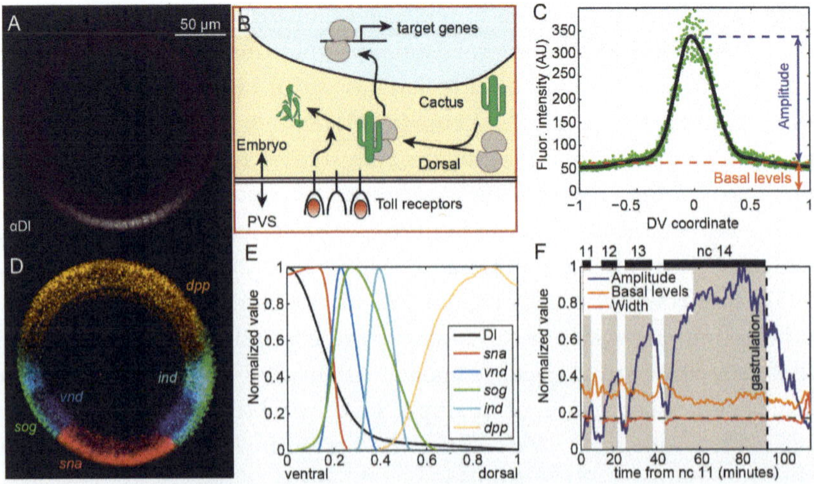

Fig. 1 Overview of the Dl gradient system. (A) Cross section of nc 14 embryo fluorescently immunostained against Dl. (B) Schematic of Toll/Dl/Cact interactions that occur on the ventral side of the embryo. (C) Example of quantification of a Dl gradient from a nc 14 embryo. Gradient amplitude and basal levels are defined on the graph. (D) Cross section of nc 14 embryo that has been hybridized with probes against DV target genes. (E) Quantification of DV target gene expression during nc 14. Each gene expression curve is the average of greater than 10 embryos. (F) Quantification of the dynamics of the Dl gradient. *Panels (A, D): Reproduced/modified with permission from Reeves, G. T., & Stathopoulos, A. (2009). Graded dorsal and differential gene regulation in the Drosophila embryo.* Cold Spring Harbor Perspectives in Biology, 1, *a000836. Panels (E, F): Reproduced/modified with permission from Reeves, G. T., Trisnadi, N., Truong, T. V, Nahmad, M., Katz, S., & Stathopoulos, A. (2012). Dorsal-ventral gene expression in the Drosophila embryo reflects the dynamics and precision of the dorsal nuclear gradient.* Developmental Cell, 22, *544–557.*

(Biemar et al., 2006; Ferguson & Anderson, 1991; Sandmann et al., 2007; Zeitlinger et al., 2007). As the Dl gradient extends just less than halfway around the embryo (Liberman, Reeves, & Stathopoulos, 2009; Reeves et al., 2012), the majority of patterning takes place in the steeply-sloped portion of the gradient (between ~20% and ~45% DV position, where 0% is the ventral midline) (Fig. 1D, E; see also Section 3.1; Reeves & Stathopoulos, 2009; Stathopoulos & Levine, 2005).

Fluorescent measurements in both live and fixed embryos have allowed for the quantification of the Dl gradient and its target genes. Initial quantification efforts found the Dl gradient empirically displays a Gaussian-like shape, with a smooth peak near the ventral midline, and a slowly declining tail past ~40% DV position (Fig. 1C; Liberman et al., 2009; Reeves et al., 2012). Nuclear Dl levels were measured in live embryos

(DeLotto, DeLotto, Steward, & Lippincott-Schwartz, 2007; Reeves et al., 2012), and it was found that both the gradient amplitude as well as the basal levels appear to oscillate in a "saw-tooth" pattern that is connected with the nuclear cycles (Fig. 1C, F). Overlaid on top of these oscillations is a continuously-growing gradient amplitude (see Section 5.2). In contrast, the gradient width remains constant (Fig. 1F).

Subsequent to the discovery of the salient biophysical features of the Dl/Cact/Toll system, formulation of mechanistic models of the Dl gradient has become possible (see Section 6; Kanodia et al., 2009). Modeling work has confirmed the saw-tooth pattern of the gradient amplitude and has made several predictions, including the presence of Cact in the nucleus with Dl and the facilitated diffusion of Dl by Cact (Carrell et al., 2017; O'Connell & Reeves, 2015). Dl-dependent gene expression has also been modeled (see Section 7; Kanodia et al., 2012; O'Connell & Reeves, 2015; Reeves et al., 2012; Zinzen, Senger, Levine, & Papatsenko, 2006).

This review is structured to be modular, so that the sections and subsections are roughly independent of each other, with cross-referencing to aid in the readers' finding further information if needed. We begin with descriptions of the components and targets of the Dl network, both in oogenesis and embryogenesis (Sections 2 and 3). In Section 4, we review the regulation of the Dl gradient. We continue with what we consider to be the main contribution of the review: a discussion of the quantitative and computational studies of the Dl gradient (Sections 5–7). These studies have largely taken place following a pioneering study in 2007 that focused on Dl-GFP (DeLotto et al., 2007). Finally, we conclude with a discussion of future directions of studies of the Dl gradient.

2. Components of the Dl morphogen network

2.1 Patterning events in oogenesis

While Dl is ultimately the transcription factor that establishes dorsal-ventral polarity in the *Drosophila* embryo, the signaling pathways upstream of Dl are crucial to the initiation of the Dl gradient itself. The initiation steps are a long and involved process, beginning in the stage 8 oocyte when the oocyte nucleus migrates to the dorsal-anterior border of the oocyte (Fig. 2A; Neuman-Silberberg & Schüpbach, 1996; Roth, Jordan, & Karess, 1999; Roth & Lynch, 2009). The TGFα ligand, Gurken, which is tightly associated with the oocyte nucleus, is secreted from this region of the oocyte and diffuses in a gap between the oocyte and the layer of follicle cells that

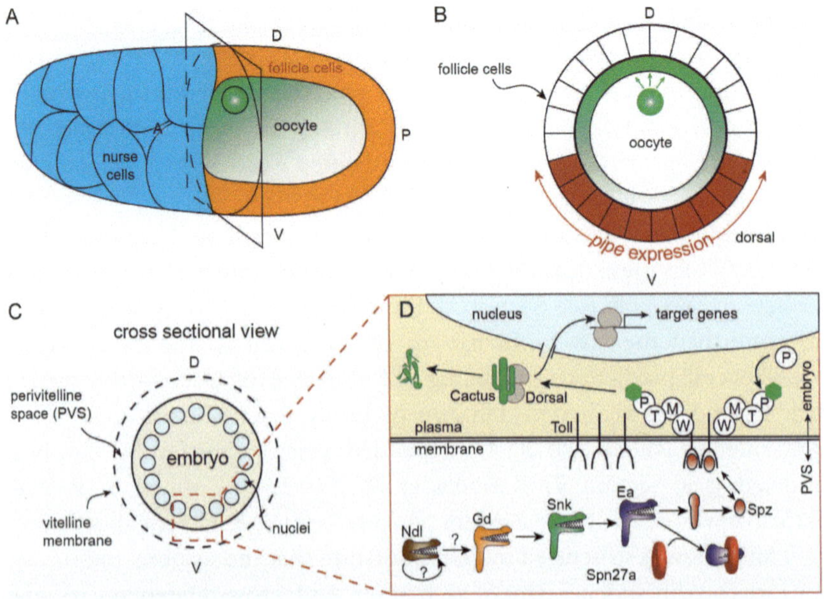

Fig. 2 Overview of the components of the Dl morphogen network. (A) Illustration of Stage 10 *Drosophila* egg chamber. The oocyte nucleus is found at the dorsal-anterior border of the oocyte. Gurken diffuses from the oocyte nucleus (green). (B) A cross section of the Stage 10 egg chamber. The diffusion of Gurken (green) establishes a gradient in the gap between oocyte and follicle cell layer and begins the cascade that represses *pipe* expression (brown) in the ventral region. (C) Illustration of a cross section of the syncytial blastoderm embryo. The protease cascade is initiated in the perivitelline space (PVS), which is between the embryo and vitelline membrane. (D) Illustration of the protease cascade and Toll signaling. In the perivitelline space, the protease cascade of Nudel (Ndl), Gastrulation defective (Gd), Snake (Snk), and Easter (Ea) is initiated. These proteases progressively process one after the other before activating Spätzle (Spz), the ligand for the Toll receptor. Ea is inhibited by Spn27a, preventing processing of Spz beyond the ventral region. Weckle (W) interacts with the cytoplasmic tail of activated Toll, then passes the signal through dMyd88 (M), Tube (T), and Pelle (P) before Pelle phosphorylates Cact, causing the degradation of Cact and the nuclear translocation of the transcription factor Dl. *Figure reproduced/modified with permission from Reeves, G. T., & Stathopoulos, A. (2009). Graded dorsal and differential gene regulation in the Drosophila embryo.* Cold Spring Harbor Perspectives in Biology, 1, *a000836. Copyright to Cold Spring Harbor Laboratory Press.*

surround the oocyte (Fig. 2B; Neuman-Silberberg & Schüpbach, 1993, 1996; Schüpbach, 1987). The range of secreted Gurken covers half of the oocyte, activating its cognate receptor, the epidermal growth factor receptor (EGFR), which is expressed in the follicle cells (Chang et al., 2008; Goentoro et al., 2006; Pai, Barcelo, & Schüpbach, 2000; Price, Clifford, & Schüpbach, 1989; Schejter & Shilo, 1989; Schüpbach, 1987).

This signaling module is the beginning of the establishment of distinct dorsal and ventral regions. EGFR is responsible for the inactivation of the HMG-box protein Capicua, which is in turn responsible for the repression of the homeodomain factor Mirror (Andreu et al., 2012). This progression of activation and repression controls where *pipe* is expressed, which defines the future ventral side of the embryo (see Fig. 2B). Inactive Capicua allows Mirror to be expressed, which then goes on to dorsally repress *pipe* (Andreu et al., 2012). EGFR expression tapers off in the lateral region of the follicle cells; as a result, Capicua is active ventrally, Mirror is repressed, and *pipe* is expressed (Andreu et al., 2012).

Pipe itself is a protein similar to heparan sulfate 2-O-sulfotransferase (Sen, Goltz, Stevens, & Stein, 1998). Pipe is responsible for the sulfonation of particular proteins involved in the establishment of the dorsal-ventral axis. Based on its similarity to other sulfotransferases, it is thought to interact with and transfer sulfate from a sulfate donor molecule called 3′-phosphoadenosine 5′-phosphosulfate (PAPS) (Robbins & Lipmann, 1957; Sen et al., 1998). This idea is supported by work that has shown the elimination of PAPS synthetase, the enzyme responsible for PAPS production, results in dorsalized embryos (Zhu, Stevens, Stein, Schüpbach, & Stein, 2007). The protein Slalom transports PAPS to the Golgi for Pipe activity (Luders et al., 2003). While Pipe is an important component linking DV polarity in the egg chamber and the Dl gradient, it in and of itself does not form a gradient. Instead, it is expressed uniformly across ~40% of ventral side of the follicular epithelium (Fig. 2B; Sen et al., 1998; Stein & Stevens, 2014). Its expression defines a region on the ventral side of the egg where a protease cascade is initiated, which eventually causes the Dl gradient to form (Andreu et al., 2012; Peri, Technau, & Roth, 2002; Sen et al., 1998). Introduction of uniformly expressed Pipe results in ventralized embryos, while Pipe mutants become dorsalized (Sen et al., 1998). At this stage, proper patterning is also reliant on appropriate subcellular localization of Pipe to the Golgi apparatus (Sen, Goltz, Konsolaki, Schüpbach, & Stein, 2000). This localization is dependent on a protein called Windbeutel (Sen et al., 2000). In the absence of Windbeutel, Pipe is retained in the endoplasmic reticulum (Sen et al., 2000) and the resulting embryos completely lack DV polarity (Konsolaki & Schüpbach, 1998; Nilson & Schu, 1998).

2.2 The protease cascade

As referenced earlier, a protease cascade consisting of a series of proenzymes—enzymes that require cleavage by an upstream protease to acquire activity—is a crucial step in formation of the Dl gradient during embryogenesis. The final step of the cascade is the processing of the Toll-ligand Spätzle (Spz; DeLotto & DeLotto, 1998). Connections between some steps in the cascade are not well known, but all of the components are vitally important to successful DV patterning. This cascade begins with Nudel (Ndl; Fig. 2D), a large trypsin-type serine protease (Hong & Hashimoto, 1995). Ndl is secreted into the perivitelline space (i.e., the space between the outer vitelline membrane and the embryo; see Fig. 2C) and is found in all follicle cells (LeMosy, Kemler, & Hashimoto, 1998; LeMosy, Leclerc, & Hashimoto, 2000). Mechanistically, Ndl does not appear to be the substrate of Pipe, which is upstream of Ndl (Stein, Cho, Zhang, & Stevens, 2008), nor do downstream proteases appear to be substrates of Ndl. However, the protease domain of Ndl is known to be integral in setting up the ventral signal leading to the activation of the rest of the protease cascade (LeMosy et al., 1998, 2000). Mutations of this domain lead to processing failures in two other protease cascade members: Gastrulation defective and Snake (Cho, Stevens, & Stein, 2010). It has been suggested that Ndl may generally potentiate protein activity in the perivitelline environment, which can influence overall activity of the protease cascade (Stein & Stevens, 2014).

Gastrulation defective (Gd), a protein similar to mammalian serine proteases C2 and B, is the next component of the protease cascade (DeLotto, 2001; Fig. 2D). Like the rest of the cascade, Gd localizes to the perivitelline space, where it serves two functions: processing of Snake (Snk) and facilitating the cleavage of Easter (Ea) by Snk (Cho, Stevens, Sieverman, Nguyen, & Stein, 2012; LeMosy, Tan, & Hashimoto, 2001). Snk, a trypsin-like serine protease, goes on to process Ea (Fig. 2D; Cho et al., 2010; DeLotto & Spierer, 1986; LeMosy et al., 2001). Pipe, along with Gd, are both required for the processing of Ea to be restricted to the ventral side of the embryo. Evidence suggests that Gd interacts with molecules sulfated by Pipe that are found in the ventral vitelline membrane (Cho et al., 2012). This interaction both enhances cleavage of Ea and restricts this cleavage to the ventral side of the embryo.

Ea is another trypsin-like serine protease that is directly responsible for processing the Toll-ligand Spz (Fig. 2D; Chasan & Anderson, 1989; DeLotto & DeLotto, 1998). While the production of activated Ea occurs on the ventral side of the embryo, it can still diffuse to lateral and dorsal regions. This is prevented through inhibition of the processed protein by Serpin27A (Spn27a; Hashimoto et al., 2003;

Ligoxygakis, Roth, & Reichhart, 2003). In this way, Spn27A helps confine Ea's activity to the ventral side of the embryo through rapid binding of processed Ea (Hashimoto et al., 2003; Ligoxygakis et al., 2003). Finally, processed Spz is released into the perivitelline space where it can then activate Toll (DeLotto & DeLotto, 1998).

2.3 Toll signaling components

Toll, a transmembrane receptor, is responsible for transmitting the ventral signal from the perivitelline space into the embryo (Fig. 2D; Hashimoto, Hudson, & Anderson, 1988). In wild-type embryos, activation of Toll leads to degradation of Cact along the ventral axis, predicated by the extracellular protease cascade (Belvin et al., 1995). Cact is the *Drosophila* homolog of vertebrate IκB, which is the protein responsible for regulation of NF-κB in vertebrates (Geisler, Bergmann, Hiromi, & Nüsslein-Volhard, 1992). Like these inhibitor molecules in vertebrates, Cact acts to inhibit Dl by sequestering it to the cytoplasm (Fig. 2D; Roth et al., 1991). Upon its degradation, Dl can then enter the nucleus. Given the region in which Toll is active, this results in a cytoplasmic gradient of Cact where there is a low concentration on the ventral side and high concentration on the dorsal side (Belvin et al., 1995; Bergmann et al., 1996; Reach et al., 1996). Embryos lacking maternal *cact* have Dl present in the nucleus all around the DV axis, while embryos with mutations in dorsal group proteins leading up to Cact do not receive a degradation signal and therefore Dl remains isolated in the cytoplasm (Roth et al., 1991). Beyond this, Cact and Dl have their own set of interactions that adds more complexity to the end of this signaling cascade (see Section 2.4).

While Toll provides the signal necessary to break down Cact, this signal is actually transduced through a signaling complex consisting of several proteins (Fig. 2D). This complex begins with Weckle, a zinc finger adaptor that localizes to the plasma membrane (Chen et al., 2006). Weckle acts as a bridge between Toll and the next part of the complex, dMyd88, while also helping dMyd88 to localize to the membrane (Chen et al., 2006). dMyd88 experiences a strong interaction with adaptor protein Tube via its death domain, continuing to pass along the Toll signal (Charatsi, Luschnig, Bartoszewski, Nüsslein-Volhard, & Moussian, 2003; Sun, Towb, Chiem, Foster, & Wasserman, 2004). Tube then interacts with the serine/threonine protein kinase Pelle (Galindo, Edwards, Gillespie, & Wasserman, 1995; Grosshans, Bergmann, Haffter, & Nüsslein-Volhard, 1994;

Shelton & Wasserman, 1993), which goes on to phosphorylate Cact, leading to its degradation (Daigneault, Klemetsaune, & Wasserman, 2013). Eliminate any of these components of the complex and the signal from Toll will not successfully be passed to Cact (Charatsi et al., 2003; Chen et al., 2006; Sun et al., 2004).

2.4 Dorsal/cactus interactions

Dl is a transcription factor with homology to mammalian NF-κB and c-rel through its rel-homology domain (Steward, 1987). The *dl* gene itself was originally discovered in a mutagenesis screen that targeted maternal effect genes (Nusslein-Volhard, 1979). It was subsequently discovered through antibody staining that Dl protein exhibits a gradient in the nuclei of blastoderm-stage embryos, and this nuclear localization is required for function (Rushlow et al., 1989; Steward, 1989; Steward et al., 1988). In 1989, it was confirmed that the nuclei in the early embryo respond to Dl in a concentration-dependent fashion (Roth et al., 1989). These two facts—that Dl exhibits a concentration gradient, and that cells respond in a concentration-dependent fashion—solidified its place with Bicoid (Driever & Nüsslein-Volhard, 1988a,b) as one of the only two confirmed morphogens at the time (Roth et al., 1989).

It initially appeared that Dl being freed from Cact was sufficient for the gradient to form. However, Toll signaling results in phosphorylation not only of Cact, but of Dl as well (Gillespie & Wasserman, 1994; Whalen & Steward, 1993), which is key for nuclear import of Dl (Drier, Huang, & Steward, 1999). By itself, Cact is an unstable protein with a high turnover rate (Belvin et al., 1995; Bergmann et al., 1996). When bound to Dl, Cact is protected from signal-independent degradation. In the absence of Dl, Cact levels are undetectable (Belvin et al., 1995; Whalen & Steward, 1993). However, there is recent evidence that signal-independent degradation of Cact may result in a Cact fragment refractory to Toll signaling (Cardoso et al., 2017; Fontenele et al., 2013). If that is the case, it is unclear how to interpret the reduction of Cact when Dl levels are reduced.

3. Dl target gene expression

3.1 Domains of gene expression

At a coarse-grain level, Dl target genes seem to cluster into three groups, or "Types," based on the location of their expression borders (Fig. 3A;

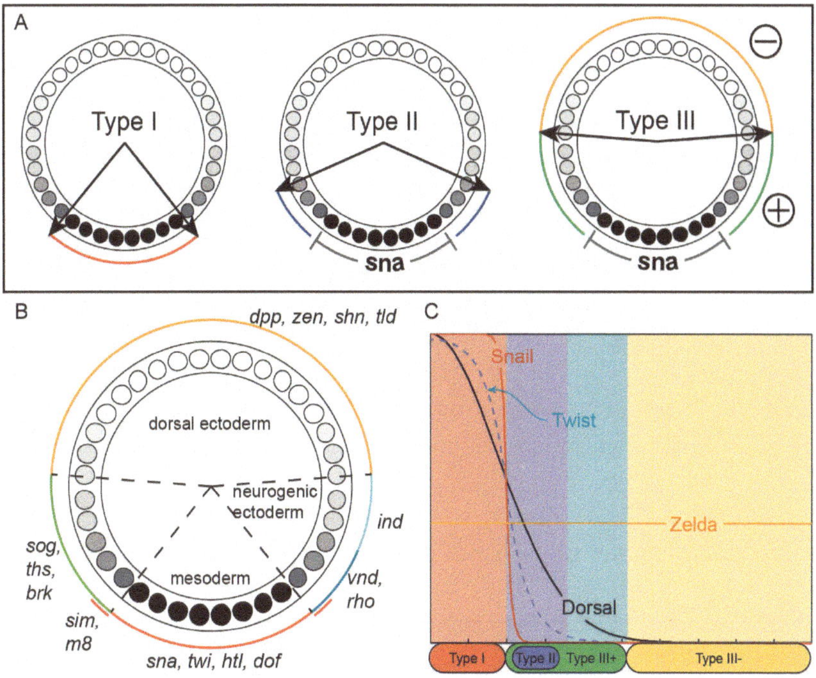

Fig. 3 Dl-dependent gene expression. (A) Illustration of the spatial extent of the Types of Dl-dependent genes. (B) Fate-map with non-exclusive lists of the types of genes found in each domain. (C) Illustration of the major transcriptional regulators present across the DV axis. *Figure reproduced/modified with permission from Reeves, G. T., & Stathopoulos, A. (2009). Graded dorsal and differential gene regulation in the Drosophila embryo.* Cold Spring Harbor Perspectives in Biology, 1, *a000836. Copyright to Cold Spring Harbor Laboratory Press.*

Chopra & Levine, 2009; Reeves & Stathopoulos, 2009; Stathopoulos & Levine, 2004). Type I genes are expressed in the ventral-most regions of the embryo with sharp borders placed at roughly 20% from ventral to dorsal (20% DV; Fig. 3A; Reeves & Stathopoulos, 2009; Stathopoulos & Levine, 2004). In particular, *sna* and *twi* activity determine the borders of the mesoderm (Fig. 3B; Alberga, Boulay, Kempe, Dennefeld, & Haenlin, 1991; Kosman, Ip, Levine, & Arora, 1991; Leptin, 1991). Type II genes have dorsal borders (less sharp than those of Type I genes) at roughly 32–37% DV, and ventrally are repressed by Sna (Fig. 3A; Reeves & Stathopoulos, 2009; Reeves et al., 2012; Stathopoulos & Levine, 2004). Genes expressed in this region include *rhomboid* (*rho*) and *ventral neuroblasts defective* (*vnd*; Fig. 3B), the latter of which is a transcription factor responsible for the specification of neuroblasts in the ventral neuroectoderm. Type III genes are defined by having graded borders that occur beyond 45% DV (Reeves & Stathopoulos, 2009; Reeves et al., 2012; Stathopoulos & Levine, 2004). However, Type III genes are further broken down into Type III + and Type III − groupings depending on the effect Dl has over their expression. Type III + genes, such as *sog*, *brk*, and *ths*, are activated by Dl and repressed by Sna. On the other hand, Type III − genes, such as *dpp*, *tld*, and *zen*, are repressed by Dl and are thus restricted to the dorsal half of the embryo (Fig. 3B). Finally, a handful of genes, such as *sim* and *ind*, are regulated by a combination of Dl and other signaling pathways, and thus defy this classification (Fig. 3B; for further details, see Chopra & Levine, 2009; Reeves & Stathopoulos, 2009; Stathopoulos & Levine, 2002, 2005). It should be stressed that these spatial classifications are only rough groupings. For example, the *twi* expression extends slightly further than fellow Type I gene *sna* (Kosman et al., 1991; Zinzen et al., 2006). Similarly, *rho* and *vnd* do not perfectly co-localize (McHale et al., 2011; Zinzen et al., 2006), nor do *sog* and *ths* (Reeves et al., 2012). On the other hand, few known genes, if any, have borders midway between those of Type I and II genes, or of Type II and III.

While border location naturally groups genes into three Types, mechanistically, border placement is determined by enhancer structure, which is unique to each individual gene (Fig. 3C). Trends can be deduced that are common to the classifications, but these are based on research done for a small group of genes and would benefit from further research and clarification (for more, see Chopra & Levine, 2009; Reeves & Stathopoulos, 2009; Stathopoulos & Levine, 2002, 2005). *twi*, a Type I gene, is expressed in regions of high concentrations of nuclear Dl, and its enhancers contain relatively low affinity Dl binding sites (Jiang, Kosman, Ip, & Levine, 1991). Twi acts synergistically with Dl to activate expression of a number of other Type I genes in the ventral region (see Fig. 3C; Section 3.3; Berman et al., 2002; Leptin, 1991; Levine & Davidson, 2005; Reeves & Stathopoulos, 2009). As a Type I gene, *sna* is activated by this combination of Dl and Twi (Ip, Park, Kosman, Yazdanbakhsh, & Levine, 1992). *rho*, a Type II gene, is characterized by a combination of higher affinity Dl sites, greater synergy between Dl and Twi, and repression by Sna (Fig. 3A, C; Ip, Park, Kosman, Bier, & Levine, 1992; Jiang & Levine, 1993; Papatsenko & Levine, 2005; Reeves & Stathopoulos, 2009). Type III + have high affinity Dl binding sites that correspond with the lower levels of Dl in this lateral region, which may be potentiated by the pioneering factor Zelda (Zld; see below). Type III − genes, such as *decapentaplegic* (*dpp*), *tolloid*, and *zen*, are repressed by Dl in more ventral regions, and thus their regulatory regions tend to have silencer elements to mediate this repression

(Huang, Schwyter, Shirokawa, & Courey, 1993; Ip, Kraut, Levine, & Rushlow, 1991; Jiang, Rushlow, Zhou, Small, & Levine, 1992; Kirov, Childs, O'Connor, & Rushlow, 1994). Interestingly, there is no clear relationship between Type III− genes and high affinity binding sites: some Type III− genes have only low affinity Dl sites (Huang et al., 1993). Regardless, mutation of these elements leads to loss of Dl binding and attenuation of ventral repression. It should be noted that this model of relating binding affinity to position of gene expression fails in other systems like Bicoid and Dpp (Chen, Xu, Mei, Yu, & Small, 2012; Liang, Xu, Chuang, & Rushlow, 2012; Lin, Park, Kirov, & Rushlow, 2006; Xu, Kirov, & Rushlow, 2005). A more detailed study of the Dl system would clarify if this model is valid.

The expression domains of Dl target genes shift during the course nc 14 (Reeves et al., 2012). For instance, the expression domain of *sna* expands slightly, which has been attributed to the dynamics of the Dl gradient (Fig. 1F; Reeves et al., 2012). Shifts in gene expression patterns may also be due to diffusion of the protein products of Dl target genes, similar to what has been proposed in the Gap gene system (Jaeger et al., 2004). Indeed, a recent live imaging study suggested that nuclei that do not initially express *sna* can be recruited into the *sna* domain as a result of Sna diffusion and autoactivation (Bothma, Norstad, Alamos, & Garcia, 2018). This same mechanism may also be responsible for the sharp and straight boundary of the *sna* domain (Bothma et al., 2018).

3.2 Zelda

While the Dl gradient is the central component to DV patterning, it acts with several partners and other signaling pathways. Perhaps most importantly, Dl acts with the pioneer factor Zelda (Zld) to influence activation of nearly all Dl target genes (Liang et al., 2008; Nien et al., 2011). Zld has been detected as early as nuclear cycle 2; however, it begins to facilitate the earliest zygotic transcription around nuclear cycle 8 (Nien et al., 2011). Zld binding sites are found in large numbers in the *cis*-regulatory regions of several genes. It has been shown that most of these genes, are primarily activated by other factors in the early embryo, but depend on Zld for proper timing, spatial domain, or strength of activation (Harrison, Li, Kaplan, Botchan, & Eisen, 2011; Liang et al., 2008; Nien et al., 2011). In *zld* mutants, several zygotically expressed genes essential for early developmental processes, such as DV and AP patterning, blastoderm formation, and sex determination, are found to be downregulated (Liang et al., 2008).

In *zld* mutants, the expression of Type I Dl target genes such as *twi* and *sna* are delayed by about two nuclear cycles but recover by nc 14. On the other hand, the expression of Type III+ genes, such as *sog*, have severely reduced domains of gene expression, including a loss of expression in lateral regions where Dl concentration is near basal levels (Dufourt et al., 2018; Kanodia et al., 2012; Liang et al., 2008; Liberman & Stathopoulos, 2009; Nien et al., 2011; Yamada et al., 2019). Furthermore, the binding sites of Zld have been shown to be in close proximity to those of Dl (Nien et al., 2011). Thus, it was inferred that Zld cooperatively interacts with Dl to boost its activity, which had been mathematically shown to sufficiently explain some *zld* phenotypes (discussed further in Section 7.3; Kanodia et al., 2012). Indeed, it was found that Zld/Dl interaction also leads to an increase in local concentration of Dl near the enhancer regions (Yamada et al., 2019).

The enhancers of most genes are found in regions of chromatin wrapped around the nucleosome and not in the more open regions, perhaps to protect it from unwanted binding. Zld binding sites have been shown to strongly correlate to regions of nucleosome depletion (Nien et al., 2011). Thus, it was suggested that Zld acts early in development to increase chromatin accessibility by nucleosome depletion, increasing the probability of transcription factor binding (Mir et al., 2017; Nien et al., 2011; Sun et al., 2015; Yamada et al., 2019). Zld protein levels have been shown to decrease as time progresses, indicating that temporal dynamics of Zld may act with the spatial gradient of patterning transcription factors to ensure robust expression (Nien et al., 2011). In spite of an appreciable understanding of the phenotypes observed in *zld* mutants, the exact molecular interactions at the Zld binding sites that lead to depletion of nucleosomes and higher chromatin accessibility remain unknown. However, the significance of Zld in DV patterning, and in the development of the pre-cellular embryo in general, is undisputed, as most of the genes that are regulated by Zld are zygotic and are activated prior to cellularization (Liang et al., 2008).

3.3 Twist

A feedforward loop with the bHLH transcription factor Twist has also been implicated in many aspects of Dl-dependent gene expression (Simpson, 1983). *Twi* itself is activated by Dl during nc 12 before aiding in activation of genes in both the mesoderm and neuroectoderm (Jiang et al., 1991;

Sandler & Stathopoulos, 2016b). Concentration levels of Twi are highly determinant of cell fate in the mesoderm: the highest levels lead to somatic muscle while lower levels determine other mesoderm structures (Baylies & Bate, 1996).

However, Twi majorly acts in concert with Dl to activate its targets. It has been shown that *twi* mRNA levels are the dominant zygotic component in Dl-dependent gene expression in the mesoderm (Sandler & Stathopoulos, 2016b). As stated in Section 3.1, different regions of the DV axis also appear to require more synergy between Dl and Twi binding sites because of differing levels of Dl concentration. In the mesoderm, a combination of Dl and Twi binding sites make up *sna*'s promoter region, a transcriptional repressor. Progressive mutations in these sites results in progressively weaker *sna* expression, along with narrowing of normal DV patterning (Ip, Park, Kosman, Yazdanbakhsh, et al., 1992). By comparing transcript levels in *twi* mutants to wild type, Sandler and Stathopoulos showed that transcriptional control of mesodermal genes by Twi, increases from ~24% at the end of nc 13 to ~77% at mid-nc 14 and later reduces to ~55% at gastrulation (Sandler & Stathopoulos, 2016b). Thus, at the height of transcriptional activity in the precellular embryo, gene regulation is significantly controlled by Twi.

Twi is also important in more lateral regions as Dl concentration tapers off. This is apparent in regulation of *rho* via the neuroectoderm element, or NEE (Ip, Park, Kosman, Bier, et al., 1992). Found within this region are closely linked Dl and Twi binding sites; mutations in either Dl or Twi binding sites markedly decrease *rho* expression, corresponding to a need for cooperative binding and activation of genes in the neuroectoderm.

Szymanski and Levine demonstrated the differences in the type of synergistic interactions between Dl and Twi by analyzing the number and spatial distribution of binding sites of these proteins (Szymanski & Levine, 1995). They found that, while in the neuroectoderm, close proximity of binding sites of Dl and Twi was essential for proper positioning of gene expression, in the mesoderm, it was not. Additionally, using synthetically constructed promoters consisting exclusively of Dl and Twi binding sites, the authors showed that synergistic interaction between Dl and Twi is both necessary and sufficient to precisely specify the mesoderm.

These Dl/Twi interactions have been quantitatively modeled using a fractional site occupancy model, where the cooperativity between Dl and Twi proteins at their binding sites regulates the different patterns of gene expression in the DV axis (see Section 7.1; Zinzen et al., 2006).

4. Regulation of the Dl gradient

As discussed in previous sections, the establishment of the Dl gradient involves myriad complex biomolecular processes. However, scant little has been discovered regarding regulation via feedback or crosstalk with other pathways. Here we discuss some discoveries that have implicated regulatory pathways in the formation of the Dl gradient itself.

4.1 WntD

A Wnt family protein, named WntD for "Wnt inhibitor of Dl" (Ganguly, Jiang, & Ip, 2005), was discovered to act as a feedback regulator of the Dl gradient (Ganguly et al., 2005; Gordon, Dionne, Schneider, & Nusse, 2005; Llimargas & Lawrence, 2001). *wntD* transcript is initially found on the ventral halves of the embryonic poles during the blastoderm stage (Fig. 4A), followed shortly thereafter by weak expression in the presumptive mesoderm (Ganguly et al., 2005; Gordon et al., 2005). The timing and localization of the *wntD* expression pattern is partially explained by overlap with Dl signaling and receptor tyrosine kinase (RTK) signaling, which occurs first at the poles of the embryo, then later in the Type II domain (Gabay, Seger, & Shilo, 1997; Helman et al., 2012). However, transcript levels of *wntD* are upregulated in constitutively active Toll mutants, which may imply that strong Dl signaling alone may be sufficient for *wntD* expression (Ganguly et al., 2005; Gordon et al., 2005). Overexpression of WntD completely blocks nuclear translocation of Dl and prevents expression of Dl-dependent target genes (Ganguly et al., 2005; Gordon et al., 2005).

Even so, null mutants of *wntD* are homozygous viable (Ganguly et al., 2005; Gordon et al., 2005), which leaves open its role in development. It may serve as the crux of the cross-talk between Dl and RTK signaling (Helman et al., 2012). However, Dl gradient measurements in *wntD* and RTK signaling mutants indicate the Dl gradient is only minimally perturbed (Fig. 4B, C; Garcia et al., 2013; Helman et al., 2012). The subtlety of the phenotype may indicate that the role of *wntD* is in restorative regulation when other factors are perturbed (Rahimi et al., 2016).

The mechanism by which WntD affects the Dl gradient is unclear. It is a secreted factor and its interaction with the Frizzled-like receptor, Fz4, is required for Dl inhibition (Ching, Hang, & Nusse, 2008; McElwain et al., 2011; Rahimi et al., 2016). Cuticle preps from *cact* hypomorphic embryos overexpressing *wntD* indicate that WntD may act at the level of

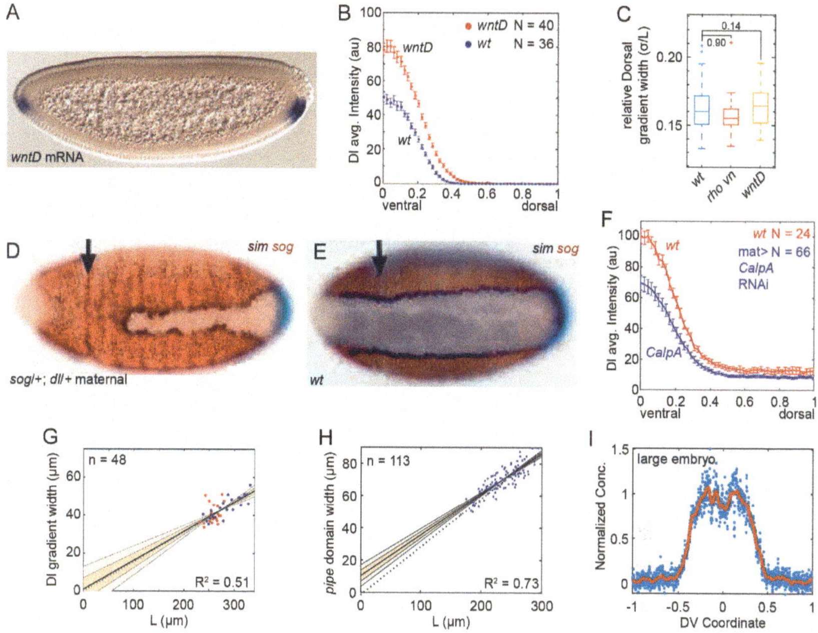

Fig. 4 Regulation of the Dl gradient. (A) Expression of *wntD* mRNA. (B) Quantification of the Dl gradient in *wntD* mutant and wild-type embryos. (C) Boxplots of Dl gradient width measurements in wild-type embryos; *rho,vn* mutants; and *wntD* mutants. (D) Embryo with reduced maternal *sog* and *dl*, co-stained for probes against *sim* and *sog*. In this maternal genotype the Type I domain collapses (loss of *sna*) and lateral genes invade the ventral-most region of the embryo. *sim* is normally expressed at the boundary of the Type I domain. (E) Wild-type embryo co-stained for probes against *sim* and *sog*. (F) Measurements of the Dl gradient in control (wild-type) embryos and embryos maternally overexpressing RNAi against *CalpA*. (G) The Dl gradient width scales perfectly with embryo size. (H) The domain of *pipe* in follicle cells scales with egg chamber size. (I) Embryos with abnormally large diameters suffer from a Dl gradient phenotype similar to those with known shuttling defects. *Panel (A): Reproduced/modified with permission from Ganguly, A., Jiang, J., & Ip, Y. T. (2005). Drosophila WntD is a target and an inhibitor of the Dorsal/Twist/Snail network in the gastrulating embryo.* Development, 132, *3419–3429. Panel (B): Reproduced/modified with permission from Helman, A., Lim, B., Andreu, M. J., Kim, Y., Shestkin, T., Lu, H., et al. (2012). RTK signaling modulates the Dorsal gradient.* Development, 139, *3032–3039. Panels (C, G, I): Reproduced/modified with permission from Garcia, M., Nahmad, M., Reeves, G. T., & Stathopoulos, A. (2013). Size-dependent regulation of dorsal–ventral patterning in the early Drosophila embryo.* Developmental Biology, 381, *286–299. Panels (D, E): Reproduced/modified with permission from Araujo, H., & Bier, E. (2000). sog and dpp exert opposing maternal functions to modify Toll signaling and pattern the dorsoventral axis of the Drosophila embryo.* Development, 127, *3631–3644. Panel (F): Reproduced/modified with permission from Fontenele, M., Lim, B., Oliveira, D., Buffolo, M., Perlman, D. H., Schupbach, T., et al. (2013). Calpain A modulates Toll responses by limited cactus/IκB proteolysis.* Molecular Biology of the Cell, 24, *2966–2980. Panel (H): Data kindly provided by Lea Goentoro, Trudi Schüpbach, and Stanislav Shvartsman. Methods describing how the data were gathered can be found in Goentoro, L. A., Reeves, G. T., Kowal, C. P., Martinelli, L., Schüpbach, T., & Shvartsman, S. Y. (2006). Quantifying the Gurken morphogen gradient in Drosophila oogenesis.* Developmental Cell, 11, *263–272.*

Cact or downstream (Gordon et al., 2005). However, a conflicting report, in which the Dl gradient is measured in *cact* RNAi embryos that overexpress *wntD*, strongly suggests that *cact* knockdown is epistatic to the overexpression of *wntD* (Rahimi et al., 2016). Furthermore, from an analysis of *Toll* mutants, it was concluded that WntD/Fz4 interfere with Toll signaling at the level of the Toll extracellular domain (Rahimi et al., 2016). Further studies to uncover details of the WntD mechanism, and its functional role in shaping the Dl gradient, are necessary for a mechanistic understanding.

4.2 BMP signaling

Maternal BMP signaling may affect the distribution of the Dl gradient (Araujo & Bier, 2000; Cardoso et al., 2017; Carneiro et al., 2006; Fontenele et al., 2009, 2013; Schloop, Carrell-Noel, & Reeves, 2019). BMP signaling components, such as the BMP ligand Dpp and the BMP inhibitor Short Gastrulation (Sog; a *Drosophila* homolog of Chordin) are maternally loaded into the egg during oogenesis and act during early embryonic patterning (Araujo & Bier, 2000; Carneiro et al., 2006). Increasing maternal BMP activity, such as by overexpressing *dpp* maternally, or reducing maternal levels of *sog*, alters DV gene expression patterns in a manner consistent with a flattening of the Dl gradient (Carneiro et al., 2006). Type I genes are reduced or lost, while Type II genes shift ventrally, and sometimes have expanded widths. This effect is particularly strong in the sensitized background of maternal *dl* heterozygotes (Araujo & Bier, 2000; Carneiro et al., 2006; Fig. 4D; compare to wild type in Fig. 4E). Furthermore, maternally overexpressing BMP signal transducers Mad (*Drosophila* Smad1) or Medea (*Drosophila* Smad4; Raftery, Twombly, Wharton, & Gelbart, 1995) results in a highly expanded, and sometimes flat-topped or double-peaked Dl gradient (Schloop et al., 2019). It should be noted that perturbing zygotic expression of *dpp* or *sog* does not produce the above phenotypes.

Maternal BMP signaling stabilizes Cact by downregulating mRNA levels of *calpainA* (*calpA*), which encodes a calcium-dependent protease that degrades Cact in a Toll-signal-independent pathway (Fontenele et al., 2009, 2013). Thus, a mutated Cact that lacks the PEST domain, which is required for signal-independent degradation, is not affected by maternal BMP signaling, nor by perturbations to CalpA activity (Fontenele et al., 2009). Maternally expressing RNAi against *calpA* results in gene expression phenotypes similar to those from maternal overexpression of *dpp* (Fontenele et al., 2009). Furthermore, measurements of the Dl gradient also suggest a flattened Dl gradient (Fontenele et al., 2013; Fig. 4F). However, it is unclear whether a gradient of calcium ions in the early embryo is consistent with these findings (Créton, Kreiling, & Jaffe, 2000).

The biophysical mechanism that links maternal BMP signaling, calcium signaling, the stability of Cact, and the Dl gradient shape remains to be determined. However, a series of biochemical and genetic data suggests that signal-independent degradation of Cact, through CalpA, may create a Cact fragment resistant to Toll-dependent degradation (Fontenele et al., 2013). Furthermore, the manner in which the Dl gradient expands upon overexpression of Mad or Medea (Schloop et al., 2019), the flattened gradient phenotype (Carneiro et al., 2006; Fontenele et al., 2009, 2013), and the sensitivity to maternal *dl* dosage (Carneiro et al., 2006; Fontenele et al., 2009, 2013), are consistent with a defect in shuttling (see Sections 5.2 and 6.3). On the other hand, neither BMP-driven stabilization of Cact, nor the decrease in production of a Cact fragment refractory to the Toll signal would be expected to cause a defect in shuttling. Further investigation of these mechanisms is required.

4.3 Size-dependent scaling of the Dl gradient

The question of how developing tissues, in particular their gene expression patterns, scale with size is one of the unmet challenges of developmental biology. Both the Dl gradient, as well as the expression patterns of its target genes, were shown to exhibit at least partial scaling with respect to the size of the embryo's DV axis (Garcia et al., 2013). In particular, the width of the Dl gradient scales perfectly (Fig. 4G), as does the border of *sna*. In contrast, the Type II gene *vnd* and the pseudo-Type III+ gene *ind* correlate with the DV axis, but not perfectly. The data also suggest the width of the *sna* domain correlates perfectly with the Dl gradient itself, while the widths of *vnd* and *ind* are nearly independent of the Dl gradient width (Garcia et al., 2013).

The mechanism for scaling the Dl gradient width is unclear. And given that some Dl target genes scale perfectly, while others exhibit overscaling (de Lachapelle & Bergmann, 2010; Garcia et al., 2013), the mechanism to scale gene expression may depend on each gene's mode of activation by the Dl gradient. One possibility for scaling of the Dl gradient itself is that the active Toll gradient itself scales with respect to size. There is evidence that, during oogenesis, the *pipe* expression domain on the ventral side of the egg chambers scales with egg chamber size (Goentoro et al., 2006; see Fig. 4H). Another piece of the puzzle may involve shuttling of Dl by Cact (see Sections 5.2 and 6.3; Carrell et al., 2017). Embryos with very large diameters have Dl gradients that are double-peaked (Garcia et al., 2013; Fig. 4I), which is a hallmark phenotype of compromised shuttling (Carrell et al., 2017). As with many tissue patterning systems, the question of size-dependent scaling of the Dl gradient remains largely open.

5. Quantitative measurements of the Dl gradient

The precise shape and spatiotemporal distribution of morphogen gradients are central to proper gene expression in developing tissues (previously reviewed in Rushlow & Shvartsman, 2012). Hence, a quantitative understanding of the features of the Dl morphogen gradient is needed. Quantitative fluorescent imaging and live, GFP-based experiments have greatly advanced the field of developmental patterning in general, and the Dl gradient network in particular, over the past dozen years (Carrell et al., 2017; Chung et al., 2011; DeLotto et al., 2007; Liberman et al., 2009; Reeves et al., 2012). In particular, our understanding of the biophysics of the Dl system has used those tools to greatly advance and test novel hypotheses.

5.1 Spatial extent of the Dl gradient

The spatial range of a morphogen gradient determines its ability to pattern distinct domains of gene expression within a tissue. The earliest quantitative measurements of the nuclear Dl gradient suggested that it has limited spatial range (Liberman et al., 2009; see also Zinzen et al., 2006). Confocal z-stacks were taken to a depth of approximately 100 microns, or roughly halfway through the diameter of the embryo (Liberman et al., 2009; Fig. 5A). Image analysis was performed to identify the Dl intensity in each nucleus in the z-stack, and to transform the x,y,z coordinate of each nucleus into a position along the DV axis (Fig. 5B). The resulting quantitative plots of the Dl nuclear intensity suggested the gradient becomes essentially flat roughly 130–150 μm from the ventral midline (Fig. 5C), which falls just short of the extent it would need to place the boundaries of Type III genes, such as such as *sog*, *dpp*, and *zen*.

From these measurements, the nuclear Dl gradient was empirically found to have roughly a bell shape (see Figs. 1C and 5C):

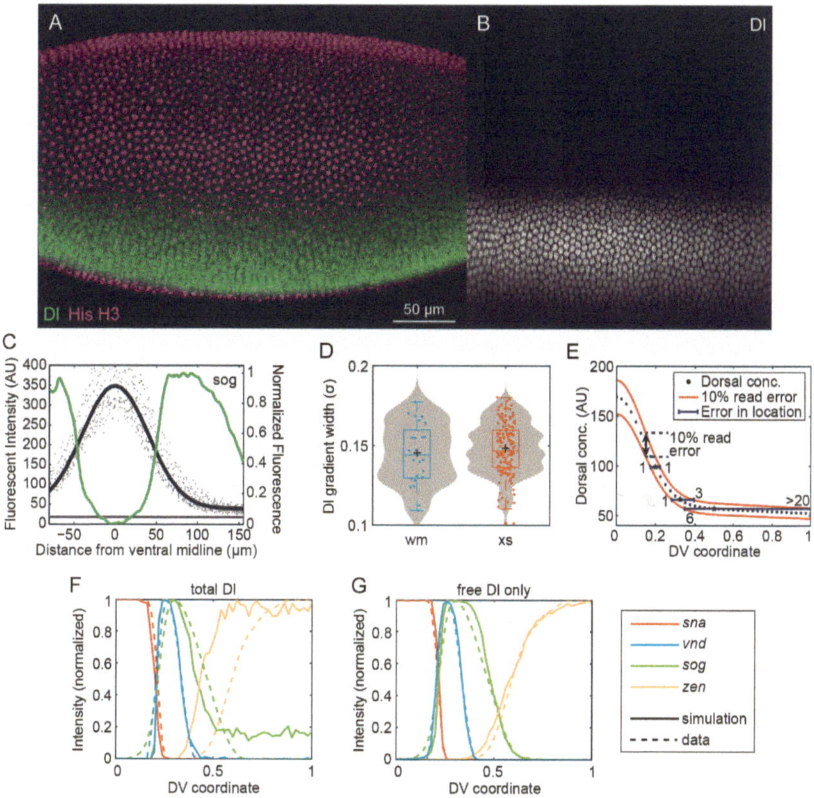

Fig. 5 Quantitative measurements of the Dl gradient width. (A) Max-intensity projection of embryo fluorescently immunostained against Dl and Histone H3. (B) Computationally flattened image of the Dl channel from embryo in (A). (C) Quantified Dl gradient and *sog* gene expression pattern. (D) Box-and-violin plot of Dl gradient widths from measurements from wholemount (wm) embryos and cross-sectioned (xs) embryos. (E) Illustration that adding 10% intrinsic noise to the Dl gradient results in drastic errors in gene expression boundary placement. (F) Simulation of gene expression using the total Dl levels, which correspond to direct fluorescent readouts, but may contain a Dl/Cact component. (G) Simulation of gene expression using only the deconvolved, free Dl levels. *Panels (A–C): Reproduced/modified with permission from Liberman, L. M., Reeves, G. T., & Stathopoulos, A. (2009). Quantitative imaging of the Dorsal nuclear gradient reveals limitations to threshold-dependent patterning in Drosophila.* Proceedings of the National Academy of Sciences of the United States of America, 106, *22317–22322. Panels (D, E): Reproduced/modified with permission from Reeves, G. T., Trisnadi, N., Truong, T. V, Nahmad, M., Katz, S., & Stathopoulos, A. (2012). Dorsal-ventral gene expression in the Drosophila embryo reflects the dynamics and precision of the dorsal nuclear gradient.* Developmental Cell, 22, *544–557. Panels (F, G): Reproduced/modified with permission from O'Connell, M. D., & Reeves, G. T. (2015). The presence of nuclear cactus in the early Drosophila embryo may extend the dynamic range of the dorsal gradient.* PLoS Computational Biology, 11, *e1004159.*

$$Dl\ conc(x) \approx A \exp\left(-\frac{(x-\mu)^2}{2\sigma^2}\right) + B,$$

where B denotes the basal levels of the gradient (i.e., the intensity of nuclear Dl at the dorsal midline), A signifies the amplitude of the gradient (above basal levels), and μ is the location of the ventral midline (within the image). For spatial range, the key parameter is the gradient width, σ, which measured to be $45 \pm 5\,\mu m$ (Liberman et al., 2009); for comparison, the spacing between nuclei is roughly 7–8 μm. The curvature of the embryos in the z-stacks suggested the embryo circumference was roughly 600 μm, which implies $\sigma \approx 0.15 \pm 0.02$ in relative units (Fig. 5D). To put that in context, as a bell-shaped curve decays to roughly zero at about three times σ, this implied the Dl gradient becomes flat roughly 45% of the way from the ventral midline to the dorsal midline, again suggesting the Dl gradient cannot specify the borders of the Type III genes. However, this discovery was in conflict with earlier genetic studies that showed the Dl gradient does in fact specify these borders (Jiang & Levine, 1993). One solution was the possibility that the spatial range of the Dl gradient changes over time, and that earlier time points carried the information to specify the Type III borders. However, measurements across nc 10–14 suggested the width remained constant (Liberman et al., 2009), a finding that was later confirmed by live imaging (see red curve in Fig. 1F; Reeves et al., 2012). While another possibility was that the fixation process confounded Dl gradient quantification, GFP-tagging of Dl caused the gradient to become wider (see also Section 6.3; Carrell et al., 2017; Kanodia et al., 2009; Liberman et al., 2009; Reeves et al., 2012). It was also possible that the computational methods used to analyze wholemount confocal z-stacks (Liberman et al., 2009) systematically missed the intensity of the gradient tail, which would affect estimates of the spatial range.

To attempt to resolve this problem, a study was performed in which the Dl gradients of more than 100 fixed embryos were imaged (Reeves et al., 2012). The embryos were cross sectioned manually (Carrell & Reeves, 2015), so that the entire DV axis could be imaged without need for spatial reconstruction (Fig. 1A, C). The Dl nuclear gradient was confirmed to be bell-shaped with a gradient width of $\sigma \approx 0.15 \pm 0.02$, which is the same as previously measured in wholemount embryos (Fig. 5D; Liberman et al., 2009). However, a shallow slope to the gradient was revealed, which persisted in the gradient tail past 45% DV, where the bell-shaped portion was otherwise flat (Fig. 1C; Reeves et al., 2012). The mechanism that ensures this shallow slope is unknown. The spatial range of the gradient was measured by examining a combination of the local slope and an estimate of the noise in the Dl nuclear concentration readout (Fig. 5E; Gregor, Tank, Wieschaus, & Bialek, 2007; Kanodia et al., 2011; Reeves et al., 2012). Even with the shallow slope, the positional information carried by the Dl gradient became negligible in the region where Type III gene borders are placed. Indeed, the Dl levels in a nucleus located at 50% DV could not be distinguished from those in a nucleus located at 100% DV (i.e., at the dorsal midline; Fig. 5E; Reeves et al., 2012).

The conclusion of limited spatial range was supported by noisy simulations of gene expression that used the measured Dl gradient as input (see also Section 7.2). Type I genes, in a region of the embryo where the Dl gradient could deliver precise positional information, were able to be simulated rather well. Type II genes were less well simulated, and Type III genes could not be properly simulated at all (Fig. 5F; Reeves et al., 2012; see also O'Connell & Reeves, 2015). A modeling study suggested the fluorescence measurements of the Dl gradient represented not just the free nuclear Dl, but also Dl/Cact complex (see Section 6.2 for more details of the Dl/Cact model; O'Connell & Reeves, 2015). Theoretically, this could resolve the conflict between measurements of the spatial range of the Dl gradient and the genetic evidence that directly shows that Dl regulates the borders of Type III genes. In particular, if Dl/Cact complex were also present in the nuclei, then some of the Dl gradient measurements would arise from a transcriptionally inactive pool of Dl molecules. Computational deconvolution of free nuclear Dl from the inactive Dl/Cact nuclear levels revealed an active Dl nuclear gradient with extended spatial range (O'Connell & Reeves, 2015). Furthermore, gene expression simulations using the deconvolved active Dl gradient as input were able to reproduce the proper positioning of Type III genes (Fig. 5G; for more details of gene expression simulations, see Section 7.2; O'Connell & Reeves, 2015). Further studies that attempt to detect Cact in the nuclei would be beneficial to test this hypothesis.

5.2 Dynamics of the Dl gradient

Live imaging of GFP-tagged Dl has revolutionized our understanding of the Dl nuclear concentration gradient. A pioneering study in 2007 showed that Dl is dynamically transported between the nucleus and its associated cytoplasmic region (DeLotto et al., 2007). Fluorescence recovery after photobleaching (FRAP) experiments showed nuclear export of Dl-GFP occurs on the rough time scale of a minute (Fig. 6A, B; DeLotto et al., 2007;

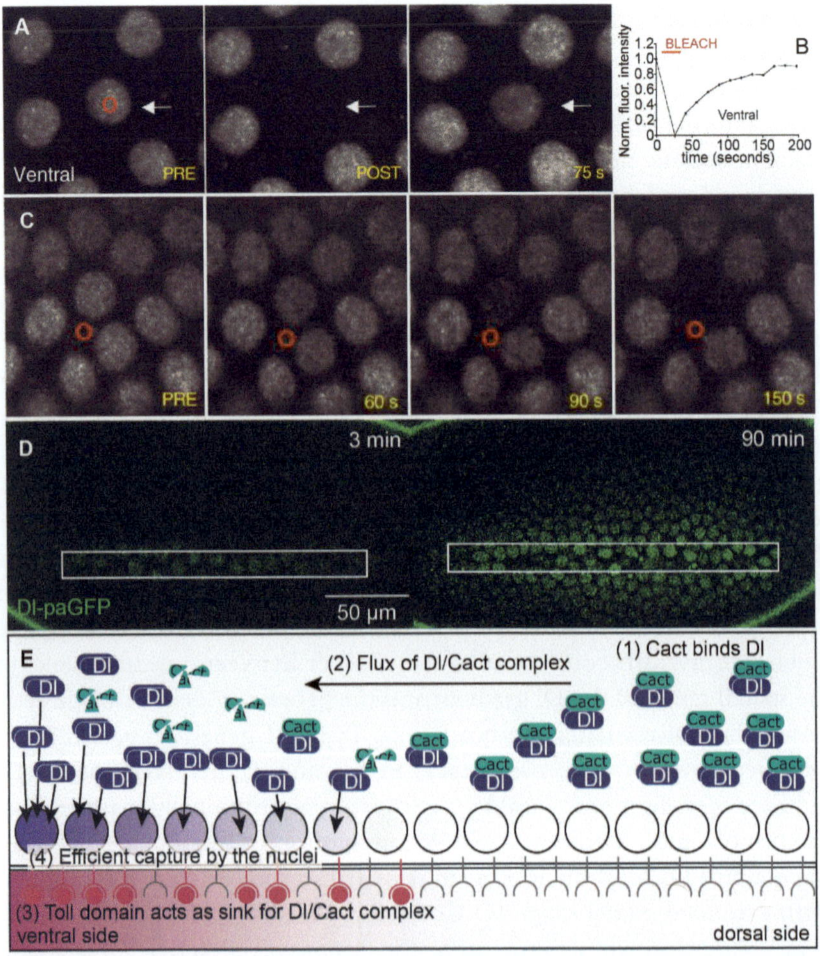

Fig. 6 The dynamics of the Dl gradient. (A) Example of FRAP experiment with a Dl-GFP. (B) Quantification of fluorescence recovery. (C) Bleaching the cytoplasm between several nuclei bleaches only one nucleus. (D) Spread of photo-activatable GFP-tagged Dl over 90 min. (E) Illustration of the processes necessary for shuttling of Dl by Cact. *Panels (A–C): Reproduced/modified with permission from DeLotto, R., DeLotto, Y., Steward, R., & Lippincott-Schwartz, J. (2007). Nucleocytoplasmic shuttling mediates the dynamic maintenance of nuclear Dorsal levels during Drosophila embryogenesis.* Development, 131, *4233–4241. Panels (D, E): Reproduced/modified with permission from Carrell, S. N., Connell, M. D. O., Jacobsen, T., Pomeroy, A. E., Hayes, S. M., & Reeves, G. T. (2017). A facilitated diffusion mechanism establishes the Drosophila Dorsal gradient.* Development, 144, *4450–4461.*

see also Al Asafen, Clark, Jacobsen, Sozzani, & Reeves, 2018; Carrell et al., 2017). Furthermore, photobleaching revealed barriers to Dl-GFP diffusion within the cytoplasm, even in the syncytial embryo (Fig. 6C). Each nucleus is surrounded by a small "island" of cytoplasm within which Dl-GFP is relatively well-mixed on the time scale of a few seconds or less (Daniels, Rikhy, Renz, Dobrowsky, & Lippincott-Schwartz, 2012; DeLotto et al., 2007). In contrast, the movement of Dl-GFP between cytoplasmic compartments occurs on a slower time scale: on the order of a minute (Carrell et al., 2017; DeLotto et al., 2007). Indeed, using a photoactivatable (pa) GFP-tagged Dl, it was shown that Dl-paGFP could move 7–10 nuclear diameters over the span of 90 min (Fig. 6D; Carrell et al., 2017), which translates to a time scale of roughly 90 s to cross one nuclear diameter.

The same study showed the Dl intensity in the ventral-most cells exhibited an oscillatory pattern in conjunction with the nuclear cycles (see also Fig. 1F; DeLotto et al., 2007), in a similar manner to that found for Bicoid in the same year (Gregor, Wieschaus, McGregor, Bialek, & Tank, 2007). Fixed embryo analysis showed that, on average, the gradient amplitude increased from nc 10 through nc 14, while the basal levels decreased (Liberman et al., 2009). Subsequent live imaging of an mVenus-tagged Dl confirmed these trends across nc 10–14, while also showing (1) the same trend of increasing amplitude and decreasing basal levels existed within individual nc's, and (2) during mitosis, the gradient amplitude would drastically decrease, while basal levels would abruptly increase (Fig. 1F; Reeves et al., 2012). It was initially suggested that after mitosis, newly-formed nuclei on the ventral side would begin with Dl levels below equilibrium, so that moving toward equilibrium during interphase caused the gradient amplitude to increase (Reeves et al., 2012). The abrupt decrease in amplitude at the onset of mitosis was caused by mixing of nuclear and cytoplasmic concentrations of Dl. In contrast, the decreasing basal levels during interphase was suggested to be caused by the dorsal-most nuclei beginning with Dl levels above equilibrium (Reeves et al., 2012). Subsequently, mathematical modeling suggested the declining basal levels were the result of the nuclei beginning interphase with high levels of *Dl/Cact complex*, which then decline over time as the complex is exported from the nuclei (see Section 6.2; O'Connell & Reeves, 2015).

It is unclear whether these complex dynamics of the gradient amplitude and basal levels have functional significance to the embryo. It has been suggested that a dynamically changing Dl gradient results in a slight shift over time in the borders of gene expression in the embryo (Reeves et al., 2012).

In particular, Type I genes, such as *sna*, expand slightly in their gene expression domain throughout nc 14, in correlation with the increasing gradient amplitude. On the other hand, the domains of Type III+ genes (e.g., *sog*), which may be subject to the decreasing basal levels, are known to contract toward the end of nc 14. Whether these correlations are caused by the dynamics of the Dl gradient, and in particular whether the dynamic gene expression boundaries are required for proper fitness, remains to be seen.

The dynamics of the Dl gradient themselves may be the result of a functionally significant mechanism called facilitated diffusion, in which Cact "shuttles" Dl ventrally, and which results in the overall accumulation of Dl on the ventral side of the embryo at the expense of the dorsal side (Carrell et al., 2017). Shuttling has been found to occur in other systems in *Drosophila* and vertebrates (Ashe & Levine, 1999; Dorfman & Shilo, 2001; Eldar et al., 2002; Haskel-Ittah et al., 2012; Holley et al., 1996; Marqués et al., 1997; Shilo, Haskel-Ittah, Ben-Zvi, Schejter, & Barkai, 2013; Shimmi, Umulis, Othmer, & O'Connor, 2005; Umulis, Serpe, O'Connor, & Othmer, 2006; Wang & Ferguson, 2005; Wharton & Serpe, 2013). This mechanism requires four biophysical processes to take place (Fig. 6E; Carrell et al., 2017; Shilo et al., 2013): (1) Cact must bind to Dl, (2) the Dl/Cact complex must be diffusible across the entire DV axis (possible because the embryo is a syncytium; see Fig. 6D), (3) Dl/Cact complex must be degraded on the ventral side, and (4) free Dl must be "captured" by the nuclei (to prevent counter-diffusion). It is clear each of these processes occur in the embryo, and previous work showed they are balanced such that significant shuttling occurs. It is also possible similar processes act to shuttle the Spz ligand within the perivitelline space (Haskel-Ittah et al., 2012).

As the shuttling mechanism causes Dl to move, on average, from the dorsal side to the ventral side, it explains the gradual decrease of the basal levels of the gradient and the accompanying increase in the gradient amplitude. Shuttling has been directly implicated in imparting robustness to the gene expression boundaries with respect to changes in overall maternal *dl* levels (Al Asafen, Bandodkar, Carrell-Noel, & Reeves, 2019; Carrell et al., 2017). In addition, the shuttling mechanism may also be required to reach the proper, peak levels of Dl required for Type I gene expression (Cardoso et al., 2017).

6. Mathematical modeling of the Dl gradient

Quantifying the shape and distribution of the Dl morphogen gradient was the first step toward gaining a mechanistic understanding of the DV patterning network. However, a systems-level understanding required computational modeling. Such modeling efforts rely on both the foundational knowledge of the components and interactions (reviewed in Section 2) for model formulation, as well as the quantitative data (reviewed in Section 5) for model constraints. In this section, we discuss several mathematical models that have successfully explained mechanisms behind known phenotypes, generated testable hypotheses, and driven further experimental discovery.

6.1 Pioneering models of the Dl gradient

In 2009, Kanodia et al. were the first to mathematically model the dynamics of the Dl gradient (Kanodia et al., 2009). They considered the DV axis to be a single dimension in space. The nuclei were modeled as compartments within well-mixed cytoplasmic regions. With every nuclear cycle, the number of nuclei, and their associated cytoplasmic compartments, increased by a factor of the square root of two, and their volume adjusted to keep the total volume constant (see also Coppey, Berezhkovskii, Kim, Boettiger, & Shvartsman, 2007). As shown in Fig. 7A (i), the model consisted of three cytoplasmic species (Dl, Cact and Dl/Cact complex), and one nuclear species (Dl). The cytoplasmic species were considered well-mixed within a cytoplasmic region and could diffuse slowly between neighboring regions. Nuclear import and export were permitted only for Dl, and the import/export rates were dependent on the surface area of the nucleus (see also Gregor, Wieschaus, et al., 2007). In the cytoplasm, Cact could reversibly bind to Dl to form Dl/Cact complex. The forward binding reaction was modeled as second order and the reverse (dissociation) reaction as first order. The reverse reaction, which was dissociation of the Dl/Cact complex, was considered catalyzed by active Toll signaling complexes, the distribution of which depended on DV position.

The model had nine dimensionless parameters that the authors constrained using data obtained from a snapshot of the Dl gradient distribution in live embryos 15 min into nc 14. As seen in Fig. 7A (ii, iii), the Dl gradient was found to be dynamic in nature, with an increasing amplitude and a constant shape. In fact, during the interphase of all nuclear cycles, the nuclear levels of Dl were found to increase monotonically at all points on the DV axis, and dropped only during mitosis. While the total amount of Dl was assumed to be constant, the continued action of Toll in the dissociation of Dl/Cact complex, resulted in increased production of free Dl that could migrate into the nucleus, thus increasing nuclear Dl levels with every nuclear

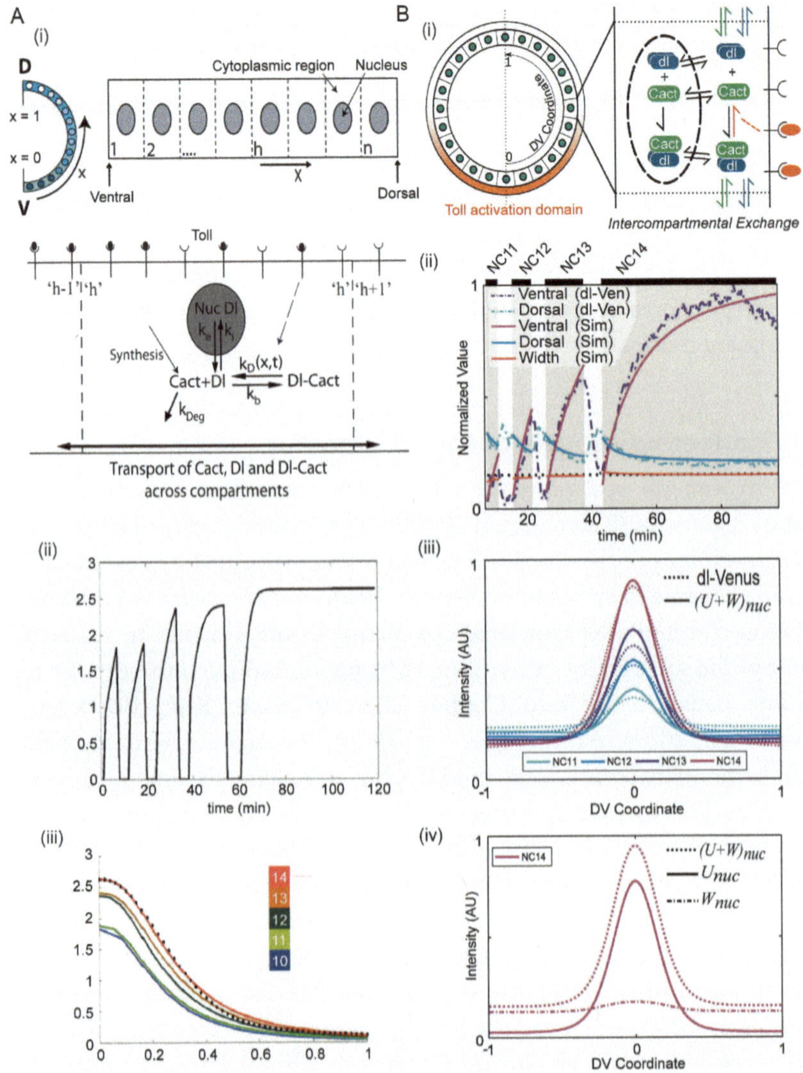

Fig. 7 Mathematical modeling of the Dl gradient. (A) Description of Kanodia model. (i) Modeling assumptions of the Kanodia model. Here the nuclei in the DV cross section are modeled as spheres in periodically arranged compartments. In each compartment, Dl reversibly reacts with Cact in the presence of Toll in the cytoplasmic region. Under the model specification, only Dl was allowed to enter and leave the nucleus. (ii) Temporal distribution of Dl at the ventral-most point as predicted by the model. (iii) Spatial distribution of Dl at the end of each nuclear cycle as predicted by the model, superimposed on experimental Dl-GFP data. (B) Description of the O'Connell deconvolution model. (i) Modeling assumptions of the deconvolution model. This is model is similar to the Kanodia model, but all three species Dl, Cact and Dl/Cact were allowed to enter and leave the nucleus. (ii) Temporal distribution of Dl at the ventral-most and dorsal-most points, as predicted by the model, superimposed on experimental Dl-Venus data from Reeves et al. (2012). (iii) Spatial distribution of Dl at the end of each nuclear cycle as predicted by the model, considering the fluorescence data to be sum of nuclear Dl, represented by U and nuclear Dl/Cact represented by W, superimposed on experimental Dl-Venus data. (iv) Spatial distribution profiles of the deconvolved components nuclear Dl and nuclear Dl/Cact complex as predicted by the model for nuclear cycle 14.

cycle. Strikingly, however, the shape of the gradient was practically unaffected by nuclear divisions. Thus, even though the slope of the gradient was relatively constant through successive nuclear divisions at all points in the DV axis, the regulatory regions of genes regulated by Dl would be exposed to rapidly escalating levels of the morphogen, raising the question of robustness of gene expression (discussed in Section 7.3).

The Dl morphogen is conserved across Drosophilids, which produce eggs of different sizes. However, in these species the Dl gradient has distinct shapes, which in turn leads to germ layer domains that do not scale linearly with size (Chahda, Sousa-Neves, & Mizutani, 2013). In 2014, Ambrosi et al. used the Kanodia model, which was developed for wild-type *D. melanogaster*, to identify parameters which could be responsible for the distinct shapes across species (Ambrosi, Chahda, Koslen, Chiel, & Mizutani, 2014). They hypothesized that modifying the nuclear size and density, and embryo geometry in the model may be sufficient to explain the differences in Dl gradient shape found between related species. First, the authors selected a representative parameter set from the ensemble of parameter sets identified by Kanodia et al. and modified those parameters that could minimally explain Dl gradient shapes in *D. melanogaster* mutant embryos. Parameters associated with Dl diffusion and export rates, degradation of Cact and embryo geometry were modified to obtain good fits with Dl gradients from these mutants. However, the modified set of parameters failed to reproduce the shape of the Dl gradient in related species. Then, the authors modified parameters associated spatially varying Dl/Cact dissociation rate constant and obtained adequate fits with experimental data from these species. They found that the species that were most distantly related (*D. sechellia* and *D. melanogaster*) have identical distribution of Dl gradient and species that were closest (*D. simulans* and *D. sechellia*) have completely different shapes. The Toll domain is broader in *D. simulans* and *D. sechellia* compared to *D. melanogaster*, while *D. simulans* produces larger eggs. The divergence in shape of the Dl gradient between *D. simulans* and *D. sechellia*, the authors argue, occurs because of similar Toll signaling dynamics. Thus, Ambrosi et al. were able to corroborate the Kanodia model by extending it for variations in Dl gradient shapes and sizes, as seen both within and across species.

6.2 Deconvolution of nuclear Dl and nuclear Dl/Cact complex

After the first mechanistic model of Dl/Cact interactions was published (Kanodia et al., 2009), the dynamics of the Dl nuclear gradient from nc 11 to 14 were measured in live embryos using a Venus-tagged Dl (Reeves et al., 2012). When the Dl gradient model was fit to these measurements, it was discovered that the model could only become consistent with the live imaging data if Dl/Cact complex was permitted to enter the nucleus (O'Connell & Reeves, 2015). The live imaging data showed that the dorsal-most nuclei began interphase with relatively high levels of Venus fluorescence, which then decreased as interphase progressed (Reeves et al., 2012). However, the previously-discussed models assumed all nuclei began interphase empty (i.e., with no fluorescence; Kanodia et al., 2009; O'Connell & Reeves, 2015). In order to have nuclei that began interphase with non-zero fluorescence, the most straightforward assumption was that, at the beginning of interphase, the nucleoplasm would have the same composition as the cytoplasm. Thus, the model was extended to include two additional nuclear species: Dl/Cact complex and free Cact, and their associated parameters (Fig. 7B (i); O'Connell & Reeves, 2015).

The attempt to align the model with the live imaging data—by assuming Dl/Cact complex was present in the nucleus—carried a further implication. The live measurements of Dl-Venus nuclear intensity were likely composed of both free Dl and Dl/Cact complex (O'Connell & Reeves, 2015). Therefore, the model parameters were constrained using the added sum of the concentrations of nuclear Dl and nuclear Dl/Cact from the model fit to experimentally derived Dl-Venus data obtained from (Fig. 7B (ii–iv); Reeves et al., 2012). If correct, this implication affects not only the live measurements, but all fluorescent measurements of the nuclear Dl gradient. This in turn implies the "active" Dl gradient, or nuclear gradient of free Dl, cannot be measured directly using fluorescent microscopy.

One possible method to infer the active Dl gradient from fluorescent intensity measurements is through computational modeling. Indeed, the authors used the model, with parameters fit to the fluorescent measurements, to deconvolve free Dl from Dl/Cact complex (O'Connell & Reeves, 2015). The computed active Dl gradient had several desirable properties not possessed by the gradient of nuclear fluorescent intensity, each of which can plausibly be attributed to the active Dl gradient having nearly zero basal levels. First, the computed active Dl gradient extended further into the DV axis (see Section 5.1; O'Connell & Reeves, 2015). This was demonstrated by simulated gene expression being superior as well (see Sections 5.1 and 7.2; O'Connell & Reeves, 2015; Reeves et al., 2012). Second, the computed active Dl gradient was more robust to noise fluctuations at the DNA level (O'Connell & Reeves, 2015;

see also Gregor, Tank, et al., 2007). Third, the computed active Dl gradient was recently shown to be more robust to perturbations to maternal dl dosage (Al Asafen et al., 2019). Given these advantages solve several theoretical issues with the Dl gradient, it is tempting to assume Dl/Cact complex is indeed in the nucleus. However, it remains to be seen whether Cact will be experimentally detected in the nucleus with Dl.

6.3 Facilitated diffusion of Dl by Cact and Toll saturation

Studies of Dl gradient dynamics showed the Dl gradient concentrations at the ventral midline (i.e., the gradient amplitude) increases on average with each nuclear cycle (see Sections 1 and 5.2; Fig. 1F; DeLotto et al., 2007; Kanodia et al., 2009; Liberman et al., 2009; Reeves et al., 2012). In 2017, Carell et al. showed that Dl accumulation on the ventral side of the embryo could be explained by facilitated diffusion of Dl by Cact from dorsal to ventral regions (Carrell et al., 2017).

The authors used a simplified model, similar to the original Kanodia model (Kanodia et al., 2009), consisting of three cytoplasmic species—Dl, Cact and Dl/Cact and one nuclear species—Dl. They modified the model to account for the possibility that Toll-mediated degradation of Dl/Cact complex could be limiting, by using Michaelis-Menten kinetics to describe the reaction. The model predicted that when shuttling rates were lowered, a hallmark series of phenotypes would be observed. Depending on the degree to which shuttling was compromised, the Dl gradient could become wider, flat-topped, or split-peaked.

To test the model predictions, the authors used GFP-tagged Dl to change the diffusion constant of Dl by varying the oligomer composition of GFP. They found that lowering the diffusion constant of Dl results in widened gradients, as predicted by the shuttling model. Also, when dosage of Dl was reduced, the phenotype showed widened gradients with flat tops and potentially split peaks (see also Ambrosi et al., 2014; Liberman et al., 2009), once again replicating the series of hallmark phenotypes of shuttling. When the Toll domain was widened, there was a split in the peak of Dl. When both dosage and mobility were varied simultaneously, the effects were exacerbated, resulting in unviable embryos. The authors showed that lowering dosage or widening the Toll domain would compromise shuttling (and thus, display the hallmark series of phenotypes) only when active Toll sites become saturated. The shuttling mechanism and Toll saturation, the authors asserted, complemented each other to ensure robustness to perturbations in development.

7. Modeling of Dl-dependent gene expression

7.1 Steady-state thermodynamic model of Dl target gene expression

Given that several enhancers of DV genes have been well-characterized (for example, see Fig. 8A (i)), a steady-state thermodynamic model of gene expression was devised (Zinzen et al., 2006). First, embryos in the early to mid-nuclear cycle 14 were imaged for Dl, Sna and Twi (proteins), as well as *rho* and *vnd* (mRNAs) (Fig. 8A (ii)). Both target genes, *rho* and *vnd*, are expressed in the ventral NeuroEctoderm (vNE) and have mostly overlapping domains, with *vnd* being slightly narrower than *rho*. Next, the protein expression data was used as an input to a site occupancy model, to predict the mRNA expression profiles. The model assumed that binding

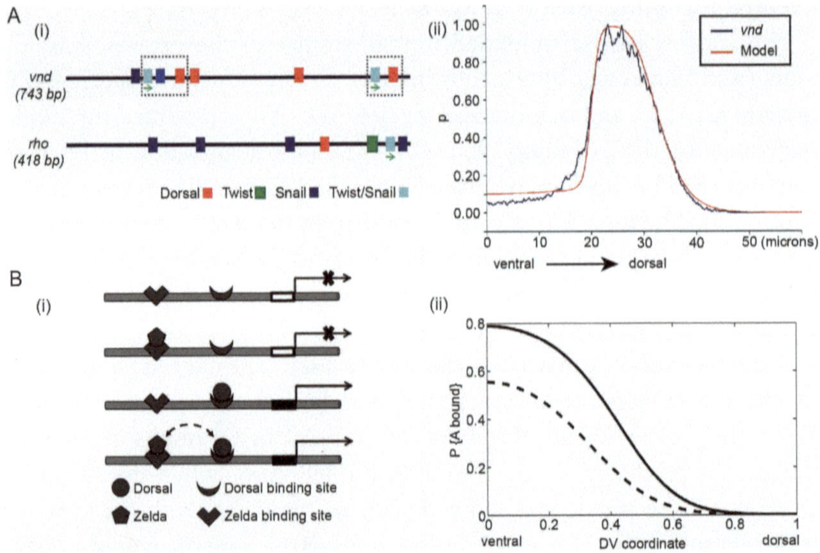

Fig. 8 Modeling of Dl-dependent gene expression. (A) Description of Zinzen thermodynamic model of Dl/Twi/Sna interactions. (i) Enhancer structures observed for *vnd* and *rho*. The dotted boxes correspond to elements with Dl and overlapping Twi/Sna binding sites and the green arrows mark the orientation of optimal Twi sites. The single-toned colored boxes indicate high scoring binding site motifs and two-toned colored boxes indicate low scoring binding motifs. (ii) The spatial expression profile of *vnd*, observed experimentally which is then fitted by the model. (B) Description of the Kanodia thermodynamic model of Dl/Zld interactions. (i) Site occupancy model for gene regulation with one binding site for Zld, indicated by the pentagon, and one for Dl, indicated by a circle. (ii) Spatial expression pattern of the activity of the regulatory region in the presence (shown by solid line) and absence (shown by dotted line) of Zld.

of at least one molecule of Dl and at least one molecule of Twi, to the enhancer region of the vNE genes would result in successful transcription, as long as Sna was not bound. It was also assumed all proteins bind with similar affinities. The model further assumed that protein binding is the rate limiting step in producing the expression patterns of *rho* and *vnd*. Most importantly, the model accounted for cooperativity effects between proximally bound proteins. The parameters of the model—namely the affinities, cooperative interactions between proteins, and number of binding sites—were constrained by fitting steady-state expression profiles to experimental measurements. The probability of successful transcription, for an element containing one Dl site and one Twi/Sna site was given by

$$p = \frac{C_{Dl-Twi}K_{Dl}[Dl]\ K_{Twi}[Twi]}{1 + K_{Dl}[Dl] + K_{Twi}[Twi] + K_{Sna}[Sna] + C_{Dl-Twi}K_{Dl}[Dl]K_{Twi}[Twi] + K_{Dl}[Dl]K_{sna}[Sna]}$$

where K_i represents binding affinity of species i, $[i]$ represents concentration of species i, and C_{ij} represents cooperative interactions between species i and j.

It was found that cooperativity between Dl/Twi, Twi/Twi and Sna/Sna proteins at binding sites was indispensable not only for making accurate model predictions, but also to explain the slight shift in the dorsal border of *rho*, as compared to *vnd*. For *rho* enhancers, the model predicted higher Dl-Twi and Twi-Twi cooperativity values and for *vnd* enhancers, it predicted higher number of Dl-Twi/Sna binding sites. Differences in cooperativity values, the authors suggest, may arise not only from direct protein interactions but indirectly by interactions with other parts of transcriptional machinery such as mediator complexes. While the number of binding sites may be directly verified by probing the enhancer elements of both genes, the predictions of cooperativity is more difficult to verify experimentally. Using principles of statistical thermodynamics, this model demonstrated at a mechanistic scale how high threshold targets, such as Twi, could act in conjunction with Dl to regulate gene expression in a cascade. However, since temporal resolution was lacking, a more complete analysis of a feed forward loop was not possible.

7.2 Simulating spatiotemporal dynamics of gene expression using the Dl gradient as input

Live imaging of the Dl gradient raised the question of how a highly dynamic, yet spatially restricted gradient could specify stable gene expression profiles of all types of Dl target genes (Reeves et al., 2012). Therefore, a simulated Dl gradient consistent with data from both live and fixed tissues—including a narrow, static gradient width; dynamic amplitude and basal levels; and a shallow slope in the gradient tail—was used as input to a model of gene expression (Reeves et al., 2012). The model focused on profiles of a spectrum of genes at different locations in the DV axis—*sna*, *vnd*, *sog*, and *zen*—during nuclear cycle 11–14. The model equation for normalized gene expression levels was as follows:

$$\frac{d[mRNA]_i}{dt} = \frac{1}{\tau_i}\left(f_i - [mRNA]_i\right),$$

where $[mRNA]_i$ is the concentration of mRNA (i=*sna*, *vnd*, *sog*, or *zen*), τ_i is the lifetime of mRNA i, and f_i is a hard threshold function of Dl to model the production rate of mRNA i. The model parameters (mRNA lifetimes and threshold values of Dl required for transcription) were constrained using experimentally obtained expression profiles of the target genes for nc 14. The authors added 10% variation to the Dl gradient input. Their reasoning was that part of the goal of the model was to determine whether the gradient was too spatially restricted to predict Type III gene expression, and a deterministic model can use even a miniscule slope to place boundaries to arbitrarily high precision (see also Gregor, Tank, et al., 2007).

Model fits obtained for earlier nuclear cycles were generally poor, due to the fact that the model was constrained by nc 14 gene expression, and the nc 10–12 Dl gradient is weak compared to the nc 13–14 gradient (see Fig. 1F). Therefore, the authors focused on the dynamics of gene expression domains in nc 14; the success of the fits varied with DV axis location. Even with noise, the model could successfully place a sharp boundary located at 20% DV, consistent with the profiles of Type I genes. Simulation of Type II genes also agreed fairly well with experimental data. However, the model could not properly match experimental data of Type III patterns. The domain of *sog*, a Type III+ gene, was predicted to expand up to the dorsal-most nucleus (see result of similar model in Fig. 5F; O'Connell & Reeves, 2015). In an attempt to eliminate the predicted *sog* expression on the dorsal half of the embryo, the threshold of Dl required to realize *sog* transcription was raised; however, this resulted in the *sog* domain collapsing to that of a Type II gene. Furthermore, the domain of Type III− gene *zen* was restricted mostly in lateral-dorsal regions. The ability of Dl to carry positional information to the boundaries of Type III genes has been in dispute since 2009 (see Section 5.1; Kanodia et al., 2009; Liberman et al., 2009). The authors suggested that a shallow gradient, plus noise, explained the inability of Dl to specify the Type III domains, and that other factors such as Zld, may also play a role. Indeed, Zld has been recently shown to enhance Dl concentration at the *sog* locus (Yamada et al., 2019). Alternatively, it is possible the assumption of Dl concentration thresholds is invalid. However, the model successfully predicted other aspects of the nc 14 expression patterns, such as expansion of the domain of Type I genes (*sna*), the static patterns of Type II genes (*vnd*) and contraction in the domain of Type III genes (*sog* and *zen*). Thus, the model satisfactorily predicted expression patterns in regions where the Dl levels were high and the gradient steep but failed in regions where Dl levels were low and the gradient shallow.

In 2015, O'Connell and Reeves used a similar threshold-dependent model for the same set of genes, but used the outputs of a model of Dl/Cact interactions as input to the gene expression simulations (see also Section 5.1; O'Connell & Reeves, 2015). Two distinct descriptions of the Dl gradient were investigated as input: (1) the modeled sum of nuclear Dl and nuclear Dl/Cact complex and (2) the modeled free nuclear Dl. The former was hypothesized to represent a direct readout of measured fluorescence (see Section 6.2), and the model produced mediocre fits for Type III genes, while making acceptable predictions for other gene expression borders (Fig. 5F), similar to what was seen previously (Reeves et al., 2012). The placement of the *sog* and *zen* boundaries were highly susceptible to perturbations in the parameters of the model describing the Dl gradient, since Dl fluorescence essentially flattens near the lateral regions. In the second case, gene expression simulations adequately fit all genes, including Type III genes (Fig. 5G). Thus, when the fluorescence data were computationally deconvolved into free nuclear Dl and nuclear Dl/Cact complex, and the active gradient was interpreted as the intensity profile of only free Dl, the gene expression simulations were robust. The robustness observed in the computationally deconvolved case can be attributed to the higher spatial range of nuclear Dl (see Section 6.2). However, even the deconvolved case had difficulty predicting gene expression of early nuclear cycles, likely due to the fact that the Dl gradient amplitude increases on average from nc 10 to 14.

7.3 Modeling the effect of Zelda on gene expression

Computational and experimental studies have shown that each nucleus in the DV axis is exposed to rapidly varying concentrations of Dl between nc 10 and 14 (DeLotto et al., 2007; Kanodia et al., 2009; Liberman et al., 2009; Reeves et al., 2012). Therefore, it was hypothesized that the large variations in the Dl concentrations could be stabilized by interaction with other regulators (Kanodia et al., 2009; Liberman et al., 2009; Reeves et al., 2012), in particular, the ubiquitous maternal protein, Zld (see Section 3.2).

Zld levels steadily decrease with time during nc 10–14, which, combined with the rising levels of nuclear Dl, could theoretically determine stable gene expression domains, even at early nuclear cycles. To test this hypothesis, a thermodynamic equation describing Zld/Dl cooperativity was analyzed theoretically (Kanodia et al., 2012). The authors used a two-site model, one each for Zld and Dl, and assumed that transcription could only occur when Dl was bound (Fig. 8B (i)). Close proximity of the binding sites of Dl and Zld would result in cooperative interactions between Zld and Dl, such that it enhances the activity of the regulatory region (Fig. 8B (ii)). These assumptions resulted in the following equation for the probability of Dl being bound:

$$P\{A\ bound\} = \frac{C_A}{K_A f(\omega, C_B, K_B) + C_A},$$

where C_A, C_B are the concentrations of Dl and Zld, respectively; K_A, K_B are the DNA binding affinities of Dl and Zld, respectively; ω is a measure of cooperativity; and $f = (1 + C_B/K_B)/(1 + \omega C_B/K_B)$. According to the equation, an increase in the binding strength of Zld or in its cooperativity with Dl could be inferred as an effective increase in the binding strength of Dl. With a simple theoretical analysis of the model equation, the authors were able to draw a few general conclusions regarding the role of Zld in Dl target gene expression. First, for a constant value of the binding strength of Dl, increasing either the binding strength of Zld or its cooperativity with Dl, increases the activity of a gene regulatory region. Second, varying the binding strength and/or cooperativity leads to corresponding changes in gene expression domains. This means that Zld could not only enhance the activity of the morphogen but could also influence patterning by controlling expression domains of its target genes.

To test these predictions experimentally, the authors specifically focused on the dorsal border of *sog*. The equation predicted a lower probability of *sog* expression in *zld* mutants than in wild type, which was seen experimentally as narrow and patchy domains of *sog*. According to the equation, in *zld* mutant embryos, *sog* expression will increase over time, as Dl is the only transcriptional activator, and its concentration steadily increases. On the other hand, in wild type, the effect of an increasing concentration of Dl is neutralized by a decreasing concentration of Zld, resulting in a broad, stable expression pattern. These dynamic effects of *zld* mutants and wild-type embryos were verified experimentally (Kanodia et al., 2012). However, the thermodynamic equation does not address how a narrow Dl gradient can specify the *sog* border. As no computation was performed in the paper, it is not clear whether the equation, using measured Dl nuclear fluorescence as input, could properly predict *sog* expression in wild-type embryos.

In conclusion, modeling efforts of Dl-dependent gene expression have each focused on one or two challenges to understanding DV patterning. From the evidence presented in this and previous sections, the expression profile of Dl alone will not be sufficient to describe detailed dynamics of gene expression: one must take into account spatiotemporal dynamics, the activity of Twi and Zld, and possibly the presence of Cact in the nuclei. A thermodynamic model accounted for Twi, but failed to account for dynamics or spatial extent of the gradient (in that there was no attempt to model Type III gene expression; Zinzen et al., 2006). Models using detailed dynamics of the Dl gradient revealed the need for accounting for Cact in the nuclei to extend the spatial range to model Type III gene expression (O'Connell & Reeves, 2015; Reeves et al., 2012). However, these models failed to account for Zld, which may have resulted in their inability to account for early gene expression, and for Twi, which acts in a coherent feed forward loop with Dl and may be responsible for dynamic stability of gene expression. It may be beneficial to model Dl-dependent gene expression in light of the efforts in each of the three sections reviewed here.

8. Conclusions and future perspectives

There remain several unanswered questions in the Dl gradient network, including the novel roles for Cact that have been identified in the last five years. For example, recent evidence suggests that signal-independent degradation of Cact produces a fragment that binds Dl yet is refractory to the ventral signal (Fontenele et al., 2013). If so, an entire species of Cact cleavage products, which has not been fully biochemically characterized, may play a major role in gradient formation. Furthermore, the signal-independent degradation may be regulated by calcium-dependent signaling, which in turn may be regulated by the maternal BMP pathway (see Section 4.2; Araujo & Bier, 2000; Carneiro et al., 2006; Fontenele et al., 2009).

Precise, quantitative measurements of Cact distribution, in either live or fixed tissues, could answer many questions. For example, visualization of Cact would answer whether it is present in the nucleus, and to what extent. However, even if it is present in the nucleus, it is unclear whether it plays a functional role there, as its homolog IκB does in mammalian systems. Indeed, mammalian IκB acts in a negative feedback loop (Cheng et al., 1994; Chiao, Miyamoto, & Verma, 1994; Scott, Fujita, Liou, Nolan, & Baltimore, 1993; Sun, Ganchi, Ballard, & Greene, 1993), and must enter the nuclei to attenuate NF-κB signaling (Arenzana-Seisdedos et al., 1995, 1997; Fagerlund et al., 2015), possibly by enhancing NF-κB dissociation from DNA (Bergqvist et al., 2009; Zabel & Baeuerle, 1990). It remains to be seen whether these processes also operate in the Dl/NF-κB network in the early *Drosophila* embryo.

The Dl gradient is an ideal system for quantitative and computational studies. Modeling work has been successful in describing the system, matching data, making predictions, and testing hypotheses (Ambrosi et al., 2014; Carrell et al., 2017; Kanodia et al., 2009; O'Connell & Reeves, 2015). However, model parameters largely remain unknown. The combination of ease of *Drosophila* transgenesis and the simple geometry of the blastoderm provide the opportunity to devise quantitative experiments to isolate and measure biophysical parameters, such as the nuclear import/export rate, the effective diffusivity, and fraction bound to DNA (Al Asafen et al., 2018; Carrell et al., 2017; DeLotto et al., 2007). Given the need for deconvolution of the nuclear fluorescent intensity of free Dl from that of Dl/Cact complex, a realistic set of model parameters may be needed for a proper understanding of the Dl activity gradient.

The input/output map between morphogen concentration and target gene expression represents the holy grail of tissue patterning. However, gene regulation within the Dl network is also dependent on other transcriptional effectors, in particular Zld and Twi. For the latter, quantitative measurements of the spatiotemporal dynamics of the Twi gradient are especially lacking (Bothma et al., 2018; Chung et al., 2011), as are models of the effect of Twi dynamics on gene expression patterns in the context of a feedforward loop. And while quantitative measurements of Zld influence on early gene expression have been prominent, we are only just beginning to plumb the depths of its ability to buffer against globally dynamic morphogen gradients (Kanodia et al., 2012), and to locally increase transcription factor concentration at the enhancer (Mir et al., 2017; Yamada et al., 2019). Ongoing research to meet these further challenges will ensure the Dl network remains at the forefront of quantitatively understanding the formation, interpretation, and regulation of morphogen gradients, tissue patterning, and gene regulation.

References

Al Asafen, H. Y., Bandodkar, P. U., Carrell-Noel, S., & Reeves, G. T. (2019). Robustness of the Dorsal morphogen gradient with respect to morphogen dosage. *BioRxiv*, 1–33. https://doi.org/10.1101/739292.

Al Asafen, H. Y., Clark, N. M., Jacobsen, T., Sozzani, R., & Reeves, G. T. (2018). Dorsal/NF-κB exhibits a dorsal-to-ventral mobility gradient in the Drosophila embryo. *BioRxiv*, 1–22. https://doi.org/10.1101/320754.

Alberga, A., Boulay, J. L., Kempe, E., Dennefeld, C., & Haenlin, M. (1991). The snail gene required for mesoderm formation in Drosophila is expressed dynamically in derivatives of all three germ layers. *Development*, *111*, 983–992.

Ambrosi, P., Chahda, J. S., Koslen, H. R., Chiel, H. J., & Mizutani, C. M. (2014). Modeling of the Dorsal gradient across species reveals interaction between embryo morphology and Toll signaling pathway during evolution. *PLoS Computational Biology*, *10*, e1003807.

Anderson, K. V. (1987). Dorsal—Ventral embryonic pattern genes of Drosophila. *Trends in Genetics*, *3*, 91–97.

Anderson, K. V., Jürgens, G., & Nüsslein-Volhard, C. (1985). Establishment of dorsal-ventral polarity in the Drosophila embryo: Genetic studies on the role of the Toll gene product. *Cell*, *42*, 779–789.

Anderson, K. V., & Nüsslein-Volhard, C. (1984). Information for the dorsal–ventral pattern of the Drosophila embryo is stored as maternal mRNA. *Nature*, *311*, 223–227.

Andreu, M. J., González-Pérez, E., Ajuria, L., Samper, N., González-Crespo, S., Campuzano, S., et al. (2012). Mirror represses pipe expression in follicle cells to initiate dorsoventral axis formation in Drosophila. *Development*, *139*, 1110–1114.

Araujo, H., & Bier, E. (2000). sog and dpp exert opposing maternal functions to modify Toll signaling and pattern the dorsoventral axis of the Drosophila embryo. *Development*, *127*, 3631–3644.

Arenzana-Seisdedos, F., Thompson, J., Rodriguez, M. S., Bachelerie, F., Thomas, D., & Hay, R. T. (1995). Inducible nuclear expression of newly synthesized I kappa B alpha negatively regulates DNA-binding and transcriptional activities of NF-kappa B. *Molecular and Cellular Biology*, *15*, 2689–2696.

Arenzana-Seisdedos, F., Turpin, P., Rodriguez, M., Thomas, D., Hay, R. T., Virelizier, J. L., et al. (1997). Nuclear localization of I kappa B alpha promotes active transport of NF-kappa B from the nucleus to the cytoplasm. *Journal of Cell Science*, *110*, 369–378.

Ashe, H. L., & Levine, M. (1999). Local inhibition and long-range enhancement of Dpp signal transduction by Sog. *Nature*, *398*, 427–431.

Baylies, M. K., & Bate, M. (1996). twist: A myogenic switch in Drosophila. *Science*, 272, 1481–1484.

Belvin, M. P., & Anderson, K. V. (1996). A conserved signaling pathway: The Drosophila Toll-dorsal pathway. *Annual Review of Cell and Developmental Biology*, *12*, 393–416.

Belvin, M. P., Jin, Y., & Anderson, K. V. (1995). Cactus protein degradation mediates Drosophila dorsal-ventral signaling. *Genes & Development*, *9*, 783–793.

Bergmann, A., Stein, D., Geisler, R., Hagenmaier, S., Schmid, B., Fernandez, N., et al. (1996). A gradient of cytoplasmic Cactus degradation establishes the nuclear localization gradient of the dorsal morphogen in Drosophila. *Mechanisms of Development*, *60*, 109–123.

Bergqvist, S., Alverdi, V., Mengel, B., Hoffmann, A., Ghosh, G., & Komives, E. A. (2009). Kinetic enhancement of NF-kappaB·DNA dissociation by IkappaB. *Proceedings of the National Academy of Sciences of the United States of America, 106*, 19328–19333.

Berman, B. P., Nibu, Y., Pfeiffer, B. D., Tomancak, P., Celniker, S. E., Levine, M., et al. (2002). Exploiting transcription factor binding site clustering to identify cis-regulatory modules involved in pattern formation in the Drosophila genome. *Proceedings of the National Academy of Sciences of the United States of America, 99*, 757–762.

Biemar, F., Nix, D. a., Piel, J., Peterson, B., Ronshaugen, M., Sementchenko, V., et al. (2006). Comprehensive identification of Drosophila dorsal-ventral patterning genes using a whole-genome tiling array. *Proceedings of the National Academy of Sciences of the United States of America, 103*, 12763–12768.

Bothma, J. P., Norstad, M. R., Alamos, S., & Garcia, H. G. (2018). LlamaTags: A versatile tool to image transcription factor dynamics in live embryos. *Cell, 173*, 1810–1822.e16.

Cardoso, M. A., Fontenele, M., Lim, B., Bisch, P. M., Shvartsman, S. Y., & Araujo, H. M. (2017). A novel function for the IκB inhibitor Cactus in promoting Dorsal nuclear localization and activity in the Drosophila embryo. *Development, 144*, 2907–2913.

Carneiro, K., Fontenele, M., Negreiros, E., Lopes, E., Bier, E., & Araujo, H. (2006). Graded maternal short gastrulation protein contributes to embryonic dorsal-ventral patterning by delayed induction. *Developmental Biology, 296*, 203–218.

Carrell, S. N., Connell, M. D. O., Jacobsen, T., Pomeroy, A. E., Hayes, S. M., & Reeves, G. T. (2017). A facilitated diffusion mechanism establishes the Drosophila Dorsal gradient. *Development, 144*, 4450–4461.

Carrell, S. N., & Reeves, G. T. (2015). Imaging the dorsal-ventral axis of live and fixed *Drosophila melanogaster* embryos. In C. M. Nelson (Ed.), *Tissue morphogenesis* (pp. 63–78). New York: Springer.

Chahda, J. S., Sousa-Neves, R., & Mizutani, C. M. (2013). Variation in the dorsal gradient distribution is a source for modified scaling of germ layers in Drosophila. *Current Biology, 23*, 710–716.

Chang, W.-L., Liou, W., Pen, H.-C., Chou, H.-Y., Chang, Y.-W., Li, W.-H., et al. (2008). The gradient of Gurken, a long-range morphogen, is directly regulated by Cbl-mediated endocytosis. *Development, 135*, 1923–1933.

Charatsi, I., Luschnig, S., Bartoszewski, S., Nüsslein-Volhard, C., & Moussian, B. (2003). Krapfen/dMyd88 is required for the establishment of dorsoventral pattern in the Drosophila embryo. *Mechanisms of Development, 120*, 219–226.

Chasan, R., & Anderson, K. V. (1989). The role of easter, an apparent serine protease, in organizing the dorsal-ventral pattern of the Drosophila embryo. *Cell, 56*, 391–400.

Chen, L.-Y., Wang, J.-C., Hyvert, Y., Lin, H.-P., Perrimon, N., Imler, J.-L., et al. (2006). Weckle is a zinc finger adaptor of the toll pathway in dorsoventral patterning of the Drosophila embryo. *Current Biology, 16*, 1183–1193.

Chen, H., Xu, Z., Mei, C., Yu, D., & Small, S. (2012). A system of repressor gradients spatially organizes the boundaries of bicoid-dependent target genes. *Cell, 149*, 618–629.

Cheng, Q., Cant, C. A., Moll, T., Hofer-Warbinek, R., Wagner, E., Birnstiel, M. L., et al. (1994). NK-kappa B subunit-specific regulation of the I kappa B alpha promoter. *The Journal of Biological Chemistry, 269*, 13551–13557.

Chiao, P. J., Miyamoto, S., & Verma, I. M. (1994). Autoregulation of I kappa B alpha activity. *Proceedings of the National Academy of Sciences of the United States of America, 91*, 28–32.

Ching, W., Hang, H. C., & Nusse, R. (2008). Lipid-independent secretion of a Drosophila Wnt protein. *The Journal of Biological Chemistry, 283*, 17092–17098.

Cho, Y. S., Stevens, L. M., Sieverman, K. J., Nguyen, J., & Stein, D. (2012). A ventrally localized protease in the Drosophila egg controls embryo dorsoventral polarity. *Current Biology, 22*, 1013–1018.

Cho, Y. S., Stevens, L. M., & Stein, D. (2010). Pipe-dependent ventral processing of Easter by snake is the defining step in Drosophila embryo DV axis formation. *Current Biology*, *20*, 1133–1137.
Chopra, V. S., & Levine, M. (2009). Combinatorial patterning mechanisms in the Drosophila embryo. *Briefings in Functional Genomics & Proteomics*, *8*, 243–249.
Chung, K., Kim, Y., Kanodia, J. S., Gong, E., Shvartsman, S. Y., & Lu, H. (2011). A microfluidic array for large-scale ordering and orientation of embryos. *Nature Methods*, *8*, 171–176.
Coppey, M., Berezhkovskii, A. M., Kim, Y., Boettiger, A. N., & Shvartsman, S. Y. (2007). Modeling the bicoid gradient: Diffusion and reversible nuclear trapping of a stable protein. *Developmental Biology*, *312*, 623–630.
Créton, R., Kreiling, J. A., & Jaffe, L. F. (2000). Presence and roles of calcium gradients along the dorsal-ventral axis in Drosophila embryos. *Developmental Biology*, *217*, 375–385.
Daigneault, J., Klemetsaune, L., & Wasserman, S. A. (2013). The IRAK homolog Pelle is the functional counterpart of IκB kinase in the Drosophila Toll pathway. *PLoS One*, *8* e75150.
Daniels, B. R., Rikhy, R., Renz, M., Dobrowsky, T. M., & Lippincott-Schwartz, J. (2012). Multiscale diffusion in the mitotic Drosophila melanogaster syncytial blastoderm. *Proceedings of the National Academy of Sciences of the United States of America*, *109*, 8588–8593.
de Lachapelle, A. M., & Bergmann, S. (2010). Precision and scaling in morphogen gradient read-out. *Molecular Systems Biology*, *6*, 351.
DeLotto, R. (2001). Gastrulation defective, a complement factor C2/B-like protease, interprets a ventral prepattern in Drosophila. *EMBO Reports*, *2*, 721–726.
DeLotto, Y., & DeLotto, R. (1998). Proteolytic processing of the Drosophila Spätzle protein by easter generates a dimeric NGF-like molecule with ventralising activity. *Mechanisms of Development*, *72*, 141–148.
DeLotto, R., DeLotto, Y., Steward, R., & Lippincott-Schwartz, J. (2007). Nucleocytoplasmic shuttling mediates the dynamic maintenance of nuclear Dorsal levels during Drosophila embryogenesis. *Development*, *134*, 4233–4241.
DeLotto, R., & Spierer, P. (1986). A gene required for the specification of dorsal-ventral pattern in Drosophila appears to encode a serine protease. *Nature*, *323*, 688–692.
Dev, A., Iyer, S., Razani, B., & Cheng, G. (2010). NF-κB and innate immunity. In M. Karin (Ed.), *NF-KB in health and disease* (pp. 115–143). Berlin Heidelberg: Springer.
Dorfman, R., & Shilo, B.-Z. (2001). Biphasic activation of the BMP pathway patterns the Drosophila embryonic dorsal region. *Development*, *128*, 965–972.
Driever, W., & Nüsslein-Volhard, C. (1988a). A gradient of bicoid protein in Drosophila embryos. *Cell*, *54*, 83–93.
Driever, W., & Nüsslein-Volhard, C. (1988b). The bicoid protein determines position in the Drosophila embryo in a concentration-dependent manner, *Cell*, *54*, 95–104.
Drier, E. A., & Steward, R. (1997). The dorsoventral signal transduction pathway and the Rel-like transcription factors in Drosophila. *Seminars in Cancer Biology*, *8*, 83–92.
Drier, E. A., Huang, L. H., & Steward, R. (1999). Nuclear import of the *Drosophila* rel protein dorsal is regulated by phosphorylation, *Genes & Development*, *13*, 556–568.
Dufourt, J., Trullo, A., Hunter, J., Fernandez, C., Lazaro, J., Dejean, M., et al. (2018). Temporal control of gene expression by the pioneer factor Zelda through transient interactions in hubs. *Nature Communications*, *9*, 5194.
Dushay, M. S., Asling, B., & Hultmark, D. (1996). Origins of immunity: Relish, a compound Rel-like gene in the antibacterial defense of Drosophila. *Proceedings of the National Academy of Sciences of the United States of America*, *93*, 10343–10347.

Eldar, A., Dorfman, R., Weiss, D., Ashe, H., Shilo, B.-Z., & Barkai, N. (2002). Robustness of the BMP morphogen gradient in Drosophila embryonic patterning. *Nature, 419*, 304–308.

Fagerlund, R., Behar, M., Fortmann, K. T., Lin, Y. E., Vargas, J. D., & Hoffmann, A. (2015). Anatomy of a negative feedback loop: The case of IκBα. *Journal of the Royal Society, Interface, 12* 20150262.

Ferguson, E. L., & Anderson, K. V. (1991). 2 Dorsal—ventral pattern formation in the Drosophila embryo: The role of zygotically active genes. *Current Topics in Developmental Biology, 25*, 17–43.

Foe, V. E., & Alberts, B. M. (1983). Studies of nuclear and cytoplasmic behaviour during the five mitotic cycles that precede gastrulation in Drosophila embryogenesis. *Journal of Cell Science, 61*, 31–70.

Fontenele, M., Carneiro, K., Agrellos, R., Oliveira, D., Oliveira-Silva, A., Vieira, V., et al. (2009). The Ca^{2+}-dependent protease Calpain A regulates Cactus/IκB levels during Drosophila development in response to maternal Dpp signals. *Mechanisms of Development, 126*, 737–751.

Fontenele, M., Lim, B., Oliveira, D., Buffolo, M., Perlman, D. H., Schupbach, T., et al. (2013). Calpain A modulates Toll responses by limited cactus/IκB proteolysis. *Molecular Biology of the Cell, 24*, 2966–2980.

Gabay, L., Seger, R., & Shilo, B.-Z. (1997). MAP kinase in situ activation atlas during Drosophila embryogenesis. *Development, 124*, 3535–3541.

Galindo, R. L., Edwards, D. N., Gillespie, S. K., & Wasserman, S. A. (1995). Interaction of the pelle kinase with the membrane-associated protein tube is required for transduction of the dorsoventral signal in Drosophila embryos. *Development, 121*, 2209–2218.

Ganguly, A., Jiang, J., & Ip, Y. T. (2005). Drosophila WntD is a target and an inhibitor of the Dorsal/Twist/Snail network in the gastrulating embryo. *Development, 132*, 3419–3429.

Garcia, M., Nahmad, M., Reeves, G. T., & Stathopoulos, A. (2013). Size-dependent regulation of dorsal–ventral patterning in the early Drosophila embryo. *Developmental Biology, 381*, 286–299.

Geisler, R., Bergmann, A., Hiromi, Y., & Nüsslein-Volhard, C. (1992). cactus, a gene involved in dorsoventral pattern formation of Drosophila, is related to the I kappa B gene family of vertebrates. *Cell, 71*, 613–621.

Ghosh, S., Gifford, A. M., Riviere, L. R., Tempst, P., Nolan, G. P., & Baltimore, D. (1990). Cloning of the p50 DNA binding subunit of NF-κB: Homology to rel and dorsal. *Cell, 62*, 1019–1029.

Gillespie, S. K., & Wasserman, S. A. (1994). Dorsal, a *Drosophila* Rel-like protein, is phosphorylated upon activation of the transmembrane protein Toll, *Molecular and Cellular Biology, 14*, 3559–3568.

Goentoro, L. A., Reeves, G. T., Kowal, C. P., Martinelli, L., Schüpbach, T., & Shvartsman, S. Y. (2006). Quantifying the Gurken morphogen gradient in Drosophila oogenesis. *Developmental Cell, 11*, 263–272.

Gordon, M. D., Dionne, M. S., Schneider, D. S., & Nusse, R. (2005). WntD is a feedback inhibitor of Dorsal/NF-kappaB in Drosophila development and immunity. *Nature, 437*, 746–749.

Govind, S. (1999). Control of development and immunity by rel transcription factors in Drosophila. *Oncogene, 18*, 6875–6887.

Gregor, T., Tank, D. W., Wieschaus, E. F., & Bialek, W. (2007). Probing the limits to positional information. *Cell, 130*, 153–164.

Gregor, T., Wieschaus, E. F., McGregor, A. P., Bialek, W., & Tank, D. W. (2007). Stability and nuclear dynamics of the Bicoid morphogen gradient. *Cell, 130*, 141–152.

Grosshans, J., Bergmann, A., Haffter, P., & Nüsslein-Volhard, C. (1994). Activation of the kinase Pelle by Tube in the dorsoventral signal transduction pathway of Drosophila embryo. *Nature, 372*, 563–566.

Harrison, M. M., Li, X.-Y., Kaplan, T., Botchan, M. R., & Eisen, M. B. (2011). Zelda binding in the early Drosophila melanogaster embryo marks regions subsequently activated at the maternal-to-zygotic transition. *PLoS Genetics*, 7 e1002266.

Hashimoto, C., Gerttula, S., & Anderson, K. V. (1991). Plasma membrane localization of the Toll protein in the syncytial Drosophila embryo: Importance of transmembrane signaling for dorsal-ventral pattern formation. *Development, 111*, 1021–1028.

Hashimoto, C., Hudson, K. L., & Anderson, K. V. (1988). The Toll gene of Drosophila, required for dorsal-ventral embryonic polarity, appears to encode a transmembrane protein. *Cell, 52*, 269–279.

Hashimoto, C., Kim, D. R., Weiss, L. A., Miller, J. W., Morisato, D., & Haven, N. (2003). Spatial regulation of developmental signaling by a serpin. *Developmental Cell, 5*, 945–950.

Haskel-Ittah, M., Ben-Zvi, D., Branski-Arieli, M., Schejter, E. D., Shilo, B.-Z., & Barkai, N. (2012). Self-organized shuttling: Generating sharp dorsoventral polarity in the early Drosophila embryo. *Cell, 150*, 1016–1028.

Helman, A., Lim, B., Andreu, M. J., Kim, Y., Shestkin, T., Lu, H., et al. (2012). RTK signaling modulates the Dorsal gradient. *Development, 139*, 3032–3039.

Hetru, C., & Hoffmann, J. a. (2009). NF-kappaB in the immune response of Drosophila. *Cold Spring Harbor Perspectives in Biology, 1* a000232.

Hoesel, B., & Schmid, J. a. (2013). The complexity of NF-κB signaling in inflammation and cancer. *Molecular Cancer, 12*, 86.

Holley, S. A., Neul, J. L., Attisano, L., Wrana, J. L., Sasai, Y., O'Connor, M. B., et al. (1996). The Xenopus dorsalizing factor noggin ventralizes Drosophila embryos by preventing DPP from activating its receptor. *Cell, 86*, 607–617.

Hong, C. C., & Hashimoto, C. (1995). An unusual mosaic protein with a protease domain, encoded by the nudel gene, is involved in defining embryonic dorsoventral polarity in Drosophila. *Cell, 82*, 785–794.

Huang, J. D., Schwyter, D. H., Shirokawa, J. M., & Courey, A. J. (1993). The interplay between multiple enhancer and silencer elements defines the pattern of decapentaplegic expression. *Genes & Development*, 7, 694–704.

Ip, Y. T., Kraut, R., Levine, M., & Rushlow, C. A. (1991). The dorsal morphogen is a sequence-specific DNA-binding protein that interacts with a long-range repression element in Drosophila. *Cell, 64*, 439–446.

Ip, Y. T., Park, R. E., Kosman, D., Bier, E., & Levine, M. (1992). The dorsal gradient morphogen regulates stripes of rhomboid expression in the presumptive neuroectoderm of the Drosophila embryo. *Genes & Development, 6*, 1728–1739.

Ip, Y. T., Park, R. E., Kosman, D., Yazdanbakhsh, K., & Levine, M. (1992). dorsal-twist interactions establish snail expression in the presumptive mesoderm of the Drosophila embryo. *Genes & Development, 6*, 1518–1530.

Ip, Y. T., Reach, M., Engstrom, Y., Kadalayil, L., Cai, H., González-Crespo, S., et al. (1993). Dif, a dorsal-related gene that mediates an immune response in Drosophila. *Cell, 75*, 753–763.

Jaeger, J., Surkova, S., Blagov, M., Janssens, H., Kosman, D., Kozlov, K. N., et al. (2004). Dynamic control of positional information in the early Drosophila embryo. *Nature, 430*, 368–371.

Jiang, J., Kosman, D., Ip, Y. T., & Levine, M. (1991). The dorsal morphogen gradient regulates the mesoderm determinant twist in early Drosophila embryos. *Genes & Development, 5*, 1881–1891.

Jiang, J., & Levine, M. (1993). Binding affinities and cooperative interactions with bHLH activators delimit threshold responses to the dorsal gradient morphogen. *Cell*, *72*, 741–752.

Jiang, J., Rushlow, C. A., Zhou, Q., Small, S., & Levine, M. (1992). Individual dorsal morphogen binding sites mediate activation and repression in the Drosophila embryo. *The EMBO Journal*, *11*, 3147–3154.

Kanodia, J. S., Kim, Y., Tomer, R., Khan, Z., Chung, K., Storey, J. D., et al. (2011). A computational statistics approach for estimating the spatial range of morphogen gradients. *Development*, *138*, 4867–4874.

Kanodia, J. S., Liang, H.-L., Kim, Y., Lim, B., Zhan, M., Lu, H., et al. (2012). Pattern formation by graded and uniform signals in the early Drosophila embryo. *Biophysical Journal*, *102*, 427–433.

Kanodia, J. S., Rikhy, R., Kim, Y., Lund, V. K., DeLotto, R., Lippincott-Schwartz, J., et al. (2009). Dynamics of the Dorsal morphogen gradient. *Proceedings of the National Academy of Sciences of the United States of America*, *106*, 21707–21712.

Kidd, S. (1992). Characterization of the Drosophila cactus locus and analysis of interactions between cactus and dorsal proteins. *Cell*, *71*, 623–635.

Kieran, M., Blank, V., Logeat, F., Vandekerckhove, J., Lottspeich, F., Le Bail, O., et al. (1990). The DNA binding subunit of NF-κB is identical to factor KBF1 and homologous to the rel oncogene product. *Cell*, *62*, 1007–1018.

Kirov, N., Childs, S., O'Connor, M., & Rushlow, C. (1994). The Drosophila dorsal morphogen represses the tolloid gene by interacting with a silencer element. *Molecular and Cellular Biology*, *14*, 713–722.

Konsolaki, M., & Schüpbach, T. (1998). windbeutel, a gene required for dorsoventral patterning in Drosophila, encodes a protein that has homologies to vertebrate proteins of the endoplasmic reticulum. *Genes & Development*, *12*, 120–131.

Kosman, D., Ip, Y. T., Levine, M., & Arora, K. (1991). Establishment of the mesoderm-neuroectoderm boundary in the Drosophila embryo. *Science*, *254*, 118–122.

Lemaitre, B., Meister, M., Govind, S., Georgel, P., Steward, R., Reichhart, J. M., et al. (1995). Functional analysis and regulation of nuclear import of dorsal during the immune response in Drosophila. *The EMBO Journal*, *14*, 536–545.

LeMosy, E. K., Kemler, D., & Hashimoto, C. (1998). Role of Nudel protease activation in triggering dorsoventral polarization of the Drosophila embryo. *Development*, *125*, 4045–4053.

LeMosy, E. K., Leclerc, C. L., & Hashimoto, C. (2000). Biochemical defects of mutant nudel alleles causing early developmental arrest or dorsalization of the Drosophila embryo. *Genetics*, *154*, 247–257.

LeMosy, E. K., Tan, Y.-Q., & Hashimoto, C. (2001). Activation of a protease cascade involved in patterning the Drosophila embryo. *Proceedings of the National Academy of Sciences*, *98*, 5055–5060.

Leptin, M. (1991). twist and snail as positive and negative regulators during Drosophila mesoderm development. *Genes & Development*, 1568–1576.

Levine, M., & Davidson, E. H. (2005). Gene regulatory networks for development. *Small*, *102*, 4936–4942.

Liang, H.-L., Nien, C.-Y., Liu, H.-Y., Metzstein, M. M., Kirov, N., & Rushlow, C. (2008). The zinc-finger protein Zelda is a key activator of the early zygotic genome in Drosophila. *Nature*, *456*, 400–403.

Liang, H.-L., Xu, M., Chuang, Y.-C., & Rushlow, C. (2012). Response to the BMP gradient requires highly combinatorial inputs from multiple patterning systems in the Drosophila embryo. *Development*, *139*, 1956–1964.

Liberman, L. M., Reeves, G. T., & Stathopoulos, A. (2009). Quantitative imaging of the Dorsal nuclear gradient reveals limitations to threshold-dependent patterning in Drosophila. *Proceedings of the National Academy of Sciences of the United States of America*, *106*, 22317–22322.

Liberman, L. M., & Stathopoulos, A. (2009). Design flexibility in cis-regulatory control of gene expression: Synthetic and comparative evidence. *Developmental Biology, 327*, 578–589.

Ligoxygakis, P., Roth, S., & Reichhart, J. (2003). A serpin regulates dorsal-ventral axis formation in the Drosophila embryo. *Current, 13*, 2097–2102.

Lin, M., Park, J., Kirov, N., & Rushlow, C. (2006). Threshold response of C15 to the Dpp gradient in Drosophila is established by the cumulative effect of Smad and Zen activators and negative cues. *Development, 133*, 4805–4813.

Liu, T., Zhang, L., Joo, D., & Sun, S.-C. (2017). NF-κB signaling in inflammation. *Signal Transduction and Targeted Therapy, 2* 17023.

Llimargas, M., & Lawrence, P. A. (2001). Seven Wnt homologues in Drosophila: A case study of the developing tracheae. *Proceedings of the National Academy of Sciences of the United States of America, 98*, 14487–14492.

Luders, F., Segawa, H., Stein, D., Selva, E. M., Perrimon, N., Turco, S. J., et al. (2003). slalom encodes an adenosine 3′-phosphate 5′-phosphosulfate transporter essential for development in Drosophila. *The EMBO Journal, 22*, 3635–3644.

Marqués, G., Musacchio, M., Shimell, M. J., Wünnenberg-Stapleton, K., Cho, K. W., & O'Connor, M. B. (1997). Production of a DPP activity gradient in the early Drosophila embryo through the opposing actions of the SOG and TLD proteins. *Cell, 91*, 417–426.

McElwain, M. A., Ko, D. C., Gordon, M. D., Fyrst, H., Saba, J. D., & Nusse, R. (2011). A suppressor/enhancer screen in Drosophila reveals a role for Wnt-mediated lipid metabolism in primordial germ cell migration. *PLoS One, 6* e26993.

McHale, P., Mizutani, C. M., Kosman, D., MacKay, D. L., Belu, M., Hermann, A., et al. (2011). Gene length may contribute to graded transcriptional responses in the Drosophila embryo. *Developmental Biology, 360*, 230–240.

Minakhina, S., & Steward, R. (2006). Nuclear factor-kappa B pathways in Drosophila. *Oncogene, 25*, 6749–6757.

Mir, M., Reimer, A., Haines, J. E., Li, X., Stadler, M., Garcia, H., et al. (2017). Dense Bicoid hubs accentuate binding along the morphogen gradient. *Genes & Development, 31*, 1784–1794.

Morisato, D., & Anderson, K. V. (1995). Signaling pathways that establish the dorsal-ventral pattern of the Drosophila embryo. *Annual Review of Genetics, 29*, 371–399.

Moussian, B., & Roth, S. (2005). Dorsoventral axis formation in the Drosophila embryo—Shaping and transducing a morphogen gradient. *Current Biology, 15*, R887–R899.

Neuman-Silberberg, F. S., & Schüpbach, T. (1993). The Drosophila dorsoventral patterning gene gurken produces a dorsally localized RNA and encodes a TGFα-like protein. *Cell, 75*, 165–174.

Neuman-Silberberg, F. S., & Schüpbach, T. (1996). The Drosophila TGF-α-like protein Gurken: Expression and cellular localization during Drosophila oogenesis. *Mechanisms of Development, 59*, 105–113.

Nien, C.-Y., Liang, H.-L., Butcher, S., Sun, Y., Fu, S., Gocha, T., et al. (2011). Temporal coordination of gene networks by Zelda in the early Drosophila embryo. *PLoS Genetics, 7* e1002339.

Nilson, L. A., & Schu, T. (1998). Localized requirements for windbeutel and pipe reveal a dorsoventral prepattern within the follicular epithelium of the Drosophila ovary. *Cell, 93*, 253–262.

Nusslein-Volhard, C. (1979). Maternal effect mutations that alter the spatial coordinates of the embryo of Drosophila melanogaster. In *Determinants of spatial organization* (pp. 185–211). Elsevier.

Nüsslein-Volhard, C., Lohs-Schardin, M., Sander, K., & Cremer, C. (1980). A dorso-ventral shift of embryonic primordia in a new maternal-effect mutant of Drosophila. *Nature, 283*, 474–476.

O'Connell, M. D., & Reeves, G. T. (2015). The presence of nuclear cactus in the early Drosophila embryo may extend the dynamic range of the dorsal gradient. *PLoS Computational Biology*, *11* e1004159.

Pai, L.-M., Barcelo, G., & Schüpbach, T. (2000). D-cbl, a negative regulator of the Egfr pathway, is required for dorsoventral patterning in Drosophila oogenesis. *Cell*, *103*, 51–61.

Papatsenko, D., & Levine, M. (2005). Quantitative analysis of binding motifs mediating diverse spatial readouts of the Dorsal gradient in the Drosophila embryo. *Proceedings of the National Academy of Sciences of the United States of America*, *102*, 4966–4971.

Peri, F., Technau, M., & Roth, S. (2002). Mechanisms of Gurken-dependent pipe regulation and the robustness of dorsoventral patterning in Drosophila. *Development*, *2975*, 2965–2975.

Price, J. V., Clifford, R. J., & Schüpbach, T. (1989). The maternal ventralizing locus torpedo is allelic to faint little ball, an embryonic lethal, and encodes the Drosophila EGF receptor homolog. *Cell*, *56*, 1085–1092.

Raftery, L. A., Twombly, V., Wharton, K., & Gelbart, W. M. (1995). Genetic screens to identify elements of the decapentaplegic signaling pathway in Drosophila. *Genetics*, *139*, 241–254.

Rahimi, N., Averbukh, I., Haskel-Ittah, M., Degani, N., Schejter, E. D., Barkai, N., et al. (2016). A WntD-dependent integral feedback loop attenuates variability in Drosophila Toll Signaling. *Developmental Cell*, *36*, 401–414.

Reach, M., Galindo, R. L., Towb, P., Allen, J. L., Karin, M., & Wasserman, S. A. (1996). A gradient of cactus protein degradation establishes dorsoventral polarity in the Drosophila embryo. *Developmental Biology*, *180*, 353–364.

Reeves, G. T., & Stathopoulos, A. (2009). Graded dorsal and differential gene regulation in the Drosophila embryo. *Cold Spring Harbor Perspectives in Biology*, *1* a000836.

Reeves, G. T., Trisnadi, N., Truong, T. V., Nahmad, M., Katz, S., & Stathopoulos, A. (2012). Dorsal-ventral gene expression in the Drosophila embryo reflects the dynamics and precision of the dorsal nuclear gradient. *Developmental Cell*, *22*, 544–557.

Robbins, P. W., & Lipmann, F. (1957). Isolation and identification of active sulfate. *The Journal of Biological Chemistry*, *229*, 837–851.

Roth, S. (2003). The origin of dorsoventral polarity in Drosophila. *Philosophical Transactions of the Royal Society of London Series B, Biological Sciences*, *358*, 1317–1329.

Roth, S., Hiromi, Y., Godt, D., & Nüsslein-Volhard, C. (1991). cactus, a maternal gene required for proper formation of the dorsoventral morphogen gradient in Drosophila embryos. *Development*, *112*, 371–388.

Roth, S., Jordan, P., & Karess, R. (1999). Binuclear Drosophila oocytes: Consequences and implications for dorsal-ventral patterning in oogenesis and embryogenesis. *Development*, *126*, 927–934.

Roth, S., & Lynch, J. A. (2009). Symmetry breaking during Drosophila oogenesis. *Cold Spring Harbor Perspectives in Biology*, *1* a001891.

Roth, S., Stein, D., & Nüsslein-Volhard, C. (1989). A gradient of nuclear localization of the dorsal protein determines dorsoventral pattern in the Drosophila embryo. *Cell*, *59*, 1189–1202.

Rushlow, C. A., Han, K., Manley, J. L., & Levine, M. (1989). The graded distribution of the dorsal morphogen is initiated by selective nuclear transport in Drosophila. *Cell*, *59*, 1165–1177.

Rushlow, C. A., & Shvartsman, S. Y. (2012). Temporal dynamics, spatial range, and transcriptional interpretation of the Dorsal morphogen gradient. *Current Opinion in Genetics & Development*, 1–5.

Sandler, J. E., & Stathopoulos, A. (2016a). Stepwise progression of embryonic patterning. *Trends in Genetics*, *32*, 432–443.

Sandler, J. E., & Stathopoulos, A. (2016b). Quantitative single-embryo profile of Drosophila genome activation and the dorsal-ventral patterning network. *Genetics*, *202*, 1575–1584.

Sandmann, T., Girardot, C., Brehme, M., Tongprasit, W., Stolc, V., & Furlong, E. E. M. (2007). A core transcriptional network for early mesoderm development in Drosophila melanogaster. *Genes & Development*, *21*, 436–449.

Schejter, E. D., & Shilo, B. Z. (1989). The Drosophila EGF receptor homolog (DER) gene is allelic to faint little ball, a locus essential for embryonic development. *Cell*, *56*, 1093–1104.

Schloop, A. E., Carrell-Noel, S., & Reeves, G. T. (2019). Shuttling of Dorsal by Cactus: Mechanism and implications. *BioRxiv*, 1–21. https://doi.org/10.1101/739284.

Schneider, D. S., Hudson, K. L., Lin, T. Y., & Anderson, K. V. (1991). Dominant and recessive mutations define functional domains of Toll, a transmembrane protein required for dorsal-ventral polarity in the Drosophila embryo. *Genes & Development*, *5*, 797–807.

Schüpbach, T. (1987). Germ line and soma cooperate during oogenesis to establish the dorsoventral pattern of egg shell and embryo in Drosophila melanogaster. *Cell*, *49*, 699–707.

Scott, M. L., Fujita, T., Liou, H. C., Nolan, G. P., & Baltimore, D. (1993). The p65 subunit of NF-kappa B regulates I kappa B by two distinct mechanisms. *Genes & Development*, *7*, 1266–1276.

Sen, R., & Baltimore, D. (1986). Multiple nuclear factors interact with the immunoglobulin enhancer sequences. *Cell*, *46*, 705–716.

Sen, J., Goltz, J. S., Konsolaki, M., Schüpbach, T., & Stein, D. (2000). Windbeutel is required for function and correct subcellular localization of the Drosophila patterning protein pipe. *Development*, *127*, 5541–5550.

Sen, J., Goltz, J. S., Stevens, L., & Stein, D. (1998). Spatially restricted expression of pipe in the Drosophila egg chamber defines embryonic dorsal–ventral polarity. *Cell*, *95*, 471–481.

Shelton, C. A., & Wasserman, S. A. (1993). pelle encodes a protein kinase required to establish dorsoventral polarity in the Drosophila embryo. *Cell*, *72*, 515–525.

Shilo, B.-Z., Haskel-Ittah, M., Ben-Zvi, D., Schejter, E. D., & Barkai, N. (2013). Creating gradients by morphogen shuttling. *Trends in Genetics*, *29*, 339–347.

Shimmi, O., Umulis, D., Othmer, H., & O'Connor, M. B. (2005). Facilitated transport of a Dpp/Scw heterodimer by Sog/Tsg leads to robust patterning of the Drosophila blastoderm embryo. *Cell*, *120*, 873–886.

Simpson, P. (1983). Maternal-zygotic gene interactions during formation of the dorsoventral pattern in Drosophila embryos. *Genetics*, *105*, 615–632.

Stathopoulos, A., & Levine, M. (2002). Dorsal gradient networks in the Drosophila embryo. *Developmental Biology*, *246*, 57–67.

Stathopoulos, A., & Levine, M. (2004). Whole-genome analysis of Drosophila gastrulation. *Current Opinion in Genetics & Development*, *14*, 477–484.

Stathopoulos, A., & Levine, M. (2005). Genomic regulatory networks and animal development. *Developmental Cell*, *9*, 449–462.

Stathopoulos, A., Van Drenth, M., Erives, A., Markstein, M., & Levine, M. (2002). Whole-genome analysis of dorsal-ventral patterning in the Drosophila embryo. *Cell*, *111*, 687–701.

Stein, D. D., Cho, Y. S. Y., Zhang, Z., & Stevens, L. L. M. (2008). No requirement for localized nudel protein expression in Drosophila embryonic axis determination. *Fly (Austin)*, *2*, 220–228.

Stein, D., Roth, S., Vogelsang, E., & Nüsslein-Volhard, C. (1991). The polarity of the dorsoventral axis in the Drosophila embryo is defined by an extracellular signal. *Cell*, *65*, 725–735.

Stein, D. S., & Stevens, L. M. (2014). Maternal control of the Drosophila dorsal-ventral body axis. *Wiley Interdisciplinary Reviews: Developmental Biology*, *3*, 301–330.

Steward, R. (1987). Dorsal, an embryonic polarity gene in Drosophila, is homologous to the vertebrate proto-oncogene, c-rel. *Science*, *238*, 692–694.
Steward, R. (1989). Relocalization of the dorsal protein from the cytoplasm to the nucleus correlates with its function. *Cell*, *59*, 1179–1188.
Steward, R., Ambrosio, L., & Schedl, P. (1985). Expression of the dorsal gene. *Cold Spring Harbor Symposia on Quantitative Biology*, *50*, 223–228.
Steward, R., & Govind, S. (1993). Dorsal-ventral polarity in the Drosophila embryo. *Current Opinion in Genetics & Development*, *3*, 556–561.
Steward, R., McNally, F. J., & Schedl, P. (1984). Isolation of the dorsal locus of Drosophila. *Nature*, *311*, 262–265.
Steward, R., Zusman, S. B., Huang, L. H., & Schedl, P. (1988). The dorsal protein is distributed in a gradient in early drosophila embryos. *Cell*, *55*, 487–495.
Sun, S., Ganchi, P., Ballard, D., & Greene, W. (1993). NF-kappa B controls expression of inhibitor I kappa B alpha: Evidence for an inducible autoregulatory pathway. *Science*, *259*, 1912–1915.
Sun, Y., Nien, C. Y., Chen, K., Liu, H. Y., Johnston, J., Zeitlinger, J., et al. (2015). Zelda overcomes the high intrinsic nucleosome barrier at enhancers during Drosophila zygotic genome activation. *Genome Research*, *25*, 1703–1714.
Sun, H., Towb, P., Chiem, D. N., Foster, B. A., & Wasserman, S. A. (2004). Regulated assembly of the toll signaling complex drives Drosophila dorsoventral patterning. *The EMBO Journal*, *23*, 100–110.
Szymanski, P., & Levine, M. (1995). Multiple modes of dorsal-bHLH transcriptional synergy in the Drosophila embryo. *The EMBO Journal*, *14*, 2229–2238.
Tornatore, L., Thotakura, A. K., Bennett, J., Moretti, M., & Franzoso, G. (2012). The nuclear factor kappa B signaling pathway: Integrating metabolism with inflammation. *Trends in Cell Biology*, *22*, 557–566.
Umulis, D. M., Serpe, M., O'Connor, M. B., & Othmer, H. G. (2006). Robust, bistable patterning of the dorsal surface of the Drosophila embryo. *Proceedings of the National Academy of Sciences of the United States of America*, *103*, 11613–11618.
Wang, Y.-C., & Ferguson, E. L. (2005). Spatial bistability of Dpp-receptor interactions during Drosophila dorsal-ventral patterning. *Nature*, *434*, 229–234.
Whalen, A. M., & Steward, R. (1993). Dissociation of the dorsal-cactus complex and phosphorylation of the dorsal protein correlate with the nuclear localization of dorsal. *The Journal of Cell Biology*, *123*, 523–534.
Wharton, K. A., & Serpe, M. (2013). Fine-tuned shuttles for bone morphogenetic proteins. *Current Opinion in Genetics & Development*, *23*, 374–384.
Whitham, S., Dinesh-Kumar, S. P., Choi, D., Hehl, R., Corr, C., & Baker, B. (1994). The product of the tobacco mosaic virus resistance gene N: Similarity to toll and the interleukin-1 receptor. *Cell*, *78*, 1101–1115.
Xia, Y., Shen, S., & Verma, I. M. (2014). NF-κB, an active player in human cancers. *Cancer Immunology Research*, *2*, 823–830.
Xu, M., Kirov, N., & Rushlow, C. (2005). Peak levels of BMP in the Drosophila embryo control target genes by a feed-forward mechanism. *Development*, *132*, 1637–1647.
Yamada, S., Whitney, P. H., Huang, S.-K., Eck, E. C., Garcia, H. G., & Rushlow, C. A. (2019). The Drosophila pioneer factor Zelda modulates the nuclear microenvironment of a dorsal target enhancer to potentiate transcriptional output. *Current Biology*, 29, 1387–1393.e5.
Zabel, U., & Baeuerle, P. A. (1990). Purified human IκB can rapidly dissociate the complex of the NF-κB transcription factor with its cognate DNA. *Cell*, *61*, 255–265.

Zeitlinger, J., Zinzen, R. P., Stark, A., Kellis, M., Zhang, H., Young, R. A., et al. (2007). Whole-genome ChIP-chip analysis of Dorsal, Twist, and Snail suggests integration of diverse patterning processes in the Drosophila embryo. *Genes & Development, 21*, 385–390.

Zhu, X., Stevens, L. M., Stein, D., Schüpbach, T., & Stein, D. (2007). Synthesis of the sulfate donor PAPS in either the Drosophila germline or somatic follicle cells can support embryonic dorsal-ventral axis formation. *Development, 134*, 1465–1469.

Zinzen, R. P., Senger, K., Levine, M., & Papatsenko, D. (2006). Computational models for neurogenic gene expression in the Drosophila embryo. *Current Biology, 16*, 1358–1365.

CHAPTER SIX

The design and logic of terminal patterning in *Drosophila*

Celia M. Smits[a,b], Stanislav Y. Shvartsman[a,b,c,d,*]

[a]The Lewis-Sigler Institute for Integrative Genomics, Princeton University, Princeton, NJ, United States
[b]Department of Molecular Biology, Princeton University, Princeton, NJ, United States
[c]Department of Chemical and Biological Engineering, Princeton University, Princeton, NJ, United States
[d]Center for Computational Biology, Flatiron Institute, Simons Foundation, New York, NY, United States
*Corresponding author: e-mail address: stas@princeton.edu

Contents

Abstract

Terminal regions of the early *Drosophila* embryo are patterned by the highly conserved ERK cascade, giving rise to the nonsegmented terminal structures of the future larva. In less than an hour, this signaling event establishes several gene expression boundaries and sets in motion a sequence of elaborate morphogenetic events. Genetic studies of terminal patterning discovered signaling components and transcription factors that are involved in numerous developmental contexts and deregulated in human diseases. This review summarizes current understanding of signaling and morphogenesis during terminal patterning and discusses several open questions that can now be rigorously investigated using live imaging, omics, and optogenetic approaches. The anatomical simplicity of the terminal patterning system and its amenability to a broad range of increasingly sophisticated genetic perturbations will continue to make it a premier quantitative model for studying multiple aspects of tissue patterning by dynamically controlled cell signaling pathways.

1. Introduction

Patterning of the *Drosophila* embryo is initiated by four maternally provided inductive cues. One of these cues activates the so-called terminal patterning system, the subject of this review. In this introductory section we provide the background needed for understanding molecular and

Current Topics in Developmental Biology, Volume 137
ISSN 0070-2153
https://doi.org/10.1016/bs.ctdb.2019.11.008

morphogenetic mechanisms of terminal patterning and its interactions with the anteroposterior (AP) and dorsoventral (DV) patterning systems.

Genetic studies of pattern formation during *Drosophila* embryogenesis yielded some of the first insights into molecular and cellular mechanisms of morphogen gradient formation and interpretation. The AP patterning of the embryo is initiated by Bicoid (Bcd), a homeodomain transcription factor that forms a concentration gradient along the long axis of the embryo (Wieschaus, 2016). The formation of the Bcd gradient can be explained quantitatively using a model that involves intracellular diffusion of a protein translated from anteriorly localized maternal mRNA. Acting as a direct transcriptional activator, Bcd controls multiple gene expression boundaries along the AP axis of the embryo (Fig. 1A; reviewed in chapters "A matter of time: Formation and interpretation of the Bicoid morphogen gradient" by Huang & Saunders and "Constraints and limitations on the transcriptional response downstream of the Bicoid morphogen gradient" by Tran, Walczak, & Dostatni).

In contrast to the AP patterning cascade, which relies on localized maternal RNAs, the DV patterning network relies on localized activation of transmembrane receptor signaling. Spatial regulation in this case depends on the patterned modification of the extracellular matrix surrounding the developing embryo (Moussian & Roth, 2005). Established during oogenesis and held dormant until egg activation, this modification triggers the cascade of extracellular proteases, resulting in the localized production of a diffusible ligand that activates the Toll receptor, which is expressed throughout the plasma membrane of the embryo (Roth & Lynch, 2009). Toll signals through a network of intracellular protein interactions and posttranslational modifications, ultimately resulting in spatially restricted nuclear import of Dorsal (Dl), an NF-kB transcription factor that controls gene expression along the DV axis (Fig. 1B; reviewed in chapter "Formation, interpretation, and regulation of the *Drosophila* Dorsal/NF-κB gradient" by Schloop, Bandodkar, & Reeves). Unlike the Bcd concentration gradient, which can be explained using a relatively simple source-diffusion-degradation model, the formation of the graded distribution of nuclear Dl involves many more steps and is yet to be understood quantitatively and adequately modeled mathematically. Importantly, both of these patterning gradients are established within a highly a dynamic syncytial embryo, where nuclei are dividing rapidly and proteins can diffuse in a shared cytoplasm (Foe & Alberts, 1983).

Acting relatively independently of each other, the Bcd and Dl gradients comprise two orthogonal inductive signals that establish the segmented body plan and the three germ layers. For both of these signals, transcriptional interpretation happens mostly during the third hour of development, when nuclear

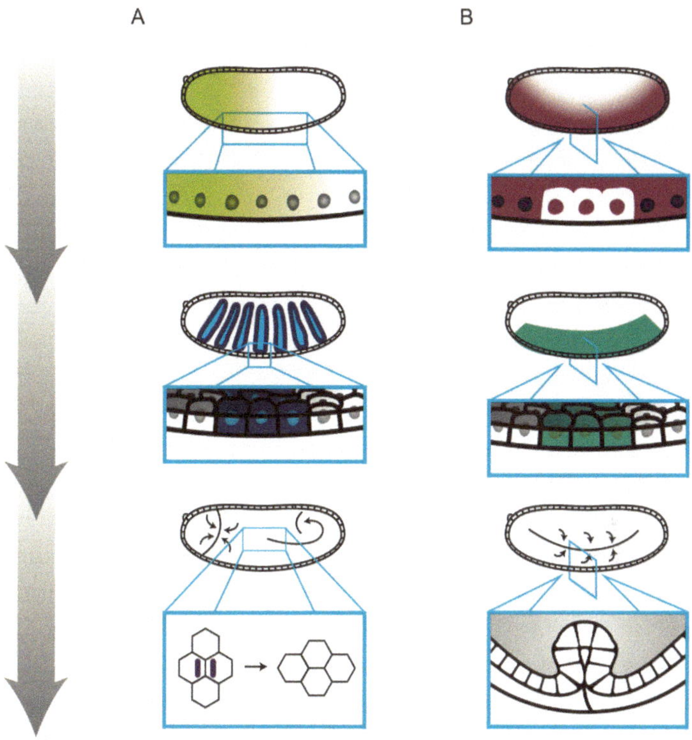

Fig. 1 AP and DV patterning and morphogenesis. (A) A concentration gradient of Bcd is established early during the syncytial blastoderm stage (top), and influences the expression of downstream genes leading to the striped expression of the pair-rule genes in the cellular blastoderm stage (middle). These genes define the areas where the cephalic furrow will form and direct the localization of Myosin to create polarized intercalation events (bottom). (B) DV patterning occurs through a graded nuclear localization of Dl at the ventral side of the syncytial blastoderm (top) leading to the ventral expression of downstream genes such as Snail in the cellular blastoderm (middle). These in turn direct the cell shape changes that lead to invagination of the ventral furrow (bottom). Insets in (B) show a cross section of the embryo, where ventral is down and lateral is to the sides. Otherwise anterior is to the left, ventral is down.

divisions have stopped and ~6000 genetically identical nuclei are tightly packed in a monolayer under a common plasma membrane (Dutta, Djabrayan, Torquato, Shvartsman, & Krajnc, 2019). During this critical time, the Bcd and Dl gradients trigger the formation of highly precise gene expression domains that both control cell fates and guide the first steps of morphogenesis. The first of these steps is the formation of two epithelial folds on the surface of newly cellularized blastoderm: the cephalic furrow, which is controlled by zygotic genes activated by the AP patterning network, and the ventral furrow, which is controlled by the DV patterning system (Fig. 1A and B).

These events depend on spatiotemporal control of actomyosin contractility. In particular, genes activated by high levels of nuclear Dl are responsible for secretion of an extracellular ligand that triggers Rho kinase signaling, causing pulsatile activity of actomyosin networks on apical cell surfaces. This in turn leads to progressive reduction of apical cell areas and causes invagination of a rectangular ventral band of cells that will form the future mesoderm (Fig. 1B). The next major morphogenetic event is germband extension, which commences shortly after mesoderm invagination and is controlled by the AP patterning system. In contrast to mesoderm invagination, which proceeds without cell rearrangements, germband extension relies on spatially and temporally ordered in-plane cell rearrangements that result in progressive elongation the embryo (Fig. 1A). These rearrangements are guided by time-dependent contractions of both apical and junction actomyosin networks. Dynamics of these networks relies on patterned expression of multiple cell-cell adhesion receptors and can be ultimately traced to the Bcd gradient and the gene network it regulates (Paré et al., 2014).

Several of the morphogenetic effects of the AP and DV patterning systems, such as segmentation and mesoderm formation, are suppressed at the terminal regions of the embryo. This antagonizing action is provided by the terminal system, which was discovered in genetic screens looking for patterning mutants (Schüpbach & Wieschaus, 1986). In cuticle preparations of this class of mutants, the cuticles of larva displayed normal segmentation patterns, but completely lacked larval head skeleton and tail structures, such as the filzkorper and posterior spiracles. Mutants of the opposite phenotype, where head and tail structures were expanded at the expense of the segmented trunk of the embryo, were also isolated. These mutations were eventually linked to loss or gain-of-function in a receptor tyrosine kinase (RTK), Torso (Goyal, Schüpbach, & Shvartsman, 2018; Li, 2005; Sprenger, Stevens, & Nusslein-Volhard, 1989).

Further genetic studies linked Torso signaling at the poles to the activation of the canonical Ras/ERK cascade. These studies revealed that the pattern of Torso activation functions as a maternal cue that induces cell fates of head and tail-specific structures while suppressing mesodermal and trunk cell fates. Normal terminal signaling is also important for proper progression of morphogenetic programs specified by the AP and DV systems. For instance, the process of germband extension is disrupted in the absence of Torso signaling (Bailles et al., 2019; Collinet, Rauzi, Lenne, & Lecuit, 2015; Irvine & Wieschaus, 1994).

Studies of the terminal patterning system continue to provide fundamental insights into molecular and systems-level mechanisms of developmental RTK signaling. These studies started with analyses of larval cuticle preparations and have over the years progressed to increasingly quantitative investigations of molecular and cellular events in live embryos. Most recently, the spatiotemporal resolution of these studies has been greatly enhanced by genome editing approaches that enable fluorescent tagging of multiple components involved in terminal patterning and by optogenetic perturbations of signaling through the Ras/ERK pathway.

The Bcd protein is used only once in the lifetime of a fruit fly: during the AP patterning of the early embryo. And although Toll signaling is a critical regulator of innate immunity, the use of Toll signaling for tissue patterning is also restricted to the early embryo. In contrast, RTK signaling that is employed by the terminal system is used recurrently during multiple stages of development in *Drosophila* and other animals. In most of these cases, locally activated RTKs provide spatial and temporal control of transcription, just as it happens during patterning of the nonsegmented terminal regions of the fly embryo. Examples of this highly conserved signaling scenario include developmental contexts as diverse as patterning of the compound insect eye and areal patterning of the mammalian neocortex (Patel & Shvartsman, 2018).

Patterning and morphogenetic outcomes of RTK signaling can be very sensitive to quantitative variations in the spatial distribution and temporal duration of RTK activation. This sensitivity is revealed by genetic studies in model organisms and studies of human developmental abnormalities associated with deregulated RTK activation (Jindal, Goyal, Burdine, Rauen, & Shvartsman, 2015). Thus, comprehensive understanding of developmental RTK signaling must go beyond identification of signaling events and their morphogenetic outcomes and quantify spatiotemporal limits of pathway activation that must be obeyed to avoid developmental defects.

Terminal patterning in *Drosophila*, similar to the other maternal patterning systems, involves a gradient of activated signal: signal is highest closest to the poles and gradually decreases toward the middle of the embryo. Additionally, this graded input elicits multiple transcriptional and downstream cell fate responses, which we will outline in this review. Terminal patterning in this sense conforms to the canonical concept of a morphogen. Interestingly, the gradient of terminal signaling is shaped by nuclear trapping, where import and export of ERK from nuclei changes its diffusion profile throughout the embryo (Coppey, Boettiger, Berezhkovskii, &

Shvartsman, 2008). The way that this allows the ERK gradient to define positional information in the anterior and posterior of the embryo remains an open question.

The terminal patterning system is uniquely positioned for the quantitative analysis of developmental RTK signaling. Indeed, genetic studies of terminal patterning identified many highly conserved genes with clear connections to human diseases. For example, the *corkscrew* (*csw*) gene received its name because of the twisted loss-of-function phenotype caused by reduced levels of RTK activation in the early embryo (Perkins, Larsen, & Perrimon, 1992). Subsequent studies revealed that *csw* encodes a non-receptor tyrosine phosphatase that is essential for optimal RTK signaling across species and is genetically deregulated in both developmental syndromes and cancers (Tajan, de Rocca Serra, Valet, Edouard, & Yart, 2015). Another gene discovered by studies of terminal patterning is *capicua* (*cic*), which also received its name from a loss-function-phenotype in the early embryo, has been recently associated with neurodegenerative diseases and cancers of neural origin (Jimenez, Guichet, Ephrussi, & Casanova, 2000; Simon-Carrasco, Jimenez, Barbacid, & Drosten, 2018).

While the discovery of *csw*, *cic*, and other genes critical for terminal patterning relied on genetic screens and studies with fixed tissues, recent advances in live imaging and quantitative approaches have propelled the terminal system to the position of one of the best-characterized RTK-dependent pattern formation events. In this review, we summarize our current understanding of signaling and morphogenetic mechanisms of terminal patterning and discuss some of the open questions.

2. Signal transduction mechanisms in terminal signaling

The terminal patterning system works by providing localized relief of gene repression (Furriols & Casanova, 2003). This section describes the main steps involved in converting Torso signaling to the localized expression of genes necessary for patterning and morphogenesis of the terminal structures of the future larva.

Terminal patterning is initiated by localized activation of Torso, a uniformly distributed RTK that it is presented with the active form of its ligand, Trunk, only at poles of the embryo (Fig. 2A and B). Trunk (Trk) is translated from a uniformly deposited maternal transcript, but its release in the extracellular space, which is essential for providing its access to Torso, depends on

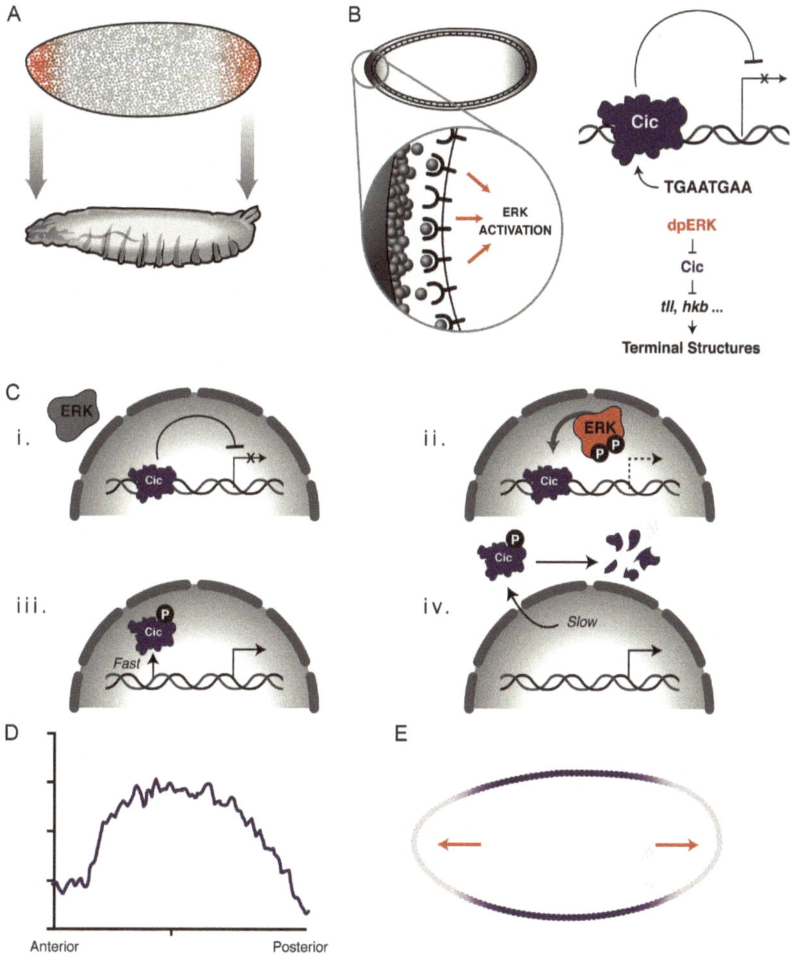

Fig. 2 Ras/ERK signaling through Capicua at the embryonic termini. (A) Ras/ERK pathway is activated only at the embryonic poles, triggering the expression of genes that specify the head and tail structures of the future larva. (B) ERK phosphorylates Capicua (Cic), a repressor which binds sites in the enhancers of patterning genes. (C) ERK regulation of Cic works on two different time scales: repression is relieved when Cic is still in the nucleus, nuclear export operates on a longer timescale, solidifying the fast response. (D) Active ERK induces nuclear export and cytoplasmic degradation of Cic. This explains why Cic is absent at the poles: the protein is translated from uniformly deposited maternal RNA and exported from the nuclei in response to signaling at the poles. (E) Cic pattern visualized in an embryo with a GFP-tagged Cic protein.

steps that are spatially restricted to the embryonic termini (Mineo, Furriols, & Casanova, 2018). The spatial regulation is provided by Torsolike (Tsl), encoded by a gene expressed in two distinct populations of the somatic cells in the epithelium that surrounds the developing cluster of the oocyte and supporting nurse cells (Furriols, Ventura, & Casanova, 2007). Localized expression of Tsl during oogenesis results in its localized incorporation at the anterior and posterior regions of the vitelline membrane that surrounds the embryo. After egg activation Tsl translocates from the vitelline membrane to the embryonic plasma membrane at the poles (Mineo, Furriols, & Casanova, 2017).

The mechanism of Tsl action is still unclear. Until recently, the predominant view has been that Tsl is involved in proteolytic processing of Trk. This model was based on the fact that the N-terminal fragment of Trk accumulates in the extracellular space at the poles and also on the fact that the C-terminal region of Trk is sufficient for Torso activation in the absence of Tsl (Casanova, Furriols, McCormick, & Struhl, 1995). More recently, an alternative model has been proposed. According to this model, Tsl controls not proteolytic cleavage, but localized secretion of Trk. Motivated by the fact that Tsl has a membrane attack complex/perforin domain that is also present in proteins involved in the formation of pores in the plasma membrane, this model correctly predicts that localized piercing of the embryonic plasma membrane should activate Torso even in the absence of Tsl (Mineo, Fuentes, Furriols, & Casanova, 2018).

In addition to transducing the signal provided by locally provided ligand, Torso limits the spatial range of Trk diffusion, thereby preventing Torso activation in the central region of the embryo (Casanova & Struhl, 1993; Sprenger & Nusslein-Volhard, 1992). This dual effect of Torso was revealed by experiments with injections of torso mRNA into embryos that lacked Torso. Specifically, when Torso mRNA was injected in central regions of the embryo, some of the terminal structures could be induced close to the site of injection. Importantly, ectopic induction of terminal structures was observed only in embryos that lack maternally provided Torso, demonstrating that locally produced active ligand is normally trapped by uniformly expressed receptors. Since the discovery of this ligand trapping effect of Torso, it is now appreciated as a general property of developmental signaling pathways activated by secreted ligands, including multiple instances of tissue patterning by RTKs.

Torso signals through the Ras pathway, a cascade of protein/protein interactions and enzymatic reactions that culminate in the dual phosphorylation and activation of the extracellular signal regulated kinase (ERK) (Futran, Link, Seger, & Shvartsman, 2013). ERK is a conserved serine-threonine kinase with hundreds of substrates involved in multiple aspects of cell regulation in different cell types, including transcription, which is the main target of Ras signaling in the early *Drosophila* embryo (Basken et al., 2018; Santini et al., 2019). Interestingly, only a handful of ERK substrates have been identified by genetic studies of terminal patterning (Kim et al., 2010). Among these, the HMG-box transcriptional repressor Capicua (Cic) emerged as a critical sensor of ERK activation in the early embryo (Jimenez, Shvartsman, & Paroush, 2012). In the absence of ERK signaling, Cic binds to highly conserved octamers in the regulatory regions of multiple genes involved in patterning of both the termini and main body of the embryo (Fig. 2B) (Ajuria et al., 2011; Fores et al., 2017).

Genetic and biochemical studies provided a detailed understanding of the ERK/Cic interaction and have started to probe the mechanisms by which ERK antagonizes gene repression by Cic (Astigarraga et al., 2007; Futran, Kyin, Shvartsman, & Link, 2015). In particular, experiments with purified proteins established that phosphorylation of ERK strongly increases its binding affinity for Cic, which ensures high specificity of the ERK/Cic interaction in the syncytial embryo where components of the nuclear and cytoplasmic compartments can be mixed during cleavage division cycles. It is also known that Cic is phosphorylated by ERK, although the phosphorylation sites essential for the ERK-dependent relief of transcriptional repression remain to be determined. Furthermore, experiments with optogenetically controlled ERK activation, time-resolved chromatin immunoprecipitation, and live imaging of transcriptional activity of gene regulatory regions with Cic binding sites revealed that repression by Cic is relieved within minutes of ERK activation and is accompanied by a rapid loss of Cic binding to its target sites across the genome (Fig. 2C) (Johnson et al., 2017; Lim et al., 2013). On a longer time scale, Cic is exported from the nucleus and degraded in the cytoplasm, which explains why Cic protein levels are downregulated at the poles (Fig. 2D and E) (Grimm et al., 2012).

The mechanisms underlying each of these effects remain to be determined. For instance, it is unclear whether rapid loss of Cic/DNA binding is caused by phosphorylation-dependent reduction in Cic/DNA affinity or more indirect effects, such as changes in Cic binding partners. Studies in cultured cells suggest that phosphorylation of Cic can result in its binding to the 14-3-3 protein, which interferes with Cic/DNA binding (Dissanayake et al., 2011). The same studies proposed that phosphorylation of Cic masks its interactions with importins responsible for the nuclear import of Cic. There are two 14-3-3 proteins in the *Drosophila* genome, D14-3-3ε and Leonardo, both of which are expressed maternally and zygotically in the early embryo (Acevedo, Tsigkari, Grammenoudi, & Skoulakis, 2007). Interestingly, Leonardo is involved in controlling transcriptional effects of Torso signaling (Li, Skoulakis, Davis, & Perrimon, 1997). However, since 14-3-3 proteins have numerous binding partners, it is still unclear whether the effects of Leonardo on terminal patterning are related to its potential binding to phosphorylated Cic.

Cic-dependent repression in the early embryo depends on Groucho (Gro), a co-repressor protein that acts by recruiting histone deacetylases (Turki-Judeh & Courey, 2012). Consistent with this mechanism, studies of Cic in human cancers revealed that genes repressed by Cic are characterized by high levels of histone deacetylation (Weissmann et al., 2018). The essential role of Gro is supported by the similarity of the phenotypes caused by genetic loss of Cic and Gro and by the Cic-dependent recruitment of Gro to the enhancers of Cic target genes (Paroush, Wainwright, & Ish-Horowicz, 1997). Gro is also phosphorylated by ERK, and this was proposed to antagonize its co-repressor function. In contrast to Cic, which is exported from the nucleus and degraded in response to ERK activation, phosphorylated Gro is stable and remains in the nucleus (Cinnamon et al., 2008). The functional role of Gro phosphorylation was proposed based on overexpression of either phosphomimetic or unphosphorylatable forms of Gro in the embryo (Helman et al., 2011). Recent advances in gene editing should enable more rigorous tests of this model. It will be also important to test whether rapid loss of Cic binding is accompanied with corresponding changes of Gro recruitment and histone acetylation.

ERK signaling during the terminal patterning of the embryo appears to work exclusively by antagonizing repression, enabling gene induction by ERK-dependent transcriptional activators. These include Zelda, a uniformly distributed activator of early zygotic transcription, and nonuniformly distributed Bcd and Dl (Samee et al., 2015). One open question is whether the spatiotemporal gradient of ERK activation, which reflects the localized pattern of Torso signaling, is needed for establishing the distinct boundaries of Cic target genes. For example, two of the best-characterized transcriptional targets of ERK signaling in the terminal system are *tailless* (*tll*) and *huckebein* (*hkb*), which encode transcription factors that play essential roles in the specification and morphogenesis of terminal structures. Based on their expression domains, the induction of *tll* and *hkb* requires distinct thresholds of ERK signaling: a high threshold for *hkb* and a lower threshold for *tll* (Greenwood & Struhl, 1997). In theory, these differential responses may be explained by the differences in the binding site composition of the regulatory regions of these genes, such as differences in the number and affinities of Cic binding sites and/or different activators. However, recent optogenetic experiments revealed that a full complement of terminal cell fates can be established by a pattern of ERK activation that is significantly more restricted than the pattern provided by endogenous Torso signaling (Johnson, Shvartsman, & Toettcher, 2019). These experiments used a restricted light pulse to optically activate ERK signaling at the poles, and showed that this discrete (rather than graded) signal is sufficient to rescue the patterning defects in Torso mutant embryos. Thus, more work is needed to understand how a spatiotemporal pulse of ERK activation is converted into nested arrangement of multiple cell fates at the poles.

In addition to inducing terminal structures, transcriptional targets of Torso signaling repress alternative cell fates and morphogenetic processes. For instance, Tll represses Knirps, a gap gene involved in AP patterning of the trunk of the embryo (Moran & Jimenez, 2006). At the same time, Hkb represses Snail, a key gene within the mesoderm specification network. As a consequence of these repressive effects, the processes of segmentation and mesoderm formation are excluded from the poles (Reuter & Leptin, 1994). A new layer of regulation has been recently identified by studies of the interactions between Torso signaling and the DV patterning system, which establishes a graded activation of Toll and graded distribution of the nuclear Dl. One of direct Dl targets, WntD, is repressed by Cic, which results in WntD expression in regions where the activation domains of Toll and Torso overlap (Helman et al., 2012; Rahimi et al., 2016). WntD is a secreted inhibitor of Toll signaling, which downregulates Toll signaling at the poles and provides an additional mechanism for antagonizing mesoderm specification by high levels of nuclear Dl. Thus, the terminal system influences DV patterning by modulating the transcriptional effects of the Dl gradient and by shaping the formation of this gradient.

3. Morphogenesis

The major morphogenetic event directly linked to terminal patterning is invagination of the posterior midgut (PMG). Additionally, acting in concert with Bcd, the terminal system patterns the area of the embryo that will give rise to head structures in the larva. In this section, we summarize what is known about these morphogenetic outcomes of Torso signaling.

The genetic pathway that functions to link Torso signaling to PMG morphogenesis is similar to the pathway that links DV patterning and ventral furrow formation (Fig. 3A). The posterior midgut, shaped like a disc at the extreme posterior pole of the blastoderm stage embryo, forms a cup-like

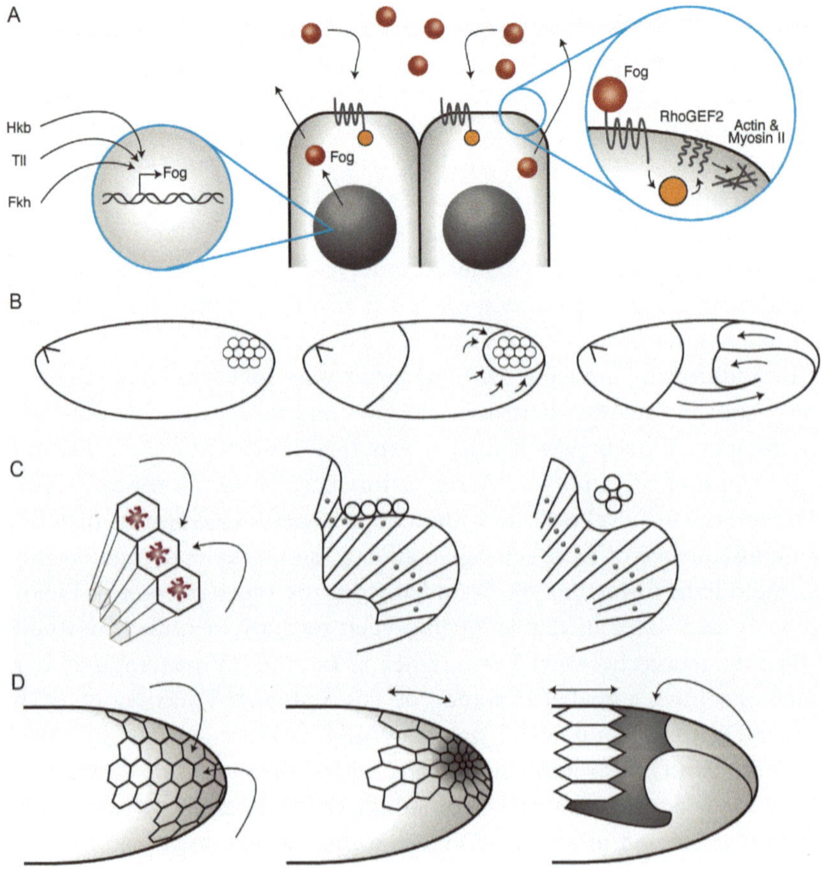

Fig. 3 Posterior midgut formation. (A) Cells that receive signaling from the terminal patterning system express Huckebein (hkb), tailless (tll) and forkhead (fkh), which together induce expression of Fog (left). This extracellular ligand is secreted outside of the cell (middle) where it locally binds to the G-protein coupled receptor Mist (right). This induces a signaling cascade through activation of the G-alpha subunit Concertina (Cta), leading to the apically localized activation of RhoGEF2 and eventually medial/apical actomyosin networks. (B) Tissue level flows that lead to internalization of the posterior midgut and pole cells, in concert with germband extension. (C) Cell shape changes that lead to internalization—apical constriction (left) of cells surrounding the pole cells leads to a ledge-shaped intermediate (middle) before isotropic contraction creates a cup-shaped pit (right). (D) Cells directly receiving Fog signal constrict first, more so on the dorsal side of the embryo (left, arrows depict force of constriction). This induces stretching of the dorsal cells, which causes apical localization of myosin in the area dorsal/anterior to the invagination pit (middle). These cells then begin to constrict, and this constriction propagates as a mechanical wave toward the anterior of the embryo (left).

invagination during the early stages of gastrulation (Sweeton, Parks, Costa, & Wieschaus, 1991). This area is specified by genes controlled most directly by Torso signaling, *hkb* and *tll*. These two genes interact with posterior patterning factors to induce another transcriptional regulator, *forkhead*. Together, these transcription factors define the PMG and activate expression of *fog* (Wu & Lengyel, 1998). Fog, a G-protein coupled receptor ligand, binds to its receptor Mist and activates the Gα subunit Concertina (Costa, Wilson, & Wieschaus, 1994; Manning, Peters, Peifer, & Rogers, 2013). This induces a cascade of protein modifications that lead to the apical accumulation of RhoGEF2, and consequent activation and recruitment of non-muscle Myosin II (hereafter "myosin") at medial-apical cell surfaces (Fig. 3A) (Gilmour, Rembold, & Leptin, 2017). Active myosin becomes stabilized through connections to the actomyosin networks of other cells via transmembrane adhesions, and this mesh-like quality of myosin in the PMG allows for the tissue-scale morphogenetic movements that follow (Fig. 3B).

The cell shape changes that lead to invagination of the PMG are also similar to those that drive ventral furrow invagination. Like cells in the future ventral furrow, cells in the posterior of the embryo are some of the first to finish cellularization, before the dorsal and lateral cells. Immediately afterward, these cells flatten along their apices and elongate along their apical/basal axis. At this stage the pole cells (which will give rise to the germline) are sitting on top of the cells that will become the PMG, and have shifted slightly dorsally. The slow phase of contraction then begins—cells undergo stochastic fluctuations in apical area, led by pulses of medial/apical actomyosin accumulation (Chanet et al., 2017; Sweeton et al., 1991). These dynamics are thought to drive basal migration of nuclei, which accompanies the cell shape changes that occur during this stage. Interestingly, the area where the pole cells sit on top of the PMG invagination site does not undergo apical constriction or basal nuclear displacement during the initial stage of PMG invagination. As a consequence, the initial invagination forms a "ledge-shaped" intermediate, in contrast to a cup-shaped pit that would be expected if all cells constricted isotropically (Fig. 3C). When pole cells are genetically ablated, this area acts the same as the rest of the PMG and forms a cup-shaped invagination (Sweeton et al., 1991). This indicates that the pole cells communicate with the underlying soma to block the program of cell shape changes that leads to invagination. The mechanism by which this occurs remains unclear; it could involve either chemical cues or mechanical blocking of apical constriction.

Following the formation of this ledge-shape intermediate, the fast phase of invagination begins, when apical area oscillations increase in amplitude. During this phase, the pole cells detach from the somatic cells, and every cell in the presumptive PMG begins to constrict (Fig. 3C). The cells dorsal to the primary site of invagination are the first to constrict, followed by the lateral and ventral cells. These cell shape changes, along with the forces of germband extension, guide the PMG up to the dorsal side of the embryo and push it toward the anterior (Sweeton et al., 1991). As the PMG invagination reaches the dorsal side and begins to move anteriorly, the ventral furrow expands up into the invagination, folding it over to create a crease along the midline that will persist through germband extension (Fig. 3B).

Recent studies have moved from fixed to live imaging to investigate the biophysical and cell biological mechanisms that lead to the cell shape changes discussed above. Like the ventral furrow, medial and apical pulses of actomyosin lead to the apical constriction of cells of the future PMG. Apical constriction of this region of the embryo occurs isotropically due to the geometry of the embryo, so that a pit-like invagination is favored over a more rectangular invagination like the ventral furrow (Chanet et al., 2017). After the initial invagination occurs, myosin becomes active in the cells adjacent to the invagination to the anterior. This myosin activation then progresses as a wave traveling toward the anterior of the embryo, and is driven by the mechanical strain induced by the invaginating pit. Mechanical strain creates pulses of active Rho1 in the medial-apical region of cells, which in turn activates myosin and directs constriction of cells toward the anterior of the initial pit (Fig. 3D) (Bailles et al., 2019). It will be interesting in the future to determine the mechanism by which this mechanotransduction occurs.

Although the genetic hierarchy that leads to apical constriction is relatively well-defined, the exact molecular mechanisms by which the transcription factors listed above elicit their actions remain unknown. For example, the combinatorial nature of Fog activation relies on Hkb and the posterior patterning factor Caudal (Wu & Lengyel, 1998), but it is unknown how this is accomplished at the molecular scale. Additionally, there may be other factors that influence the specification of cell fate and behavior in the posterior of the embryo.

Compared to the posterior of the embryo, dissecting the contributions of Torso signaling at the anterior pole is complicated by the interactions between ERK signaling and the anterior patterning system driven by Bcd. Like Cic, Bcd is also phosphorylated by ERK. Competing with Cic for access to active ERK, Bcd effectively reduces the amount of Cic down-regulation, thereby attenuating the effect of terminal signaling in the anterior-most cells (Kim et al., 2010). These competitive effects can be discerned at the level of gene expression: for instance, the expression domains of *tll* and *hkb* at the posterior pole are entirely overlapping, whereas the *tll* expression domain at the anterior is shifted toward the posterior, forming a broad stripe positioned slightly posterior to the terminus. An interesting consequence of these shifted expression domains is that many of the morphogenetic movements and structures specified by anterior terminal signaling are in the areas of the blastoderm exposed to intermediate levels of ERK activation.

One of the key experimental methods used to determine which positional cues in the anterior of the embryo relate to certain head structures in later developmental time is fate mapping. This approach enabled the identification of areas in early blastoderm embryos that relate to specific structures formed later in development, and characterized some of the intermediate steps. Many fate maps of the *Drosophila* embryo were constructed in the 1970s and 80s using fixed specimens, and researchers became particularly interested in fate mapping the head region of the developing *Drosophila* embryo (Campos-Ortega & Hartenstein, 1985). One of the most comprehensive of these fate maps was constructed by Jurgens et al., who used a UV laser to specifically ablate cells in a certain area, then analyzed both late-stage embryos and larvae to determine which structures were missing (Fig. 4) (Jurgens, Lehmann, Schardin, & Nusslein-Volhard, 1986). These studies established an accurate fate map of the blastoderm, which was subsequently cross-referenced with the spatial patterns of gene expression and information from mutant phenotypes (Finkelstein & Perrimon, 1991).

The anterior-most tip of the fly embryo, the area most exposed to both Bcd and ERK signaling, is fated to become the stomodeum, an epithelial structure that forms a tube at the base of the anterior midgut and eventually functions during the process of head involution. On the ventral side of the embryo, just posterior to the area giving rise to the stomodeum, is the precursor population to the anterior midgut, followed by the mesoderm to its posterior. The anterior midgut is an endodermal cell population that is initially internalized with the mesoderm, but later adopts different invagination characteristics that allow it to form a tube and connect with the posterior midgut and hindgut. Dorsal to the anterior tip of the embryo, and posteriorly adjacent to the stomodeum, lies the area giving rise to the labrum, which will eventually form the labral organs and epipharynx in the larva.

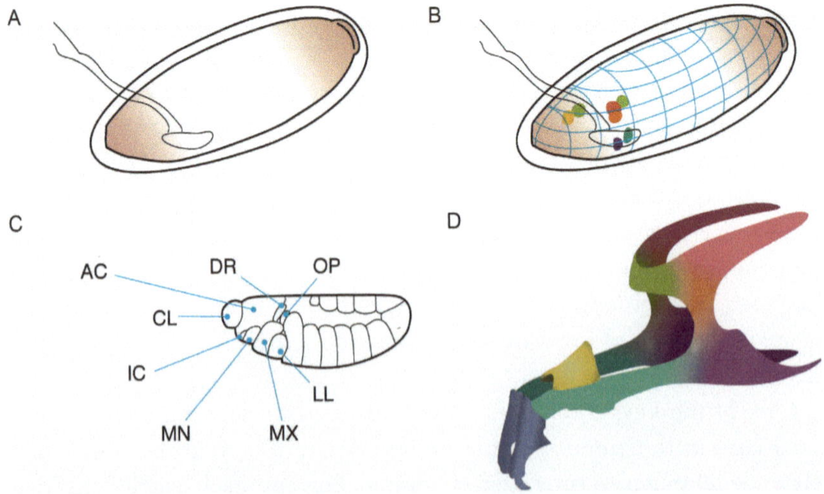

Fig. 4 Fate map of *Drosophila* anterior. (A) Torso signaling (red) informs cell fates in the earliest stages of embryogenesis. (B) Different areas of the blastoderm that receive varying amounts of signal adopt distinct cell fates. (C) Head segments are formed at the end of gastrulation and represent intermediate structures between the blastoderm (A,B) and the larval head skeleton. (D) Larval head skeleton color coded based on the areas in B that are fated to form each structure. AC, acron; DR, dorsal ridge; OP, optic lobe; LL, labial lobe; MX, maxillary lobe; MN, mandibular lobe; IC, intercalary segment; CL, clypeolabrum.

A large area on the dorsolateral side of the head, just posterior to the stomodeum and labrum, is a nonsegmental area called the acron. A portion of this area will give rise to the brain, and the rest forms a large portion of the larval head skeleton. Directly ventral to the acron lies the intercalary segment, followed to the posterior by the gnathal segments, the mandibular, maxillary, and labial (from anterior to posterior). Lying above the gnathal segments (to the dorsal side) is a small area that comprises the antennal segment precursors, which will eventually become the anterior-most structures in the larva (after head involution). This area seems to overlap heavily with posterior-most side of the acron and the dorsal sides of the gnathal segment precursors (Fig. 4).

Interestingly, many of the structures described above depend on the joint effects of all three patterning systems. For instance, the anterior expression of *tll* is induced by Torso and refined to a slightly posterior and dorsal area by interactions with AP and DV patterning factors, respectively (Pignoni, Steingrimsson, & Lengyel, 1992). Another layer of complexity is added to the system when considering that the spatial domains of gene expression vary over time—in the case of *tll*, expression initially starts as a cap at both the anterior and posterior poles, and is refined to its final position only during cellularization (Pignoni et al., 1992). Additionally, because of the high degree of interaction between patterning systems, head structures are deleted or perturbed in many different mutant conditions. These layers of complexity make the head region of the embryo a perfect system to study the downstream effects of gene expression networks; the dynamics and systematic specification of cell fates and morphogenetic behaviors promises to be a fruitful area for future research.

4. Discussion

The terminal patterning system has long been used as a model to study inductive signaling and pattern formation. However, several key questions related to molecular and systems-level mechanisms of terminal patterning remain unresolved. This section highlights some of these questions and places them in a more general context of developmental signaling.

The first set of open questions is related to molecular interpretation of the temporal pulse of ERK activation at the poles. What are the relevant substrates of active ERK during terminal patterning? Are they limited to Cic and Gro, the DNA-binding repressor and transcriptional co-repressor that control well-known ERK targets such as *tll*, *hkb*, and *wntD*? Does the ERK-dependent phosphorylation of other regulators of early embryogenesis, such as Bcd and Hb, play a role in their transcriptional effects? What are the downstream consequences of any given ERK-dependent phosphorylation event in terminal patterning? For instance, what are the mechanisms that connect Cic phosphorylation to the fast downregulation of its repressor function? Such questions can be systematically addressed using proteomics and phosphoproteomics approaches to the large-scale discovery of direct binding partners and substrates of ERK. When applied to ERK activation in mammalian cells, these approaches have revealed dozens of ERK substrates (Basken et al., 2018; Santini et al., 2019). Although only a few of these substrates are well known to produce terminal patterning defects, these defects may not fully encapsulate the functions that ERK performs in the early embryo. For example, genetic redundancy could mask the role of other ERK substrates whose functions are in maintaining robustness in terminal patterning, the effects of certain substrates may be essential for embryonic viability before the larval stage (such that they do not give rise to readily scorable patterning defects) or the effects of other ERK substrates may require more sensitive metrics to understand. The early *Drosophila* embryo provides an excellent opportunity to test these approaches in a highly controllable developmental setting and promises to reveal new connections between ERK signaling and gene regulation.

The second set of open questions is related to the quantitative properties of ERK signaling and its effects on pattern formation. Considering a pulse of ERK activation by Torso, one can explore the functional consequences of changing the starting point and the duration of this pulse, its spatial extent, and its amplitude. Studies with complete loss of function and strong gain of function mutants demonstrated what happens when the amplitude of the pulse is set to zero or when the spatial extent of Torso activation is not restricted to the poles. New tools for optogenetic activation of ERK make it possible to design perturbations that are much more diverse and, in this way, establish requirements that are both necessary and sufficient for normal patterning (Johnson et al., 2017). Optogenetic stimulation of ERK in embryos that lack endogenous Torso activation revealed that different effects of ERK signaling, such as the formation of filzkorper or repression of mesoderm fates, require different levels of ERK activation. This result could have been expected based on the graded wild-type pattern of Torso-dependent ERK activation, which establishes at least two distinct gene expression borders. Fine-tuned control of ERK activation and live imaging of transcriptional responses are needed for a mechanistic understanding of these borders and their effects on the formation of the full complement of terminal cell fates.

ERK signaling is used many times in the development of a diverse range of organisms, as well as re-used in different contexts in the same organism (Patel & Shvartsman, 2018). The cellular effects of ERK signaling depend on the cellular context in which the signaling occurs. For instance, even though the levels of Torso signaling at the two poles are the same, transcriptional effects of Torso signaling at the anterior and posterior poles are clearly different, which can be most readily appreciated by the differences in the anterior and posterior expression domains of *tll*. Based on mathematical modeling and genetic perturbations, these differences can be attributed to both ERK substrate competition and combinatorial control of gene expression (Kim et al., 2013). In addition, several of the known targets of Torso, including *tll* and *hkb*, are concurrently regulated by multiple enhancers (Hong, Hendrix, & Levine, 2008). While recent studies have started to probe signaling responses of isolated enhancers, a deeper understanding requires analysis of transcriptional responses in the normal genomic context. This applies to both terminal patterning and other ERK-dependent developmental contexts.

As previously mentioned, the terminal signaling system appears to generate a morphogenetic gradient of ERK activity that is read out into distinct cell responses in a concentration dependent manner (Greenwood & Struhl, 1997). Many of the questions outlined above involve the mechanisms by which the terminal signaling, acting as a morphogen gradient, is decoded in space and in time to elicit different cellular responses. The answers to these questions are likely to be applicable to many different patterning systems, particularly those in other organisms that function through conserved cell signaling pathways. Many of the insights gained in this system are likely to be relevant, including the mechanisms that link graded gene expression to cell behaviors and morphogenesis. Indeed, even though the method of signaling in terminal patterning is distinct from the other maternal gradients in *Drosophila*, the mechanisms that link patterning to morphogenesis follow similar paradigms. For example, cells are able to decode information like the frequency, amplitude and overall duration of the signals they receive, then modulate their response based on these quantitative parameters. The cell (or embryo) thus acts in a similar way to a computer: it receives a complex input, breaks it down into pieces of information that it can interpret, then responds with some behavioral output (Goglia & Toettcher, 2019). Most research to date has focused on understanding how patterning factors, such as morphogens and inductive signals, are read out by cells to modulate gene expression. In a number of different cases, it has been shown that through combinations of different quantitative parameters, a single input can elicit a wide range of possible outcomes (Toettcher, Weiner, & Lim, 2013). Interestingly, however, these transcriptional outputs are still funneled back into a small number of cellular behaviors (Stooke-Vaughan & Campas, 2018). Thus the programs from gene expression to morphogenesis appear to have a neural network-like architecture, in which a small number of signaling inputs (Fig. 5A) are broken down and reassembled into a large number of transcriptional outputs (Fig. 5B, teal), which can then further influence gene expression (Fig. 5B, blue). This transcriptional landscape is then read out as a cell behavioral response, such as proliferation, apoptosis, movement, or shape change (Fig. 5C).

Specifically, in the case of AP patterning, Bcd acts as a maternal gradient in the first layer of the network (Fig. 5A, left, green), and induces the expression of gap genes, the second layer (Fig. 5A, right, teal). Gap genes then induce expression of pair-rule genes in their characteristic stripes, which allows the expression of other proteins that eventually localize myosin to AP cell-cell interfaces (Fig. 5A, right, blue). This third layer of the network

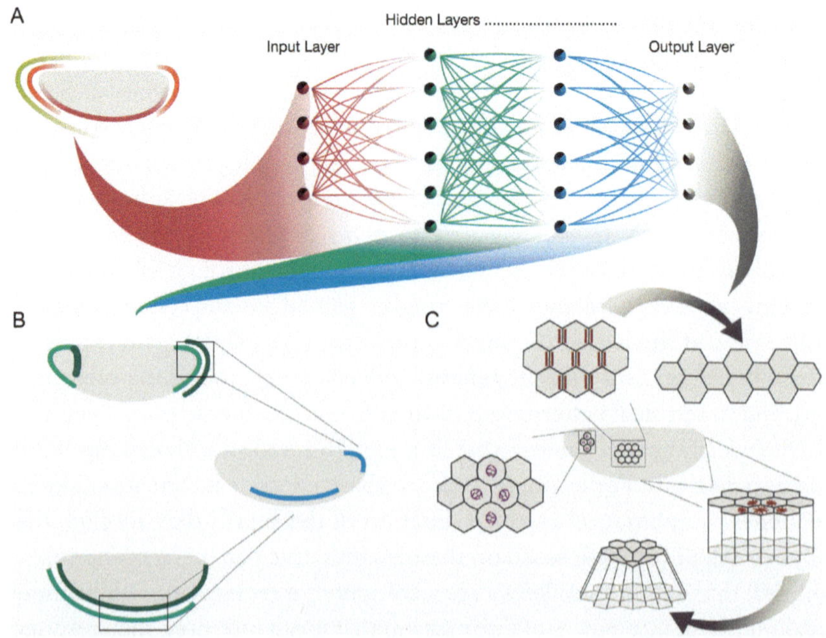

Fig. 5 Neural network-like architecture of patterning and morphogenesis in the early *Drosophila* embryo. (A) A small number of maternal patterning factors (left) act as the inputs to the neural network-like architecture (right) of signaling pathways in early *Drosophila* development. The patterning factors depicted here are the Dorsal gradient along the ventral side of the embryo (purple), the Ras/ERK gradient through activation of Torso at the poles (red), and the Bicoid gradient at the anterior (green). (B) The major patterning factors shown in (A) act to influence the expression patterns of a large number of downstream genes, which are themselves transcriptional regulators. The transcription factors downstream of the terminal patterning system (top; Huckebein in teal, Tailless in dark green), and the dorsal-ventral patterning system (bottom, Twist in dark green and Snail in teal) both induce the expression of the GPCR ligand Folded Gastrulation (Fog, blue), which binds to its receptor at the apical sides of cells and recruits contractile actomyosin networks to cell apices. (C) Groups of cells specified by different cues undergo a small number of specific cell behavioral responses to physically shape the developing tissues. These include cell intercalation (top), invagination (bottom right) and mitosis (left). All embryos are drawn to show their lateral sides, with anterior to the left and ventral on the bottom.

is less well understood relative to the other layers, but the overall effect of anisotropic myosin localization leads to the behavior of directed cell intercalation and germband extension, the final layer of the network (Fig. 5C). In the case of DV patterning, maternally supplied factors induce nuclear localization of Dl in a ventral to dorsal gradient (first layer; Fig. 5A, purple), which allows for the action of transcription factors Twist and Snail (second layer; Fig. 5B, teal). These together modulate expression of other downstream genes including localized secretion of Fog (third layer; Fig. 5B, blue), and this leads to apical localization of myosin and tissue invagination (final layer; Fig. 5C). The terminal patterning system follows a similar trend in the posterior of the embryo, with locally activated Torso at the posterior pole (first layer; Fig. 5A, red) inducing the expression of downstream factors like Hkb (second layer; Fig. 5B, top left). These also induce expression of Fog and other factors (third layer; Fig. 5B, blue), leading to pulses of apical myosin and invagination (Fig. 5C). A similar mechanism can control Torso-dependent morphogenesis at the anterior pole.

What is the utility of such layered architecture of morphogenetic networks and how does it influence development? Neural network-like architecture could introduce redundancy within the information transfer system, and thereby increase robustness. Indeed, it has been shown that redundancy at the genetic level allows organisms to adapt to mutations and environmental perturbations (Hong et al., 2008; Osterwalder et al., 2018) and that redundancy at the level of biophysical mechanisms additionally allows for small perturbations or rapid dynamic changes to the mechanical (cytoskeletal) networks without overt structural consequences (Naganathan et al., 2018; Yevick, Miller, Dunkel, & Martin, 2019). It will be a very interesting to fully map the network that leads from terminal patterning to morphogenetic movements and cell fates, and to determine the role of this network topology in *Drosophila* development. This will require a combination of techniques, spanning molecular biology, biophysical measurements, and computational modeling. Overall, the terminal patterning system will continue to be a highly tractable system for the formulation and rigorous testing of increasingly quantitative models of developmental dynamics.

Acknowledgments

We would like to thank Trudi Schüpbach, Eric Wieschaus and members of the Shvartsman lab for discussions. We also like to thank Jeremy Guay, Nareg Djabrayan, Gael McGill, and Shannon Keenan for help with figure design preparation. Some panels in Fig. 2 were created with BioRender.

References

Acevedo, S. F., Tsigkari, K. K., Grammenoudi, S., & Skoulakis, E. M. (2007). In vivo functional specificity and homeostasis of Drosophila 14-3-3 proteins. *Genetics*, *177*(1), 239–253.

Ajuria, L., Nieva, C., Winkler, C., Kuo, D., Samper, N., Andreu, M. J., et al. (2011). Capicua DNA-binding sites are general response elements for RTK signaling in Drosophila. *Development*, *138*(5), 915–924.

Astigarraga, S., Grossman, R., Diaz-Delfin, J., Caelles, C., Paroush, Z., & Jimenez, G. (2007). A MAPK docking site is critical for downregulation of Capicua by Torso and EGFR RTK signaling. *The EMBO Journal*, *26*(3), 668–677.

Bailles, A., Collinet, C., Philippe, J. M., Lenne, P. F., Munro, E., & Lecuit, T. (2019). Genetic induction and mechanochemical propagation of a morphogenetic wave. *Nature*, *572*(7770), 467–473.

Basken, J., Stuart, S. A., Kavran, A. J., Lee, T., Ebmeier, C. C., Old, W. M., et al. (2018). Specificity of phosphorylation responses to mitogen activated protein (MAP) kinase pathway inhibitors in melanoma cells. *Molecular & Cellular Proteomics*, *17*(4), 550–564.

Campos-Ortega, J. A., & Hartenstein, V. (1985). *The embryonic development of Drosophila melanogaster*. Berlin: Springer-Verlag.

Casanova, J., Furriols, M., McCormick, C. A., & Struhl, G. (1995). Similarities between trunk and spatzle, putative extracellular ligands specifying body pattern in Drosophila. *Genes & Development*, *9*(20), 2539–2544.

Casanova, J., & Struhl, G. (1993). The torso receptor localizes as well as transduces the spatial signal specifying terminal body pattern in Drosophila. *Nature*, *362*(6416), 152–155.

Chanet, S., Miller, C. J., Vaishnav, E. D., Ermentrout, B., Davidson, L. A., & Martin, A. C. (2017). Actomyosin meshwork mechanosensing enables tissue shape to orient cell force. *Nature Communications*, *8*, 15014.

Cinnamon, E., Helman, A., Ben-Haroush Schyr, R., Orian, A., Jimenez, G., & Paroush, Z. (2008). Multiple RTK pathways downregulate Groucho-mediated repression in Drosophila embryogenesis. *Development*, *135*(5), 829–837.

Collinet, C., Rauzi, M., Lenne, P. F., & Lecuit, T. (2015). Local and tissue-scale forces drive oriented junction growth during tissue extension. *Nature Cell Biology*, *17*(10), 1247–1258.

Coppey, M., Boettiger, A. N., Berezhkovskii, A. M., & Shvartsman, S. Y. (2008). Nuclear trapping shapes the terminal gradient in the Drosophila embryo. *Current Biology*, *18*(12), 915–919.

Costa, M., Wilson, E. T., & Wieschaus, E. (1994). A putative cell signal encoded by the folded gastrulation gene coordinates cell shape changes during Drosophila gastrulation. *Cell*, *76*(6), 1075–1089.

Dissanayake, K., Toth, R., Blakey, J., Olsson, O., Campbell, D. G., Prescott, A. R., et al. (2011). ERK/p90(RSK)/14-3-3 signalling has an impact on expression of PEA3 Ets transcription factors via the transcriptional repressor capicua. *The Biochemical Journal*, *433*(3), 515–525.

Dutta, S., Djabrayan, N. J., Torquato, S., Shvartsman, S. Y., & Krajnc, M. (2019). Self-similar dynamics of nuclear packing in the early Drosophila embryo. *Biophysical Journal*, *117*(4), 743–750.

Finkelstein, R., & Perrimon, N. (1991). The molecular genetics of head development in Drosophila melanogaster. *Development*, *112*(4), 899–912.

Foe, V. E., & Alberts, B. M. (1983). Studies of nuclear and cytoplasmic behaviour during the five mitotic cycles that precede gastrulation in Drosophila embryogenesis. *Journal of Cell Science*, *61*, 31–70.

Fores, M., Simon-Carrasco, L., Ajuria, L., Samper, N., Gonzalez-Crespo, S., Drosten, M., et al. (2017). A new mode of DNA binding distinguishes Capicua from other HMG-box factors and explains its mutation patterns in cancer. *PLoS Genetics*, *13*(3), e1006622.

Furriols, M., & Casanova, J. (2003). In and out of Torso RTK signalling. *The EMBO Journal*, *22*(9), 1947–1952.

Furriols, M., Ventura, G., & Casanova, J. (2007). Two distinct but convergent groups of cells trigger Torso receptor tyrosine kinase activation by independently expressing torso-like. *Proceedings of the National Academy of Sciences of the United States of America*, *104*(28), 11660–11665.

Futran, A. S., Kyin, S., Shvartsman, S. Y., & Link, A. J. (2015). Mapping the binding interface of ERK and transcriptional repressor Capicua using photocrosslinking. *Proceedings of the National Academy of Sciences of the United States of America*, *112*(28), 8590–8595.

Futran, A. S., Link, A. J., Seger, R., & Shvartsman, S. Y. (2013). ERK as a model for systems biology of enzyme kinetics in cells. *Current Biology*, *23*(21), 972.

Gilmour, D., Rembold, M., & Leptin, M. (2017). From morphogen to morphogenesis and back. *Nature*, *541*(7637), 311–320.

Goglia, A. G., & Toettcher, J. E. (2019). A bright future: Optogenetics to dissect the spatiotemporal control of cell behavior. *Current Opinion in Chemical Biology*, *48*, 106–113.

Goyal, Y., Schüpbach, T., & Shvartsman, S. Y. (2018). A quantitative model of developmental RTK signaling. *Developmental Biology*, *442*(1), 80–86.

Greenwood, S., & Struhl, G. (1997). Different levels of Ras activity can specify distinct transcriptional and morphological consequences in early Drosophila embryos. *Development*, *124*(23), 4879–4886.

Grimm, O., Sanchez Zini, V., Kim, Y., Casanova, J., Shvartsman, S. Y., & Wieschaus, E. (2012). Torso RTK controls Capicua degradation by changing its subcellular localization. *Development*, *139*(21), 3962–3968.

Helman, A., Cinnamon, E., Mezuman, S., Hayouka, Z., Von Ohlen, T., Orian, A., et al. (2011). Phosphorylation of Groucho mediates RTK feedback inhibition and prolonged pathway target gene expression. *Current Biology*, *21*(13), 1102–1110.

Helman, A., Lim, B., Andreu, M. J., Kim, Y., Shestkin, T., Lu, H., et al. (2012). RTK signaling modulates the dorsal gradient. *Development*, *139*(16), 3032–3039.

Hong, J. W., Hendrix, D. A., & Levine, M. S. (2008). Shadow enhancers as a source of evolutionary novelty. *Science*, *321*(5894), 1314.

Irvine, K. D., & Wieschaus, E. (1994). Cell intercalation during Drosophila germband extension and its regulation by pair-rule segmentation genes. *Development*, *120*(4), 827–841.

Jimenez, G., Guichet, A., Ephrussi, A., & Casanova, J. (2000). Relief of gene repression by torso RTK signaling: Role of Capicua in Drosophila terminal and dorsoventral patterning. *Genes & Development*, *14*(2), 224–231.

Jimenez, G., Shvartsman, S. Y., & Paroush, Z. (2012). The Capicua repressor—A general sensor of RTK signaling in development and disease. *Journal of Cell Science*, *125*(Pt. 6), 1383–1391.

Jindal, G. A., Goyal, Y., Burdine, R. D., Rauen, K. A., & Shvartsman, S. Y. (2015). RASopathies: Unraveling mechanisms with animal models. *Disease Models & Mechanisms*, *8*(8), 769–782.

Johnson, H. E., Goyal, Y., Pannucci, N. L., Schüpbach, T., Shvartsman, S. Y., & Toettcher, J. E. (2017). The spatiotemporal limits of developmental Erk signaling. *Developmental Cell*, *40*(2), 185–192.

Johnson, H. E., Shvartsman, S. Y., & Toettcher, J. E. (2019). Optogenetic rescue of a developmental patterning mutant. *bioRxiv* Retrieved fromhttps://www.biorxiv.org/content/10.1101/776120v2.article-info.

Jurgens, G., Lehmann, R., Schardin, M., & Nusslein-Volhard, C. (1986). Segmental organisation of the head in the embryo of Drosophila melanogaster: A blastoderm fate map of the cuticle structures of the larval head. *Roux's Archives of Developmental Biology*, *195*(6), 359–377.

Kim, Y., Coppey, M., Grossman, R., Ajuria, L., Jimenez, G., Paroush, Z., et al. (2010). MAPK substrate competition integrates patterning signals in the Drosophila embryo. *Current Biology*, *20*(5), 446–451.

Kim, Y., Iagovitina, A., Ishihara, K., Fitzgerald, K. M., Deplancke, B., Papatsenko, D., et al. (2013). Context-dependent transcriptional interpretation of mitogen activated protein kinase signaling in the Drosophila embryo. *Chaos*, *23*(2) 025105.

Li, W. X. (2005). Functions and mechanisms of receptor tyrosine kinase Torso signaling: Lessons from Drosophila embryonic terminal development. *Developmental Dynamics: An Official Publication of the American Association of the Anatomists*, *232*(3), 656–672.

Li, W., Skoulakis, E. M., Davis, R. L., & Perrimon, N. (1997). The Drosophila 14-3-3 protein Leonardo enhances Torso signaling through D-Raf in a Ras 1-dependent manner. *Development*, *124*(20), 4163–4171.

Lim, B., Samper, N., Lu, H., Rushlow, C., Jimenez, G., & Shvartsman, S. Y. (2013). Kinetics of gene derepression by ERK signaling. *Proceedings of the National Academy of Sciences of the United States of America*, *110*(25), 10330–10335.

Manning, A. J., Peters, K. A., Peifer, M., & Rogers, S. L. (2013). Regulation of epithelial morphogenesis by the G protein-coupled receptor mist and its ligand fog. *Science Signaling*, *6*(301), ra98.

Mineo, A., Fuentes, E., Furriols, M., & Casanova, J. (2018). Holes in the plasma membrane mimic torso-like perforin in torso tyrosine kinase receptor activation in the Drosophila embryo. *Genetics*, *210*(1), 257–262.

Mineo, A., Furriols, M., & Casanova, J. (2017). Transfer of dorsoventral and terminal information from the ovary to the embryo by a common group of eggshell proteins in Drosophila. *Genetics*, *205*(4), 1529–1536.

Mineo, A., Furriols, M., & Casanova, J. (2018). The trigger (and the restriction) of Torso RTK activation. *Open Biology*, *8*(12) 180180.

Moran, E., & Jimenez, G. (2006). The tailless nuclear receptor acts as a dedicated repressor in the early Drosophila embryo. *Molecular and Cellular Biology*, *26*(9), 3446–3454.

Moussian, B., & Roth, S. (2005). Dorsoventral axis formation in the Drosophila embryo—Shaping and transducing a morphogen gradient. *Current Biology*, *15*(21), 887.

Naganathan, S. R., Furthauer, S., Rodriguez, J., Fievet, B. T., Julicher, F., Ahringer, J., et al. (2018). Morphogenetic degeneracies in the actomyosin cortex. *eLife7*, e37677. https://doi.org/10.7554/eLife.37677.

Osterwalder, M., Barozzi, I., Tissieres, V., Fukuda-Yuzawa, Y., Mannion, B. J., Afzal, S. Y., et al. (2018). Enhancer redundancy provides phenotypic robustness in mammalian development. *Nature*, *554*(7691), 239–243.

Paré, A. C., Vichas, A., Fincher, C. T., Mirman, Z., Farrell, D. L., Mainieri, A., et al. (2014). A positional toll receptor code directs convergent extension in Drosophila. *Nature*, *515*(7528), 523–527.

Paroush, Z., Wainwright, S. M., & Ish-Horowicz, D. (1997). Torso signalling regulates terminal patterning in Drosophila by antagonising Groucho-mediated repression. *Development*, *124*(19), 3827–3834.

Patel, A. L., & Shvartsman, S. Y. (2018). Outstanding questions in developmental ERK signaling. *Development*, *145*(14), dev143818. https://doi.org/10.1242/dev.143818.

Perkins, L. A., Larsen, I., & Perrimon, N. (1992). Corkscrew encodes a putative protein tyrosine phosphatase that functions to transduce the terminal signal from the receptor tyrosine kinase torso. *Cell*, *70*(2), 225–236.

Pignoni, F., Steingrimsson, E., & Lengyel, J. A. (1992). Bicoid and the terminal system activate tailless expression in the early Drosophila embryo. *Development*, *115*(1), 239–251.

Rahimi, N., Averbukh, I., Haskel-Ittah, M., Degani, N., Schejter, E. D., Barkai, N., et al. (2016). A WntD-dependent integral feedback loop attenuates variability in Drosophila Toll signaling. *Developmental Cell*, *36*(4), 401–414.

Reuter, R., & Leptin, M. (1994). Interacting functions of snail, twist and huckebein during the early development of germ layers in Drosophila. *Development*, *120*(5), 1137–1150.

Roth, S., & Lynch, J. A. (2009). Symmetry breaking during Drosophila oogenesis. *Cold Spring Harbor Perspectives in Biology*, *1*(2) a001891.

Samee, M. A., Lim, B., Samper, N., Lu, H., Rushlow, C. A., Jimenez, G., et al. (2015). A systematic ensemble approach to thermodynamic modeling of gene expression from sequence data. *Cell Systems*, *1*(6), 396–407.

Santini, C. C., Longden, J., Schoof, E. M., Simpson, C. D., Jeschke, G. R., Creixell, P., et al. (2019). Global view of the RAF-MEK-ERK module and its immediate downstream effectors. *Scientific Reports*, *9*(1), 10865.

Schüpbach, T., & Wieschaus, E. (1986). Maternal-effect mutations altering the anterior-posterior pattern of the Drosophila embryo. *Roux's Archives of Developmental Biology*, *195*(5), 302–317.

Simon-Carrasco, L., Jimenez, G., Barbacid, M., & Drosten, M. (2018). The Capicua tumor suppressor: A gatekeeper of Ras signaling in development and cancer. *Cell Cycle*, *17*(6), 702–711.

Sprenger, F., & Nusslein-Volhard, C. (1992). Torso receptor activity is regulated by a diffusible ligand produced at the extracellular terminal regions of the Drosophila egg. *Cell*, *71*(6), 987–1001.

Sprenger, F., Stevens, L. M., & Nusslein-Volhard, C. (1989). The Drosophila gene torso encodes a putative receptor tyrosine kinase. *Nature*, *338*(6215), 478–483.

Stooke-Vaughan, G. A., & Campas, O. (2018). Physical control of tissue morphogenesis across scales. *Current Opinion in Genetics & Development*, *51*, 111–119.

Sweeton, D., Parks, S., Costa, M., & Wieschaus, E. (1991). Gastrulation in Drosophila: The formation of the ventral furrow and posterior midgut invaginations. *Development*, *112*(3), 775–789.

Tajan, M., de Rocca Serra, A., Valet, P., Edouard, T., & Yart, A. (2015). SHP2 sails from physiology to pathology. *European Journal of Medical Genetics*, *58*(10), 509–525.

Toettcher, J. E., Weiner, O. D., & Lim, W. A. (2013). Using optogenetics to interrogate the dynamic control of signal transmission by the Ras/Erk module. *Cell*, *155*(6), 1422–1434.

Turki-Judeh, W., & Courey, A. J. (2012). Groucho: A corepressor with instructive roles in development. *Current Topics in Developmental Biology*, *98*, 65–96.

Weissmann, S., Cloos, P. A., Sidoli, S., Jensen, O. N., Pollard, S., & Helin, K. (2018). The tumor suppressor CIC directly regulates MAPK pathway genes via histone deacetylation. *Cancer Research*, *78*(15), 4114–4125.

Wieschaus, E. (2016). Positional information and cell fate determination in the early Drosophila embryo. *Current Topics in Developmental Biology*, *117*, 567–579.

Wu, L. H., & Lengyel, J. A. (1998). Role of caudal in hindgut specification and gastrulation suggests homology between Drosophila amnioproctodeal invagination and vertebrate blastopore. *Development*, *125*(13), 2433–2442.

Yevick, H. G., Miller, P. W., Dunkel, J., & Martin, A. C. (2019). Structural redundancy in supracellular actomyosin networks enables robust tissue folding. *Developmental Cell*, *50*(5), 586–598.e3.

CHAPTER SEVEN

Dynamic positional information: Patterning mechanism versus precision in gradient-driven systems

Johannes Jaeger[a,b,*], **Berta Verd**[c]
[a]Complexity Science Hub (CSH), Vienna, Austria
[b]Department of Molecular Evolution & Development, University of Vienna, Vienna, Austria
[c]Department of Genetics, University of Cambridge, Cambridge, United Kingdom
*Corresponding author: e-mail address: yoginho@gmail.com

Contents

Abstract

There is much talk about information in biology. In developmental biology, this takes the form of "positional information," especially in the context of morphogen-based pattern formation. Unfortunately, the concept of "information" is rarely defined in any precise manner. Here, we provide two alternative interpretations of "positional information," and examine the complementary meanings and uses of each concept. Positional information defined as Shannon information helps us understand decoding and error propagation in patterning systems. General relativistic positional information, in contrast, provides a metric to assess the output of pattern-forming mechanisms. Both interpretations provide powerful conceptual tools that do not compete, but are best used in combination to gain a proper mechanistic understanding of robust patterning.

While we feel information theory is indeed a valuable tool in providing fundamental insights into the nature of communication problems, it is certainly no panacea for the communication engineer, or a fortiori, *for anyone else.*

Claude Shannon,* The Bandwagon, *1956.

Current Topics in Developmental Biology, Volume 137
ISSN 0070-2153
https://doi.org/10.1016/bs.ctdb.2019.11.017

1. Introduction

The use of the term "information" is "strikingly prominent" in contemporary biology (Godfrey-Smith, 2007). Broadly speaking, it is motivated by the idea that biological activities—such as perception, cognition, and signaling—are best understood in terms of information processing and representation (Godfrey-Smith & Sterelny, 2007). In evolutionary biology, inheritance can be treated as the flow of information from parents to offspring (Godfrey-Smith & Sterelny, 2007). In cell and developmental biology, it has come to reflect the common notion that genes exert their causal effects by carrying information about their products and the phenotypic traits that result from their expression (Godfrey-Smith, 2007; Griffiths, 2001; Maynard Smith, 2000). This view originated with Erwin Schrödinger's book "What is Life?" (1944) in which he postulates a "codescript" underlying order in biology. It became firmly entrenched with the elaboration of the genetic code (Godfrey-Smith, 2007). Since then, it has taken on some very strong meanings, such as the view that genes are "made of information," that phenotypic traits are "encoded" by genes, or that regulatory processes can be viewed as the execution of a "genetic program" (Godfrey-Smith, 2007). Through the use of such computational metaphors, "information" has become a fundamental concept in biology.

But strictly speaking, "information" only applies to the genetic code, where DNA sequences can be said to encode RNA and protein products in a well-defined sense (e.g., Griffiths, 2001).[1] All other talk about information in biology uses the term vaguely, either denoting some kind of correlation between observables, or remaining entirely at the metaphorical level. This is one of the main reasons why the use of the concept has been heavily criticized. Sarkar (1996, p. 187), for example, observes that "there is no clear, technical notion of 'information' in molecular biology. It is little more than a metaphor that masquerades as a theoretical concept and (…) leads to a misleading picture of possible explanations in molecular biology." Similarly, Griffiths (2001) calls information "a metaphor in search of a theory," while Longo, Miquel, Sonnenschein, and Soto (2012) deny outright that information is a proper observable for biology, quoting Godfrey-Smith and Sterelny (2007), which state that "enthusiasm for

[1] Alternative splicing and RNA editing complicate the picture, but do not alter the fact that there is a strong and relatively straightforward correlation between DNA and RNA sequences.

information in biology has been a serious theoretical wrong turn" as it "fosters naive genetic determinism." We will encounter this particular problem again when we discuss information in the context of developmental biology.

At the heart of this conceptual confusion lies the fact that "information" can be defined and applied in two very different ways. In its weaker and more clear-cut sense, information measures contingent but non-accidental correlations between a signal and its source. This sense of the term originates with Claude Shannon's theory of information, which is concerned with the accuracy of signals transmitted through some kind of channel (Shannon, 1948). Shannon information is sometimes (rather confusingly) called causal information (Griffiths et al., 2015). This notion of information is counter-intuitive and has several important limitations. First of all, Shannon information is highest when the signal is random. It is not meant to capture the quantity of *meaningful* information contained in a signal. Because of this, Shannon information cannot convey false information. It cannot misrepresent, as happens, for example, with the transmission of a deceitful signal, or the misexpression of a gene during development (Godfrey-Smith, 2007). Moreover, Shannon information flows both ways: we can learn as much about the source from the signal as the other way around. However, information flow is asymmetric in most biological contexts: genes are inherited from parents to offspring only, development unravels genomic information over time, and inductive signaling conveys information from the inducer to the target tissue.

This is why "information" in biology is often used in a richer semantic or intentional sense of the term. In this sense, "information" is about the content or meaning of a signal. The richer semantic concept is required if we are to move from a mere description of correlations to a causal-mechanistic understanding of biology, for instance, if we want to explain *how* a particular tissue interprets an inductive signal, *how* a particular pattern is generated during development, or *how* a character trait has originated, been modified, and inherited across generations in evolution (Godfrey-Smith & Sterelny, 2007).[2] Only semantic information can distinguish between true and false signals (e.g., honest or dishonest advertising during mating rituals), or correct (wild-type) development and aberrations induced by genetic or

[2] Even a causal-mechanistic understanding of translation and transcription—the very processes that underlie the genetic code—requires a semantic notion of "information" since many aspects of these processes (e.g., which tRNA matches which particular amino acid) are evolutionarily contingent.

environmental perturbations. In return, semantic information poses some formidable conceptual problems of its own. How do we define "meaning" in a context without intentionality or a conscious interpreter? Meaning for whom? What is the signal about? How is its normativity, its proper function, defined and how did it come about? If we are to take semantic concepts seriously in biology—and not interpret them as merely metaphorical—then we need answers to these questions.

One way to interpret meaning in biology is through a teleosemantic (or teleofunctional) approach (reviewed in Godfrey-Smith & Sterelny, 2007). It provides a reductive explanation of semantic information as derived from evolution. Genes semantically specify their wild-type products and traits because their function has been adapted to this task by natural selection.[3] In this sense, adaptive evolution results in genes that truly represent the traits they are involved in generating. Meaning depends on an etiological account of function, which derives from evolution (Wright, 1973).

Unfortunately, there are many problems with this approach. It is overly adaptationist, assuming that all traits represented by genes have a proper function. Moreover, it cannot account for exaptations, and other contingent or opportunistic features of evolution. Finally, it ignores the fact that evolution at the genetic, gene network, morphological, and behavioral levels can often be quite dissociable (Needham, 1933; see also DiFrisco & Jaeger, 2019; Von Dassow & Munro, 1999). What is functionally conserved at one level may not be at another.

An alternative and complementary way to interpret meaning and function in biology is by examining how a particular process contributes to the overall self-maintaining dynamics of a living system. This is the systemic or organizational approach to function (Cummins, 1975; Mossio, Saborido, & Moreno, 2009). In this framework, the meaning or function of a process, trait, or signal is determined by its contribution to the continued and healthy survival of an organism.

In what follows, we will focus on the specific problem of "positional information" in developmental biology and how it is used to study pattern formation by morphogen gradients (Wolpert, 1968, 1969, 1989, 1994, 1996; see, Briscoe & Small, 2015 and Green & Sharpe, 2015, for recent reviews and contextualization). We provide a short historical overview on definitions and applications of the term, and examine how they correspond

[3] This mirrors the approach which claims to use teleonomy—purposive function evolved by natural selection—to explain teleological aspects of organisms (Mayr, 1965; Pittendrigh, 1958).

to either one of the two basic meanings of information introduced above. We discuss the use of Shannon information theory to study error propagation and precision in gradient-based pattern formation. We compare this to more semantic, causal-mechanistic, approaches to the same problems, and show how the two perspectives can complement each other to yield a deeper understanding of robust developmental patterning. Finally, we discuss how discussions about function inform our view of the evolution of positional information, and provide a number of suggestions for future research directions in the field.

2. Positional information and developmental biology

Information talk first appears in the field of developmental biology through the invention of the "genetic program" metaphor, which historically originated in two convergent ways (reviewed in Peluffo, 2015). On the one hand, Jacob and Monod (1961) noted that the logic of transcriptional regulation—in bacteria, and presumably also in other organisms—provides the basis for a "co-ordinated program of protein synthesis and the means to control its execution" (Peluffo, 2015, p. 686).[4] On the other hand, Mayr (1961) stipulated that such a genetic program shaped by natural selection can explain the apparent goal-orientedness and function of the organismic traits and behaviors it determines. In this sense, informational metaphors were seen as a solution to the problem of teleology in biology (see our discussion of the teleosemantic approach in the Introduction). Primarily, however, the program metaphor represents the idea that the genome encodes a complete set of instructions for the construction of an organism under the influence of a given environment (see DiFrisco & Jaeger, 2019, for a detailed criticism of this idea).

A few years later, Apter and Wolpert (1965) review some less well-known efforts to apply information theory directly to the development of an organism. They compare a naïve approach, which literally considers development as a communication channel between embryo and adult, and a more sophisticated, program-based approach inspired by the arguments of Jacob, Monod, and Mayr. The paper focuses on the problem of preformationism: is it realistic, or even possible, that the egg contains a sufficient amount of information or instructions to determine the adult

[4] This idea closely resembles Schrödinger's (1944) "hereditary code script," but Jacob and Monod do not seem to have been aware of Schrödinger's book and do not cite it in their publication (Peluffo, 2015).

phenotype? This question is easy to answer for the naïve approach. Several authors in the late 1950s and early 1960s had attempted to estimate the (Shannon) information content of successive developmental stages, from egg to the full-grown organism. Although afflicted by much uncertainty and serious controversies over what to measure, they conclude that the information content of the adult must be orders of magnitude higher than that of the egg. Therefore, development cannot be treated directly with information theory, as Shannon information can never increase between source and signal (Apter & Wolpert, 1965). Much later, Developmental System Theory (DST) returns to this argument, and bases its criticism of informational approaches to development on this fundamental point (see, for example, Griffiths & Gray, 2001; Griffiths & Stotz, 2018; Oyama, 1985).

The case of the genetic program is more difficult to judge. Apter and Wolpert (1965) note that much of the information increase from egg to adult is due to the increasing complexity of spatial organization. Earlier authors (especially Monod) had neglected this rather obvious problem when comparing bacteria to elephants. This raises two important questions. First, how much of this information is redundant, meaning that it does not require an explicit representation in the egg? And second, how do we measure aspects of complexity, which go beyond the capabilities of Shannon's information theory? Apter and Wolpert (1965) focus on the first of these questions asking how a "program for development" could implement increases in spatial complexity.

As a consequence of this discussion, Wolpert (1969) develops the notion of "positional information" through a simple conceptual model of gradient-based patterning, the French Flag (Wolpert, 1968), which is intended to illustrate how spatial information can be "encoded" in a developmental program (Fig. 1A). The gradient-based French Flag model consists of a one-dimensional tissue, made up of a row of cells. At one end of the tissue, there is a source of diffusible morphogen; at the other end, there is a sink, where the morphogen molecule gets degraded. If there is no morphogen production or degradation in between source and sink, diffusion will generate a linear concentration gradient of the morphogen across the tissue (Fig. 1A). This gradient is then read out by its target cells. If above or below specific thresholds of morphogen concentration, a target cell will activate alternative sets of genes that determine its future fate. To distinguish it from a general inductive signal, the morphogen gradient must span at least two such thresholds, leading to three (or more) distinct territories of gene expression in the target tissue, colored blue, white, and red in case of the French

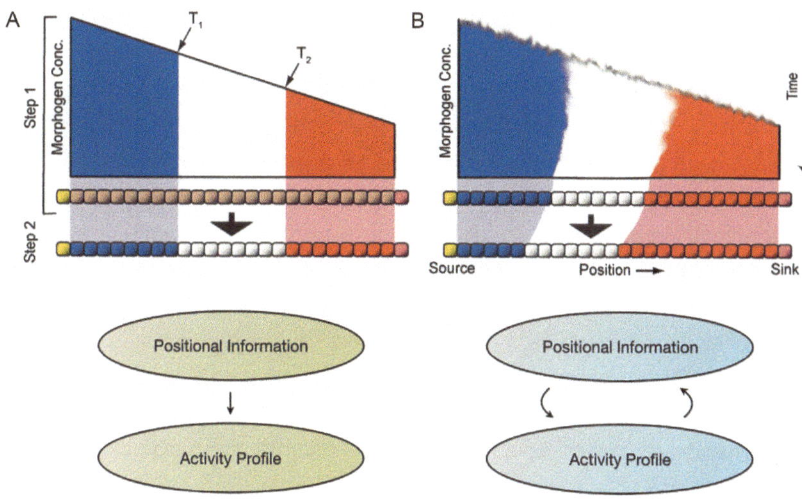

Fig. 1 The French Flag Model and Positional Information. (A) Wolpert's original gradient-based French Flag Model (1968). A morphogen diffuses from its source (green) across a tissue toward its sink (pink), where it is degraded. Concentration thresholds in the resulting spatial gradient (T1 and T2) are read out by cells in the tissue, leading to the establishment of different territories of target gene expression (blue, white, and red). This model treats development as a two-step process: first, positional information is imposed on the target tissue by the morphogen gradient (step 1). Later, this information is interpreted by cells in the tissue leading to distinct pathways of differentiation (step 2). Concentration thresholds in the gradient correspond exactly to borders of expression territories. Downstream activity profiles are determined by the morphogen in a feed-forward manner. (B) A revised French Flag, incorporating target domain shifts and increasing precision over time. Both depend on feedback regulation involving target genes. In this revised model, which is now explicitly dynamic, there is no longer a precise correspondence between concentration thresholds in the gradient and the final position of target domain boundaries. Positional information now explicitly depends on feedback regulation from downstream activity. *Simplified from Jaeger, J., Irons, D., & Monk, N. (2008). Regulative feedback in pattern formation: Towards a general relativistic theory of positional information. Development, 135, 3175–3183 and Jaeger, J. (2011). The gap gene network. Cellular and Molecular Life Science, 68, 243–274.*

Flag (Fig. 1A). The pattern scales across variable tissue sizes as long as the strength of the source and the sink are held constant. In this model, morphogen concentration is said to "encode positional information," which can be decoded by the target cells to reduce their uncertainty about where in the tissue they are located (Wolpert, 1969). In this sense, "positional information" provides a static and feed-forward coordinate system, which is imposed onto the tissue by the signal encoded in the spatial distribution of the morphogen gradient.

Wolpert (1969) argues that positional information is a "mechanism" according to which cells in a developing tissue have their position specified in relation to one or more reference points in the system (the source and the sink at the boundaries of the tissue). Cells that have their positional information specified with respect to the same points constitute a field. This reduces the classic embryological concept of the morphogenetic field to a "mechanism" represented by the genetic program in a spatial patterning context. Importantly, specification occurs before, and is disconnected from, subsequent cellular and morphological differentiation. Spatial pattern formation is seen as an essentially two-step feed-forward process. This allows Wolpert (1969) to speculate that positional information may be universal between different lineages of organisms, a view he later saw confirmed by the discovery of conserved Hox gene clusters and their expression across animal phyla (Wolpert, 1994, 1996).

The discovery of molecular gradients involved in patterning a wide variety of developmental systems led to the widespread adoption of the term "positional information" far beyond its originally intended scope. While some of its use can be precisely defined in the context of gradient-based patterning (see, for example, Briscoe & Small, 2015), most of it remains vague and metaphorical, indicating a general sense that "cells know where they are in a developing embryo." This has led to controversies and discussions analogous to those concerning the use of "information" in biology in general. Here, we will focus on two recent attempts to make the term precise: on the one hand, work on patterning precision that interprets "positional information" in terms of Shannon's theory; and on the other hand, work on the mechanisms of patterning dynamics that provides an alternative interpretation of "positional information," not as a mechanism, but as a metric for embryonic patterning. This metric is an epiphenomenon of the underlying patterning dynamics.

3. Precision in patterning: Positional information as Shannon information

Positional information, as originally defined by Wolpert in 1969, proposes that spatial asymmetry in a signal can be used by cells to determine their relative position in a tissue. Concentration thresholds in the gradient determine boundaries between distinct territories of gene expression (see Fig. 1A). 20 years later, Wolpert's model received a great boost in popularity with the discovery that a spatial gradient of the Bicoid (Bcd) protein

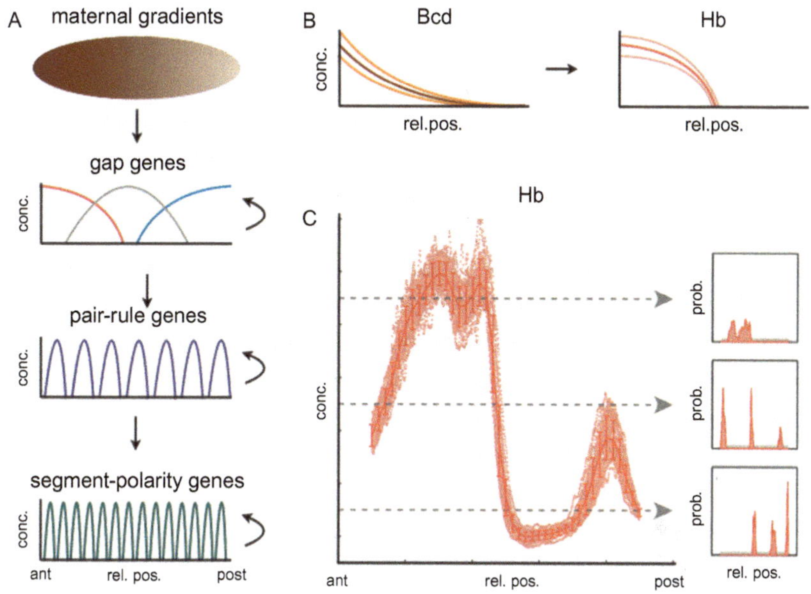

Fig. 2 Error propagation through the segmentation gene network in *Drosophila*. (A) The blastoderm-stage embryo of *Drosophila melanogaster* is patterned through hierarchical interactions among regulatory layers of the segmentation gene network. Regulatory input is provided by maternal morphogen gradients such as the Bicoid (Bcd) protein gradient (shown in brown, projected onto a schematic embryo). Maternal gradients regulate expression of the zygotic gap genes. Maternal and gap genes together then generate the periodic 7-stripe patterns of pair-rule genes which, in turn, activate the segment-polarity genes, whose 14-stripe patterns form a molecular prepattern for the process of morphological segmentation, which occurs later in development. Curved arrows indicate cross-regulation among members of each hierarchical layer in the network. (B) Bcd (brown) activates early expression of the gap gene Hunchback (Hb, red) in a concentration-dependent manner. Fluctuations in Bcd and Hb concentrations are indicated by bright lines. (C) Positional information as Shannon information: the main graph shows measured Hb expression data (dots), with the average activity profile (red line) and error bars indicating standard deviations. Dashed arrows indicate specific concentration levels of Hb. Small graphs show the probability (prob.) of being at a certain position, given a specific concentration of Hb. From these distributions, the mutual information between concentration and relative position can be calculated. In this way, positional information can be shown to correspond to Shannon information (see text). (C) Is highly simplified from Dubuis, Samanta, and Gregor (2013)). Unless indicated otherwise, graphs in all panels show relative protein concentration levels (conc.) plotted against relative position (rel. pos.) along the anteroposterior axis: anterior (ant) is to the left, posterior (post) to the right.

determines position in a concentration-dependent manner in the early blastoderm embryo of the vinegar fly *Drosophila melanogaster* (Fig. 2A) (Driever & Nüsslein-Volhard, 1988a, 1988b; Struhl, Struhl, & Macdonald, 1989). Bcd is a prototypical example of what Turing (1952) had called a morphogen (a form-giving substance; see also Jaeger & Reinitz, 2006). This experimental work established a causal link between morphogen gradient distribution and the resulting fate map of the embryo. But it did not yet provide a detailed quantitative characterization or mechanistic explanation of the intermediary developmental processes. These processes are governed by hierarchical regulatory interactions between the layered components of the segmentation network: gap, pair-rule, and segment-polarity genes (Fig. 2A) (Akam, 1987; Ingham, 1988).

Well-nigh another 20 years had to pass before the methodological advances required to start addressing these issues became available. Gregor, Wieschaus, McGregor, Bialek, and Tank (2007) provide the first detailed experimental characterization of the temporal dynamics, reproducibility, and precision of the Bcd gradient. The authors estimated the precision limit of positional information encoded by the Bcd gradient in the presence of diffusion-induced noise (Gregor, Tank, Wieschaus, & Bialek, 2007). The detected amount of spatial precision in the gradient was remarkably high, very close to the predicted theoretical limit. Even more unexpected, however, was the precision with which Bcd positions the expression domain boundary of one of its targets, the gap gene *hunchback* (*hb*) (Fig. 2B). In fact, the measured levels of precision for the *hb* domain boundary are not compatible with purely feed-forward regulation by Bcd alone. In the absence of other candidate maternal inputs, Gregor, Tank, et al. (2007) invoked the possibility that nuclei in the blastoderm embryo must be able to perform spatial averaging when measuring Bcd concentrations. This mechanism reappears in later theoretical work (Hillenbrand, Gerland, & Tkačik, 2016; Sokolowski & Tkačik, 2015), but is currently not supported by any experimental evidence.

Be that as it may, these results do suggest a surprisingly tight level of control very early in the patterning process. This is further corroborated by quantitative measurements of maternally deposited *bcd* mRNA in *Drosophila* wild-type and mutant embryos, which reveal that fluctuations in the levels of mRNA are as small as those measured for Bcd protein, and that the amount of mRNA is directly proportional to *bcd* gene dosage in the mother (Petkova, Little, Liu, & Gregor, 2014; see also Liu, Morrison, & Gregor, 2013). This raises the possibility that developmental precision and reproducibility in the *Drosophila* embryo is controlled, or at least initiated, by the mother.

One central aspect of the quantitative work discussed so far is the attempt to precisely measure the amount of positional information encoded in an observable gene expression pattern. How is this achieved? Dubuis, Samanta, et al. (2013) introduce a quantitative framework, which is further developed in Tkačik, Dubuis, Petkova, and Gregor (2015). As a first step, Dubuis, Tkačik, Wieschaus, Gregor, and Bialek (2013) generated a dataset, in which the concentrations of all four trunk gap genes—*hb*, *Krüppel* (*Kr*), *knirps* (*kni*), and *giant* (*gt*)—were measured simultaneously by immunofluorescent staining in the same embryo. By a careful dissection of experimental versus biological "errors," this allowed the authors to quantify relative expression levels as well as (co-)variation of expression patterns. To capture the positional information present in each pattern, Dubuis, Samanta, et al. (2013) assess entropy (uncertainty) reduction in each blastoderm nucleus by quantifying the mutual information between morphogen concentration and spatial position (Fig. 2C).

Mutual information is one of the central quantities in information theory (Shannon, 1948). Shannon proposed his theory as a means to quantify communication through a noisy transmission channel. As an input, a message is encoded by a transmitter. This message is then sent down the channel and decoded by the receiver, providing the read-out or output of the system. The channel is noisy: information only decreases, but never increases during transmission. In his theory, Shannon introduced the concept of entropy as a measure of uncertainty in a random variable. He defined mutual information as a measure of the statistical dependence between two random variables, in the case of the transmission channel, its input and output. More specifically, mutual information measures the reduction in the uncertainty (or entropy) of the output given a noisy measurement (or transmission) of the input.

Mutual information is extremely useful and broadly applicable. This has several reasons: first, it accounts for both linear and non-linear correlations between variables. Second, it is independent of the units of measurement, always being measured in bits (roughly, the number of yes/no questions that would need to be answered to account for the measured reduction in uncertainty/entropy). Finally, mutual information is symmetric. It does not depend on the direction of the flow of information. We can predict the state of the output, given a certain input, but we can also estimate the state of the input, given a certain output.

By formally equating positional information to mutual information between morphogen concentration and spatial position, Dubuis and colleagues enable the straightforward application of Shannon's theory to pattern formation (Fig. 2C). We can now measure the positional information contained in any observed morphogen distribution. Individual gap gene profiles, for example, contain approximately two bits of information (*hb*: 2.25 bits; *Kr*: 1.95 bits; *kni*: 1.75 bits; *gt*: 1.84bits; Dubuis, Samanta, et al., 2013). Interestingly, this is almost twice the amount of information that would be carried if gap genes were simple on/off switches: graded concentration profiles obviously matter. Together, all four gap genes encode 4.27 bits of information, evenly distributed along the anteroposterior axis of the embryo, which is sufficient to convey a unique identity to each nucleus based on its particular spatial position (Dubuis, Samanta, et al., 2013; Petkova, Tkačik, Bialek, Wieschaus, & Gregor, 2019).

Unfortunately, there is something not quite right with this simple picture of tight maternal control with subsequent feed-forward transmission of positional information. When the amplitude of the Bcd protein gradient is perturbed experimentally, the resulting changes in downstream gene expression are corrected, or canalized, toward the wild-type state (Liu et al., 2013). Generally, the precision of gap and pair-rule expression increases and becomes more independent of maternal inputs over time (Dubuis, Tkačik, et al., 2013; Petkova et al., 2019; Surkova et al., 2008). This requires cross-regulatory interactions downstream of the maternal gradients (Manu et al., 2009a), indicating that regulatory feedback among target genes is essential for the control of patterning robustness, reproducibility and ultimately precision in the *Drosophila* blastoderm.

One way to tackle this issue is to treat the gap gene network as an optimal decoder of positional information, as first suggested in Liu et al. (2013). A later study tests this suggestion explicitly by "predicting" the position of stripe positions for the pair-rule gene *even-skipped* (*eve*) under the assumption that the positional information provided by maternal gradients is optimally decoded through the gap genes (Petkova et al., 2019). The authors show that the terminal maternal system is affecting the precision of gap genes expressed in the central region of the embryo (where terminal genes are not expressed). This leads to the claim that downstream regulatory interactions integrate different maternal inputs across the embryo, which may explain why precision at late stages is higher than predicted from feed-forward regulation by Bcd alone.

Theoretical studies provide further support that integration of multiple gradients can lead to higher precision. Applied to the opposing gradients of Bcd and the posterior gradient of Caudal (Cad) protein in the Drosophila embryo, this work found that positional information, measured in terms of maximum likelihood, was highest in the middle of the embryo and reduced toward the poles (Morishita & Iwasa, 2009, 2011).

All of this leaves us with intriguing evidence that information decoding may be optimal, and with the fact that patterning is astonishingly precise already early on in the *Drosophila* blastoderm, but without any convincing mechanistic explanation of how any of this could be achieved. The focus of all the work described so far is on feed-forward error propagation from maternal gradients to their downstream targets. This (perhaps somewhat excessive) focus may explain the necessity to invoke implausible mechanistic assumptions such as spatial averaging to account for the observed levels of precision. But are gap genes really just integrating inputs from different maternal systems, or do they actively contribute to the control of patterning precision through cross-regulatory feedback over time? The evidence we have described so far simply cannot answer that.

As we have mentioned above, Shannon-style positional information is symmetrical, and therefore agnostic concerning the flow of information in the system. Another problem is that we do not even have a proper null model to tell us what levels of precision (or what kind of error propagation behavior in general) we should expect from a heavily feedback-driven regulatory system such as the segmentation gene network. Shannon's information theory is ill-suited for this context.

The problem is that the information-based approach has been expanded from its initial application to a well-defined problem—the feed-forward regulation of early *hb* expression by its only known activator Bcd (e.g., Gregor, Tank, et al., 2007; Gregor, Wieschaus, et al., 2007)—to an overall paradigm for studying patterning in the complex regulatory context of the gap and pair-rule networks (e.g., Liu et al., 2013; Petkova et al., 2019). By doing so, has this approach transgressed its proper limits? To answer this question, we need to take a closer look at the mechanisms of gene regulation underlying pattern formation in the *Drosophila* blastoderm.

4. Patterning precision versus patterning mechanism

Measurements of expression co-variation are not only useful for measuring positional information, but also enable us to make inferences about other aspects of stability in pattern formation. Of particular interest in this regard is the claim that biological systems must be in a state of criticality to be both robust and adaptable (see Mora & Bialek, 2010). As Kauffman (1993) put it, biological systems must be poised at "the edge of chaos"—stable against small perturbations yet close to a bifurcation boundary where the behavior of the system can change drastically and abruptly.

Critical systems exhibit a number of characteristic signatures. Krotov, Dubuis, Gregor, and Bialek (2014) assume that domain boundaries can be modeled by a simple two-gene network. From these simple models, they derive four predicted signatures of criticality, which they identify in the gap gene expression data from Dubuis, Tkačik, et al. (2013). This leads them to conclude that the gap gene system is in a state of criticality all along the anteroposterior axis of the embryo (Krotov et al., 2014). Unfortunately, this work is based on a number of simplifying assumptions which call this conclusion into question. Their models only consider the interactions of overlapping gap genes (and do not even specify these in any detail), ignoring the strong repression between genes with mutually exclusive spatial domains (Jaeger, 2011). Furthermore, the analysis focuses exclusively on local patterning at boundary interfaces. Finally, the authors assume that gap gene patterns are at steady-state, which is not the case (see below). Consequently, their analysis fails to consider the transient expression dynamics of gap genes, and neglects much of the regulatory complexity of the system.

Although this work may be seen as a step toward mechanistic investigation of patterning precision, it also highlights the need for dynamic models with the relevant level of detail that have been rigorously validated against experimental evidence. Only such models will yield solid mechanistic insight. Luckily, the *Drosophila* blastoderm provides a unique opportunity for such a detailed, data-driven modeling approach (see Jaeger, 2009, 2018; Jaeger, Manu, & Reinitz, 2012). Over the past two decades, quantitative data sets of segmentation gene expression have been generated (see Ashyraliyev et al., 2009; Surkova et al., 2008, among others), that enable us to fit detailed models of gap gene regulation (Ashyraliyev et al., 2009; Crombach, Wotton, Cicin-Sain, Ashyraliyev, & Jaeger, 2012; Jaeger, Blagov, et al., 2004; Jaeger, Surkova, et al., 2004; Manu et al., 2009a, 2009b; Perkins, Jaeger, Reinitz, & Glass, 2006; Verd et al., 2018; Verd, Crombach, & Jaeger, 2017; Verd, Monk, & Jaeger, 2019).[5] The solutions resulting from these fits provide a detailed representation of the complex regulatory mechanisms governing the dynamics of gap gene expression. These mechanisms are entirely consistent with the extensive experimental evidence that is available in this system (reviewed in Jaeger, 2011).

[5] This data-driven modeling approach has also been extended to non-model species of flies (Diptera), such as the moth midge *Clogmia albipunctata* (Crombach, Garcia-Solache, & Jaeger, 2014), and the scuttle fly *Megaselia abdita* (Crombach, Wotton, Jiménez-Guri, & Jaeger, 2016).

What have we learned from these mechanistic models? The first important point to note concerns the dynamics of gap gene expression. While boundaries of gap gene expression domains remain stationary over time in the anterior, they shift toward the anterior in the posterior region of the embryo, resulting in an accordion-like narrowing and sharpening of the shifting domains (Fig. 3A) (Jaeger, Blagov, et al., 2004; Jaeger, Surkova, et al., 2004; Surkova et al., 2008). These two qualitatively different types of expression dynamics are governed by different regulatory mechanisms. Stable anterior domain boundaries result from different nuclei converging toward different attractor states in a multi-stable dynamic regime (Fig. 3A) (Manu et al., 2009b; Verd et al., 2017). For example, a nucleus may converge toward a state of high *hb* expression, or toward a state of high *Kr* expression, depending on the amount of maternal morphogen it has been exposed to. In contrast, shifting domain boundaries in the posterior result from an underlying damped oscillator mechanism (Fig. 3A) (Verd et al., 2018). Each nucleus in this region of the embryo goes through part of a stereotypical succession of gap gene expression whose temporal sequence (from *Kr–kni* to *gt–hb*) is imposed by the oscillator. Different nuclei start at different positions in this sequence—they are phase-shifted with regard to each other—depending on their maternal inputs. This leads to an apparent anterior movement of posterior domains—so-called kinematic shifts[6]—although no gap protein is actually being transported across the tissue.

A second important point, which affects our understanding of the stability of the system, is that the gap gene network never reaches steady state. Gap gene dynamics remain transient throughout the blastoderm stage (Verd, Crombach, & Jaeger, 2014; Verd et al., 2017). Even though far from steady state, gap gene regulation shows canalysed behavior in the posterior region of the embryo, as first demonstrated by Manu et al. (2009a, 2009b). Canalization means that system trajectories converge toward each other long before the system approaches its attractors. In the case of the damped oscillator, it happens because of strong repressive feedback between complementary gap genes *hb*/*kni* and *Kr*/*gt* (Verd et al., 2019). This may provide a possible mechanism for the observed increase in precision of gap gene expression over time.

[6] A Mexican wave in a soccer stadium is a good example of such a kinematic phenomenon. It originates from a temporal sequence of arm movements by each spectator, and travels around the stadium without any people actually changing position.

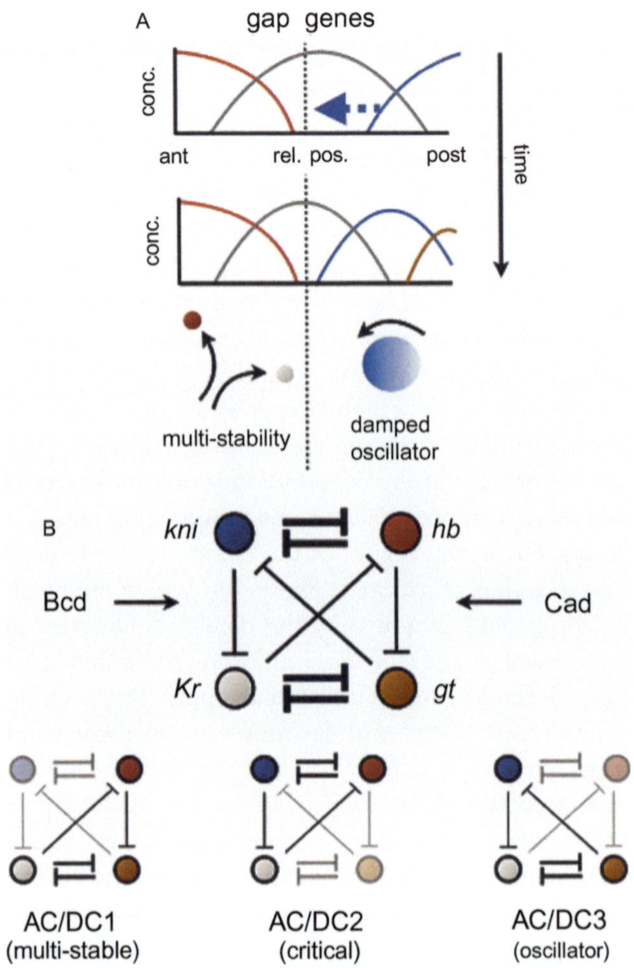

Fig. 3 Mechanisms for gap gene regulation. (A) The dynamics of gap gene expression differ between the anterior (ant) and posterior (post) halves of the *Drosophila* embryo. In the anterior, domain boundaries remain stationary while in the posterior, they shift toward the anterior over time (indicated by dashed blue arrow). Stable boundaries in the anterior are positioned by a multi-stable dynamical regime that causes different nuclei to converge toward different stable attractor states (represented as red and white circles indicating high expression of the red and white gene, respectively). Shifting boundaries in the posterior are governed by a damped oscillator mechanism, which causes nuclei to cycle through a stereotypical succession of gene expression states (only shown for white and blue here). Nuclei are phase-shifted with respect to each other, depending on their position in the embryo. This leads to the kinematic shifts of the domain boundaries in this region. (B) Schematic regulatory topology of the gap gene network (including external activating inputs by maternal gradients of Bicoid, Bcd, in the anterior, and Caudal, Cad, in the posterior). T-bars indicate repressive cross-regulatory

Our third and last point is that the gap gene system exhibits modular dynamics, even though its regulatory structure (or topology) shows no modularity at all (Fig. 3B) (Verd et al., 2019). This is because different subsets of gap genes are expressed and active in different regions of the embryo, which allows us to identify three individual subcircuits, or dynamical modules: AC/DC1 consists of *hb*, *Kr*, and *gt* and contributes to patterning through its multi-stable regime in the anterior region of the embryo; AC/DC2 consists of *hb*, *Kr*, and *kni*, and is in a critical state throughout the central region; AC/DC3 consists of *Kr*, *kni*, and *gt* and implements the damped oscillator in the posterior. Interestingly, AC/DC2 straddles the bifurcation boundary between stationary and shifting domains in the middle of the embryo (Fig. 3B). This suggests that the gap gene system is critical in the central region only, while exhibiting stable dynamics further anterior and posterior, contradicting the claim by Krotov et al. (2014) that the network is critical throughout. What kind of consequences this may have for patterning precision remains an open question.

5. General relativistic positional information (GRPI)

In contrast to statistical work based on information theory, our detailed analysis of the dynamics driven by the gap gene system provides a realistic and accurate causal-mechanistic explanation of *how* gap genes pattern the fly blastoderm. This mechanistic aspect of patterning is not in contradiction but complementary to questions of precision and error propagation. However, only the latter can be handled within the framework of Shannon's information theory. To confuse the two aspects amounts to a

interactions between gap genes *hunchback* (*hb*), *Krüppel* (*Kr*), *knirps* (*kni*), and *giant* (*gt*). Line thickness indicates relative interaction strength. The gap gene network can be subdivided into three dynamical modules: AC/DC1 is in a multi-stable regime, active in the anterior; AC/DC2 is critical, exhibiting both multi-stability and mono-stability with spiraling trajectories in the central region of the embryo; AC/DC3 is in a mono-stable regime with spiraling trajectories, active in the posterior. All subcircuits share the same network topology, but consist of distinct subsets of gap genes, with overlap between the modules. *Simplified from Verd, B., Clark, E., Wotton, K. R., Janssens, H., Jiménez-Guri, E., Crombach, A., & Jaeger, J. (2018). A damped oscillator imposes temporal order on posterior gap gene expression in Drosophila. PLoS Biology, 16, e2003174; Verd, B., Monk, N., & Jaeger, J. (2019). Modularity, criticality, and evolvability of a developmental gene regulatory network. eLife, 8, e42832. Graphs in (A) show relative protein concentration levels (conc.) plotted against relative position (rel. pos.) along the anteroposterior axis: anterior (ant) is to the left, posterior (post) to the right. See text for details.*

category error, muddling statistical with causal-mechanistic explanation (see also Calcott, Pocheville, & Griffiths, 2018). Causal-mechanistic explanation of patterning calls for a radically different, semantic, interpretation of "positional information."

The dynamic conceptual framework of "general relativistic positional information" (GRPI) (Jaeger, Irons, & Monk, 2008; Jaeger & Reinitz, 2006) provides such a semantic interpretation[7] (Fig. 1B). It treats positional specification in strictly processual terms: there is no particular point in time at which a cell becomes specified in terms of its position within the tissue. Instead, specification is seen as a dynamic process that ends with stable determination of cell fate. There is no invariable coordinate system as in the French Flag anymore. More importantly, there is no one-to-one correspondence between morphogen concentration and cell fate. Instead, the interpretation of the signal depends on the state of the target cell, dynamically changing over time (Jaeger et al., 2008). This is very close to Waddington's interpretation of morphogens as "evocators," exerting their effect dependent on the "competence" of the target cells (Jaeger & Reinitz, 2006; Waddington, 1940, 1956). Here, the role of the receiving cell is one of active, dynamic, and context-dependent interpretation. The cell needs to be ready (or "competent") for the signal to evoke an appropriate response. Therefore, the relevant content of the signal becomes fundamentally semantic and context-dependent. Moreover, the shape of the morphogen gradient itself is often altered in the process (Jaeger et al., 2008).

Instead of a statically imposed coordinate system, we now have a dialectic interaction between gradient and target tissue (Fig. 1B) that establishes a dynamic positional metric, which is not only actively interpreted but also actively altered by the receiving cells. Conceptually, this is analogous to the difference between absolute space and time in Newtonian versus dynamically malleable spacetime in Einstein's theory of general relativity (Jaeger et al., 2008). Just as spacetime geometry emerges from the interplay between massive objects with their environment, so does GRPI emerge from reciprocal interactions between gradient and tissue. In this sense, it is not a "mechanism" in the sense of Wolpert (1968, 1969), but a mere reflection, an epiphenomenon, of the underlying regulatory dynamics.

[7] Since its original publication, there has been some debate on whether GRPI should be seen as a type of "information" at all. In what follows, we will argue that it should indeed, as it can be smoothly integrated into a systemic or organizational account of function and semantic information.

Still, GRPI is a type of semantic information, since it conveys a "meaning" to the cell, namely, about its time- and context-dependent relative position in a developing embryo. Its role within the organism is to establish the proportions and relative positions of different parts and organs. This is a central contribution to the continued survival of the organism: without it, a living system could not mature to its adult stage and reproduce. Therefore, it has a clear systemic or organizational function (see Cummins, 1975; Mossio et al., 2009). Note again that this interpretation is fundamentally different to the information-theoretic analysis of error propagation, which asks how precise a pattern can be. Here, we ask instead, *how* the specific pattern of an individual organism comes to be in the first place. The first question makes no sense in the absence of the second. But only answering both questions together yields a complete understanding of how development proceeds. Yet, surprisingly, there are only very few studies that have tried to integrate the two approaches.

6. Mechanisms for patterning precision

One of the few studies that integrates patterning mechanisms and precision in the way suggested above examines morphogen-based patterning in the vertebrate neural tube (Zagorski et al., 2017). Different populations of neurons develop at different dorso-ventral positions in this growing tissue (Fig. 4A). Cell specification depends on two antagonistic morphogen gradients: one of Sonic Hedgehog (Shh) emanating from the floor plate and ventral neural tube, and another one of bone morphogenetic protein (BMP) emanating from the dorsal end of the tissue (Fig. 4A). As the neural tube grows, cells experience changing concentrations of both morphogens. As in the case of the gap gene network, cross-repressive interactions among target genes alter the response of the receiving cells over time. At this level of abstraction, patterning in the fly blastoderm embryo and the vertebrate neural tube follow very similar regulatory principles (see Briscoe & Small, 2015).

Zagorski et al. (2017) ask a very simple question: what kind of regulatory network provides the most accurate fit to the observed target gene expression patterns given an antagonistic arrangement of contrapolar morphogen gradients of Shh and BMP? What is special about this study is that the authors fit their model to both expression patterns and a decoding map based on a maximum-likelihood analysis of the system (based on the methodology of Tkačik et al., 2015). This type of analysis predicts error propagation patterns based on the assumption that the system achieves optimal decoding of the

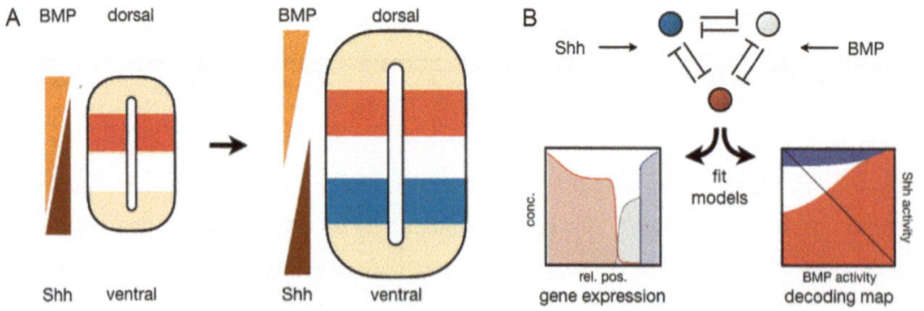

Fig. 4 Patterning and optimal decoding in the vertebrate neural tube. (A) The vertebrate neural tube is patterned by two antagonistic morphogen gradients: Sonic Hedgehog (Shh) and bone morphogenetic protein (BMP). These gradients induce different states of target gene expression (shown in red, white, and blue). As the tissue grows over time, cells experience changing concentrations of morphogens. Cross-regulatory feedback between target genes further modifies the boundary positions of expression territories. (B) Models of the response network implement various combinations of target gene interactions drawn from the fully connected topology shown here. These models are fit to both spatio-temporal gene expression data and a decoding map derived from maximum likelihood under an optimal decoding assumption. This forces fitting solutions to reproduce both the dynamics and fluctuations of gene expression correctly. The best-fitting solution across all models is achieved with a fully connected network topology. See text for details. *Panel (A) simplified from Briscoe, J. & Small, S. (2015). Morphogen rules: Design principles of gradient-mediated embryo patterning. Development, 142, 3996–4009.; Panel (B) simplified from Zagorski, M., Tabata, Y., Brandenberg, N., Lutolf, M. P., Tkačik, G., Bollenbach, T., Briscoe, J., & Kicheva, A. (2017). Decoding of position in the developing neural tube from antiparallel morphogen gradients. Science, 356, 1379–1383.*

positional information provided by the upstream morphogenetic gradients. Dynamical models, implementing various topologies of target gene networks, were fit to both gene expression data and the predicted decoding map (Fig. 4B). This approach evaluates which network topology gets closest to optimal decoding while faithfully reproducing the observed patterns of gene expression.

The network topology that emerges from this dual-aspect fitting procedure is a fully connected network with reciprocal repressive interactions between all target genes (Fig. 4B) (Zagorski et al., 2017). As long as we cannot constrain the fit with more detailed evidence on individual regulatory interactions, it may not be surprising that this is the case. One reason is methodological: the more interactions in a model, the more free parameters there are, and hence degrees of freedom for the fit. Another reason is dynamic: feedback regulation generally allows for tighter and more fine-grained control. Lastly, we cannot entirely exclude the possibility that there may be a missed alternative topology that would fit the data and decoding map better, since the search of parameter space is not exhaustive, and the convergence to the best solution never guaranteed for such complex model-fitting procedures.

But despite all these caveats, there is progress here. First of all, this study shows, for the first time, that it is *possible* to achieve optimal decoding in a morphogen-driven patterning system, while closely reproducing the observed dynamics of gene expression. And second, it introduces an integrative methodology of mechanistic modeling combined with error-propagation analysis that should be used much more widely in the study of developmental regulation, not only in the context of morphogen-based pattern formation.

Indeed, such an integrative approach will be necessary to reconcile causal-mechanistic and information-based studies of the segmentation gene system of *Drosophila*. As it stands, our knowledge of these two complementary aspects of gene regulation in the fly blastoderm, are based on incompatible assumptions.

How patterning occurs in the gap gene system has been studied with data-driven models of the network that fit the dynamics of the observed (averaged) patterns of gene expression very closely (Ashyraliyev et al., 2009; Jaeger, Blagov, et al., 2004; Jaeger, Surkova, et al., 2004; Manu et al., 2009a, 2009b; Verd et al., 2018, 2017, 2019). These models are fully compatible with the available experimental evidence on gap gene regulation (reviewed in Jaeger, 2011), but do not reproduce the observed patterns of expression fluctuations in the system very accurately. In contrast, our knowledge of error propagation and optimal decoding in the system is based on extremely precise, quantitative measurements of expression (co-)variation, in the absence of realistic mechanistic models of the underlying regulatory dynamics (Dubuis, Samanta, et al., 2013; Dubuis, Tkačik, et al., 2013; Krotov et al., 2014; Liu et al., 2013; Petkova et al., 2019). These measurements indicate that decoding must be near optimal in the gap gene system, but we do not know *how* this is achieved at the level of regulatory mechanisms. As already mentioned, these two approaches do not compete, but represent two different sides of the same coin. The sobering truth is that, as long as they remain in contradiction, we cannot claim to truly understand all important aspects of patterning in this most carefully studied model system for pattern formation.

7. Conclusions

Here, we have argued that information-based and causal-mechanistic approaches to pattern formation seek complementary types of explanations,

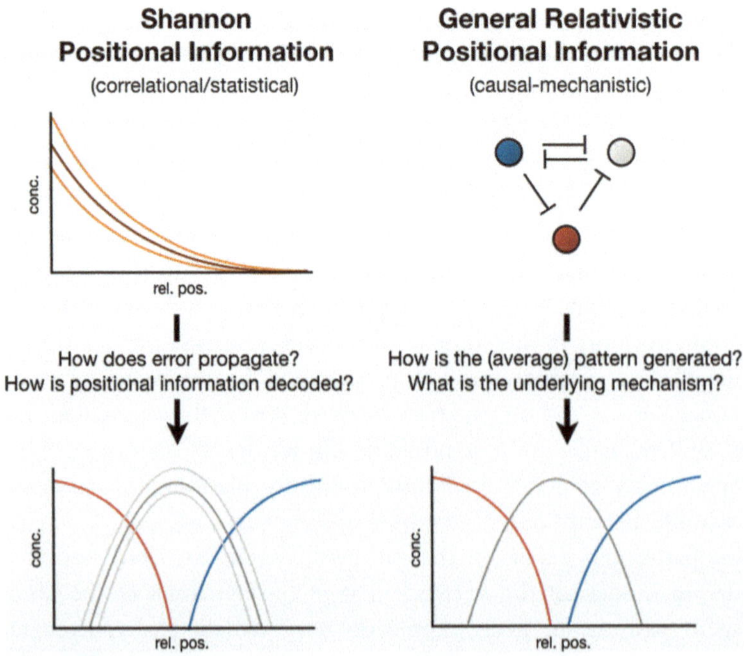

Fig. 5 Two different kinds of explanation. Shannon positional information and General Relativistic Positional Information (GRPI) are two alternative ways in which the concept of information can be made precise in the context of pattern formation. The represent two entirely different kinds of explanation, two different types of perspectives, on the same underlying pattern-forming system. See text for details.

which are of a fundamentally different nature (Fig. 5). Shannon's information theory provides powerful tools to investigate precision, error propagation and decoding of positional information in patterning systems. If we interpret positional information *sensu* Shannon, however, it no longer represents an instructive program or mechanism of pattern formation as argued by Wolpert (1968, 1969). Information-based explanations are correlational (statistical), while mechanistic explanations must be framed in causal terms. The latter require detailed experimental evidence on the relevant interactions among components of the underlying regulatory network, as well as a detailed and accurate dynamic model of the system to show how these interactions synergize to generate the observed patterning output (see DiFrisco & Jaeger, 2019).

In contrast, information-based explanations are independent of mechanistic detail (Tkačik et al., 2015). On the one hand, this can be advantageous if we are interested in a broad theory of patterning precision, which aims at predicting rather than explaining features of regulation. For instance, an information-theoretic framework enables us to test the hypothesis that transmission of information is close to optimal decoding in different classes of regulatory networks, and asks what types of target gene expression patterns indicate such optimal decoding (Hillenbrand et al., 2016; Tkačik, Callan, & Bialek, 2008; Tkačik et al., 2015; Tkačik, Walczak, & Bialek, 2009, 2012; Sokolowski & Tkačik, 2015; Walczak, Tkačik, & Bialek, 2010; reviewed in Tkačik and Walczak (2011). This is certainly a fascinating question to pursue. Preliminary evidence suggests that optimal decoding may indeed apply both in the *Drosophila* blastoderm and the vertebrate neural tube (Dubuis, Samanta, et al., 2013; Gregor, Tank, et al., 2007; Petkova et al., 2019; Zagorski et al., 2017).

However, it remains to be seen whether this insight can be further generalized. In particular, it is not at all clear whether the essential underlying assumption of optimal decoding holds in a broader developmental and evolutionary context. Model organisms are often chosen because of their short generation times and fast development. As a consequence, these creatures often show genetically hardwired modes of development, while slower developers are likely to rely more heavily on a combination of genetic and cellular-physical mechanisms of morphogenesis (see, for example, Love, 2018; Newman, 2008, 2012; Newman & Bhat, 2009; Salazar-Ciudad, 2010). This may introduce a bias toward model systems exhibiting high levels of decoding optimality. It will be challenging but worthwhile to gain the evidence on patterning in slow-developing non-model organisms required to resolve this issue.

On the other hand, the lack of specificity of information-based approaches prevents us from learning anything particular about the mechanisms underlying any given developmental system. This is a serious limitation, because details *do* matter in biology. Evidence from the few developmental systems for which we have suitable experimental and modeling-based data indicate that biological pattern formation is heavily feedback-driven (e.g., Briscoe & Small, 2015; Jaeger, 2018; Jaeger et al., 2008). Unfortunately, we do not yet know how the flow of positional information is regulated in feedback-heavy systems. Shannon's original theory is not applicable in the presence of regulatory feedback, and efforts to extend it have not gone beyond very simple examples of genetic auto-regulation (Tkačik et al., 2012). An integrative approach, combining information-based as well as causal-mechanistic analysis and modeling with detailed experimental evidence will be required to transcend this fundamental limitation.

There is one last philosophical point to make: our argument reveals that information-based and causal-mechanistic approaches provide different perspectives on the same underlying reality. Both perspectives deal with the same physico-chemical processes, the same regulatory systems composed of material entities and the dynamic interactions between them. Depending on our interpretation, these interactions result in an overall flow of information, or an overall mechanistic flow of cause-and-effect. Therefore, information theory provides us with an alternative angle to the problem of pattern formation. But information is not a thing. It is not a substance. Genes cannot be "made of information" and the "genetic program" remains a metaphor (Godfrey-Smith, 2007).

Instead, information is a conceptual tool, to be employed with a clear understanding of its meaning. Unfortunately, there is much loose talk about information in biology, which leads to confusion and unnecessary arguments. As Shannon (1956) himself noted, "information is no panacea." It must be used in clearly specified ways. We have shown that there are two alternative ways of precisely defining "positional information" in the context of pattern formation. Shannon information helps us to understand statistical patterns of error propagation and decoding in pattern-forming systems. General relativistic positional information (GRPI), in contrast, is a semantic metric for cells to "read" and "interpret" their relative position in a growing embryo. Both of these conceptual tools are complementary. They are most practical and powerful when used together with a clear understanding of their domains and limits of application.

Acknowledgments

This contribution is dedicated to all researchers out there who use the term positional information without having the faintest clue what it may mean. An early draft of this argument was presented in the context of the workshop "Causal Foundations of Biological Information," organized by Paul E. Griffiths and the recently deceased Karola Stotz, at the Konrad Lorenz Institute for Evolution and Cognition Research in Klosterneuburg. A late but heartfelt thank you to both for the invitation to talk and the discussions at the workshop. J.J. thanks Murillo Pagnotta for our fascinating and inspiring exchanges about information in biology.

References

Akam, M. (1987). The molecular basis for metameric pattern in the *Drosophila* embryo. *Development, 101*, 1–22.

Apter, M. J., & Wolpert, L. (1965). Cybernetics and development: I. Information theory. *Journal of Theoretical Biology, 8*, 244–257.

Ashyraliyev, M., Siggens, K., Janssens, H., Blom, J., Akam, M., & Jaeger, J. (2009). Gene circuit analysis of the terminal gap gene *huckebein*. *PLoS Computational Biology, 5,* e1000548.

Briscoe, J., & Small, S. (2015). Morphogen rules: Design principles of gradient-mediated embryo patterning. *Development, 142,* 3996–4009.

Calcott, B., Pocheville, A., & Griffiths, P. (2018). Signals that make a difference. *The British Journal for the Philosophy of Science,* axx022. https://doi.org/10.1093/bjps/axx022.

Crombach, A., Garcia-Solache, M. A., & Jaeger, J. (2014). Evolution of early development in dipterans: Reverse-engineering the gap gene network in the moth midge *Clogmia albipunctata* (Psychodidae). *Biosystems, 123,* 74–85.

Crombach, A., Wotton, K. R., Cicin-Sain, D., Ashyraliyev, M., & Jaeger, J. (2012). Efficient reverse-engineering of a developmental gene regulatory network. *PLoS Computational Biology, 8,* e1002589.

Crombach, A., Wotton, K. R., Jiménez-Guri, E., & Jaeger, J. (2016). Gap gene regulatory dynamics evolve along a genotype network. *Molecular Biology and Evolution, 33,* 1293–1307.

Cummins, R. (1975). Functional analysis. *Journal of Philosophy, 72,* 741–765.

DiFrisco, J., & Jaeger, J. (2019). Beyond networks: Mechanism and process in evo-devo. *Biology & Philosophy, 34*(6), 54.

Driever, W., & Nüsslein-Volhard, C. (1988a). A gradient of *bicoid* protein in Drosophila embryos. *Cell, 54,* 83–93.

Driever, W., & Nüsslein-Volhard, C. (1988b). The *bicoid* protein determines position in the Drosophila embryo in a concentration-dependent manner. *Cell, 54,* 95–104.

Dubuis, J. O., Samanta, R., & Gregor, T. (2013). Accurate measurements of dynamics and reproducibility in small genetic networks. *Molecular Systems Biology, 9,* 639.

Dubuis, J. O., Tkačik, G., Wieschaus, E. F., Gregor, T., & Bialek, W. (2013). Positional information, in bits. *Proceedings of the National Academy of Sciences of the United States of America, 110,* 16301–16308.

Godfrey-Smith, J. (2007). Information in biology. In D. Hull & M. Ruse (Eds.), *The Cambridge companion to the philosophy of biology* (pp. 103–119). Cambridge: Cambridge University Press.

Godfrey-Smith, J., & Sterelny, K. (2007). *Biological information. In E. N. Zalta (Ed.), The Standford encyclopedia of philosophy.* (Summer 2016 Edition), URL. https://plato.stanford.edu/archives/sum2016/entries/information-biological.

Green, J. B. A., & Sharpe, J. (2015). Positional information and reaction-diffusion: Two big ideas in developmental biology combine. *Development, 142,* 1203–1211.

Gregor, T., Tank, D. W., Wieschaus, E. F., & Bialek, W. (2007). Probing the limits to positional information. *Cell, 130,* 153–164.

Gregor, T., Wieschaus, E. F., McGregor, A. P., Bialek, W., & Tank, D. W. (2007). Stability and nuclear dynamics of the Bicoid morphogen gradient. *Cell, 130,* 141–152.

Griffiths, P. E. (2001). Genetic information: A metaphor in search of a theory. *Philosophy of Science, 68,* 394–412.

Griffiths, P. E., & Gray, R. D. (Eds.), (2001). *Cycles of contingency: Developmental systems and evolution.* Cambridge, MA: MIT Press.

Griffiths, P. E., Pocheville, A., Calcott, B., Stotz, K., Kim, H., & Knight, R. (2015). Measuring causal specificity. *Philosophy of Science, 82,* 529–555.

Griffiths, P., & Stotz, K. (2018). Developmental systems theory as a process theory. In D. J. Nicholson & J. Dupré (Eds.), *Everything flows: Towards a processual philosophy of biology* (pp. 225–245). Oxford: Oxford University Press.

Hillenbrand, P., Gerland, U., & Tkačik, G. (2016). Beyond the French flag model: Exploiting spatial and gene regulatory interactions for positional information. *PLoS One, 11,* e0163628.

Ingham, P. W. (1988). The molecular genetics of embryonic pattern formation in *Drosophila*. *Nature*, *335*, 25–34.
Jacob, F., & Monod, J. (1961). Genetic regulatory mechanisms in the synthesis of proteins. *Journal of Molecular Biology*, *3*, 318–356.
Jaeger, J. (2009). Modelling the *Drosophila* embryo. *Molecular BioSystems*, *5*, 1549–1568.
Jaeger, J. (2011). The gap gene network. *Cellular and Molecular Life Science*, *68*, 243–274.
Jaeger, J. (2018). Shift happens: The developmental and evolutionary dynamics of the gap gene system. *Current Opinion in Systems Biology*, *11*, 65–73.
Jaeger, J., Blagov, M., Kosman, D., Kozlov, K. N., Manu, Myasnikova, E., et al. (2004). Dynamical analysis of regulatory interactions in the gap gene system of *Drosophila melanogaster*. *Genetics*, *167*, 1721–1737.
Jaeger, J., Irons, D., & Monk, N. (2008). Regulative feedback in pattern formation: Towards a general relativistic theory of positional information. *Development*, *135*, 3175–3183.
Jaeger, J., Manu, & Reinitz, J. (2012). *Drosophila* blastoderm patterning. *Current Opinion in Genetics & Development*, *22*, 533–541.
Jaeger, J., & Reinitz, J. (2006). On the dynamic nature of positional information. *BioEssays*, *28*, 1102–1111.
Jaeger, J., Surkova, S., Blagov, M., Janssens, H., Kosman, D., Kozlov, K. N., et al. (2004). Dynamic control of positional information in the early *Drosophila* embryo. *Nature*, *430*, 368–371.
Kauffman, S. A. (1993). *The origins of order: Self-organization and selection in evolution*. Oxford: Oxford University Press.
Krotov, D., Dubuis, J. O., Gregor, T., & Bialek, W. (2014). Morphogenesis at criticality. *Proceedings of the National Academy of Sciences of the United States of America*, *111*, 3683–3688.
Liu, F., Morrison, A. H., & Gregor, T. (2013). Dynamic interpretation of maternal inputs by the *Drosophila* segmentation gene network. *Proceedings of the National Academy of Sciences of the United States of America*, *110*, 6724–6729.
Longo, G., Miquel, P.-A., Sonnenschein, C., & Soto, A. M. (2012). *Progress in Biophysics and Molecular Biology*, *109*, 108–114.
Love, A. C. (2018). Developmental mechanisms. In S. Glennan & P. Illari (Eds.), *The Routledge handbook of mechanisms and mechanical philosophy* (pp. 332–347). London: Taylor & Francis.
Manu, Surkova, S., Spirov, A. V., Gursky, V. V., Janssens, H., Kim, A.-R., et al. (2009a). Canalization of gene expression in the *Drosophila* blastoderm by gap gene cross regulation. *PLoS Biology*, 7, e1000049.
Manu, Surkova, S., Spirov, A. V., Gursky, V. V., Janssens, H., Kim, A.-R., et al. (2009b). Canalization of gene expression and domain shifts in the *Drosophila* blastoderm by dynamical attractors. *PLoS Computational Biology*, *5*, e1000303.
Maynard Smith, J. (2000). The concept of information in biology. *Philosophy of Science*, *67*, 177–194.
Mayr, E. (1961). Cause and effect in biology: Kinds of causes, predictability, and teleology are viewed by a practicing biologist. *Science*, *134*, 1501–1506.
Mayr, E. (1965). Cause and effect in biology. In D. Lerner (Ed.), *Cause and effect* (pp. 33–50). New York: Free Press.
Mora, T., & Bialek, W. (2010). Are biological systems poised at criticality? *Journal of Statistical Physics*, *144*, 268–302.
Morishita, Y., & Iwasa, Y. (2009). Accuracy of positional information provided by multiple morphogen gradients with correlated noise. *Physical Review E*, *79*, 061905.
Morishita, Y., & Iwasa, Y. (2011). Coding design of positional information for robust morphogenesis. *Biophysical Journal*, *101*, 2324–2335.

Mossio, M., Saborido, C., & Moreno, A. (2009). *The British Journal for the Philosophy of Science, 60*, 813–841.
Needham, J. (1933). On the dissociability of the fundamental processes in ontogenesis. *Biological Reviews, 8*, 180–223.
Newman, S. A. (2008). Dynamical patterning modules: Physico-genetic determinants of morphological development and evolution. *Physical Biology, 5*, 015008.
Newman, S. A. (2012). Physico-genetic determinants in the evolution of development. *Science, 338*, 217–219.
Newman, S. A., & Bhat, R. (2009). Dynamical patterning modules: A "pattern language" for development and evolution of multicellular form. *International Journal of Developmental Biology, 53*, 693–705.
Oyama, S. (1985). *The ontogeny of information: Developmental systems and evolution*. Cambridge: Cambridge University Press.
Peluffo, A. E. (2015). The "genetic program": Behind the genesis of an influential metaphor. *Genetics, 200*, 685–696.
Perkins, T. J., Jaeger, J., Reinitz, J., & Glass, L. (2006). Reverse engineering the gap gene networks of *Drosophila melanogaster*. *Genetics, 167*, 1721–1737.
Petkova, M. D., Little, S. C., Liu, F., & Gregor, T. (2014). Maternal origins of developmental reproducibility. *Current Biology, 24*, 1283–1288.
Petkova, M. D., Tkačik, G., Bialek, W., Wieschaus, E. F., & Gregor, T. (2019). Optimal decoding of cellular identities in a genetic network. *Cell, 176*, 1–12.
Pittendrigh, C. S. (1958). Adaptation, natural selection, and behavior. In A. Roe & G. G. Simpson (Eds.), *Behavior and evolution* (pp. 390–416). New Haven: Yale University Press.
Salazar-Ciudad, I. (2010). Morphological evolution and embryonic developmental diversity in metazoa. *Development, 137*, 531–539.
Sarkar, S. (1996). Biological information: A sceptical look at some central dogmas of molecular biology. In S. Sarkar (Ed.), *The philosophy and history of molecular biology: New perspectives* (pp. 187–232). Dordrecht: Kluwer Academic Publishers.
Schrödinger, E. (1944). *What is life?* Cambridge: Cambridge University Press.
Shannon, C. E. (1948). A mathematical theory of communication. *The Bell System Technical Journal, 27*, 379–423.
Shannon, C. E. (1956). The bandwagon. *IRE Transactions on Information Theory, 2*, 3.
Sokolowski, T. R., & Tkačik, G. (2015). Optimizing information flow in small genetic networks. IV. Spatial coupling. *Physical Reviews E, 91,* 062710.
Struhl, G., Struhl, K., & Macdonald, P. M. (1989). The gradient morphogen *bicoid* is a concentration-dependent transcriptional activator. *Cell, 57*, 1259–1273.
Surkova, S., Kosman, D., Kozlov, K., Manu, Myasnikova, E., Samsonova, A. A., et al. (2008). Characterization of the *Drosophila* segment determination morphome. *Developmental Biology, 313*, 844–862.
Tkačik, G., Callan, C. G., & Bialek, W. (2008). Information flow and optimization in transcriptional regulation. *Proceedings of the National Academy of Sciences of the United States of America, 105*, 12265–12270.
Tkačik, G., Dubuis, J. O., Petkova, M. D., & Gregor, T. (2015). Positional information, positional error, and readout precision in morphogenesis: A mathematical framework. *Genetics, 199*, 39–59.
Tkačik, G., & Walczak, A. M. (2011). Information transmission in genetic regulatory network: A review. *Journal of Physics: Condensed Matter, 23*, 153102.
Tkačik, G., Walczak, A. M., & Bialek, W. (2009). Optimizing information flow in small genetic networks. *Physical Review E, 80*, 031920.
Tkačik, G., Walczak, A. M., & Bialek, W. (2012). Optimizing information flow in small genetic networks. III. A self-interacting gene. *Physical Review E, 85,* 041903.

Turing, A. M. (1952). The chemical basis of morphogenesis. *Philosophical Transactions of the Royal Society of London B*, *237*, 37–72.

Verd, B., Clark, E., Wotton, K. R., Janssens, H., Jiménez-Guri, E., Crombach, A., et al. (2018). A damped oscillator imposes temporal order on posterior gap gene expression in *Drosophila*. *PLoS Biology*, *16*, e2003174.

Verd, B., Crombach, A., & Jaeger, J. (2014). Classification of transient behaviours in a time-dependent toggle switch model. *BMC Systems Biology*, *8*, 43.

Verd, B., Crombach, A., & Jaeger, J. (2017). Dynamic maternal gradients control timing and shift-rates of *Drosophila* gap gene expression. *PLoS Computational Biology*, *13*, e1005285.

Verd, B., Monk, N., & Jaeger, J. (2019). Modularity, criticality, and evolvability of a developmental gene regulatory network. *eLife*, *8*, e42832.

Von Dassow, G., & Munro, E. (1999). Modularity in animal development and evolution: Elements of a conceptual framework for EvoDevo. *Journal of Experimental Zoology*, *285*, 307–325.

Waddington, C. H. (1940). *Organisers and genes*. Cambridge: Cambridge University Press.

Waddington, C. H. (1956). *Principles of embryology*. London: George Allen & Unwin.

Walczak, A. M., Tkačik, G., & Bialek, W. (2010). Optimizing information flow in small genetic networks. II. Feed-forward interactions. *Physical Review E*, *81*, 041905.

Wolpert, L. (1968). The French flag problem: A contribution to the discussion on pattern development and regulation. In C. H. Waddington (Ed.), *Towards a theoretical biology: Vol. 1* (pp. 125–133). Edinburgh: Edinburgh University Press.

Wolpert, L. (1969). Positional information and the spatial pattern of cellular differentiation. *Journal of Theoretical Biology*, *25*, 1–47.

Wolpert, L. (1989). Positional information revisited. *Development*, *107*(Suppl), 3–12.

Wolpert, L. (1994). Positional information and pattern formation in development. *Developmental Genetics*, *15*, 485–490.

Wolpert, L. (1996). One hundred years of positional information. *Trends in Genetics*, *12*, 359–364.

Wright, L. (1973). Functions. *The Philosophical Review*, *82*, 139–168.

Zagorski, M., Tabata, Y., Brandenberg, N., Lutolf, M. P., Tkačik, G., Bollenbach, T., et al. (2017). Decoding of position in the developing neural tube from antiparallel morphogen gradients. *Science*, *356*, 1379–1383.

CHAPTER EIGHT

Intracellular morphogens: Specifying patterns at the subcellular scale

Lars Hubatsch[a], Nathan W. Goehring[b,c,d,*]

[a]Max Planck Institute for the Physics of Complex Systems, Dresden, Germany
[b]The Francis Crick Institute, London, United Kingdom
[c]Institute for the Physics of Living Systems, University College London, London, United Kingdom
[d]MRC Laboratory for Molecular Cell Biology, University College London, London, United Kingdom
[*]Corresponding author: e-mail address: nate.goehring@crick.ac.uk

Contents

Abstract

The notion that graded distributions of signals underlie the spatial organization of biological systems has long been a central pillar in the fields of cell and developmental biology. During morphogenesis, morphogens spread across tissues to guide development of the embryo. Similarly, a variety of dynamic gradients and pattern-forming networks have been discovered that shape subcellular organization. Here we discuss the principles of intracellular pattern formation by these *intracellular morphogens* and relate them to conceptually similar processes operating at the tissue scale. We will specifically review mechanisms for generating cellular asymmetry and consider how intracellular patterning networks are controlled and adapt to cellular geometry. Finally, we assess the general concept of intracellular gradients as a mechanism for positional control in light of current data, highlighting how the simple readout of fixed concentration thresholds fails to fully capture the complexity of spatial patterning processes occurring inside cells.

Current Topics in Developmental Biology, Volume 137
ISSN 0070-2153
https://doi.org/10.1016/bs.ctdb.2019.11.006

1. Introduction

Roughly a century of work has identified the morphogen gradient as a central pillar in the development of multicellular life. Produced by local populations of signaling cells, morphogens spread through developing tissues resulting in concentration gradients that can be used by cells to guide morphogenesis and developmental programs (Briscoe & Small, 2015; Rogers & Schier, 2011). With the advent of tools to peer deep into the inner workings of cells, it has become apparent that individual cells also rely on spatial gradients of what are sometimes referred to as *intracellular morphogens* to define their internal organization (Table 1). Intracellular gradients are implicated in processes as diverse as regulation of cell division (Afonso, Figueiredo, & Maiato, 2017; Rowlett & Margolin, 2015; Wood & Nurse, 2015), assembly of the mitotic spindle (Weaver & Walczak, 2015), cell polarity (Chiou, Balasubramanian, & Lew, 2017; Goehring, 2014), and cell fate specification (Wu & Griffin, 2017).

2. The challenge of intracellular gradients—A consideration of length and time scales

While gradients across tissues and those inside cells can share conceptual similarities, details can differ. One obvious contrast is the difference in the length over which gradients operate. Classically, morphogen gradients operate on a scale of tens to hundreds of microns, equivalent to a few to tens of cell diameters (Crick, 1970). By contrast, gradients in cells typically operate on a scale of 1–10 μm. These length scales place constraints on how gradients form.

For a gradient formed by localized production, diffusion and uniform degradation, the gradient will tend to take the form of a simple exponential

$$C(x) = C_0 e^{-x/\lambda}$$

where λ is the characteristic length scale of the gradient, defined by the distance over which the concentration is reduced to approx. 36% ($1/e$) of its initial value. λ is itself given by

$$\lambda = \sqrt{D\tau} \tag{1}$$

Table 1 Intracellular patterning systems.

	System	Mechanism/function	Key references
Source sink	**Ran**	**Chromosome localized GTP exchange factor RCC1 with uniform GTPase activating protein RanGAP**	Carazo-Salas et al. (1999), Kalab et al. (2002), Maresca et al. (2009), Oh et al. (2016). **Review:** Kalab and Heald (2008)
	Eukaryotes	*Regulates spindle assembly*	
	Pom1	**Polar-localized phosphatase promotes loading at cell tips. Dissociation through autophosphorylation**	Martin and Berthelot-Grosjean (2009), Moseley et al. (2009), Saunders et al. (2012), Pan et al. (2014), Facchetti et al. (2019). **Reviews:** Howard (2012), Wood and Nurse (2015)
	S. pombe	*Midcell specification and potential cell size sensor*	
Differential diffusion	**MEX-5, POS-1**	**Mobility restriction by RNA binding. RNA binding is suppressed by polarized kinase activity, shifting molecules to a high mobility state**	Lipkow and Odde (2008), Daniels et al. (2009), Daniels, Dobrowsky, Perkins, Sun, and Wirtz (2010), Griffin et al. (2011), Wu, Zhang, et al. (2015), Han et al. (2018). **Review:** Griffin (2015)
	C. elegans	*Asymmetric inheritance of fate markers during asymmetric division*	
	ParA	**Mobility restriction by bacterial DNA binding. Released from DNA by plasmid segregation complex to create a poleward shifting gradient**	Lim et al. (2014), Vecchiarelli et al. (2014). **Review:** Hu et al. (2017)
	Bacteria	*Plasmid segregation by a "diffusion ratchet"*	
Self-organizing reaction-diffusion	**MinCDE**	**Cooperative assembly of MinD onto membrane is coupled to delayed recruitment and displacement by MinE, yielding bipolar oscillations**	Corbin et al. (2002), Loose et al. (2008), Zieske and Schwille (2014), Wu, van Schie, et al. (2015). **Reviews:** Rowlett and Margolin (2015), Wettmann and Kruse (2018)

Continued

Table 1 Intracellular patterning systems—Cont'd

	System	Mechanism/function	Key references
	E. coli	*Proper positioning of the division machinery at midcell*	
	Cdc42	**Autocatalytic Cdc42 accumulation is linked to global depletion of cytoplasmic pool of rapidly diffusing positive regulators**	Wedlich-Soldner et al. (2003), Goryachev and Pokhilko (2008), Kozubowski et al. (2008), Howell et al. (2009), Wu, Chiou, et al. (2015). **Review:** Chiou et al. (2017)
	S. cerevisiae	*Orientation of bud site and cell growth*	
	PAR proteins	**Segregation of PAR proteins into domains through reciprocal negative feedback with domain size limited by depletion of a cytoplasmic pool**	Goehring et al. (2011), Sailer, Anneken, Li, Lee, and Munro (2015), Gross et al. (2019), Hubatsch et al. (2019). **Review:** Goehring (2014)
	Metazoans	*Asymmetric cell division and polarized cell architecture*	
Trigger waves	**Cdk1 activation**	**Local diffusion and bistable reaction kinetics promote long-range signal propagation, similar to the conduction of nerve impulses**	Chang and Ferrell (2013). **Reviews:** Gelens et al. (2014), Deneke and Di Talia (2018)
	Xenopus, Drosophila	*Coordination of cell cycle progression in large cells*	

where D is the diffusion coefficient and τ is the typical lifetime of the morphogen (Fig. 1A). Given this relationship, for typical protein lifetimes of 100 min and diffusivities of 10 $\mu m^2/s$, we obtain a characteristic length scale of 250 μm, very reasonable for defining a gradient across a tissue. However, obtaining a gradient of 5 μm for the same morphogen would instead imply unrealistic protein lifetimes of only 2.5 s. Only if diffusivity is reduced to 0.01 $\mu m^2/s$ in this case, more than two orders of magnitude below what

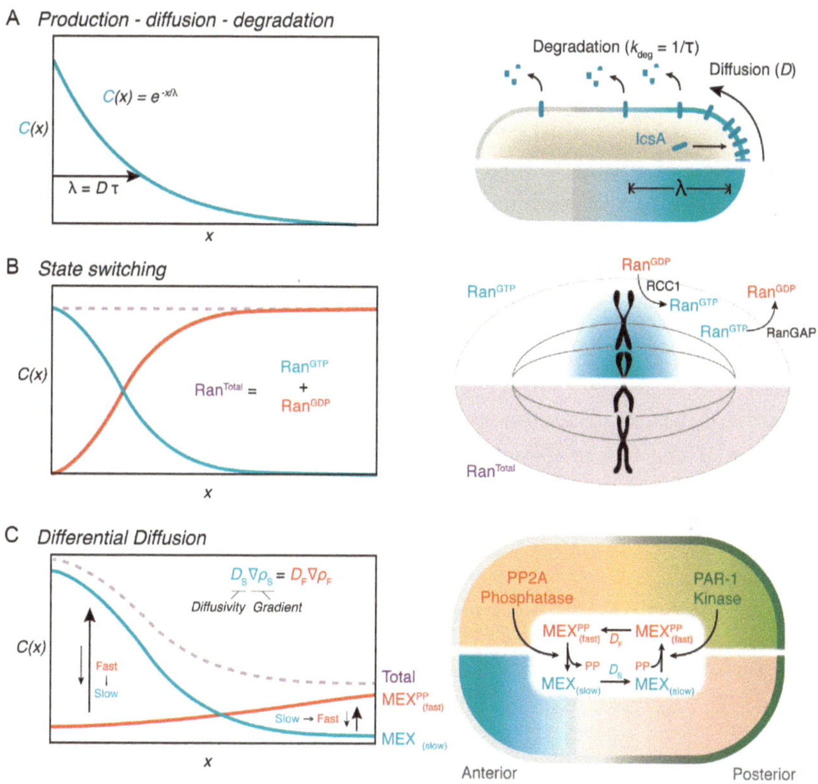

Fig. 1 Building intracellular gradients. (A) Profile of a typical source-sink gradient. Such a gradient is formed in *Shigella flexneri* through local insertion of IcsA at the posterior, followed by lateral spread and degradation. (B) An activity gradient formed by local state switching. Opposing gradients of active and inactive species are formed, but total levels of protein remain uniform. A RanGTP gradient is formed by local conversion of RanGDP to GTP by chromosome-associated RCC1, followed by diffusion and GTP-GDP switching induced by the uniform RanGAP. (C) A protein gradient formed through differential diffusion. Similar to (B), but one species diffuses slower than the other. This results in a steeper gradient of the slow species and thus a gradient in protein concentration across the cell. In *C. elegans*, PAR-1 induces a switch to a fast MEX state through phosphorylation, which is reversed by a uniform phosphatase PP2A, leading to overall enrichment in the anterior.

is typical for a cytoplasmic protein, can we sustain the gradient while maintaining reasonable protein lifetimes (~42 min).

This logic therefore suggests that the establishment of gradients via production and degradation, while ubiquitous in developing tissues, would be rather rare in cells. The most famous example of an intracellular gradient is that of Bicoid, but it must be noted that this cell is 500 μm in length! One of the few examples of such an intracellular gradient in a more typically sized cell is the polar gradient of the bacterial virulence factor IcsA in *Shigella* (Fig. 1A). IcsA drives local actin nucleation at one pole of the bacteria, resulting in unidirectional actin comet-based motility of the pathogen (Goldberg & Theriot, 1995). IcsA is secreted preferentially at the pole and incorporated into the bacterial outer membrane, diffuses laterally, and is degraded in a uniform fashion along the bacterial membrane (Robbins, Monack, McCallum, et al., 2001; Steinhauer, Agha, Pham, Varga, & Goldberg, 1999). Achieving an IcsA gradient that decays over ~1 μm is likely only possible due to severe restriction of diffusion in the membrane potentially due to corralling by large outer membrane protein clusters (Chavent, Duncan, Rassam, et al., 2018) or high levels of molecular crowding giving rise to glass-like behavior (Munguira, Casuso, Takahashi, et al., 2016).

3. State switching allows spatial activity gradients

It is far more typical for cells to create gradients through state switching (Brown & Kholodenko, 1999; Howard, 2012; Kholodenko, 2009; Peglion & Goehring, 2019). Rather than localized production and degradation, a molecule can be switched, for example, between active and inactive states. Cells have many tools at their disposal for regulating protein activity, such as changes in nucleotide-bound state or post-translational modifications, which are less energetically demanding and more rapid than synthesis and degradation. Although the mathematics remain similar, state switching converts pattern formation to a process of redistribution within the cell, with relevant timescales now constrained by the enzymatic rate constants that drive state switching rather than protein synthesis and degradation.

A canonical example is the gradient of the small diffusible GTPase Ran (Fig. 1B) (Kalab & Heald, 2008; Kalab, Weis, & Heald, 2002). Ran activity is tightly linked to the state of its bound nucleotide. RanGTP is produced from RanGDP by the GTP exchange factor RCC1 (Carazo-Salas et al., 1999). RanGTP is converted back to RanGDP through its inherent GTPase activity, stimulated by the GTPase activating protein RanGAP, completing the GTP-GDP cycle. Cells are able to leverage this cycle for local regulation by spatially decoupling these two processes. During interphase, RCC1 is bound to chromatin and RanGAP is cytoplasmic, generating a concentration gradient of RanGTP across the nuclear envelope, which can be used to drive protein import into and export from the nucleus. During mitosis, RCC1 remains concentrated at chromosomes, but as the nuclear envelope breaks down, RanGAP enters the nucleus converting the RanGTP gradient across the nuclear envelope into a spatially extended gradient emanating from the condensing chromosomes. This gradient in RanGTP in turn promotes local activation of microtubule regulators to drive assembly of the mitotic spindle. Conceptually, this process is similar to a source-sink mechanism: RanGTP diffuses away from a pool of chromosome-associated RCC1 (the source) and is subject to a constant rate of "degradation" by RanGAP (a uniform sink).

Similar activity gradients, though reliant on phosphorylation rather than nucleotide state govern activity of the MAP kinase Fus3, which controls polarized growth of the yeast mating projection known as a schmoo (Maeder et al., 2007), and the mitotic kinase Aurora B, which extends from midcell to spatially regulate spindle microtubules, chromosome decondensation, mitotic exit and cytokinesis (Afonso et al., 2014; Fuller, Lampson, Foley, et al., 2008; Wang, Ballister, & Lampson, 2011).

4. Molecular gradients through differential diffusion

It is important to emphasize that in the above cases of state switching, cells are not creating gradients of total protein concentration, but gradients of an active species. Because the molecules in the two states are otherwise equivalent, at steady-state, there must be a balance in the flux of molecules across the cell: Net diffusion of active molecules from the source to the sink must be balanced by net diffusion of inactive species back to the source, where they provide the substrate for generating active species. Because net flux must balance and diffusion rates of active and inactive species are the same, total concentration must remain uniform across the cell (Kholodenko, 2009) (Fig. 1B).

This changes, however, if the diffusion coefficients of the active and inactive species are different. Theoretical work has demonstrated how diffusion differences of the relevant species in such systems could lead to concentration gradients across a cell (Kholodenko, 2009; Lipkow & Odde, 2008). Diffusive flux is given by: $J = -D\nabla\rho$, where D is diffusivity and ρ is the concentration gradient. Because flux must still balance, if diffusivity of the two species is different, the decreased diffusivity of the slow species will be compensated by accumulation of a steeper concentration gradient, ultimately leading to asymmetric accumulation of total protein (slow + fast) (Fig. 1C).

The *C. elegans* zygote has proved to be a prototypical model to understand how mobility switches enable the formation of polarized gradients. Gradients of fate determinants are induced by asymmetric activity of PAR polarity proteins (discussed in detail below), allowing them to be segregated asymmetrically upon division of the zygote. Gradients rely on asymmetric PAR-1 kinase activity, which induces a gradient of the RNA binding protein MEX-5 through a phosphorylation-dependent switch in MEX-5 mobility (Fig. 1C). Phosphorylated MEX-5 is displaced from cytoplasmic RNAs, converting it into a freely diffusing form, with dephosphorylation by a uniform phosphatase returning it to a slow RNA-bound state (Griffin, Odde, & Seydoux, 2011). MEX-5 then induces segregation of a variety of downstream molecules, again through regulation of their mobility, which in at least one case also involves asymmetric regulation of RNA binding by phosphorylation (Brangwynne, Eckmann, Courson, et al., 2009; Han et al., 2018; Wu, Zhang, & Griffin, 2015).

One set of molecules downstream of MEX-5 are those that make up the germ line P granules. P granules are biomolecular condensates that result from liquid-liquid phase separation (LLPS) (Brangwynne, Eckmann, et al., 2009). During LLPS, molecules condense into droplets, becoming highly concentrated relative to the surrounding cytoplasm in a process that is similar to the separation of oil from water (Banani, Lee, Hyman, & Rosen, 2017). MEX-5 locally suppresses LLPS by sequestering cytoplasmic RNA. Thus as MEX-5 becomes concentrated in the anterior, the pool of available RNA is increased in the posterior allowing P granules to form (Saha, Weber, Nousch, et al., 2016; Smith et al., 2016). Although individual components of P granules can rapidly exchange between P granules and the cytoplasm, and can diffuse internally within both compartments, molecules in the condensed phase are partitioned within P granules that, due to their size, are minimally-diffusive on the relevant cellular time and length scales.

More generally, any spatially regulated mechanism to bias the reversible conversion between high and low mobility states can support asymmetric distributions. In bacteria, a family of molecules related to the plasmid-partitioning protein ParA associate with the bacterial chromosome and use the resulting patterns to spatially regulate diverse processes, including plasmid segregation (ParA, *Caulobacter crescentus* (Hu, Vecchiarelli, Mizuuchi, Neuman, & Liu, 2017; Lim et al., 2014; Vecchiarelli,

Neuman, & Mizuuchi, 2014)), spacing of organelles (McdA, *Synechococcus elongatus* (MacCready, Hakim, Young, et al., 2018)), and cell division (PomZ, *Myxococcus xanthus* (Schumacher, Bergeler, Harms, et al., 2017)). Another common strategy is to reversibly bind to the plasma membrane or other internal membrane system, which is typically associated with order of magnitude reductions in diffusivity and may be further constrained by corralling or membrane crowding effects (Jacobson, Liu, & Lagerholm, 2019; Kusumi, Fujiwara, Chadda, et al., 2012; Schavemaker, Boersma, & Poolman, 2018). Thus, cells have multiple tools to locally modulate the mobility of proteins.

5. Toward self-organization

We have so far shown how spatially segregated sources and sinks can give rise to patterning inside the cell. In these cases, the output pattern is largely set by the geometry imposed by the source and sink. However, if molecules are able to influence their own production and degradation or conversion between states, patterns can become self-organizing.

This notion of self-organization, and in fact the term "morphogen" itself was put forth in 1952 by an unlikely author: British mathematician Alan Turing. In seminal work, he provided a theoretical demonstration that the diffusion of molecules, when coupled to the appropriate biochemical feedback pathways, could give rise to spontaneously forming patterns through the amplification of stochastic fluctuations (Turing, 1952). Later formulated into a more intuitive form by Gierer and Meinhardt, the central features of these so-called reaction-diffusion (RD) systems were the ability of a morphogen to locally amplify itself, for example by promoting its own production, while at the same time generating a long-range signal to restrict production to these local regions of self-activation (Gierer & Meinhardt, 1972). In the simplest form, this can be achieved if (i) morphogen production is autocatalytic, thereby allowing fluctuations to be amplified, (ii) morphogen production is coupled to production of an inhibitor of morphogen production, and (iii) the diffusion of the inhibitor is faster than that of the morphogen (Fig. 2A). However, numerous variants involving various numbers of species, interaction networks and patterns of mobility are possible. For example, in activator-depletion systems (Fig. 2B), depletion of a fast diffusing substrate required for morphogen production serves as the long-range inhibitory signal. Depending on the precise implementation, parameter choices and time delays, the types of patterns can vary enormously, from spots and stripes to traveling waves and oscillations. These

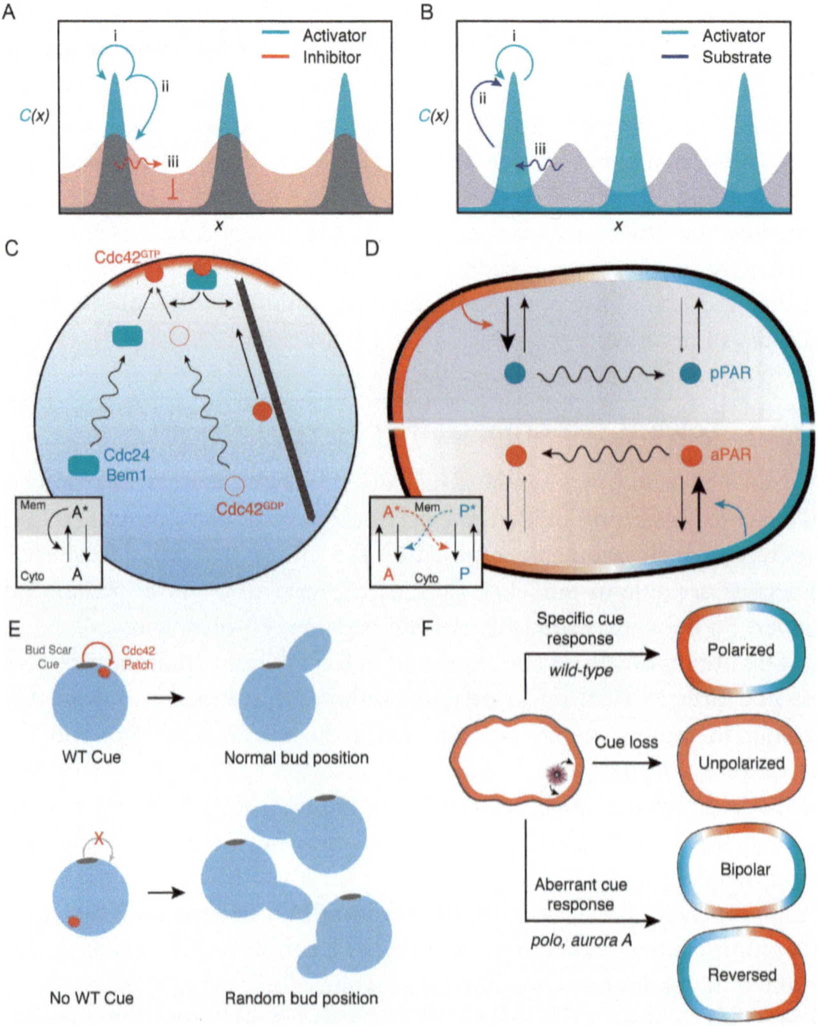

Fig. 2 Self-organizing patterning networks. (A) An activator-inhibitor network in which the activator induces its own production (i) and a fast diffusing inhibitor (ii), which spreads beyond the activator peak to suppress additional peaks over a characteristic length scale (iii). (B) An activator-depletion network in which an activator induces its own production (i) from a diffusible substrate (ii). Diffusion of the substrate converts local consumption into long-range depletion, again suppressing additional peaks over a characteristic length scale (iii). (C) Polarity establishment in budding yeast by a self-amplifying positive feedback loop. Cdc42(GTP) induces its own accumulation via stimulating actin cable-based transport of Cdc42(GTP)-containing vesicles to the polarity patch and via recruitment of a GEF complex (Cdc24/Bem1). Long-range inhibition is believed to occur via depletion of freely diffusible GEF complex. (D) Polarity establishment in the *C. elegans* zygote through a double negative feedback loop. On one side of

networks can also be integrated with mechanics, growth, and shape change, greatly expanding the complexity and range of patterning capabilities available (Bois, Jülicher, & Grill, 2011; Diego, Marcon, Müller, & Sharpe, 2018; Goehring & Grill, 2013; Green & Sharpe, 2015; Halatek & Frey, 2018; Howard, Grill, & Bois, 2011; Schweisguth & Corson, 2019).

Compared to the source-sink models, self-organizing networks have a number of attractive properties. Most fundamentally, self-organizing systems allow patterns to emerge spontaneously from an initially un-patterned state without the need to invoke upstream spatial inputs. Moreover, even in cases where spatial signals are present, the nature of these systems allows them to amplify and lock-in persistent patterns in response to weak or transient cues (Jilkine & Edelstein-Keshet, 2011). By contrast, source-sink gradients can only persist as long as upstream cues are present—the distribution of RanGTP becomes uniform as soon as the source is switched off.

One of the best studied intracellular self-organizing networks is the budding yeast cell polarity pathway (Fig. 2C). In budding yeast, cell polarity is required to guide reproduction, both in organizing the process of budding by which new daughter cells are formed and "bud off" from mother cells, and for directed growth in response to pheromones from compatible neighboring yeast cells during the mating process (Chiou et al., 2017). Polarization involves formation of a single stable patch of GTP-bound Cdc42 at the cell membrane, which spatially focuses downstream pathways, including the biosynthetic machinery for cell wall synthesis. Polarization of Cdc42 is achieved through a Turing-like activation-depletion model of self-organization in which Cdc42(GTP) promotes its own recruitment and activation through two reinforcing feedback loops (Goryachev & Pokhilko, 2008; Kozubowski et al., 2008). One involves recruitment of a Cdc42 GTP exchange factor (GEF), Cdc24p, and the scaffold Bem1 from the cytoplasm, which in turn promotes further local production of Cdc42(GTP) (Goryachev & Pokhilko, 2008; Kozubowski et al., 2008). The second relies

the embryo, anterior PARs (A, aPAR) exclude posterior PARs (P, pPARs), while the reverse holds on the other side. Domain size is limited by depletion of the respective pools of cytoplasmic PAR proteins. (E) In wild-type budding yeast, signals from the bud scar bias accumulation of Cdc42. When this cue is absent, spontaneous symmetry-breaking occurs, leading to random bud placement. (F) In the *C. elegans* zygote, the centrosome triggers symmetry-breaking at the posterior pole. In the absence of this cue, polarity fails. However, in *polo* and *aurora A* mutants, the system responds to aberrant cues, resulting in polarization that occurs at either or both poles.

on Cdc42p-dependent stimulation of actin cables and delivery of a vesicle-associated pool of Cdc42p (Wedlich-Soldner, Altschuler, Wu, & Li, 2003), though details remain controversial (Chiou et al., 2017). The restriction of Cdc42 to a single polarity site relies on rapid depletion of cytoplasmic components (e.g., Bem1) as they are sequestered within the growing polarity domain (Howell, Savage, Johnson, et al., 2009; Wu, Chiou, Minakova, et al., 2015). This depletion serves a long-range inhibitory function, as depleted material is not available to form a peak elsewhere. Importantly, like in the above systems that give rise to gradients of total concentration rather than activity, differential diffusion plays a critical role: A slow membrane-associated state of Cdc42p and its associated polarity factors allows them to accumulate within the micron scale patch, while the fast cytoplasmic state allows local depletion by the growing polarity patch to propagate rapidly across the cell to suppress formation of additional polarity sites (Woods & Lew, 2019). While this network is sufficient to allow spontaneous polarization, under normal circumstances, polarity is under the direct control of spatial cues (Chiou et al., 2017).

A similar self-organizing network underlies polarity in animal cells. This network, known as the *PAR-titioning defective* or PAR network, relies on the ability of its components to form asymmetric membrane-associated domains (Fig. 2D). In the *C. elegans* zygote, these domains are formed through the segregation of two sets of PAR proteins into opposing domains (Rose & Gönczy, 2014). One set (aPAR) marks the anterior pole of the embryo and the other (pPAR) marks the posterior, which together spatially regulate downstream pathways to yield the asymmetric division of the zygote. Rather than positive feedback, the segregation of aPAR and pPAR proteins into mutually exclusive domains relies on reciprocal, double negative feedback in which each set of PAR proteins excludes the opposing set from the plasma membrane (Goehring, 2014). However, the result is similarly autocatalytic: an initial advantage for one set of proteins will locally reduce the concentration of and antagonism by the second set, allowing more of the first to accumulate. Similar to budding yeast, long-range inhibition is thought to occur through cytoplasmic depletion (Goehring, Trong, Bois, et al., 2011). As PAR domains grow, they incorporate more PAR proteins from a freely diffusing cytoplasmic pool, reducing the pool available to sustain growth of the domain and formation of additional domains elsewhere in the cell. Mathematical modeling based on quantitative measurements in live embryos is consistent with the ability of this reaction-diffusion network to sustain stable, polarized states

(Dawes & Munro, 2011; Goehring et al., 2011). However, unlike yeast, the feedback circuits responsible for self-organization are thought to be subcritical (Goehring et al., 2011; Gross, Kumar, Goehring, et al., 2019; Trong, Nicola, Goehring, Kumar, & Grill, 2014). In other words, the feedback is not strong enough to amplify stochastic fluctuations to allow spontaneous breaking of symmetry. Consequently, not only do spatial cues guide pattern formation, but are required to induce patterns from an initially homogeneous state.

6. Controlling self-organization in space and time

Why might self-organizing systems rely on spatial cues if they are, at least in principle, capable of spontaneous polarization? While advantageous for amplifying spatial cues and enabling the de novo emergence of patterns, self-organization can be a liability in a developmental system, where uncontrolled and spontaneously emerging patterns can wreak havoc on morphogenetic processes. For example, cell polarity plays crucial roles in defining the ability of cells to migrate or grow in response to environmental cues. Cell polarity is also critical for defining whether a cell divides asymmetrically and if so, ensuring that it divides in the right orientation to establish the correct tissue geometry. Thus, not only must cells polarize, but must do so at the right time and place.

In budding yeast, these dual requirements are manifest in the tight control over the orientation of budding events. For reasons that are not entirely clear, haploid cells form buds near previous events, while in diploid cells, daughter cells preferentially undergo budding at the pole opposite to the bud scar (Chiou et al., 2017). If spatial cues are absent or compromised, the feedback is sufficient to drive spontaneous symmetry-breaking, potentially as part of a failsafe mechanism. Thus, the Cdc42p network must be carefully controlled to bias the system toward spatial cues. Several processes appear to be at play. First, as we noted, fast diffusion of limiting Cdc42 regulators ensures that it is rapidly depleted throughout the cell as the initial patch forms, ensuring the establishment of a single polarity axis (Wu, Chiou, et al., 2015). Second, there are additional negative feedback pathways that destabilize Cdc42p patches as they form to help ensure stable polarization at a single dominant site (Howell et al., 2012). Finally, the wiring of the network is under cell cycle control, becoming more permissive to symmetry-breaking over time, potentially giving the network a chance to respond to cues if they are present before allowing it to undergo spontaneous symmetry-breaking (Moran, Kang, Araujo, et al., 2019; Witte, Strickland, & Glotzer, 2017).

In the *C. elegans* zygote, the induction of polarity is normally restricted to a single cue, the paternal centrosome, which acts through two semi-redundant pathways (Rose & Gönczy, 2014). In the dominant pathway, local centrosomal signals induce long-range flows of cortical actin, which sweep aPAR proteins into the nascent anterior and exclude them from the posterior (Goehring et al., 2011; Munro, Nance, & Priess, 2004). A second, concurrent mechanism invokes a role for centrosome-nucleated microtubules in locally inhibiting aPARs in the posterior, allowing pPARs to load onto the membrane (Motegi et al., 2011; Tsai & Ahringer, 2007; Wallenfang & Seydoux, 2000). Together these cues enable robust polarization over the course of several minutes, after which the self-organizing properties of the network take over to maintain the polarized state in the zygote. Simultaneous disruption of both cues prevents proper symmetry-breaking, consistent with the subcritical nature of the PAR feedback circuits (Gross et al., 2019; Motegi et al., 2011). The centrosome itself is temporally regulated, only becoming active after two meiotic divisions. Thus, the zygote is positioned to undergo symmetry-breaking precisely at the onset of the first mitotic division and sets up a single axis of polarity in response to a single specific cue (Cowan & Hyman, 2004, 2006). The specificity of this response, however, is no accident. Disruption of either of two cell cycle kinases, the polo-like kinase homolog PLK-1 and the Aurora A kinase homolog AIR-1, prematurely activates the self-organizing capability of the PAR network, allowing it to respond to aberrant cues that are normally ignored, resulting in zygotes with reversed polarity or bipolar pPAR caps on both poles (Klinkert, Levernier, Gross, et al., 2019; Reich, Hubatsch, Illukkumbura, et al., 2019) see also (Kotak & Kapoor, 2018; Zhao, Teng, Tantirimudalige, et al., 2019). Thus, like budding yeast, the *C. elegans* embryo actively modulates the responsiveness of the PAR network to enhance specificity to dedicated spatial cues and ensure the proper timing and geometry of patterning by the PAR network (Reich et al., 2019).

7. Geometry sensing by patterning networks

Ultimately, to be useful, patterns must be correctly coordinated with the geometry of the cell. As we have seen, this can be achieved by the provision of dedicated cues, which override natural tendencies of the underlying patterning network. At the same time, the ability of networks to amplify even small spatial imbalances allows direct geometry sensing by networks themselves (Haupt & Minc, 2018).

One way this is achieved is through the ability of molecules or molecular assemblies to directly sense local geometry, such as membrane curvature (McMahon & Boucrot, 2015). By seeding the local recruitment of network components, they can serve to anchor an otherwise unstable pattern geometry. Bacteria have several molecules that recognize features of the poles (Laloux & Jacobs-Wagner, 2014). One of these, DivIVA localizes to regions of negative curvature to direct cell division inhibitors to the poles and form the septum in *Bacillus subtilis* (Lenarcic, Halbedel, Visser, et al., 2009; Ramamurthi & Losick, 2009). In eukaryotes, curvature sensing BAR proteins have been implicated in traveling waves on the cell cortex, which emerge through positive feedback between the ability of BAR proteins to locally recruit actin nucleating factors to regions of particular curvature and the generation of curvature by the resulting actin networks (Wu, Su, Tong, Wu, & Liu, 2018).

Transport processes can also render pattern-forming networks sensitive to local and global geometry (Fig. 3A and B). Such behaviors can arise through effective confinement due to limits on transport processes, typically diffusion. For example, local increases in surface to volume ratio in regions of high curvature or in thin plasma membrane extensions, can result in reactions becoming diffusion-limited, leading to spatial biases (Meyers, Craig, & Odde, 2006; Rangamani, Lipshtat, Azeloglu, et al., 2013; Schmick & Bastiaens, 2014; Thalmeier, Halatek, & Frey, 2016), while effective compartmentalization of membrane regions can enable locally distinct behaviors in reaction-diffusion systems (Hansen et al., 2019). In some cases, simple transitions from spherical to ellipsoid or flattened can be enough to induce spatial inhomogeneity in concentrations or globally shift the balance of reaction networks.

More global changes in surface to volume ratio accompanying size or shape changes can also tune the reaction parameters of a system. The effect of changing geometry has been explored in the context of organelle scaling by surface to volume ratios (Brownlee & Heald, 2019) and activation of the zygotic genome at the midblastula transition by reduced volume to DNA ratios during embryonic cleavage divisions (Amodeo, Jukam, Straight, & Skotheim, 2015; Chen, Einstein, Little, & Good, 2019). Such mechanisms open the door to size- and shape-dependent regulation and scaling of patterns to geometry.

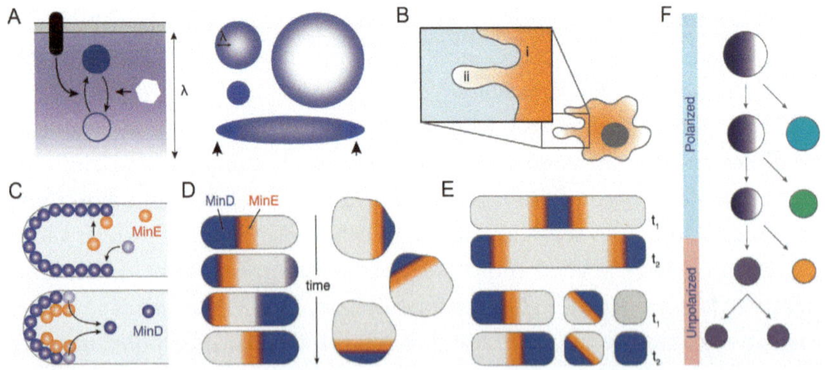

Fig. 3 Geometric effects on patterning networks. (A) Local activation of a molecule by a membrane-localized activator (e.g., kinase) coupled to a uniformly acting inactivator (e.g., phosphatase) leads to a gradient of active species (solid) extending into the cytoplasm. Because of the fixed length scale of this gradient, changes in size and shape will alter the ratio of active vs inactive species, which for non-spherical cells will vary in space. (B) Spatial biases can arise from changes in local membrane-cytoplasm ratio, for example, due to local depletion of a cytoplasmic signaling molecule (orange) by membrane-associated receptors, which here results in distinct local concentrations in regions (i) and (ii). (C) MinD oscillations arise through the combination of cooperative membrane association of MinD and the delayed accumulation of MinE which triggers MinD release. (D) Schematic of Min behavior in normal and round *E. coli* mutants. (E) Min patterns in *E. coli* cells confined in microwells of different size. Multiple stripes form in long cells, while in small cells, alignment shifts to the diagonal to maximize the oscillation axis or is lost completely. (F) A model for size dependent regulation of polarity and asymmetric division modeled on PAR polarity in *C. elegans*. In cleavage divisions, cells do not grow and thus become smaller at each division. If a polarity network exhibits a fixed length scale, at some point cells will become too small to polarize and will be unable to divide asymmetrically.

Geometry effects can also emerge from interactions between the favored length scale of a patterning network and the geometry of the system in which it operates. One of the best studied examples in this context is the bacterial MinCDE system, initially discovered in the rod-shaped bacteria *E. coli* (de Boer, Crossley, & Rothfield, 1989). The Min proteins undergo dynamic pole-to-pole oscillations, which guide assembly of the cell division machinery at midcell (Rowlett & Margolin, 2015) (Fig. 3C). MinD and MinE comprise the pattern-forming unit. MinD is an ATPase that undergoes cooperative binding to the plasma membrane and recruits the division inhibitor MinC. MinD(ATP) assembles at the membrane and recruits MinE. MinE then triggers ATP hydrolysis by MinD, yielding MinD(ADP) which is displaced into the cytoplasm. Following nucleotide exchange, MinD(ATP) undergoes a new cycle of membrane association at the opposite pole. The dynamic oscillations result in time-averaged gradients of MinC that are highest at the cell poles and lowest at midcell, thereby restricting assembly of the division apparatus to the center of the cell.

Experiments in cell shape mutants provided the first insight into geometry sensing. Min proteins generally aligned with the long axis of the cell if this axis was clear, but moved between multiple sites in round or aberrantly sized cells (Corbin, Yu, & Margolin, 2002) (Fig. 3D). Microfabrication techniques have enabled precise control of parameters by growing bacteria in compartments of defined size and shape. For rectangular cells of fixed, narrow width, these experiments revealed an optimal length scale of roughly 3–6 μm, with a robust ability to oscillate along the long axis (Wu, van Schie, Keymer, & Dekker, 2015).

However, while long axis specification is the general rule in *E. coli*, as one moves beyond preferred length scales, qualitative changes in the pattern can be observed (Fig. 3E). As *E. coli* cells grow longer, bipolar Min oscillations transition into oscillating stripes. Conversely, as cells are confined to smaller sizes, oscillations shift to the diagonal, maximizing the oscillation axis. Below a critical length, the pattern is lost and the Min proteins revert to stochastic fluctuations (Wu, van Schie, et al., 2015).

This notion of a minimum system size has been noted in models for cell polarity networks, which generically exhibit minimal length scales below which pattern formation is not possible (Hubatsch, Peglion, Reich, et al., 2019; Jilkine & Edelstein-Keshet, 2011; Trong et al., 2014). In the case of the PAR polarity network in the *C. elegans* embryo, this size threshold has been linked to a developmental transition between asymmetric and symmetric division in the germ lineage, raising the possibility for a direct role of pattern length scales in regulating developmental events in response to cell size (Hubatsch et al., 2019) (Fig. 3F).

8. Patterning and signal propagation across length scales

In the systems discussed up to this point, we have focused on length scales that are dominated by diffusion as the fundamental mode of morphogen transport. However, the efficiency of diffusive transport drops rapidly as length scales increase because diffusion time increases with the distance squared ($\tau = \lambda^2/D$ from Eq. 1, Fig. 4A). This is not a problem in small cells. Given a typical cytoplasmic protein ($D = 10\,\mu m^2/s$), the time to traverse the length of a bacterium like *E. coli* (~3 μm) is less than a second. For the

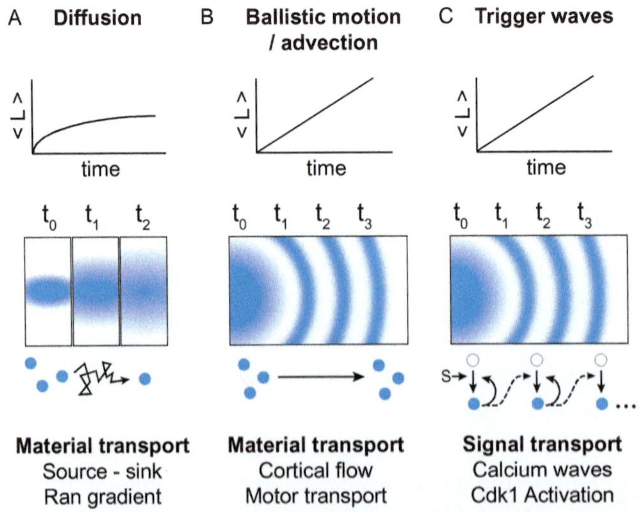

Fig. 4 Modes of signal transport. (A) Diffusive transport arises from the random motion of molecules which will yield net transport down a concentration gradient. The random nature of motion leads to a square root relationship between time and average distance traveled ($L \propto \sqrt{t}$). (B) For ballistic motion or advection, e.g., motor-driven transport, molecules are transported unidirectionally, with the time required for molecular transport increasing linearly with distance ($L \propto \sqrt{t}$). (C) In the case of trigger waves, the time required for a signal to spread also increases linearly with time. However, in this case, information (e.g., the local state of a network) rather than molecules are transported: a signal induces a local switch in the state of molecules. As these molecules diffuse to adjacent regions and induce switching of additional molecules, they trigger a chain reaction that results in a propagating wave of signal across the system. Because molecules only need to spread short distances to relay the signal, this wave can travel much faster than would be possible by diffusion alone.

C. elegans zygote (50 μm), the time increases to 25 s. However, for very large cells, like *Xenopus* or Zebrafish oocytes, which can reach up to 0.5–1.2 mm, diffusion times reach into hours to tens of hours, making them unsuited for governing intracellular processes operating on minute timescales, such as mitosis.

One solution for large cells is directed transport—also known as ballistic motion, which in cells typically involves molecular motors. Molecules be moved along cytoskeletal tracks or through the generation of large scale cytoskeletal or cytoplasmic flows, which transport molecules by bulk material flow in a process known as advection. Because ballistic and advective timescales increase linearly with distance, they are much more effective than diffusion in large cells (Fig. 4B). Transport by molecular motors is the rule for transport along axons which can reach up to several meters in length

(Sheetz, Steuer, & Schroer, 1989). The *C. elegans* zygote is also an interesting case in that it can be polarized either by cortical flows (advection) or biochemical, diffusion-limited processes (microtubule cue). However, the timescales of polarization differ dramatically. Whereas the PAR reaction-diffusion model alone is sufficient to drive the system to a stable polarized state with a boundary at midcell if provided with a local biochemical stimulus (e.g., local protection of pPAR proteins), the system approaches that steady-state slowly (>30 min). By contrast, cortical flows can actively push the PAR boundary toward midcell, allowing a fully polarized state to be achieved in <10 min (Goehring et al., 2011). This process is facilitated by the clustering of aPAR proteins into multimolecular assemblies, reducing their diffusive motion and maximizing net transport (Dickinson, Schwager, Pintard, Gotta, & Goldstein, 2017; Rodriguez, Peglion, Martin, et al., 2017; Wang et al., 2017). Given that the first cell cycle is only about 20 min, advection is critical for polarization to occur in time. A similar role for advection has been proposed in Drosophila embryos to allow the gradient of the morphogen Bicoid to more rapidly reach its steady-state length scale in the syncytial embryo (Hecht, Rappel, & Levine, 2009).

Advection can also drive active mixing, allowing molecules to sample space much more rapidly than by diffusion alone. In this case, local, randomly oriented ballistic motion appears as an increase in effective diffusion on the scale of the cell, enabling diffusion-driven patterning processes such as diffusion and capture that would otherwise occur far too slowly (Brangwynne, Koenderink, MacKintosh, & Weitz, 2009; Lu et al., 2018; Quinlan, 2016).

Another way to enable rapid communication across a cell is not to transport molecules themselves, but to transmit information through so-called trigger waves (Deneke & Di Talia, 2018; Gelens, Anderson, & Ferrell, 2014). Trigger waves are traveling waves of chemical activity that rely on bistable feedback circuits. In these systems, molecules will locally tend to remain in one state, e.g., inactive. However, once the local concentration of active molecules reaches a threshold, positive feedback will amplify the activation process to convert the region into active state. As molecules from this active region diffuse, they can trigger activation in neighboring regions, leading to a propagating wave (Fig. 4C). Importantly, molecules need only diffuse sufficiently far to relay information to immediately adjacent regions and thus, unlike a diffusive signal, trigger waves propagate linearly with time. Such trigger waves are critical for coordinating mitotic entry and microtubule nucleation across large eggs and embryos (Chang & Ferrell, 2013; Ishihara et al., 2014), waves of actin polymerization and protrusion across the surface of cells (Weiner, Marganski, Wu, Altschuler, & Kirschner, 2007), propagation of action potentials in neurons (Hodgkin & Huxley, 1952), oocyte activation (Gilkey, Jaffe, Ridgway, & Reynolds, 1978), and coordination of multicellular behaviors via cAMP and calcium signaling waves (Tomchik & Devreotes, 1981).

9. Gradient precision, noise, and readout strategies

The ultimate output of intracellular morphogens is a pattern of spatially varying concentrations that can be used to influence downstream processes. This notion of a concentration-dependent response is a central feature of the morphogen hypothesis as put forth by Wolpert, who envisioned that a cell could tell how far it was from a morphogen source by reading out local morphogen concentration from the gradient (Wolpert, 1969). However, an intrinsic problem for models that rely on concentration-dependent thresholds is that they are expected to be highly sensitive to both the magnitude and length scale of the gradient (Fig. 5). This sensitivity suggests systems must either possess mechanisms to tightly control gradients or their readout in the face of fluctuations and noise, or downstream pathways must be robust to changes in the gradient.

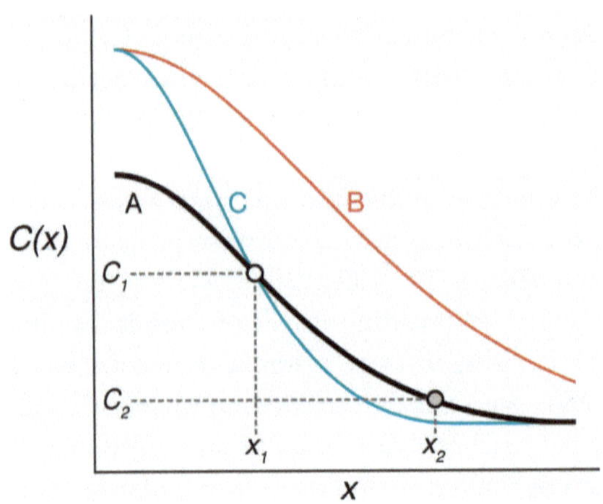

Fig. 5 Gradient adaptation and interpretation. In a French Flag Model as proposed by Wolpert, consider a hypothetical gradient (A) with a single characteristic length scale that triggers specific outputs at defined thresholds C_1 and C_2 at positions x_1 and x_2, respectively. (B) If the peak concentration increases, but the length scale is held constant, neither position is accurately specified. (C) If gradient length is a function of peak concentration, the gradient can accurately specify one position (e.g., x_1), but will invariably fail to specify the second, in this case x_2.

In an intracellular context, these concepts have been best explored in the context of the Pom1 kinase gradient in fission yeast. Pom1 was first identified for its role in the regulation of mitotic entry and positioning of the cytokinetic ring, which it achieves through phosphorylation of the cell division promoting kinase Cdr2 (Bähler & Pringle, 1998; Martin & Berthelot-Grosjean, 2009; Moseley, Mayeux, Paoletti, & Nurse, 2009). Pom1 is loaded at the cell poles in a dephosphorylated form, diffuses toward midcell, and eventually dissociates through autophosphorylation (Hachet et al., 2011), giving rise to a bipolar gradient with the lowest concentration at midcell. If this gradient is fixed, midcell Pom1 concentrations will drop as a function of cell size, suggesting it could act as a ruler to set the size at which cells enter mitosis: In short cells, the gradient is sufficient to keep levels of Pom1 at midcell high enough to suppress Cdr2. Growth progressively separates the peaks of the bipolar Pom1 gradient, reducing midcell Pom1 levels until they drop sufficiently to allow activation of Cdr2 and mitotic entry (Martin & Berthelot-Grosjean, 2009; Moseley et al., 2009).

It is clear that at some level, cells care about the extent of the Pom1 gradient: Gradient steepness is altered as a function of concentration, maintaining Pom1 levels at midcell relatively constant despite several-fold variation in cellular Pom1 levels (Hersch, Hachet, Dalessi, et al., 2015; Saunders, Pan, Angel, et al., 2012). On the other hand, although a key ingredient in the ruler model of size control is the length-dependent reduction in midcell Pom1 concentration, whether midcell concentrations are meaningfully reduced in long cells remains controversial (Allard, Opalko, & Moseley, 2019; Gerganova, Floderer, Archetti, et al., 2019; Pan, Saunders, Flor-Parra, Howard, & Chang, 2014). Moreover, although bulk measurement of diffusion and membrane lifetimes from FRAP experiments are consistent with the observed gradient length scale, single molecule measurements reveal a more complex picture, suggesting that the molecular dynamics underlying gradient formation are more complicated than currently envisioned (Allard et al., 2019; Gerganova et al., 2019; Pan et al., 2014). In this light, it is striking that the biggest effects of the Pom1 gradient are seen in small cells: Midcell levels drop significantly as cells increase beyond 8 μm, but less so thereafter. Given that cells divide when they are approximately 14 μm in length, it remains an open question as to how precisely midcell Pom1 concentrations reflect cell size at the time that the decision to enter mitosis is made. Indeed, recent work has questioned this notion of a Pom1 ruler. Notably, *pom1* mutants exhibit unexpectedly good size control (Wood & Nurse, 2013, 2015) and quantitative analysis indicates that cell area, rather than size, appears to be the key determinant of mitotic entry with Cdr1 constituting the core sizing mechanism (Facchetti, Knapp, Flor-Parra, Chang, & Howard, 2019; Pan et al., 2014).

Thus, while Pom1 clearly impacts the function of Cdr1 and effectively suppresses division in small cells, the dominant role of Pom1 may actually be similar to the bacterial Min system—i.e., suppression of polar divisions. Indeed, polar Pom1 enrichment is known to help focus the division machinery at the cell center (Padte, Martin, Howard, & Chang, 2006; Rincon, Bhatia, Bicho, et al., 2014). Compared to accurately reading out position from a noisy gradient, defining a broad region in the cell center would be much less sensitive to variation in the gradient. In such a model, the precision of positioning and timing of mitotic entry would arise from subsequent refinement by downstream pathways. This is also analogous to the Min system in which Min proteins prevent division at the poles, but the precision of midcell positioning arises through the combined effects of additional pathways (Männik, Wu, Hol, et al., 2012; Mulder & Woldringh, 1989).

A similar situation can be envisioned for the Ran gradient. Notably, spindle organization, including spindle size, is largely insensitive to changes in either the magnitude or length of the Ran gradient (Maresca et al., 2009; Oh, Yu, & Needleman, 2016). Moreover, it is not RanGTP itself that drives microtubule nucleation, but RanGTP in complex with various spindle assembly factors (SAFs) which can act at a substantial distance away from the peak of the Ran gradient (Kalab & Heald, 2008; O'Connell, Loncarek, Kaláb, & Khodjakov, 2009). The distribution of SAF complexes is influenced in turn by the spindle microtubule dynamics, resulting in a feedback network that defines spindle size (Oh et al., 2016). Thus, the primary function of RanGTP in spindle assembly may be to create a local source of nucleation factors to guide initial steps in spindle assembly, which can then be refined by the many mechanisms available for controlling spindle size (Oh et al., 2016; Rieckhoff, Ishihara, & Brugués, 2019).

Finally, despite the formation of reproducible and reliably positioned boundary gradients between PAR protein domains in polarized *C. elegans* zygotes, there is little evidence that the zygotes care. The size asymmetry of the daughter cells is largely defined by the spindle midzone with contributions from cortical contractility that set the division plane (Bringmann & Hyman, 2005; Pacquelet, Uhart, Tassan, & Michaux, 2015). Moreover, asymmetric division is not particularly sensitive to the position of the PAR boundary or relative PAR domain size (Schenk, Bringmann, Hyman, & Cowan, 2010; Schonegg, Constantinescu, Hoege, & Hyman, 2007).

In fact, at division the PAR boundary often corrects to match the cleavage furrow rather than vice versa. Thus, while PAR domains set up the gross asymmetry of the zygote to orient the asymmetric division, this asymmetry is refined by downstream pathways to define the ultimate precision with which asymmetric division occurs, including the unequal size of the two daughter cells and the relevant inheritance of fate determinants.

Thus, in nearly all the cases we have discussed here, it is unclear whether the precision and extent of the spatial gradient really matters to the cell. The few exceptions to this rule are cases such as the Min and PAR systems in which there is a mismatch between the length scale of the patterning network and the system in which it operates. However, here, the problem is less that the gradient is providing incorrect positional information, but that the correct pattern fails to form at all (Hubatsch et al., 2019; Wu, van Schie, et al., 2015). Rather, in most cases, the key role of these patterning systems is to generate spatial landmarks that prime the spatial organization of downstream pathways that are ultimately responsible for ensuring precise positional control in the cell. These landmarks are typically characterized by two regions of distinct behavior corresponding to high and low concentrations of the intracellular morphogen. While the Pom, Min and PAR systems can generate alternating stripes of high and low concentration (e.g., pole-midcell-pole), we are not aware of any examples in which an intracellular gradient is read out at multiple concentrations to generate more than two regions with distinct properties.

The notion that intracellular morphogens may primarily act to specify the coarse-grained organization of the cell has parallels in our evolving understanding of morphogen gradients in development (see several excellent reviews: Briscoe & Small, 2015; Kicheva & Briscoe, 2015; Rogers & Schier, 2011; Sagner & Briscoe, 2017). One emergent theme is that the role of morphogen gradients may primarily be to specify tissue polarity relative to system boundaries and thereby provide a crude outline for development that is refined by subsequent regulatory networks to ensure robust and reproducible patterning in the developing embryo (Briscoe & Small, 2015).

10. Outlook

Many of the theories for patterning and positional specification emerged in an era in which we were largely blind to the molecular basis of the processes under investigation (Crick, 1970; Gierer & Meinhardt, 1972; Turing, 1952; Wolpert, 1969). Despite this limitation, these models have shown remarkable prescience. And while hypotheses have not always been borne out, they have been pivotal in defining the questions that have occupied subsequent generations of biologists. With the molecular revolution in cell and developmental biology, many of the players have been identified, laying the framework for critical assessment of model predictions. Notably thanks to recent advances in genetics, quantitative live imaging and measurement techniques, and reconstitution of systems in vitro, direct quantitative comparisons between theory and experiment are becoming increasingly tenable.

A key limitation for most models, however, remains the lack of robust, quantitative data. This includes not only the ability to measure, but to introduce quantitative perturbations of the concentration, localization, mobility and activity of molecules in living cells. New technologies will be critical. For example, in the case of the PAR polarity network, the introduction of single molecule imaging techniques (Robin, McFadden, Yao, & Munro, 2014), selective kinase inhibitors (Hannaford, Loyer, Tonelli, Zoltner, & Januschke, 2019; Rodriguez et al., 2017), and new methods for altering intracellular flows (Mittasch, Gross, Nestler, et al., 2018) has helped reveal key design principles, while the application of optogenetics to the well-studied Cdc42 network in yeast has allowed characterization of positive feedback circuits in the context of mating and cell cycle regulation (Lamas, Merlini, Vještica, Vincenzetti, & Martin, 2019; Witte et al., 2017). Reconstitution has provided an unrivaled experiment system to explore core behaviors of the Min system (Loose, Fischer-Friedrich, Ries, Kruse, & Schwille, 2008; Zieske & Schwille, 2014). Intelligent data analysis tools will likely be transformational in maximizing the extraction of relevant information (Bannon, Moen, Borba, et al., 2018; Van Valen, Kudo, Lane, et al., 2016).

Even where detailed kinetics remain inaccessible, the availability of high-quality quantitative datasets across experimental perturbations could be used to distinguish between models and estimate relevant model parameters based on fitting techniques. Such approaches will likely benefit from the use of artificial intelligence to score and optimize model selection (Maddu, Cheeseman, Sbalzarini, & Müller, 2019). Finally, new theoretical frameworks will also be necessary, including implementation of more varied and realistic geometries (Wettmann & Kruse, 2018) and, in the case of self-organizing systems, moving beyond linear stability analysis, which is typically used to understand how systems break symmetry, toward methods that allow analytical insight into the patterns that ultimately emerge (Halatek & Frey, 2018).

Acknowledgments

We would like to thank Jake Reich for fruitful discussions. Work in N.W.G.'s lab is supported by the Francis Crick Institute, which receives its core funding from Cancer Research UK (FC001086), the UK Medical Research Council (FC001086), and Wellcome Trust (FC001086).

References

Afonso, O., Figueiredo, A. C., & Maiato, H. (2017). Late mitotic functions of Aurora kinases. *Chromosoma*, *126*(1), 93–103.

Afonso, O., Matos, I., Pereira, A. J., Aguiar, P., Lampson, M. A., & Maiato, H. (2014). Feedback control of chromosome separation by a midzone Aurora B gradient. *Science*, *345*(6194), 332–336.

Allard, C. A. H., Opalko, H. E., & Moseley, J. B. (2019). Stable Pom1 clusters form a glucose-modulated concentration gradient that regulates mitotic entry. *eLife*, *8*, e46003.

Amodeo, A. A., Jukam, D., Straight, A. F., & Skotheim, J. M. (2015). Histone titration against the genome sets the DNA-to-cytoplasm threshold for the Xenopus midblastula transition. *Proceedings of the National Academy of Sciences of the United States of America*, *112*(10), E1086–E1095.

Bähler, J., & Pringle, J. R. (1998). Pom1p, a fission yeast protein kinase that provides positional information for both polarized growth and cytokinesis. *Genes & Development*, *12*(9), 1356–1370.

Banani, S. F., Lee, H. O., Hyman, A. A., & Rosen, M. K. (2017). Biomolecular condensates: Organizers of cellular biochemistry. *Nature Reviews. Molecular Cell Biology*, *18*(5), 285–298.

Bannon, D., Moen, E., Borba, E., et al. (2018). DeepCell 2.0: Automated cloud deployment of deep learning models for large-scale cellular image analysis. In *bioRxiv*.

Bois, J. S., Jülicher, F., & Grill, S. W. (2011). Pattern formation in active fluids. *Physical Review Letters*, *106*(2), 028103.

Brangwynne, C. P., Eckmann, C. R., Courson, D. S., et al. (2009). Germline P granules are liquid droplets that localize by controlled dissolution/condensation. *Science*, *324*(5935), 1729–1732.

Brangwynne, C. P., Koenderink, G. H., MacKintosh, F. C., & Weitz, D. A. (2009). Intracellular transport by active diffusion. *Trends in Cell Biology*, *19*(9), 423–427.

Bringmann, H., & Hyman, A. A. (2005). A cytokinesis furrow is positioned by two consecutive signals. *Nature*, *436*(7051), 731–734.

Briscoe, J., & Small, S. (2015). Morphogen rules: Design principles of gradient-mediated embryo patterning. *Development*, *142*(23), 3996–4009.

Brown, G. C., & Kholodenko, B. N. (1999). Spatial gradients of cellular phospho-proteins. *FEBS Letters*, *457*(3), 452–454.

Brownlee, C., & Heald, R. (2019). Importin α partitioning to the plasma membrane regulates intracellular scaling. *Cell*, *176*(4), 805–815.e8.

Carazo-Salas, R. E., Guarguaglini, G., Gruss, O. J., Segref, A., Karsenti, E., & Mattaj, I. W. (1999). Generation of GTP-bound Ran by RCC1 is required for chromatin-induced mitotic spindle formation. *Nature*, *400*(6740), 178–181.

Chang, J. B., & Ferrell, J. E. (2013). Mitotic trigger waves and the spatial coordination of the Xenopus cell cycle. *Nature*, *500*(7464), 603–607.

Chavent, M., Duncan, A. L., Rassam, P., et al. (2018). How nanoscale protein interactions determine the mesoscale dynamic organisation of bacterial outer membrane proteins. *Nature Communications*, *9*(1), 2846.

Chen, H., Einstein, L. C., Little, S. C., & Good, M. C. (2019). Spatiotemporal patterning of zygotic genome activation in a model vertebrate embryo. *Developmental Cell*, *49*(6), 852–866.e7.

Chiou, J.-G., Balasubramanian, M. K., & Lew, D. J. (2017). Cell polarity in yeast. *Annual Review of Cell and Developmental Biology*, *33*, 77–101.

Corbin, B. D., Yu, X.-C., & Margolin, W. (2002). Exploring intracellular space: Function of the Min system in round-shaped *Escherichia coli*. *The EMBO Journal*, *21*(8), 1998–2008.

Cowan, C. R., & Hyman, A. A. (2004). Centrosomes direct cell polarity independently of microtubule assembly in *C. elegans* embryos. *Nature*, *431*(7004), 92–96.

Cowan, C. R., & Hyman, A. A. (2006). Cyclin E-Cdk2 temporally regulates centrosome assembly and establishment of polarity in *Caenorhabditis elegans* embryos. *Nature Cell Biology*, *8*(12), 1441–1447.

Crick, F. (1970). Diffusion in embryogenesis. *Nature*, *225*(5231), 420–422.

Daniels, B. R., Perkins, E. M., Dobrowsky, T. M., et al. (2009). Asymmetric enrichment of PIE-1 in the *Caenorhabditis elegans* zygote mediated by binary counterdiffusion. *Journal of Cell Biology*.

Daniels, B. R., Dobrowsky, T. M., Perkins, E. M., Sun, S. X., Wirtz, D., et al. (2010). MEX-5 enrichment in the *C. elegans* early embryo mediated by differential diffusion. *Development*.

Dawes, A. T., & Munro, E. M. (2011). PAR-3 oligomerization may provide an actin-independent mechanism to maintain distinct par protein domains in the early Caenorhabditis elegans embryo. *Biophysical Journal*, *101*(6), 1412–1422.

de Boer, P. A., Crossley, R. E., & Rothfield, L. I. (1989). A division inhibitor and a topological specificity factor coded for by the minicell locus determine proper placement of the division septum in *E. coli*. *Cell*, *56*(4), 641–649.

Deneke, V. E., & Di Talia, S. (2018). Chemical waves in cell and developmental biology. *The Journal of Cell Biology*, *217*(4), 1193–1204.

Dickinson, D. J., Schwager, F., Pintard, L., Gotta, M., & Goldstein, B. (2017). A single-cell biochemistry approach reveals PAR complex dynamics during cell polarization. *Developmental Cell*, *42*(4), 416–434.e11.

Diego, X., Marcon, L., Müller, P., & Sharpe, J. (2018). Key features of turing systems are determined purely by network topology. *Physical Review X*, *8*(2), 021071.

Facchetti, G., Knapp, B., Flor-Parra, I., Chang, F., & Howard, M. (2019). Reprogramming Cdr2-dependent geometry-based cell size control in fission yeast. *Current Biology*, *29*(2), 350–358.e4.

Fuller, B. G., Lampson, M. A., Foley, E. A., et al. (2008). Midzone activation of aurora B in anaphase produces an intracellular phosphorylation gradient. *Nature*, *453*(7198), 1132–1136.

Gelens, L., Anderson, G. A., & Ferrell, J. E. (2014). Spatial trigger waves: Positive feedback gets you a long way. *Molecular Biology of the Cell*, *25*(22), 3486–3493.

Gerganova, V., Floderer, C., Archetti, A., et al. (2019). Multi-phosphorylation reaction and clustering tune Pom1 gradient mid-cell levels according to cell size. *eLife*, *8*, e45983.

Gierer, A., & Meinhardt, H. (1972). A theory of biological pattern formation. *Kybernetik*, *12*(1), 30–39.

Gilkey, J. C., Jaffe, L. F., Ridgway, E. B., & Reynolds, G. T. (1978). A free calcium wave traverses the activating egg of the medaka, *Oryzias latipes*. *The Journal of Cell Biology*, *76*(2), 448–466.

Goehring, N. W. (2014). PAR polarity: From complexity to design principles. *Experimental Cell Research*, *328*(2), 258–266.

Goehring, N. W., & Grill, S. W. (2013). Cell polarity: Mechanochemical patterning. *Trends in Cell Biology*, *23*(2), 72–80.

Goehring, N. W., Trong, P. K., Bois, J. S., et al. (2011). Polarization of PAR proteins by advective triggering of a pattern-forming system. *Science*, *334*(6059), 1137–1141.

Goldberg, M. B., & Theriot, J. A. (1995). Shigella flexneri surface protein IcsA is sufficient to direct actin-based motility. *Proceedings of the National Academy of Sciences of the United States of America*, *92*(14), 6572–6576.

Goryachev, A. B., & Pokhilko, A. V. (2008). Dynamics of Cdc42 network embodies a Turing-type mechanism of yeast cell polarity. *FEBS Letters*, *582*(10), 1437–1443.

Green, J. B. A., & Sharpe, J. (2015). Positional information and reaction-diffusion: Two big ideas in developmental biology combine. *Development*, *142*(7), 1203–1211.

Griffin, E. E. (2015). Cytoplasmic localization and asymmetric division in the early embryo of *Caenorhabditis elegans*. *Wiley Interdisciplinary Reviews: Developmental Biology*, *4*(3), 267–282.

Griffin, E. E., Odde, D. J., & Seydoux, G. (2011). Regulation of the MEX-5 gradient by a spatially segregated kinase/phosphatase cycle. *Cell*, *146*(6), 955–968.

Gross, P., Kumar, K. V., Goehring, N. W., et al. (2019). Guiding self-organized pattern formation in cell polarity establishment. *Nature Physics*, *15*(3), 293–300.

Hachet, O., Berthelot-Grosjean, M., Kokkoris, K., Vincenzetti, V., Moosbrugger, J., & Martin, S. G. (2011). A phosphorylation cycle shapes gradients of the DYRK family kinase Pom1 at the plasma membrane. *Cell*, *145*(7), 1116–1128.

Halatek, J., & Frey, E. (2018). Rethinking pattern formation in reaction-diffusion systems. *Nature Physics*, *14*(14), 507–514.

Han, B., Antkowiak, K. R., Fan, X., Rutigliano, M., Ryder, S. P., & Griffin, E. E. (2018). Polo-like kinase couples cytoplasmic protein gradients in the C. elegans zygote. *Current Biology*, *28*(1), 60–69.e8.

Hannaford, M., Loyer, N., Tonelli, F., Zoltner, M., & Januschke, J. (2019). A chemical-genetics approach to study the role of atypical Protein Kinase C in Drosophila. *Development*, *146*(2), dev170589.

Hansen, S. D., Huang, W. Y. C., Lee, Y. K., Bieling, P., Christensen, S. M., & Groves, J. T. (2019). Stochastic geometry sensing and polarization in a lipid kinase-phosphatase competitive reaction. *Proceedings of the National Academy of Sciences of the United States of America*, *116*, 15013–15022.

Haupt, A., & Minc, N. (2018). How cells sense their own shape—Mechanisms to probe cell geometry and their implications in cellular organization and function. *Journal of Cell Science*, *131*(6), jcs214015.

Hecht, I., Rappel, W.-J., & Levine, H. (2009). Determining the scale of the Bicoid morphogen gradient. *Proceedings of the National Academy of Sciences of the United States of America*, *106*(6), 1710–1715.

Hersch, M., Hachet, O., Dalessi, S., et al. (2015). Pom1 gradient buffering through intermolecular auto-phosphorylation. *Molecular Systems Biology*, *11*(7), 818.

Hodgkin, A. L., & Huxley, A. F. (1952). A quantitative description of membrane current and its application to conduction and excitation in nerve. *The Journal of Physiology*, *117*(4), 500–544.

Howard, M. (2012). How to build a robust intracellular concentration gradient. *Trends in Cell Biology*, *22*(6), 311–317.

Howard, J., Grill, S. W., & Bois, J. S. (2011). Turing's next steps: The mechanochemical basis of morphogenesis. *Nature Reviews. Molecular Cell Biology*, *12*(6), 392–398.

Howell, A. S., Jin, M., Wu, C.-F., Zyla, T. R., Elston, T. C., & Lew, D. J. (2012). Negative feedback enhances robustness in the yeast polarity establishment circuit. *Cell*, *149*(2), 322–333.

Howell, A. S., Savage, N. S., Johnson, S. A., et al. (2009). Singularity in polarization: Rewiring yeast cells to make two buds. *Cell*, *139*(4), 731–743.

Hu, L., Vecchiarelli, A. G., Mizuuchi, K., Neuman, K. C., & Liu, J. (2017). Brownian ratchet mechanisms of ParA-mediated partitioning. *Plasmid*, *92*, 12–16.

Hubatsch, L., Peglion, F., Reich, J. D., et al. (2019). A cell-size threshold limits cell polarity and asymmetric division potential. *Nature Physics*, *15*, 1078–1085.

Ishihara, K., Nguyen, P. A., Wühr, M., Groen, A. C., Field, C. M., & Mitchison, T. J. (2014). Organization of early frog embryos by chemical waves emanating from centrosomes. *Philosophical Transactions of the Royal Society of London. Series B, Biological Sciences*, *369*(1650), 20130454.

Jacobson, K., Liu, P., & Lagerholm, B. C. (2019). The lateral organization and mobility of plasma membrane components. *Cell, 177*(4), 806–819.

Jilkine, A., & Edelstein-Keshet, L. (2011). A comparison of mathematical models for polarization of single eukaryotic cells in response to guided cues. *PLoS Computational Biology, 7*(4), e1001121.

Kalab, P., & Heald, R. (2008). The RanGTP gradient—A GPS for the mitotic spindle. *Journal of Cell Science, 121*(Pt. 10), 1577–1586.

Kalab, P., Weis, K., & Heald, R. (2002). Visualization of a Ran-GTP gradient in interphase and mitotic Xenopus egg extracts. *Science, 295*(5564), 2452–2456.

Kholodenko, B. N. (2009). Spatially distributed cell signalling. *FEBS Letters, 583*(24), 4006–4012.

Kicheva, A., & Briscoe, J. (2015). Developmental pattern formation in phases. *Trends in Cell Biology, 25*(10), 579–591.

Klinkert, K., Levernier, N., Gross, P., et al. (2019). Aurora A depletion reveals centrosome-independent polarization mechanism in Caenorhabditis elegans. *eLife, 8*, e44552.

Kotak, S., & Kapoor, S. (2018). Centrosome aurora a gradient ensures a single PAR-2 polarity axis by regulating RhoGEF ECT-2 localization in C. elegans embryos. In *bioRxiv*.

Kozubowski, L., Saito, K., Johnson, J. M., Howell, A. S., Zyla, T. R., & Lew, D. J. (2008). Symmetry-breaking polarization driven by a Cdc42p GEF-PAK complex. *Current Biology, 18*(22), 1719–1726.

Kusumi, A., Fujiwara, T. K., Chadda, R., et al. (2012). Dynamic organizing principles of the plasma membrane that regulate signal transduction: Commemorating the fortieth anniversary of Singer and Nicolson's fluid-mosaic model. *Annual Review of Cell and Developmental Biology, 28*, 215–250.

Laloux, G., & Jacobs-Wagner, C. (2014). How do bacteria localize proteins to the cell pole? *Journal of Cell Science, 127*(Pt. 1), 11–19.

Lamas, I., Merlini, L., Vještica, A., Vincenzetti, V., & Martin, S. G. (2019). Optogenetics reveals Cdc42 local activation by scaffold-mediated positive feedback and Ras GTPase. In *bioRxiv*.

Lenarcic, R., Halbedel, S., Visser, L., et al. (2009). Localisation of DivIVA by targeting to negatively curved membranes. *The EMBO Journal, 28*(15), 2272–2282.

Lim, H. C., Surovtsev, I. V., Beltran, B. G., Huang, F., Bewersdorf, J., & Jacobs-Wagner, C. (2014). Evidence for a DNA-relay mechanism in ParABS-mediated chromosome segregation. *eLife, 3*, e02758.

Lipkow, K., & Odde, D. J. (2008). Model for protein concentration gradients in the cytoplasm. *Cellular and Molecular Bioengineering, 1*(1), 84–92.

Loose, M., Fischer-Friedrich, E., Ries, J., Kruse, K., & Schwille, P. (2008). Spatial regulators for bacterial cell division self-organize into surface waves in vitro. *Science, 320*(5877), 789–792.

Lu, W., Lakonishok, M., Serpinskaya, A. S., Kirchenbüechler, D., Ling, S.-C., & Gelfand, V. I. (2018). Ooplasmic flow cooperates with transport and anchorage in Drosophila oocyte posterior determination. *The Journal of Cell Biology, 217*(10), 3497–3511.

MacCready, J. S., Hakim, P., Young, E. J., et al. (2018). Protein gradients on the nucleoid position the carbon-fixing organelles of cyanobacteria. *eLife*, 7, e39723.

Maddu, S., Cheeseman, B. L., Sbalzarini, I. F., & Müller, C. L. (2019). *Stability selection enables robust learning of partial differential equations from limited noisy data.* [Preprint]. Available at: *https://arxiv.org/abs/1907.07810v1.*

Maeder, C. I., Hink, M. A., Kinkhabwala, A., Mayr, R., Bastiaens, P. I. H., & Knop, M. (2007). Spatial regulation of Fus3 MAP kinase activity through a reaction-diffusion mechanism in yeast pheromone signalling. *Nature Cell Biology, 9*(11), 1319–1326.

Männik, J., Wu, F., Hol, F. J. H., et al. (2012). Robustness and accuracy of cell division in Escherichia coli in diverse cell shapes. *Proceedings of the National Academy of Sciences of the United States of America, 109*(18), 6957–6962.
Maresca, T. J., Groen, A. C., Gatlin, J. C., Ohi, R., Mitchison, T. J., & Salmon, E. D. (2009). Spindle assembly in the absence of a RanGTP gradient requires localized CPC activity. *Current Biology, 19*(14), 1210–1215.
Martin, S. G., & Berthelot-Grosjean, M. (2009). Polar gradients of the DYRK-family kinase Pom1 couple cell length with the cell cycle. *Nature, 459*(7248), 852–856.
McMahon, H. T., & Boucrot, E. (2015). Membrane curvature at a glance. *Journal of Cell Science, 128*(6), 1065–1070.
Meyers, J., Craig, J., & Odde, D. J. (2006). Potential for control of signaling pathways via cell size and shape. *Current Biology, 16*(17), 1685–1693.
Mittasch, M., Gross, P., Nestler, M., et al. (2018). Non-invasive perturbations of intracellular flow reveal physical principles of cell organization. *Nature Cell Biology, 20*(3), 344–351.
Moran, K. D., Kang, H., Araujo, A. V., et al. (2019). Cell-cycle control of cell polarity in yeast. *The Journal of Cell Biology, 218*(1), 171–189.
Moseley, J. B., Mayeux, A., Paoletti, A., & Nurse, P. (2009). A spatial gradient coordinates cell size and mitotic entry in fission yeast. *Nature, 459*(7248), 857–860.
Motegi, F., Zonies, S., Hao, Y., Cuenca, A. A., Griffin, E., & Seydoux, G. (2011). Microtubules induce self-organization of polarized PAR domains in Caenorhabditis elegans zygotes. *Nature Cell Biology, 13*(11), 1361–1367.
Mulder, E., & Woldringh, C. L. (1989). Actively replicating nucleoids influence positioning of division sites in Escherichia coli filaments forming cells lacking DNA. *Journal of Bacteriology, 171*(8), 4303–4314.
Munguira, I., Casuso, I., Takahashi, H., et al. (2016). Glasslike membrane protein diffusion in a crowded membrane. *ACS Nano, 10*(2), 2584–2590.
Munro, E., Nance, J., & Priess, J. R. (2004). Cortical flows powered by asymmetrical contraction transport PAR proteins to establish and maintain anterior-posterior polarity in the early C. elegans embryo. *Developmental Cell, 7*(3), 413–424.
O'Connell, C. B., Loncarek, J., Kaláb, P., & Khodjakov, A. (2009). Relative contributions of chromatin and kinetochores to mitotic spindle assembly. *The Journal of Cell Biology, 187*(1), 43–51.
Oh, D., Yu, C.-H., & Needleman, D. J. (2016). Spatial organization of the Ran pathway by microtubules in mitosis. *Proceedings of the National Academy of Sciences of the United States of America, 113*(31), 8729–8734.
Pacquelet, A., Uhart, P., Tassan, J.-P., & Michaux, G. (2015). PAR-4 and anillin regulate myosin to coordinate spindle and furrow position during asymmetric division. *The Journal of Cell Biology, 210*(7), 1085–1099.
Padte, N. N., Martin, S. G., Howard, M., & Chang, F. (2006). The cell-end factor pom1p inhibits mid1p in specification of the cell division plane in fission yeast. *Current Biology, 16*(24), 2480–2487.
Pan, K. Z., Saunders, T. E., Flor-Parra, I., Howard, M., & Chang, F. (2014). Cortical regulation of cell size by a sizer cdr2p. *eLife, 3*, e02040.
Peglion, F., & Goehring, N. W. (2019). Switching states: Dynamic remodelling of polarity complexes as a toolkit for cell polarization. *Current Opinion in Cell Biology, 60*, 121–130.
Quinlan, M. E. (2016). Cytoplasmic streaming in the Drosophila oocyte. *Annual Review of Cell and Developmental Biology, 32*, 173–195.
Ramamurthi, K. S., & Losick, R. (2009). Negative membrane curvature as a cue for subcellular localization of a bacterial protein. *Proceedings of the National Academy of Sciences of the United States of America, 106*(32), 13541–13545.
Rangamani, P., Lipshtat, A., Azeloglu, E. U., et al. (2013). Decoding information in cell shape. *Cell, 154*(6), 1356–1369.

Reich, J. D., Hubatsch, L., Illukkumbura, R., et al. (2019). Regulated activation of the PAR polarity network ensures a timely and specific response to spatial cues. *Current Biology, 29*(12), 1911–1923.e5.

Rieckhoff, E. M., Ishihara, K., & Brugués, J. (2019). How to tune spindle size relative to cell size? *Current Opinion in Cell Biology, 60*, 139–144.

Rincon, S. A., Bhatia, P., Bicho, C., et al. (2014). Pom1 regulates the assembly of Cdr2-Mid1 cortical nodes for robust spatial control of cytokinesis. *The Journal of Cell Biology, 206*(1), 61–77.

Robbins, J. R., Monack, D., McCallum, S. J., et al. (2001). The making of a gradient: IcsA (VirG) polarity in Shigella flexneri. *Molecular Microbiology, 41*(4), 861–872.

Robin, F. B., McFadden, W. M., Yao, B., & Munro, E. M. (2014). Single-molecule analysis of cell surface dynamics in Caenorhabditis elegans embryos. *Nature Methods, 11*(6), 677–682.

Rodriguez, J., Peglion, F., Martin, J., et al. (2017). aPKC cycles between functionally distinct PAR protein assemblies to drive cell polarity. *Developmental Cell, 42*(4), 400–415.e9.

Rogers, K. W., & Schier, A. F. (2011). Morphogen gradients: From generation to interpretation. *Annual Review of Cell and Developmental Biology, 27*, 377–407.

Rose, L., & Gönczy, P. (2014). Polarity establishment, asymmetric division and segregation of fate determinants in early C. elegans embryos. *Wormbook: The Online Review of C Elegans Biology*, 1–43.

Rowlett, V. W., & Margolin, W. (2015). The Min system and other nucleoid-independent regulators of Z ring positioning. *Frontiers in Microbiology, 6*, 478.

Sagner, A., & Briscoe, J. (2017). Morphogen interpretation: Concentration, time, competence, and signaling dynamics. *Wiley Interdisciplinary Reviews Developmental Biology, 6*(4), e271.

Saha, S., Weber, C. A., Nousch, M., et al. (2016). Polar positioning of phase-separated liquid compartments in cells regulated by an mRNA competition mechanism. *Cell, 166*(6), 1572–1584.e16.

Sailer, A., Anneken, A., Li, Y., Lee, S., & Munro, E. (2015). Dynamic opposition of clustered proteins stabilizes cortical polarity in the *C. elegans* zygote. *Developmental Cell, 35*(1), 131–142.

Saunders, T. E., Pan, K. Z., Angel, A., et al. (2012). Noise reduction in the intracellular pom1p gradient by a dynamic clustering mechanism. *Developmental Cell, 22*(3), 558–572.

Schavemaker, P. E., Boersma, A. J., & Poolman, B. (2018). How important is protein diffusion in prokaryotes? *Frontiers in Molecular Biosciences, 5*, 93.

Schenk, C., Bringmann, H., Hyman, A. A., & Cowan, C. R. (2010). Cortical domain correction repositions the polarity boundary to match the cytokinesis furrow in C. elegans embryos. *Development, 137*(10), 1743–1753.

Schmick, M., & Bastiaens, P. I. H. (2014). The interdependence of membrane shape and cellular signal processing. *Cell, 156*(6), 1132–1138.

Schonegg, S., Constantinescu, A. T., Hoege, C., & Hyman, A. A. (2007). The Rho GTPase-activating proteins RGA-3 and RGA-4 are required to set the initial size of PAR domains in Caenorhabditis elegans one-cell embryos. *Proceedings of the National Academy of Sciences of the United States of America, 104*(38), 14976–14981.

Schumacher, D., Bergeler, S., Harms, A., et al. (2017). The PomXYZ proteins self-organize on the bacterial nucleoid to stimulate cell division. *Developmental Cell, 41*(3), 299–314.e13.

Schweisguth, F., & Corson, F. (2019). Self-organization in pattern formation. *Developmental Cell, 49*(5), 659–677.

Sheetz, M. P., Steuer, E. R., & Schroer, T. A. (1989). The mechanism and regulation of fast axonal transport. *Trends in Neurosciences, 12*(11), 474–478.

Smith, J., Calidas, D., Schmidt, H., Lu, T., Rasoloson, D., & Seydoux, G. (2016). Spatial patterning of P granules by RNA-induced phase separation of the intrinsically-disordered protein MEG-3. *eLife*, *5*, e21337.

Steinhauer, J., Agha, R., Pham, T., Varga, A. W., & Goldberg, M. B. (1999). The unipolar Shigella surface protein IcsA is targeted directly to the bacterial old pole: IcsP cleavage of IcsA occurs over the entire bacterial surface. *Molecular Microbiology*, *32*(2), 367–377.

Thalmeier, D., Halatek, J., & Frey, E. (2016). Geometry-induced protein pattern formation. *Proceedings of the National Academy of Sciences of the United States of America*, *113*(3), 548–553.

Tomchik, K. J., & Devreotes, P. N. (1981). Adenosine 3',5'-monophosphate waves in Dictyostelium discoideum: A demonstration by isotope dilution—Fluorography. *Science*, *212*(4493), 443–446.

Trong, P. K., Nicola, E. M., Goehring, N. W., Kumar, K. V., & Grill, S. W. (2014). Parameter-space topology of models for cell polarity. *New Journal of Physics*, *16*(6), 065009.

Tsai, M.-C., & Ahringer, J. (2007). Microtubules are involved in anterior-posterior axis formation in C. elegans embryos. *The Journal of Cell Biology*, *179*(3), 397–402.

Turing, A. M. (1952). The chemical basis of morphogenesis. *Philosophical Transactions of the Royal Society B: Biological Sciences*, *237*(641), 37–72.

Van Valen, D. A., Kudo, T., Lane, K. M., et al. (2016). Deep learning automates the quantitative analysis of individual cells in live-cell imaging experiments. *PLoS Computational Biology*, *12*(11), e1005177.

Vecchiarelli, A. G., Neuman, K. C., & Mizuuchi, K. (2014). A propagating ATPase gradient drives transport of surface-confined cellular cargo. *Proceedings of the National Academy of Sciences of the United States of America*, *111*(13), 4880–4885.

Wallenfang, M. R., & Seydoux, G. (2000). Polarization of the anterior-posterior axis of C. elegans is a microtubule-directed process. *Nature*, *408*(6808), 89–92.

Wang, E., Ballister, E. R., & Lampson, M. A. (2011). Aurora B dynamics at centromeres create a diffusion-based phosphorylation gradient. *The Journal of Cell Biology*, *194*(4), 539–549.

Wang, S.-C., Low, T. Y. F., Nishimura, Y., Gole, L., Yu, W., & Motegi, F. (2017). Cortical forces and CDC-42 control clustering of PAR proteins for Caenorhabditis elegans embryonic polarization. *Nature Cell Biology*, *19*(8), 988–995.

Weaver, L. N., & Walczak, C. E. (2015). Spatial gradients controlling spindle assembly. *Biochemical Society Transactions*, *43*(1), 7–12.

Wedlich-Soldner, R., Altschuler, S., Wu, L., & Li, R. (2003). Spontaneous cell polarization through actomyosin-based delivery of the Cdc42 GTPase. *Science*, *299*(5610), 1231–1235.

Weiner, O. D., Marganski, W. A., Wu, L. F., Altschuler, S. J., & Kirschner, M. W. (2007). An actin-based wave generator organizes cell motility. *PLoS Biology*, *5*(9), e221.

Wettmann, L., & Kruse, K. (2018). The Min-protein oscillations in Escherichia coli: An example of self-organized cellular protein waves. *Philosophical Transactions of the Royal Society of London Series B, Biological Sciences*, *373*(1747), 20170111.

Witte, K., Strickland, D., & Glotzer, M. (2017). Cell cycle entry triggers a switch between two modes of Cdc42 activation during yeast polarization. *eLife*, *6*, e26722.

Wolpert, L. (1969). Positional information and the spatial pattern of cellular differentiation. *Journal of Theoretical Biology*, *25*(1), 1–47.

Wood, E., & Nurse, P. (2013). Pom1 and cell size homeostasis in fission yeast. *Cell Cycle*, *12*(19), 3228–3236.

Wood, E., & Nurse, P. (2015). Sizing up to divide: Mitotic cell-size control in fission yeast. *Annual Review of Cell and Developmental Biology*, *31*, 11–29.

Woods, B., & Lew, D. J. (2019). Polarity establishment by Cdc42: Key roles for positive feedback and differential mobility. *Small GTPases*, *10*(2), 130–137.

Wu, C.-F., Chiou, J.-G., Minakova, M., et al. (2015). Role of competition between polarity sites in establishing a unique front. *eLife*, *4*, e11611.

Wu, Y., & Griffin, E. E. (2017). Regulation of cell polarity by PAR-1/MARK kinase. *Current Topics in Developmental Biology*, *123*, 365–397.

Wu, Z., Su, M., Tong, C., Wu, M., & Liu, J. (2018). Membrane shape-mediated wave propagation of cortical protein dynamics. *Nature Communications*, *9*(1), 136.

Wu, F., van Schie, B. G. C., Keymer, J. E., & Dekker, C. (2015). Symmetry and scale orient Min protein patterns in shaped bacterial sculptures. *Nature Nanotechnology*, *10*(8), 719–726.

Wu, Y., Zhang, H., & Griffin, E. E. (2015). Coupling between cytoplasmic concentration gradients through local control of protein mobility in the Caenorhabditis elegans zygote. *Molecular Biology of the Cell*, *26*(17), 2963–2970.

Zhao, P., Teng, X., Tantirimudalige, S. N., et al. (2019). Aurora-A breaks symmetry in contractile actomyosin networks independently of its role in centrosome maturation. *Developmental Cell*, *48*(5), 631–645.e6.

Zieske, K., & Schwille, P. (2014). Reconstitution of self-organizing protein gradients as spatial cues in cell-free systems. *eLife*, *3*, e03949.

CHAPTER NINE

Insights into mammalian morphogen dynamics from embryonic stem cell systems

Elena Camacho-Aguilar[a], Aryeh Warmflash[a,b,*]
[a]Department of Biosciences, Rice University, Houston, TX, United States
[b]Department of Bioengineering, Rice University, Houston, TX, United States
*Corresponding author: e-mail address: aryeh.warmflash@rice.edu

Contents

Abstract

Morphogens play an essential role in cell fate specification and patterning including in laying out the mammalian body plan during gastrulation. *In vivo* studies have shed light on the signaling pathways involved in this process and the phenotypes associated with their disruption, however, several important open questions remain regarding how morphogens function in space and time. Self-organized patterning systems based on embryonic stem cells have emerged as a powerful platform for beginning to address these questions that is complementary to *in vivo* approaches. Here we review recent progress in understanding morphogen signaling dynamics and patterning in early mammalian development by taking advantage of cutting-edge embryonic stem cell technology.

Current Topics in Developmental Biology, Volume 137
ISSN 0070-2153
https://doi.org/10.1016/bs.ctdb.2019.11.010

1. Introduction

During development, extracellular signaling molecules are essential for creating spatial patterns of cell fates. In 1952, Alan Turing introduced the term morphogen to define "a *chemical signal that could produce a form*" (as the name suggests) (Turing, 1952). It was decades before such molecules were actually discovered and, since then, much has been learned about the mechanisms by which morphogens pattern tissues. However, much remains obscure, particularly in mammals where development is difficult to observe and to manipulate.

In Turing's time, a broad definition of a morphogen was appropriate as nothing was known about the molecular underpinnings of development. In this review, we will use more recent definitions which add the requirement of a morphogen producing more than two fates, to exclude simpler cases in which the molecule acts as a simple switch, where cells adopt one fate in its absence and a different one in its presence (Dyson & Gurdon, 1998; Wolpert, 1969). Here we focus on mammalian development and only briefly summarize general concepts in morphogen gradient formation and interpretation. For more details about these processes we refer the reader to several recent reviews (Briscoe & Small, 2015; Green & Sharpe, 2015; Kondo & Miura, 2010; Rogers & Schier, 2011; Sagner & Briscoe, 2017).

The simplest models for morphogen gradient generation rely on localized production, and diffusion and degradation throughout the tissue (Crick, 1970; Lawrence, 1966; Stumpf, 1966; Wolpert, 1969). This creates a gradient that provides positional information to cells which differentiate to different fates according to its concentration (Wolpert, 1969). This synthesis-diffusion-degradation (SDD) model predicts an exponential decay in the morphogen concentration (Driever & Nusslein-Volhard, 1988; Drocco, Wieschaus, & Tank, 2012; Gregor, Wieschaus, Mcgregor, Bialek, & Tank, 2007), and has been supported by measurements of Bicoid in *Drosophila* and Sonic-Hedgehog in the murine neural tube, which produce a gradient with this property (Chamberlain, Jeong, Guo, Allen, & McMahon, 2008; Gregor, Tank, Wieschaus, & Bialek, 2007; Gregor, Wieschaus, et al., 2007).

The SDD model assumes localized production, but many developmental processes are self-organized without a preexisting source. Turing showed mathematically that, under particular conditions, diffusion causes the state in which the morphogen is homogenous in space to become unstable, and results in the development of a pattern of morphogen concentration

(Turing, 1952). This is known as a diffusion-driven or Turing instability (Kondo & Miura, 2010; Meinhardt, 1978; Turing, 1952). This pattern of morphogen could then serve as the gradient which conveys positional information to the cells (Green & Sharpe, 2015). Turing patterns can be generated through the coupling of two molecules: a short-range activator that activates itself as well as its own long-range inhibitor (Gierer & Meinhardt, 1972), and several biological examples have been demonstrated (Economou et al., 2012; Green & Sharpe, 2015; Raspopovic, Marcon, Russo, & Sharpe, 2014).

As the morphogen diffuses through the tissue, it interacts with the field of cells by binding to cell-surface receptors and initiating signaling cascades within the cells. Let us use the TGF-β pathway as an example, since we will focus on this pathway below. TGF-β ligand binding leads to the assembly of heteromeric complexes of type I and type II receptors, which causes the type II receptor to activate the type I by phosphorylation. It has been suggested that the absolute number of occupied receptors is the signal that is conveyed to the nucleus (Dyson & Gurdon, 1998), although, as we discuss below, this simple formulation ignores the time-dependence of signaling. The type I receptor activates a class of signal transducers known as receptor-associated Smads (R-Smads). R-Smads then form a complex with another signal transducer, Smad4, and this complex moves into the nucleus where it regulates transcription. There are two branches of this pathway, the BMP branch and the Activin/Nodal branch, which share some components, including Smad4, but utilize different type I receptors and R-Smads. Thus, the signaling pathway relays information from outside the cell to its nucleus.

In the following sections, we summarize our current understanding of morphogens in early mammalian development, and opportunities to make progress in ESC systems. We first review what is known from *in vivo* experiments in the mouse, and discuss remaining open questions. We then focus on progress to date, and opportunities for the future, in addressing these questions using ESCs. The ESC systems allow for experiments that are currently impossible in the embryo, but the findings must be tested to ensure that that they reflect what occurs *in vivo*. They also allow for investigation of development directly in human cells, providing a powerful complement to ongoing research using model organisms.

2. The role of morphogens in mammalian gastrulation

Gastrulation is the first differentiation event of the embryo proper, when the pluripotent epiblast differentiates into the three germ layers of the embryo: ectoderm, mesoderm, and endoderm. Its initiation relies on three signaling pathways: the two branches of the TGF-β pathway, BMP and Activin/Nodal, and the Wnt pathway. As discussed above, BMP and Activin/Nodal signal through complexes of R-Smads (Smad1/5/8 and Smad2/3, respectively) and Smad4, while Wnt signals are transduced by β-catenin. In the mouse, Nodal signaling from the epiblast promotes BMP4 expression in the extra-embryonic ectoderm which activates WNT in both the visceral endoderm and the epiblast (Ben-Haim et al., 2006; Tortelote et al., 2013; Yoon et al., 2015). Wnt, in turn, increases Nodal signaling in the epiblast. Expression of Wnt and Nodal inhibitors in the anterior visceral endoderm (AVE) restricts these signals to the posterior of the embryo, where the primitive steak, the site of gastrulation, forms (Arnold & Robertson, 2009; Shahbazi & Zernicka-Goetz, 2018). Here, we review what has been learned *in vivo* regarding the identity and function of these morphogen ligands, focusing on Nodal signaling as an example, and highlight open questions that have remained difficult to address.

The most direct evidence for the function of TGF-beta ligands as morphogens in vertebrates comes from experiments with dissected *Xenopus* animal caps. These cells will adopt dorsal ectodermal fates when cultured in isolation, however, when exposed to low doses of Activin, they adopt a ventral mesodermal fate, while high doses lead to increasing dorsalization (Green, New, & Smith, 1992). Activin is also capable of spatially patterning a tissue *in vitro*, as an Activin bead placed within an animal cap will induce localized expression of Goosecoid, which requires high Activin, and a ring of Brachyury (BRA), which requires lower signaling, at a distance from the bead (Gurdon, Harger, Mitchell, & Lemaire, 1994). Similarly, exposure of dissociated animal caps to BMP4 induces increasingly ventral fates within the ectoderm with the highest doses giving rise to epidermis (Wilson, Lagna, Suzuki, & Hemmati-Brivanlou, 1997). While these experiments demonstrate the ability of these ligands to induce multiple fates at different concentrations, they do not demonstrate that they function in this manner *in vivo*. Notably, Activin plays a limited role in embryonic patterning, while Nodal, which activates the same pathway, is central to patterning, but signals over a shorter range (Jones, Armes, & Smith, 1996). Moreover, while these experiments in which ligand is suddenly added to culture media are consistent with ligand concentration as the central factor in determining cell fate, they are also consistent with other factors, such as ligand rate of change (Heemskerk & Warmflash, 2016). Finally, as discussed below, similar experiments with mammalian pluripotent cells yield different results

(Nemashkalo, Ruzo, Heemskerk, & Warmflash, 2017; Yoney et al., 2018), calling into question whether these ligands can function as classical morphogens in these systems.

It is instructive to consider the case of Nodal signaling in mammalian embryos. While there are six Nodal ligands in *Xenopus* and two in *Zebrafish*, mammals have only a single ligand, and deletion of this ligand leads to severe defects in embryogenesis shortly after implantation (Brennan et al., 2001; Conlon et al., 1994), consistent with the roles of Nodal signaling both in maintaining epiblast pluripotency (James, Levine, Besser, & Hemmati-Brivanlou, 2005; Vallier, Alexander, & Pedersen, 2005), and primitive streak induction (Brennan et al., 2001). Understanding of the role of Nodal in the mammalian embryo has come primarily from two types of experiments, measurements of expression patterns of Nodal, its inhibitors, and its targets, and assessments of the phenotypes of modified alleles of the genes that code for these extracellular molecules or for components of the intracellular signaling pathway (Arnold & Robertson, 2009; Dunn, Vincent, Oxburgh, Robertson, & Bikoff, 2004) (Fig. 1). Nodal is initially expressed throughout the epiblast shortly after implantation, while during gastrulation its expression is highest in the proximal posterior epiblast. At late gastrulation, it becomes restricted to the node, which forms at the anterior end of the primitive streak (Zhou et al., 1993).

Intracellularly, Nodal signals are transduced from the receptor to the cell nucleus by two very similar proteins known as Smad2 and Smad3. Surprisingly, while Nodal knockout mice rarely form mesoderm (Conlon et al., 1994), in Smad2 knockout mice, the entire embryo proper is converted to mesodermal cells (Brennan et al., 2001; Waldrip, Bikoff, Hoodless, Wrana, & Robertson, 1998). This results from loss of anterior identity due to a failure to form the anterior visceral endoderm signaling center within the extraembryonic endoderm. Epiblast specific Smad2 knockouts do form this signaling center, however, they have severe defects in forming anterior portions of the embryo which originate in a failure to specify the anterior most mesoderm and endoderm in the primitive streak during gastrulation (Vincent, Dunn, Hayashi, Norris, & Robertson, 2003). This phenotype is very similar to that observed in embryos which are heterozygous at the Nodal locus with one allele representing a null mutation and the other lacking an enhancer which drives Nodal expression in the primitive streak, which led the authors to conclude that the defects in epiblast specific Smad2 knockouts result from a general reduction in Nodal signaling levels (Vincent et al., 2003). In these embryos, the node and midline structures still form, however, removal of one copy of

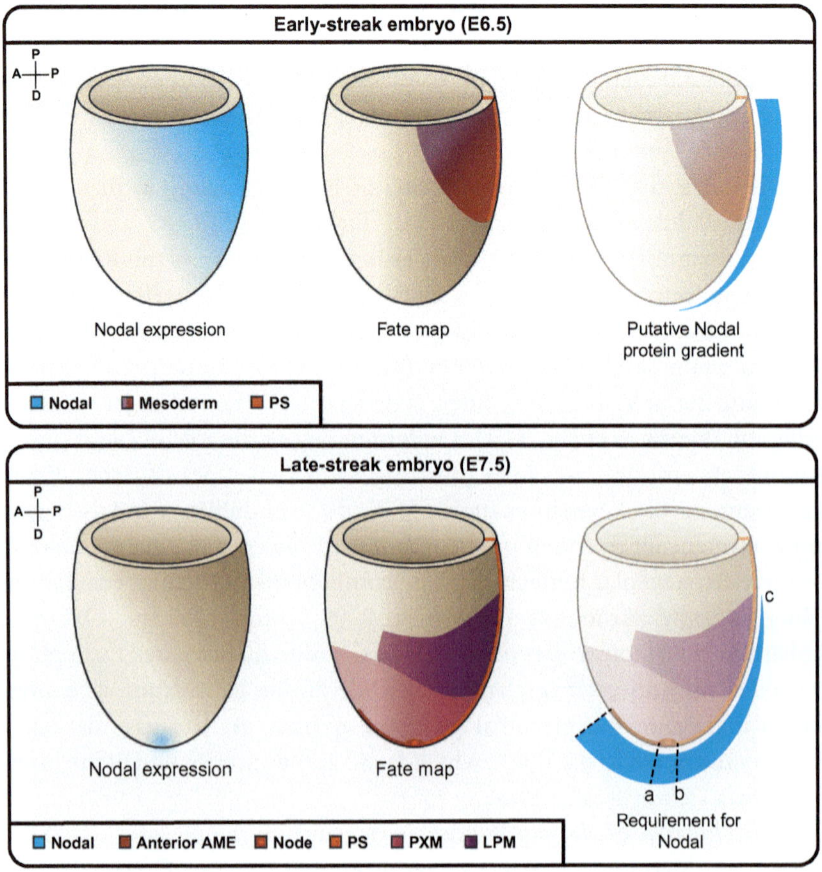

Fig. 1 Schematics of the mouse embryo at gastrulation stage. Schematics show the expression of Nodal mRNA and the fate map of cells arising from the primitive streak (PS) in the early gastrulation (E6.5, top left and center), and late gastrulation (E7.5, bottom left and center) staged mouse embryo. Also shown are the Nodal gradient at E6.5 (top right), and the requirement for Nodal signaling to differentiate to primitive streak fates as determined from genetic experiments (bottom right). The phenotypic thresholds (a), (b) and (c) correspond to gradual decrease in Smad2/3 dosage, obtained from Dunn et al. (2004). The first eliminates definitive endoderm and anterior axial mesoderm (AME), followed by the node in (b), and finally the paraxial mesoderm (PXM) and lateral plate mesoderm (LPM) in (c). *Adapted from Brennan, J., Lu, C.C., Norris, D.P., Rodriguez, T.A., Beddington, R.S.P., Robertson, E.J. 2001. Nodal signalling in the epiblast patterns the early mouse embryo.* Nature, 411, *965–969; Dunn, N.R., Vincent, S.D., Oxburgh, L., Robertson, E.J., & Bikoff, E.K. 2004. Combinatorial activities of Smad2 and Smad3 regulate mesoderm formation and patterning in the mouse embryo.* Development, 131, *1717–1728; Martinez Arias, A. & Steventon, B. 2014. On the nature and function of organizers. Development, 2018, 145; Robertson, E.J. Dose-dependent nodal/Smad signals pattern the early mouse embryo.* Seminars in Cell & Developmental Biology, 32, *73–79; Stern, C.D. 2004.* Gastrulation: From cells to embryo; *Cold Spring Harbor Laboratory Press; Tam, P.P.L., & Behringer, R.R. 1997. Mouse gastrulation: The formation of a mammalian body plan.* Mechanisms of Development, 68, *3–25; Zhou, X., Sasaki, H., Lowe, L., Hogan, B.L.M., & Kuehn, M.R. 1993. Nodal is a novel TGF-β-like gene expressed in the mouse node during gastrulation.* Nature, 361, *543–547.*

Smad3 in this context causes a loss of both structures, while embryos lacking both Smad2 and 3 fail to form mesoderm altogether (Dunn et al., 2004; Vincent et al., 2003). Taken together, these results have been interpreted to indicate a dose-dependent patterning of the mesendoderm by Nodal signals (Robertson, 2014).

There remain, however, several points of confusion. First, there is not a clear correspondence between the expression patterns of Nodal and the phenotypes induced. During gastrulation, Nodal is highest in the posterior proximal primitive streak, while reduction in Nodal levels results in a loss of the anterior most derivatives (Fig. 1). Similarly, Nodal expression is widespread at earlier stages so it is unclear what restricts the position of the DVE/AVE signaling center to the distal and anterior side, although it has been suggested to be due to a balance between Nodal and BMP signaling (Yamamoto et al., 2009). Second, we largely lack measurements of signaling activity, such as the amount of phosphorylated, nuclear-localized Smad proteins, and so interpretations of phenotypes must assume what the effect on signaling has been, and these can often be counterintuitive. The case of global knockout of Smad2 discussed above provides a good example. Unexpectedly, mesoderm is drastically expanded due to the loss of the AVE and its inhibitors. As the pathway contains multiple levels of feedback, effects such as this may be common. Further, some perturbations may change the spatial or temporal patterns of signaling in a manner which is difficult to predict. Well-studied examples include changes in the expression of molecules which affect the range of signaling gradients such as inhibitors or receptors (Wang & Ferguson, 2005). Changing expression of molecules to reduce diffusion can have the effect of sharpening the gradient—increasing levels close to the source while reducing those at a distance. Existing measurements of ligand and inhibitor expression, while informative, are largely non-quantitative and interpreting these requires determining how signaling activity is determined as a function of both activating and inhibiting influences. In the next section, we will return to these questions and discuss how *in vitro* experiments may provide an inroad into these issues.

3. ESC studies: Insights and opportunities

The difficulty in addressing questions about how signaling pathways pattern the early mammalian embryo in space and time has led to the development of ESC cultures that are complementary to *in vivo* models. These are particularly critical for studying early human development as ethical and practical considerations limit availability of embryos between approximately 2 weeks, when the 14-day rule takes effect, and 5 weeks when the earliest tissue from elective terminations is available. In this section, we briefly review ESC-based systems for studying early embryonic patterning. The basic features of most of these systems have been covered in more detail in previous reviews (Heemskerk & Warmflash, 2016; Shahbazi, Siggia, & Zernicka-Goetz, 2019; Siggia & Warmflash, 2018). In this review, we will not discuss organoids, which recapitulate some of the patterning of particular organs at later developmental stages. The reader is directed to recent reviews that focus on these (Brassard & Lutolf, 2019; Simunovic & Brivanlou, 2017).

3.1 ESC systems for studying gastrulation

Organized two-dimensional ESC patterns were first reported in Warmflash, Sorre, Etoc, Siggia, and Brivanlou (2014). That study took advantage of micropatterning technology to control the geometry of hESC cultures and showed that this was sufficient to overcome the hurdles of inconsistency and lack of reproducibility in regular culture. This micropatterning technology creates complementary patterns of adhesive and non-adhesive territories on the culture surface. This can be accomplished by coating the surface with a non-adhesive "passivating" material such as PLL-PEG, and then burning this material off the surface in defined regions using a mask and UV light (Azioune, Storch, Bornens, Th, & Piel, 2009). Alternative approaches based on stamping are also possible (Théry & Piel, 2009). Remarkably, simple confinement and stimulation with BMP4 led to the emergence of patterns which contained all three germ layers and extra embryonic cells. In what follows, we refer to these micropatterned colonies treated with BMP4 as "2D gastruloids". Treatment of micropatterned colonies with Wnt ligands also produces cell fate patterns, but without the extra-embryonic population at the edge (Martyn, Brivanlou, & Siggia, 2019).

While it is impossible to directly compare these micropatterns with human embryos, a recent study developed an analogous platform with mESCs, and compared the resulting patterns to the mouse embryo (Morgani, Metzger, Nichols, Siggia, & Hadjantonakis, 2018). They showed that the patterning process of Epiblast like cells (EpiLCs) grown in two-dimensional micropatterns mimics that observed *in vivo* (E5.0–E8.0), with cell differentiation following similar developmental paths. Induction of EpiLCs by FGF, BMP, Wnt and Nodal (signals observed at the posterior end of the embryo) resulted in a pattern comprised of posterior cell types including posterior epiblast, primitive streak, mesoderm, and extra-embryonic mesoderm. In agreement with *in vivo* observations, if BMP is not present in the induction media, the pattern is composed of anterior and distal cell types, such as anterior epiblast, anterior primitive streak, and definitive endoderm. These findings support the idea that organized two-dimensional ESC systems can mimic some of the pattern formation that occurs *in vivo*.

Two-dimensional systems are inherently limited in their ability to reproduce the morphogenesis and symmetry breaking that is observed *in vivo*. Several three-dimensional ESC systems to mimic early development have been devised. The closest analogue of the 2D systems are three-dimensional gastruloids which are created from mESCs. These systems are built by pulsing aggregates of mESCs of a defined size with a WNT activator 2 days into their differentiation (Turner et al., 2017; Van Den Brink et al., 2014). They subsequently develop an anterior-posterior axis that elongates, and even shows sequential and regionalized activation of Hox genes (Beccari et al., 2018). Other systems are generated by combining multiple cell types of the early embryo together. This is most easily done with mouse cells, because there are defined cell lines for all three lineages of the early embryo: epiblast, visceral endoderm (known as XEN cells), and trophoblast. Combining mESCs with trophoblast stem cells (TSCs) can produce configurations which look remarkably similar to the mammalian blastocyst (mouse E3.5) (Rivron et al., 2018), while in a slightly different configuration it can also lead to the development of structures resembling the egg cylinder staged mouse embryo (E5) (Harrison, Sozen, Christodoulou, Kyprianou, & Zernicka-Goetz, 2017), known as ETS embryos. These ETS embryos go on to show some early signs of gastrulation including regionalized induction of BRA close to the boundary between the ESCs and TSCs. Combining XEN cells with trophoblasts and ESCs forms structures known as an ETX embryos, and further improves this process (Sozen et al., 2018), with more robust symmetry breaking and primitive streak differentiation.

For human cells, it has been shown that culturing hESCs in Matrigel on top of a soft gel leads to differentiation to a hollow sphere of amniotic ectodermal cells (Shao, Taniguchi, Gurdziel, et al., 2017). Interestingly, in some fraction of cases, the initial spherical symmetry is broken and only half of the aggregate becomes amnion while the rest remains pluripotent epiblast, a configuration which resembles the amniotic sac *in vivo* (Shao, Taniguchi, Townshend, et al., 2017). By varying the gel composition, colonies of hESCs can also be grown as three-dimensional spheres of pluripotent cells. When treated with low doses of BMP, these also break symmetry, but instead divide into SOX2 expressing ectoderm, and BRA expressing mesoderm, which could be indicative of an anterior-posterior axis (Simunovic et al., 2019).

3.2 Do ligands function as morphogens in mammalian gastrulation?

BMP signal is essential for mammalian gastrulation (Arnold & Robertson, 2009; Winnier, Blessing, Labosky, & Hogan, 1995) and has been shown to act as a morphogen in other organisms. In particular, Dpp, the BMP4 homolog in *Drosophila*, was shown to function as a morphogen to control dorsal-ventral patterning of the early embryo and anterior-posterior patterning and growth of the imaginal disk (Ferguson & Anderson, 1992; Lecuit et al., 1996; Matsuda, Harmansa, & Affolter, 2016; Nellen, Burke, Struhl, & Basler, 1996). Treating dissociated cells from the Xenopus animal cap with increasing doses of BMP4 causes them to adopt progressively more ventral fates (Wilson et al., 1997). Simple treatment of mammalian ESCs by BMP ligands leads to mixtures of trophectodermal (Horii et al., 2016; Li et al., 2013; Xu et al., 2002), extra-embryonic (Morgani et al., 2018), and mesodermal fates (Etoc et al., 2016; Warmflash et al., 2014), while as noted above, in the 2D gastruloid system, BMP4 treatment causes ordered differentiation to all three germ layers.

While some studies have suggested that the patterns in 2D gastruloids reflect cells reading levels of BMP4 to give rise to multiple different fates (Tewary et al., 2017), there is evidence to suggest that this is not the case. First, measuring the time-dependent response to BMP treatment shows that BMP initially activates SMAD1/5/8 throughout the colony and this later becomes restricted to the edge (Etoc et al., 2016; Heemskerk et al., 2019). This response is binary rather than graded along the radial axis: cells at the edge show a sustained response and differentiate to extra-embryonic fates, while the remainder of the cells show a transient response. Termination of BMP signaling at the colony center depends on both secreted inhibitors and polarization of BMP receptors to the basal side of the cells (Etoc et al., 2016; Zhang, Zwick, Loew, Grimley, & Ramanathan, 2019), and is highly synchronous between cells (Heemskerk et al., 2019). Thus, there is not sufficient diversity in the direct response to BMP to explain the cell fates that arise. Further, a separate study used isolated colonies containing fewer than 10 cells to show that there is a sharp transition between pluripotency and extra-embryonic fates as the BMP concentration is raised, and BMP alone never gives rise to intermediate fates

(Nemashkalo et al., 2017). Finally, in larger colonies, application of intermediate BMP concentrations does yield mesoderm differentiation, however, this effect is lost upon either WNT or NODAL inhibition (Nemashkalo et al., 2017). Similarly, in micropatterned colonies, inhibiting endogenous WNT or NODAL signals results in a loss of the rings of mesoderm and endoderm (Chhabra, Liu, Goh, Kong, & Warmflash, 2019; Martyn, Kanno, Ruzo, Siggia, & Brivanlou, 2018; Tewary et al., 2017; Warmflash et al., 2014). Thus, evidence from stem cells suggests that in this context, BMP does not function as a morphogen, but rather as a switch which both activates extra-embryonic differentiation and WNT/NODAL activity, and it is the combinatorial effect of these pathways that gives the appearance of a morphogen effect and patterns the 2D gastruloids.

We illustrate how two signals could lead to an apparent morphogen effect as well as pattern the radial axis of 2D gastruloids with a mathematical model. We will compare this two-signal model with the classic model in which the concentration of a single signal is utilized for patterning (Fig. 2A). Details of both models can be found in Appendix. In both cases, the concentration of the signal is interpreted by a network of three mutually repressive transcription factors, which for simplicity we label CDX2, BRA, SOX2, well established markers of the trophoectodermal, mesodermal, and pluripotent states, respectively. We note that this choice is not meant to model actual interactions between these genes, but, as we are focused on signaling, this is the simplest gene regulatory model that implements a three-way decision. A gene regulatory network with a similar architecture has been shown to govern patterning of the neural tube in response to Sonic hedgehog (Balaskas et al., 2012) and the anterior-posterior axis of the drosophila embryo (Chen, Xu, Mei, Yu, & Small, 2012; Jaeger et al., 2004). In the classic model in which BMP acts as a morphogen, the cell remains in a default SOX2-positive pluripotent state in the absence of signal (Fig. 2A–C). When the concentration of the morphogen is sufficiently high, CDX2 is strongly induced and represses both SOX2 and BRA (Fig. 2A–C). However, at intermediate morphogen concentrations, CDX2 expression is not sufficient to repress BRA, and once BRA is activated, it suppresses both CDX2 and SOX2 (Fig. 2A–C). Thus, in this model, the concentration of a single morphogen yields three different fates as in the French Flag model (Fig. 2C).

We asked now if two signals, neither of which acts as a morphogen on its own, could produce the same phenomena. In this model, the gene regulatory network remains the same but, the upstream signaling molecule, BMP,

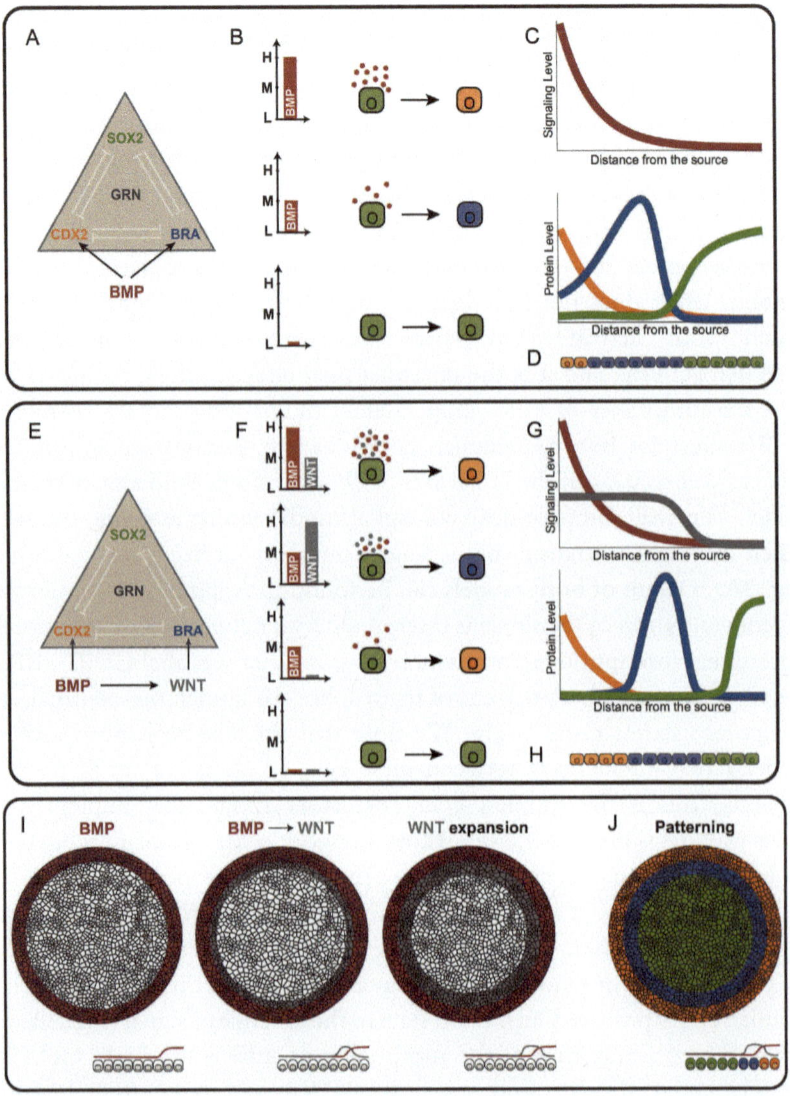

Fig. 2 Mathematical models for signal induction. (A) Classic model for morphogen induction: three mutually repressive transcription factors (CDX2, BRA, SOX2) are activated at different rates by the morphogen ligand, in this case BMP. (B) Schematic representation of cell fate dependence on BMP signal concentration. High BMP signal induces pluripotent (green) cells to adopt CDX2 (yellow) fate (top). At intermediate concentrations of the morphogen, BRA (blue) expression is induced (middle). However, if BMP concentration is low, SOX2 (green) expression remains active (bottom). (C) Morphogen signaling gradient (top) and the corresponding TF expression profile (bottom) as functions of the distance from the source. For example, a cell that is at a

(Continued)

would activate both the expression of the transcription factor CDX2, and the production of a secondary signaling molecule, WNT, which activates the second transcription factor BRA (Fig. 2D–F). At low concentrations of BMP, WNT is also not expressed, and cells remain in the default (pluripotent) state expressing SOX2. At intermediate concentrations, BMP activates the WNT signal, but does not override it, and cells differentiate to mesoderm. When BMP is sufficiently high, WNT is expressed but CDX2 is expressed too strongly and the cells differentiate to trophectoderm. A similar effect can occur in space when BMP activates a diffusible WNT signal (Fig. 2I and J). Thus, an identical transcriptional network could interpret either a single signal *via* a classic morphogen effect or two signals, in which cells do not sense the absolute concentration of either one. These interactions could also explain the pattern observed 2D gastruloids (Fig. 2I), where the BMP signal is restricted to the edge but induces WNT signaling that propagates toward the center of the colony (Chhabra et al., 2019), resulting in a ring of mesendoderm differentiation.

It is important to note that while the two models have very similar networks, the logic at the level of signaling is different, and they make different predictions regarding experiments in which ligand levels are varied. In the single-signal model, cells autonomously decide their fate regardless of

Fig. 2—Cont'd distance from the source will be exposed to an intermediate BMP level and will express a high level of BRA (blue). (D) Pattern created in a row of cells exposed to the gradient signal showed in (C). (E) Transcriptional network mimicking morphogen induction by induction of a secondary signal: three mutually repressive transcription factors (CDX2, BRA, SOX2) are controlled by a balance of an upstream signal (BMP) and a secondary signal (WNT). (F) Schematic representation of cell fate dependence on the combination of BMP and WNT signals. High BMP signal induces pluripotent (green) cells to adopt CDX2 (yellow) fate (top). At intermediate BMP concentrations, the secondary signal, WNT, is induced and BRA (blue) expression is induced (middle top). If at middle concentrations of BMP, the secondary signal was inhibited, CDX2 (yellow) fate would be induced instead (middle bottom). Low BMP concentration would not affect SOX2 (green) expression, which remains active in this case (bottom). (G) Signaling levels (top) and the corresponding TF expression profile (bottom) as functions of the distance from the source of BMP. (H) Pattern created in a row of cells exposed to the signaling levels showed in (G). The blue cells are only present due to the activation of the secondary signal and not to the BMP gradient directly. (I) Sketch of the signaling dynamics of a micropattern exposed to BMP signal. First, BMP signal is restricted to the edge of the colony. Second, BMP signal activates the secondary signal WNT at a distance from the edge, which expands inwards in the colony. (J) The signaling dynamics result into a pattern with three regions: extra-embryonic (yellow), mesendodermal (blue) and ectodermal (green).

secondary signaling, while in the two-signal model, interactions between cells through the downstream pathway are required to specify the intermediate fate. Thus, if one grows cells under conditions where secondary signaling is reduced, such as in small colonies or at low density, the one-signal model predicts that the fate decision will remain the same, while the two-signal model predicts a loss of the intermediate fate. Similarly, the two-signal model predicts a loss of the intermediate fate when secondary signals are inhibited, while the one-signal model does not. All of these predictions were tested by varying BMP dose in hESC colonies of controlled size or density (Nemashkalo et al., 2017), and the results strongly support the two-signal model.

While BMP does not function directly as a morphogen in human pluripotent cells, whether WNT or NODAL can remains an open question. ACTIVIN alone is not sufficient to induce primitive streak differentiation, however, hESCs pretreated with WNT3A for 24h before ACTIVIN differentiate into mesendodermal fates (Yoney et al., 2018). Thus, whether or not these signals function as morphogens, combinatorial effects will also be essential in understanding their function during germ layer specification.

3.3 How are changing morphogen concentrations interpreted in time?

While there is a wealth of information on the phenotypes of mouse embryos when signaling components have been removed, very few direct measurements of signaling activity have been performed due to the challenge of following subcellular events in mammalian embryos. Fluorescently tagging endogenous signal transducers in ESCs and imaging them during differentiation and patterning provides a window into how cells interpret dynamic signals (Heemskerk et al., 2019; Massey et al., 2019; Nemashkalo et al., 2017; Yoney et al., 2018). These studies have led to the conclusion that each pathway performs its own unique signal processing.

hESCs respond stably to stimulation by BMP, and the sustained response is necessary for differentiation to extra-embryonic fates (Nemashkalo et al., 2017). In contrast, hESCs exposed to WNT or ACTIVIN signals show an adaptive response which decays even under constant stimulation (Heemskerk et al., 2019; Massey et al., 2019; Yoney et al., 2018). ACTIVIN treated cells show a decay on the time scale of about 4h to level slightly above the baseline independently of ligand concentration. In contrast, in response to WNT, adaptation is slower (taking around 12h to reach baseline) and dose-dependent, such that it is nearly complete at lower doses of WNT, and partial at higher doses (Massey et al., 2019). Adaptation allows the cells to read the rate of change of the signal in time instead of the absolute concentration (Heemskerk et al., 2019; Sorre, Warmflash, Brivanlou, & Siggia, 2014), and changing the ligand concentration slowly abrogates the response to ACTIVIN but not to BMP. Interestingly, the cells of the neural tube have also been shown to respond adaptively to Sonic hedgehog (Dessaud, McMahon, & Briscoe, 2008), and it will be interesting to determine whether the mechanisms by which different adaptive signals are interpreted are similar.

Stem cell differentiation protocols can be optimized based on this dynamic information. In the case of ACTIVIN, pulsing ligand is more efficient that constant exposure even though the integrated exposure is lower (Heemskerk et al., 2019). Thus, ESCs have allowed for direct measurement of signaling dynamics in pluripotent cells and in response to controlled concentrations and dynamic presentation of ligand. Whether these dynamics are relevant to the embryo remains an open question, however, insights are beginning to emerge from the patterning systems described in the next section.

3.4 How do morphogen systems result in spatial patterns of cell fates?

During development, the signaling dynamics described above lead to the formation of spatial patterns. The 2D gastruloid system has allowed for the direct observation of signaling dynamics during self-organized patterning. These results show that initially the entire colony responds to BMP before its restriction to the edge approximately 12h later (Heemskerk et al., 2019). This pattern of BMP signaling, that is high only at the edge, is then stable for the remainder of patterning. BMP signaling at the edge triggers a wave of WNT signaling which initiates around 24h after BMP treatment near the edge and moves inward at a constant rate (Fig. 3A) (Chhabra et al., 2019). Slightly later, WNT initiates NODAL signaling near the colony edge, and NODAL also expands inwards at a constant rate (Fig. 3B) (Heemskerk et al., 2019). The wave of WNT signaling is initiated earlier than the wave of NODAL, but the NODAL wave travels faster allowing it to overtake the WNT wave (Chhabra et al., 2019).

Importantly, mesodermal differentiation requires both WNT and NODAL, however, the spatial extent of differentiation does not map to any particular level of these pathways (Chhabra et al., 2019). As the signaling travels in a wave, the timing of signaling onset is much more variable than

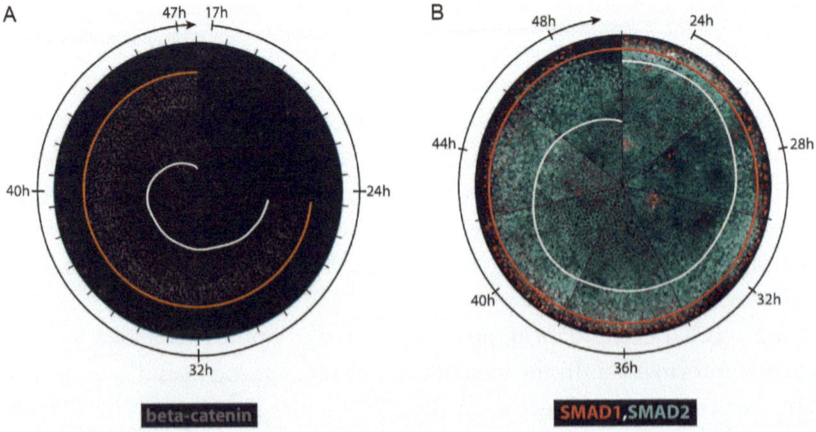

Fig. 3 Nodal and WNT signaling dynamics in BMP treated micropatterned hESC colonies. (A) Pie sections of a live-imaged micropatterned GFP-beta-catenin hESC colony stimulated with BMP4 showing a clockwise time evolution (17–47 h) of non-membrane beta-catenin expression. Each section is a snapshot from time-lapse imaging at clockwise increasing time points post BMP treatment. The outer orange circumference highlights the pinned back end of the WNT wave while the white spiral approximates the inward movement of the front end of the WNT wave. (B) Pie sections of micropatterned hESC colonies stimulated with BMP4 showing a clockwise time evolution (24–48 h) of SMAD1 (red) and SMAD2 (cyan) expression. Each section corresponds to a micropatterned colony fixed at the indicated time point post BMP treatment. The red circumference highlights that SMAD1 stays restricted to the edge at all time points. The white spiral approximates the front end of the SMAD2 wave that moves inward in time. *Panel (A): Data from reference Chhabra, S., Liu, L., Goh, R., Kong, X., & Warmflash, A. 2019. Dissecting the dynamics of signaling events in the BMP, WNT, and NODAL cascade during self-organized fate patterning in human gastruloids. PLoS Biology, 17, 2019, e3000498; Panel (B): Data from reference Heemskerk, I., Burt, K., Miller, M., Chhabra, S., Guerra, M.C., Liu, L., et al. 2019. Rapid changes in morphogen concentration control self-organized patterning in human embryonic stem cells.* eLife, 8, *e40526.*

the level of signaling, and it is possible that it is this timing what cells sense when determining their fate. With the exception of the edge cells, each cell inside the colony experiences three events: the attenuation of the initial BMP signal, the WNT signal, and the NODAL signal. We speculate that the relative timing of these events is what determines the final pattern. As noted above, differentiation in response to NODAL signal requires previous WNT signaling (Yoney et al., 2018), so one hypothesis is that the point in space where NODAL signaling overtakes WNT marks the edge of the territory of mesodermal differentiation. As this mechanism differs from previously proposed models of spatial patterning, it will be interesting to determine how the gene regulatory network, which determines cell fates downstream of signaling, functions to implement this decision.

A WNT wave also forms when hESCs colonies are treated with WNT3A, this time patterning the colonies into two regions: an outer ring of mesendodermal cells and an inner region of ectodermal cells (Martyn et al., 2019, 2018). The initial response to WNT is limited to the colony edge by cell-cell contacts through *E*-cadherin (E-CAD). The wave is formed by a *domino effect* where response to WNT ligands causes the cell to undergo EMT and downregulate E-CAD at the junctions with its neighbors, thereby freeing more cells to respond to the WNT signal. This process continues until a sufficient amount of the secreted feedback inhibitor DKK1 builds up and halts the EMT wave.

The WNT waves created by BMP or WNT3A, although similar in appearance, are mechanistically distinct. In the case of WNT3A induction, the wave created is due to a wave of competence to respond to exogenous WNT signaling, and it is not inhibited even if IWP2, an inhibitor of WNT ligand secretion, is present in the media the entire time (Martyn et al., 2019). In the case of BMP induction, the wave results from endogenous signaling, and does not start at the edge of the colony, where *E*-CAD expression is low. It also does not require EMT as the wave propagates farther into the colony than the territories where differentiation and EMT are observed (Chhabra et al., 2019).

An interesting open question is what separates the mesoderm from the endoderm. Knockout of CER1 during micropatterned differentiation, causes upregulated NODAL signaling and promotes endoderm at the expense of mesoderm differentiation (Yoney et al., 2018). However, how the NODAL signal is different upon CER1 knockout is unclear. Are its dynamics or spatial extent perturbed or only its level? Recent results with a reporter system in mouse embryos and stem cells suggest that the endoderm arises in randomly distributed cells within the anterior primitive streak and then segregates from the mesoderm (Pour et al., 2019) but, how does this observation relate to the Nodal levels or distribution? One hypothesis is that the fate choice is stochastic with a bias towards endoderm that depends on the level of Nodal signaling that a cell receives.

3.5 How are symmetries broken in the early embryo?

Two-dimensional models are very useful for studying signaling dynamics and patterning, however, they do not recapitulate the symmetry breaking or morphogenesis which is observed *in vivo*. This issue has driven the development of the 3D models described above. Most studies to date have concentrated on determining to what degree these models resemble *in vivo* embryos, however, they have provided some insights about the signals that drive patterning during early mammalian development and have set a course for future investigations.

Several of the three-dimensional models described above break symmetry from an initial spherical configuration. In the case of the human models of amnion (Shao, Taniguchi, Gurdziel, et al., 2017; Shao, Taniguchi, Townshend, et al., 2017), it is BMP signaling which separates the amnion from the epiblast. The ETS (Harrison et al., 2017) and ETX (Sozen et al., 2018) embryos, as well as 3D hESCs treated with low doses of BMP4 (Simunovic et al., 2019), localize expression of BRA to one side of the epiblast compartment. This is particularly surprising in the case of ETS embryos as they lack the visceral endoderm which places a role in positioning the primitive streak to the posterior side of the embryo. In all of these cases, symmetry breaking depends on Wnt signaling, and in the hESC model it has been shown that removing DKK1 causes the entire sphere to adopt a mesodermal fate (Simunovic et al., 2019). Some caution in interpretation is warranted, however, as the markers used such as Sox2 and Bra delineate germ layers rather than the AP axis specifically. Further, in the case of the hESC model, the final object contains one axis of rotational signaling, the same as the 2D micropatterned colonies, so it is not clear that one model represents a true AP axis more than the other (Heemskerk & Warmflash, 2016). Nonetheless, these are useful models for how morphogen pathways are deployed to break spherical symmetry, and it will be interesting to use them to study the dynamics of the BMP and WNT pathways in these events.

The mouse gastruloid system develops a more complete AP axis, including axial elongation and the temporal and spatial emergence of Hox genes (Beccari et al., 2018; Turner et al., 2017). A pulse of the Wnt agonist CHIR99021 (CHIR) is essential for these events, however, little is known about the response to this pulse or how downstream signals ultimately lead to patterning. In the future, this represents an exciting system to begin to explore mechanistically how signals pattern the AP axis.

4. Conclusions

In the previous sections we have reviewed recent insights on morphogen dynamics and patterning that have come from newly developed stem cell systems. Recent results indicate that the BMP pathway does not function as a morphogen in hESCs (Nemashkalo et al., 2017). Analogous experiments have not been performed for Wnt and Nodal, however, the mechanisms uncovered in hESC demonstrate that these pathways are not functioning as classical morphogens during self-organized fate patterning in micropatterned systems. It will be interesting to test more directly whether these pathways are capable of functioning as morphogens in mammalian ESCs.

The potential of *in vitro* systems for studying early development has only begun to be realized, and many new and exciting questions are yet to be explored. In the above sections, we highlighted several questions that are currently being addressed, including understanding precisely how the position of mesendodermal differentiation is established by the combinatorial effects of BMP, Wnt, and Nodal, how mesoderm and endoderm are separated, and what signaling dynamics lead to the symmetry breaking that underlies axis formation in the mammalian embryo. The combination of the stem cells systems and live cell reporters that have been developed should allow rapid progress on these longstanding questions. For example, DKK1 knockouts have been shown to cause defects in symmetry breaking in hESC spheroid models, and CER1 knockouts have been shown to lead to increased endoderm differentiation. Creating these knockouts directly in WNT or NODAL reporter lines, respectively, will allow for direct observation of the signaling dynamics, and how they lead to the observed phenotypes.

Here we have focused on signaling, however, much also remains to be learned about how these signals are interpreted by transcription factor networks inside of cells to implement cell fate decisions, the subject of intense study in other systems (Briscoe & Small, 2015; Driever, Thoma, & Nusslein-Volhard, 1989). The simplest model is that different binding sites have different affinities for the transcription factors induced by the morphogen (Driever et al., 1989; Struhl, Struhl, & Macdonald, 1989), however, recent studies have failed to show that there is a correlation between binding affinities and the range of morphogen induction in systems such as the vertebrate neural tube and AP patterning in fly (Hannon, Blythe, & Wieschaus, 2017; Oosterveen et al., 2012; Peterson et al., 2012). Further, it is clear that understanding the interplay between morphogen dynamics and their interpretation by the individual promoters of target genes is important for understanding patterning (Dubrulle et al., 2015). Directly studying these issues in mammalian stem cells could shed light on the gene regulatory network that separates germ layers during gastrulation.

A major remaining challenge is comparing the results from *in vitro* systems with actual embryos, particularly for human. Recently, a system for culturing post-implantation human embryos has been developed, and the different cell types present at these early stages were characterized (Deglincerti et al., 2016; Shahbazi et al., 2016), however, these are still limited by the 14-day rule, and it is also still unclear to what degree the cultured embryos mimic processes that occur *in vivo*.

Interspecies comparisons can serve as tool for validating *in vitro* systems and for understanding differences between humans and model organisms. Mouse ESC systems have begun to be compared to mouse embryos (Beccari et al., 2018; Harrison et al., 2017; Morgani et al., 2018), and validating *in vitro* systems for mouse would yield more confidence in human *in vitro* systems. There are a number of known differences between mouse and human including the geometry of the embryo (Behringer, Wakamiya, Tsang, & Tam, 2000), the positioning of the extra embryonic tissues (Dobreva et al., 2010), and molecular differences that are beginning to be probed by single cell RNA-Seq studies (Blakeley et al., 2015; Petropoulos, Edsga, Reinius, & Linnarsson, 2016). *In vitro* systems have the potential to examine these differences in more controlled settings by growing ESCs from the same developmental stages in the same conditions. Continuing to develop insights from stem cell systems and validating them *in vivo* is providing windows into previous inaccessible times during human development and into understanding developmental differences between humans and other species.

Acknowledgments

We thank Miguel Ángel Ortiz Salazar for a critical reading of the manuscript, Sapna Chhabra and Idse Heemskerk for providing data used to create Fig. 3, and all the members of the Warmflash lab for helpful feedback. This work was funded by Rice University and grants to AW from NSF (MCB-1553228), NIH (R01GM126122), and Simons Foundation (511079).

Appendix

Mathematical model for fate patterning by a single morphogen

The dynamics of a gene regulatory network comprised of three mutually repressive transcription factors (TF1, TF2 and TF3) controlled by a morphogen signal (S), as in Fig. 2A, can be described by the system of differential equations:

$$\frac{d\text{TF1}}{dt} = a_1 \frac{(\text{S}/k_{S1})^{n_{S1}}}{1 + (\text{S}/k_{S1})^{n_{S1}}} \frac{1}{1 + (\text{TF2}/k_{21})^{n_{21}} + (\text{TF3}/k_{31})^{n_{31}}} - \delta_1 \text{TF1}$$

$$\frac{d\text{TF2}}{dt} = a_2 \frac{(\text{S}/k_{S2})^{n_{S2}}}{1 + (\text{S}/k_{S2})^{n_{S2}}} \frac{1}{1 + (\text{TF1}/k_{12})^{n_{12}} + (\text{TF3}/k_{32})^{n_{32}}} - \delta_2 \text{TF2}$$

$$\frac{d\text{TF3}}{dt} = a_3 \frac{1}{1 + (\text{TF1}/k_{13})^{n_{13}} + (\text{TF2}/k_{23})^{n_{23}}} - \delta_3 \text{TF3}$$

where the variables TF1, TF2 and TF3 represent the expression values of the corresponding transcription factors. In the main text these transcription factors are CDX2, BRA and SOX2, respectively. The parameter S corresponds to the concentration of the morphogen, which in the main text is BMP. The parameters n_{ij}, k_{ij} (with $i \in \{\text{S}, 1,2,3\}, j \in \{1,2,3\}$) are the Hill coefficients and dissociation constants respectively, and a_i, δ_i with $i \in \{1,2,3\}$ are the production and degradation rates, respectively.

In Fig. 2C of the main text, the system of differential equations was simulated with initial condition TF1 = 0, TF2 = 0 and TF3 = 2, the parameter values given in Table A1, and S decreased exponentially as a function of the distance from the source.

Table A1 Table of parameter values used in the simulations shown in Fig. 2C.

Parameter	Value	Parameter	Value	Parameter	Value
a_1	8	a_2	4	a_3	2
δ_1	1	δ_2	1	δ_3	1
k_{S1}	7	k_{S2}	0.5	k_{13}	0.1
n_{S1}	1	n_{S2}	5	n_{13}	1
k_{21}	0.4	k_{12}	0.3	k_{23}	0.1
n_{21}	1	n_{12}	1	n_{23}	1
k_{31}	0.5	k_{32}	1		
n_{31}	1	n_{32}	1		

Mathematical model for patterning multiple fates with two signals

The dynamics of a gene regulatory network comprised of three mutually repressive transcription factors (TF1, TF2 and TF3), this time controlled by a balance of an upstream signal (S1) and a secondary signal (S2), as in Fig. 2E, can be described by the system of differential equations:

$$\frac{dS2}{dt} = a_{S2}S1 - \delta_{S2}S2$$
$$\frac{dTF1}{dt} = \frac{a_1 + b_1 S1}{1 + (TF2/k_{21})^{n_{21}} + (TF3/k_{31})^{n_{31}}} - \delta_1 TF1$$
$$\frac{dTF2}{dt} = \frac{a_2 + b_2 S2}{1 + (TF1/k_{12})^{n_{12}} + (TF3/k_{32})^{n_{32}}} - \delta_2 TF2$$
$$\frac{dTF3}{dt} = \frac{a_3}{1 + (TF1/k_{13})^{n_{13}} + (TF2/k_{23})^{n_{23}}} - \delta_3 TF3$$

where the variables S2, TF1, TF2 and TF3 correspond to the expression values of the secondary signal and the corresponding transcription factors. In the main text, the secondary signal is WNT and the transcription factors are CDX2, BRA and SOX2, respectively. The parameter S1 is the concentration of the upstream signal, BMP in the main text. The parameters n_{ij}, k_{ij} (with $i,j \in \{1,2,3\}$) are the Hill coefficients and dissociation constants and a_i, δ_i with $i \in \{S2,1,2,3\}$ are the production and degradation rates, respectively.

Table A2 Tables of parameter values used in the simulations shown in Fig. 2G of the main text.

Parameter	Value	Parameter	Value	Parameter	Value
a_{S2}	2	a_2	0.05	a_3	5
δ_{S2}	0.5	b_2	14	δ_3	0.1
a_1	0.7	δ_2	0.5	k_{13}	1
b_1	5	k_{12}	0.1	n_{13}	1
δ_1	0.5	n_{12}	1	k_{23}	0.01
k_{21}	0.8	k_{32}	0.5	n_{23}	1
n_{21}	1	n_{32}	1		
k_{31}	0.5				
n_{31}	1				

In Fig. 2G of the main text, the system of differential equations was simulated with initial condition S2 = 0, TF1 = 0, TF2 = 0 and TF3 = 3, the parameter values given in Table A2 and the main signal, and S1 decreased exponentially as a function of the distance from the source.

References

Arnold, S. J., & Robertson, E. J. (2009). Making a commitment: Cell lineage allocation and axis patterning in the early mouse embryo. *Nature Reviews. Molecular Cell Biology*, *10*, 91–103.

Azioune, A., Storch, M., Bornens, M., Th, M., & Piel, M. (2009). Simple and rapid process for single cell micro-patterning. *Lab on a Chip*, *9*, 1640–1642.

Balaskas, N., Ribeiro, A., Panovska, J., Dessaud, E., Sasai, N., Page, K. M., et al. (2012). Gene regulatory logic for reading the sonic hedgehog signaling gradient in the vertebrate neural tube. *Cell*, *148*, 273–284.

Beccari, L., Moris, N., Girgin, M., Turner, D. A., Baillie-Johnson, P., Cossy, A.-C., et al. (2018). Multi-axial self-organization properties of mouse embryonic stem cells into gastruloids. *Nature*, *562*, 272.

Behringer, R. R., Wakamiya, M., Tsang, T. E., & Tam, P. P. L. (2000). A flattened mouse embryo: Leveling the playing field. *Genesis*, *28*, 23–30.

Ben-Haim, N., Lu, C., Guzman-Ayala, M., Pescatore, L., Mesnard, D., Bischofberger, M., et al. (2006). The nodal precursor acting via activin receptors induces mesoderm by maintaining a source of its convertases and BMP4. *Developmental Cell*, *11*, 313–323.

Blakeley, P., Fogarty, N. M. E., Del Valle, I., Wamaitha, S. E., Hu, T. X., Elder, K., et al. (2015). Defining the three cell lineages of the human blastocyst by single-cell RNA-seq. *Development*, *142*, 3151–3165.

Brassard, J. A., & Lutolf, M. P. (2019). Engineering stem cell self-organization to build better organoids. *Cell Stem Cell*, *24*, 860–876.

Brennan, J., Lu, C. C., Norris, D. P., Rodriguez, T. A., Beddington, R. S. P., & Robertson, E. J. (2001). Nodal signalling in the epiblast patterns the early mouse embryo. *Nature*, *411*, 965–969.

Briscoe, J., & Small, S. (2015). Morphogen rules: Design principles of gradient-mediated embryo patterning. *Development*, *142*, 3996–4009.

Chamberlain, C. E., Jeong, J., Guo, C., Allen, B. L., & McMahon, A. P. (2008). Notochord-derived Shh concentrates in close association with the apically positioned basal body in neural target cells and forms a dynamic gradient during neural patterning. *Development*, *135*, 1097–1106.

Chen, H., Xu, Z., Mei, C., Yu, D., & Small, S. (2012). A system of repressor gradients spatially organizes the boundaries of bicoid-dependent target genes. *Cell*, *149*, 618–629.

Chhabra, S., Liu, L., Goh, R., Kong, X., & Warmflash, A. (2019). Dissecting the dynamics of signaling events in the BMP, WNT, and NODAL cascade during self-organized fate patterning in human gastruloids. *PLoS Biology*, *17*, e3000498.

Conlon, F. L., Lyons, K. M., Takaesu, N., Barth, K. S., Kispert, A., Herrmann, B., et al. (1994). A primary requirement for nodal in the formation and maintenance of the primitive streak in the mouse. *Development*, *120*, 1919–1928.

Crick, F. (1970). Diffusion in embryogenesis. *Nature*, *225*, 420–422.

Deglincerti, A., Croft, G. F., Pietila, L. N., Zernicka-Goetz, M., Siggia, E. D., & Brivanlou, A. H. (2016). Self-organization of the in vitro attached human embryo. *Nature*, *533*, 251–254.

Dessaud, E., McMahon, A. P., & Briscoe, J. (2008). Pattern formation in the vertebrate neural tube: A sonic hedgehog morphogen-regulated transcriptional network. *Development*, *135*, 2489–2503.

Dobreva, M., Pereira, P. N. G., Zwijsen, A., Leuven, K. U., Dobreva, M. P., Deprest, J., et al. (2010). On the origin of amniotic stem cells: Of mice and men. *The International Journal of Developmental Biology*, *54*, 761–777.

Driever, W., & Nusslein-Volhard, C. (1988). A gradient of bicoid protein in Drosophila embryos. *Cell*, *54*, 83–93.

Driever, W., Thoma, G., & Nusslein-Volhard, C. (1989). Determination of spatial domains of zygotic gene expression in the Drosophila embryo by the affinity of binding sites for the bicoid morphogen the Hb regulatory region. *Nature*, *340*, 363–367.

Drocco, J. A., Wieschaus, E. F., & Tank, D. W. (2012). The synthesis–diffusion–degradation model explains bicoid gradient formation in unfertilized eggs. *Physical Biology*, *9* 055004.

Dubrulle, J., Jordan, B. M., Akhmetova, L., Farrell, J. A., Kim, S.-H., Solnica-Krezel, L., et al. (2015). Response to nodal morphogen gradient is determined by the kinetics of target gene induction. *eLife*, *4*, e05042.

Dunn, N. R., Vincent, S. D., Oxburgh, L., Robertson, E. J., & Bikoff, E. K. (2004). Combinatorial activities of Smad2 and Smad3 regulate mesoderm formation and patterning in the mouse embryo. *Development*, *131*, 1717–1728.

Dyson, S., & Gurdon, J. (1998). The interpretation of position in a morphogen gradient as revealed by occupancy of activin receptors. *Cell*, *93*, 557–568.

Economou, A. D., Ohazama, A., Porntaveetus, T., Sharpe, P. T., Kondo, S., Basson, M. A., et al. (2012). Periodic stripe formation by a turing mechanism operating at growth zones in the mammalian palate. *Nature Genetics*, *44*, 348–351.

Etoc, F., Metzger, J., Ruzo, A., Kirst, C., Yoney, A., Ozair, M. Z., et al. (2016). A balance between secreted inhibitors and edge sensing controls gastruloid self-organization. *Developmental Cell*, *39*, 302–315.

Ferguson, E. L., & Anderson, K. V. (1992). Decapentaplegic acts as a morphogen to organize dorsal-ventral pattern in the Drosophila embryo. *Cell*, *71*, 451–461.

Gierer, A., & Meinhardt, H. (1972). A theory of biological pattern formation. *Kybernetik*, *12*, 30–39.

Green, J. B. A., New, H. V., & Smith, J. C. (1992). Responses of embryonic xenopus cells to activin and FGF are separated by multiple dose thresholds and correspond to distinct axes of the mesoderm. *Cell*, *71*, 731–739.

Green, J. B. A., & Sharpe, J. (2015). Positional information and reaction-diffusion: Two big ideas in developmental biology combine. *Development*, *142*, 1203–1211.

Gregor, T., Tank, D. W., Wieschaus, E. F., & Bialek, W. (2007). Probing the limits to positional information. *Cell*, *130*, 153–164.

Gregor, T., Wieschaus, E. F., Mcgregor, A. P., Bialek, W., & Tank, D. W. (2007). Stability and nuclear dynamics of the bicoid morphogen gradient. *Cell*, *130*, 141–152.

Gurdon, J. B., Harger, P., Mitchell, A., & Lemaire, P. (1994). Activin signalling and response to a morphogen gradient. *Nature*, *371*, 487–492.

Hannon, C. E., Blythe, S. A., & Wieschaus, E. F. (2017). Concentration dependent chromatin states induced by the bicoid morphogen gradient. *eLife*, *6*, e28275.

Harrison, S. E., Sozen, B., Christodoulou, N., Kyprianou, C., & Zernicka-Goetz, M. (2017). Assembly of embryonic and extraembryonic stem cells to mimic embryogenesis in vitro. *Science*, *356* eaal1810.

Heemskerk, I., Burt, K., Miller, M., Chhabra, S., Guerra, M. C., Liu, L., et al. (2019). Rapid changes in morphogen concentration control self-organized patterning in human embryonic stem cells. *eLife*, *8*, e40526.

Heemskerk, I., & Warmflash, A. (2016). Pluripotent stem cells as a model for embryonic patterning: From signaling dynamics to spatial organization in a dish. *Developmental Dynamics*, *245*, 976–990.

Horii, M., Li, Y., Wakeland, A. K., Pizzo, D. P., Nelson, K. K., Sabatini, K., et al. (2016). Human pluripotent stem cells as a model of trophoblast differentiation in both normal development and disease. *Proceedings of the National Academy of Sciences of the United States of America*, *113*, E3882–E3891.

Jaeger, J., Surkova, S., Blagov, M., Janssens, H., Kosman, D., Kozlov, K. H., et al. (2004). Dynamic control of positional information in the early drosophila embryo. *Nature*, *430*, 368–371.

James, D., Levine, A. J., Besser, D., & Hemmati-Brivanlou, A. (2005). TGF-beta/activin/nodal signaling is necessary for the maintenance of pluripotency in human embryonic stem cells. *Development*, *132*, 1273–1282.

Jones, C. M., Armes, N., & Smith, J. C. (1996). Signalling by TGF-beta family members: Short-range effects of Xnr-2 and BMP-4 contrast with the long-range effects of activin. *Current Biology*, *6*, 1468–1475.

Kondo, S., & Miura, T. (2010). Reaction-diffusion model as a framework for understanding biological pattern formation. *Science*, *329*, 1616–1620.

Lawrence, P. A. (1966). Gradients in the insect segment: The orientation of hairs in the milk-weed bug Oncopeltus fasciatus. *The Journal of Experimental Biology*, *44*, 607–620.

Lecuit, T., Brook, W. J., Ng, M., Calleja, M., Sun, H., & Cohen, S. M. (1996). Two distinct mechanisms for long-range patterning by Decapentaplegic in the Drosophila wing. *Nature*, *381*, 387–393.

Li, Y., Moretto-Zita, M., Soncin, F., Wakeland, A., Wolfe, L., Leon-Garcia, S., et al. (2013). BMP4-directed trophoblast differentiation of human embryonic stem cells is mediated through a ΔNp63+ cytotrophoblast stem cell state. *Development*, *140*, 3965–3976.

Martyn, I., Brivanlou, A. H., & Siggia, E. D. (2019). A wave of WNT signaling balanced by secreted inhibitors controls primitive streak formation in micropattern colonies of human embryonic stem cells. *Development*, *146*, dev172791.

Martyn, I., Kanno, T. Y., Ruzo, A., Siggia, E. D., & Brivanlou, A. H. (2018). Self-organization of a human organizer by combined Wnt and nodal signalling. *Nature*, *11*, 1–19.

Massey, J., Liu, Y., Alvarenga, O., Saez, T., Schmerer, M., & Warmflash, A. (2019). Synergy with TGFβ ligands switches WNT pathway dynamics from transient to sustained during human pluripotent cell differentiation. *Proceedings of the National Academy of Sciences*, *116*, 4989.

Matsuda, S., Harmansa, S., & Affolter, M. (2016). BMP morphogen gradients in flies. *Cytokine & Growth Factor Reviews*, *27*, 119–127.

Meinhardt, H. (1978). Space-dependent cell determination under the control of a morphogen gradient. *Journal of Theoretical Biology*, *74*, 307–327.

Morgani, S. M., Metzger, J. J., Nichols, J., Siggia, E. D., & Hadjantonakis, A.-K. (2018). Micropattern differentiation of mouse pluripotent stem cells recapitulates embryo regionalized cell fate patterning. *eLife*, *7*, e32839.

Nellen, D., Burke, R., Struhl, G., & Basler, K. (1996). Direct and long-range action of a DPP morphogen gradient. *Cell*, *85*, 357–368.

Nemashkalo, A., Ruzo, A., Heemskerk, I., & Warmflash, A. (2017). Morphogen and community effects determine cell fates in response to BMP4 signaling in human embryonic stem cells. *Development*, *144*, 3042–3053.

Oosterveen, T., Kurdija, S., Alekseenko, Z., Uhde, C. W., Bergsland, M., Sandberg, M., et al. (2012). Mechanistic differences in the transcriptional interpretation of local and long-range Shh morphogen signaling. *Developmental Cell*, *23*, 1006–1019.

Peterson, K. A., Nishi, Y., Ma, W., Vedenko, A., Shokri, L., Zhang, X., et al. (2012). Neural-specific Sox2 input and differential Gli-binding affinity provide context and positional information in Shh-directed neural patterning. *Genes & Development*, *26*, 2802–2816.

Petropoulos, S., Edsga, D., Reinius, B., & Linnarsson, S. (2016). Single-cell RNA-seq reveals lineage and x chromosome dynamics in human preimplantation embryos. *Cell*, *165*, 1–15.

Pour, M., Sampath Kumar, A., Walther, M., Wittler, L., Meissner, A., & Nachman, I. (2019). Emergence and patterning dynamics of mouse definitive endoderm. *BioRxiv*, 728642.

Raspopovic, J., Marcon, L., Russo, L., & Sharpe, J. (2014). Digit patterning is controlled by a Bmp-Sox9-Wnt turing network modulated by morphogen gradients. *Science*, *345*, 566–570.

Rivron, N. C., Frias-Aldeguer, J., Vrij, E. J., Boisset, J.-C., Korving, J., Vivié, J., et al. (2018). Blastocyst-like structures generated solely from stem cells. *Nature*, *557*, 106–111.

Robertson, E. J. (2014). Dose-dependent nodal/Smad signals pattern the early mouse embryo. *Seminars in Cell & Developmental Biology*, *32*, 73–79.

Rogers, K. W., & Schier, A. F. (2011). Morphogen gradients: From generation to interpretation. *Annual Review of Cell and Developmental Biology*, *27*, 377–407.

Sagner, A., & Briscoe, J. (2017). Morphogen interpretation: Concentration, time, competence, and signaling dynamics. *Wiley Interdisciplinary Reviews: Developmental Biology*, *6*, e271.

Shahbazi, M. N., Jedrusik, A., Vuoristo, S., Recher, G., Hupalowska, A., Bolton, V., et al. (2016). Self-organization of the human embryo in the absence of maternal tissues. *Nature Cell Biology*, *18*, 700–708.

Shahbazi, M. N., Siggia, E. D., & Zernicka-Goetz, M. (2019). Self-organization of stem cells into embryos: A window on early mammalian development. *Science*, *364*, 948–951.

Shahbazi, M. N., & Zernicka-Goetz, M. (2018). Deconstructing and reconstructing the mouse and human early embryo. *Nature Cell Biology*, *20*, 818–887.

Shao, Y., Taniguchi, K., Gurdziel, K., Townshend, R. F., Xue, X., Yong, K. M. A., et al. (2017). Self-organized amniogenesis by human pluripotent stem cells in a biomimetic implantation-like niche. *Nature Materials*, *16*, 419–425.

Shao, Y., Taniguchi, K., Townshend, R. F., Miki, T., Gumucio, D. L., & Fu, J. (2017). A pluripotent stem cell-based model for post-implantation human amniotic sac development. *Nature Communications*, *8*, 208.

Siggia, E. D., & Warmflash, A. (2018). Modeling mammalian gastrulation with embryonic stem cells. *Current Topics in Developmental Biology*, *129*, 1–23.

Simunovic, M., & Brivanlou, A. H. (2017). Embryoids, organoids and gastruloids: New approaches to understanding embryogenesis. *Development*, *144*, 976–985.

Simunovic, M., Metzger, J. J., Etoc, F., Yoney, A., Ruzo, A., Martyn, I., et al. (2019). A 3D model of a human epiblast reveals BMP4-driven symmetry breaking. *Nature Cell Biology*, *21*, 900–910.

Sorre, B., Warmflash, A., Brivanlou, A. H., & Siggia, E. D. (2014). Encoding of temporal signals by the TGF-β pathway and implications for embryonic patterning. *Developmental Cell*, *30*, 334–342.

Sozen, B., Amadei, G., Cox, A., Wang, R., Na, E., Czukiewska, S., et al. (2018). Self-assembly of embryonic and two extra-embryonic stem cell types into gastrulating embryo-like structures. *Nature Cell Biology*, *20*, 979–989.

Struhl, G., Struhl, K., & Macdonald, P. M. (1989). The gradient morphogen bicoid is a concentration-dependent transcriptional activator. *Cell*, *57*, 1259–1273.

Stumpf, H. F. (1966). Mechanism by which cells estimate their location within the body. *Nature*, *212*, 430–431.

Tewary, M., Ostblom, J., Prochazka, L., Zulueta-Coarasa, T., Shakiba, N., Fernandez-Gonzalez, R., et al. (2017). A stepwise model of reaction-diffusion and positional information governs self-organized human peri-gastrulation-like patterning. *Development*, *144*, 4298–4312.

Théry, M., & Piel, M. (2009). Adhesive micropatterns for cells: A microcontact printing protocol. *Cold Spring Harbor Protocols*, *4*, pdb.prot5255.

Tortelote, G. G., Hernández-Hernández, J. M., Quaresma, A. J. C., Nickerson, J. A., Imbalzano, A. N., & Rivera-Pérez, J. A. (2013). Wnt3 function in the epiblast is required for the maintenance but not the initiation of gastrulation in mice. *Developmental Biology*, *374*, 164–173.

Turing, A. M. (1952). The chemical basis of morphogenesis. *Philosophical Transactions of the Royal Society of London B*, *237*, 37–72.

Turner, D. A., Girgin, M., Alonso-Crisostomo, L., Trivedi, V., Baillie-Johnson, P., Glodowski, C. R., et al. (2017). Anteroposterior polarity and elongation in the absence of extra-embryonic tissues and of spatially localised signalling in gastruloids: Mammalian embryonic organoids. *Development*, *144*, 3894–3906.

Vallier, L., Alexander, M., & Pedersen, R. A. (2005). Activin/nodal and FGF pathways cooperate to maintain pluripotency of human embryonic stem cells. *Journal of Cell Science*, *118*, 4495–4509.

Van Den Brink, S. C., Baillie-Johnson, P., Balayo, T., Hadjantonakis, A.-K., Nowotschin, S., Turner, D. A., et al. (2014). Symmetry breaking, germ layer specification and axial organisation in aggregates of mouse embryonic stem cells. *Development*, *141*, 4231–4242.

Vincent, S. D., Dunn, N. R., Hayashi, S., Norris, D. P., & Robertson, E. J. (2003). Cell fate decisions within the mouse organizer are governed by graded nodal signals. *Genes & Development*, *17*, 1646–1662.

Waldrip, W. R., Bikoff, E. K., Hoodless, P. A., Wrana, J. L., & Robertson, E. J. (1998). Smad2 signaling in extraembryonic tissues determines anterior-posterior polarity of the early mouse embryo. *Cell*, *92*, 797–808.

Wang, Y.-C., & Ferguson, E. L. (2005). Spatial bistability of Dpp-receptor interactions during drosophila dorsal-ventral patterning. *Nature*, *434*, 229–234.

Warmflash, A., Sorre, B., Etoc, F., Siggia, E. D., & Brivanlou, A. H. (2014). A method to recapitulate early embryonic spatial patterning in human embryonic stem cells. *Nature Methods*, *11*, 847–854.

Wilson, P. A., Lagna, G., Suzuki, A., & Hemmati-Brivanlou, A. (1997). Concentration-dependent patterning of the xenopus ectoderm by BMP4 and its signal transducer Smad1. *Development*, *124*, 3177–3184.

Winnier, G., Blessing, M., Labosky, P. A., & Hogan, B. L. M. (1995). Bone morphogenetic protein-4 is required for mesoderm formation and patterning in the mouse. *Genes & Development*, *9*, 2105–2116.

Wolpert, L. (1969). Positional information and the spatial pattern of cellular differentiation. *Journal of Theoretical Biology*, *25*, 1–47.

Xu, R.-H., Chen, X., Li, D. S., Li, R., Addicks, G. C., Glennon, C., et al. (2002). BMP4 initiates human embryonic stem cell differentiation to trophoblast. *Nature Biotechnology*, *20*, 1261–1264.

Yamamoto, M., Beppu, H., Takaoka, K., Meno, C., Li, E., Miyazono, K., et al. (2009). Antagonism between Smad1 and Smad2 signaling determines the site of distal visceral endoderm formation in the mouse embryo. *The Journal of Cell Biology*, *184*, 323–334.

Yoney, A., Etoc, F., Ruzo, A., Carroll, T., Metzger, J. J., Martyn, I., et al. (2018). WNT signaling memory is required for ACTIVIN to function as a morphogen in human gastruloids. *eLife*, 7, e38279.

Yoon, Y., Huang, T., Tortelote, G. G., Wakamiya, M., Hadjantonakis, A.-K., Behringer, R. R., et al. (2015). Extra-embryonic Wnt3 regulates the establishment of the primitive streak in mice. *Developmental Biology*, *403*, 80–88.

Zhang, Z., Zwick, S., Loew, E., Grimley, J. S., & Ramanathan, S. (2019). Embryo geometry drives formation of robust signaling gradients through receptor localization. *Nature Communication*, *10*, 4516.

Zhou, X., Sasaki, H., Lowe, L., Hogan, B. L. M., & Kuehn, M. R. (1993). Nodal is a novel TGF-β-like gene expressed in the mouse node during gastrulation. *Nature*, *361*, 543–547.

CHAPTER TEN

Control of size, fate and time by the Hh morphogen in the eyes of flies

David G. Míguez[a,*], Diana García-Morales[b], Fernando Casares[b,*]
[a]Dept. Física de la Materia Condensada, Instituto de Física de la Materia Condensada (IFIMAC), Facultad de Ciencias, and Centro de Biología Molecular Severo Ochoa (CBMSO, CSIC-UAM), Universidad Autónoma de Madrid, Madrid, Spain
[b]CABD (Andalusian Centre for Developmental Biology), GEM-DMC2 Maria de Maeztu Unit, CSIC-Universidad Pablo de Olavide-Junta de Andalucia, Campus UPO, Seville, Spain
*Corresponding authors: e-mail address: david.miguez@uam.es; fcasfer@upo.es

Contents

Abstract

Molecules of the *hedgehog* (*hh*) family are involved in the specification and patterning of eyes in vertebrates and invertebrates. These organs, though, are of very different sizes, raising the question of how Hh molecules operate at such different scales. In this paper we discuss the strategies used by Hh to control the development of the two eye types in *Drosophila:* the large compound eye and the small ocellus. We first describe the distinct ways in which these two eyes develop and the evidence for the key role played by Hh in both; then we consider the potential for variation in the range of action of a "typical" morphogen and measure this range ("characteristic length") for Hh in different organs, including the compound eye and the ocellus. Finally, we describe how different

Current Topics in Developmental Biology, Volume 137
ISSN 0070-2153
https://doi.org/10.1016/bs.ctdb.2019.10.011

feedback mechanisms are used to extend the Hh range of action to pattern the large and even the small eye. In the ocellus, the basic Hh signaling pathway adds to its dynamics the attenuation of its receptor as cell differentiate. This sole regulatory change can result in the decoding of the Hh gradient by receiving cells as a wave of constant speed. Therefore, in the fly ocellus, the Hh morphogen adds to its spatial patterning role a novel one: patterning along a time axis.

1. The small and large eyes in *Drosophila* rely on Hh

Insects typically harbor two eye types. Two large compound eyes (CE), located laterally, and three small eyes, called ocelli (OC), on the dorsal medial head—the so-called ocellar complex (Fig. 1A). CEs are the main visual organs, responsible for color and motion vision and the detection of polarized light. OC might have different functions in different insect species, but they have been generally involved in the detection of fast changes in light intensity, such as those produced by looming objects, to trigger a fast escape reflex, or helping in the detection of the horizon during flight (Krapp, 2009). CEs and OC are of very different sizes. In *Drosophila*, each CE comprises about 16,000 cells (of which 6400 are photoreceptors ("PR") cells), while each of the three OC is formed by about 80 cells (with 40 PRs each). However, and despite their very different size and structure (the CE is formed by many small unit eyes, called ommatidia, while each CO is a single retina), their genetic makeup is very similar (Blanco, Pauli, Seimiya, Udolph, & Gehring, 2010; Brockmann, Dominguez-Cejudo, Amore, & Casares, 2011; Friedrich, 2006).

Both eyes types derive from two bilateral head primordia (also called "eye-antennal imaginal discs"; Fig. 1B; Haynie & Bryant, 1986) which fuse during the fly's metamorphosis to give rise to most of the fly head, including its major sensory organs (in addition to the CEs and OC, the antenna and the maxillary palps (the olfactory and gustatory organs, respectively) also derive from the same primordium). The eye-antennal disc is a monolayer epithelium, and in it, each organ has its own Hh source (Fig. 1B). Specifically, prior to the differentiation onset, Hh is expressed along the disc's margin in two regions: posterior, for the CE (Borod & Heberlein, 1998; Bras-Pereira, Bessa, & Casares, 2006; Dominguez & Hafen, 1997) and dorsal, for the OC (Aguilar-Hidalgo et al., 2013; Blanco, Seimiya, Pauli, Reichert, & Gehring, 2009; Dominguez-Cejudo & Casares, 2015; Royet & Finkelstein, 1996).

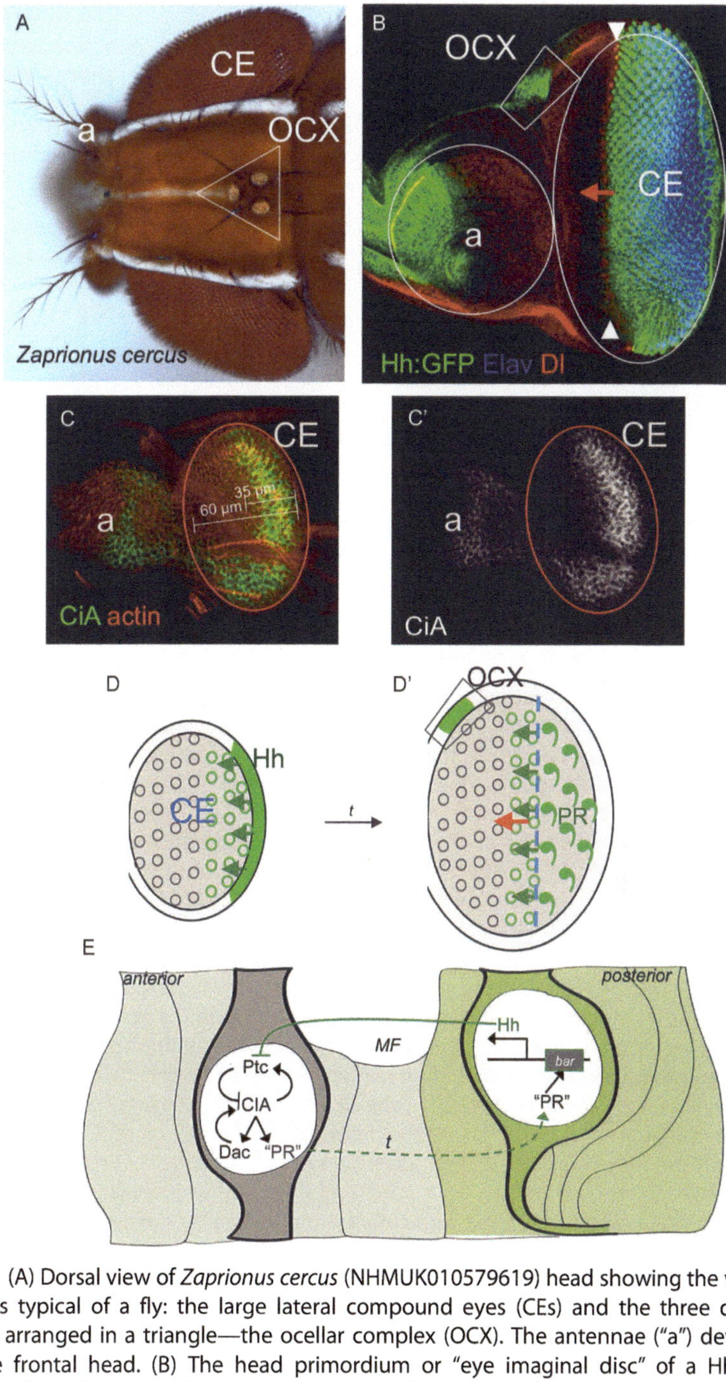

Fig. 1 (A) Dorsal view of *Zaprionus cercus* (NHMUK010579619) head showing the visual organs typical of a fly: the large lateral compound eyes (CEs) and the three dorsal ocelli, arranged in a triangle—the ocellar complex (OCX). The antennae ("a") develop in the frontal head. (B) The head primordium or "eye imaginal disc" of a Hh:GFP *Drosophila melanogaster* larva. The expression of GFP-tagged Hh is driven from a BAC *(Continued)*

This early Hh expression is necessary for the onset of retina differentiation in both eyes (Dominguez & Hafen, 1997; Ma, Zhou, Beachy, & Moses, 1993; Royet & Finkelstein, 1996). Beyond flies, the role of Hh-family morphogens in retina development is conserved in vertebrates (Neumann & Nuesslein-Volhard, 2000), which also exhibit a huge range of eye sizes in different species. Therefore, the two *Drosophila* Hh-controlled eyes, with similar genetics but very different size, represent an excellent system to explore how the Hh patterning mechanisms cope with these spatial scale differences.

2. The range of Hh action is small and constant in different organs

The onset of retina differentiation in the CE happens after the transition from the second to the third (and last) larval instar (L3). At this point, the width of the retina progenitor field is about 60 μm (Fig. 1C). The differentiation of the OC starts later, in mid-L3, and the width of the ocellar field is

Fig. 1—Cont'd that recapitulates endogenous *hh* expression. The prospective antenna, CE and OCX regions are marked. The primordium has been stained for Hh (GFP, green), the photoreceptor marker Elav (blue) and the *Notch* ligand *Delta* (*Dl*), which is used here to outline the different disc's regions. The two arrowheads mark the position of the differentiation wave-front (the so-called "morphogenetic furrow," or MF). (C,C′) Before the onset of PR differentiation, Hh produced at the posterior margin signals across a domain 30–40 μm wide, when the width of the primordium is about 60 μm. Hh signaling readout is CiA (the activator form of *cubitus interruptus, ci*). The disc has been counterstained with rhodamine phalloidin ("actin") to outline the tissue. CiA and actin signals (C) or only CiA (C′) are shown. (D) Initiation of PR differentiation depends on Hh produced at the primordium's posterior margin. Once differentiation starts (D′), PRs (inverted comas) express Hh themselves, causing a positive feedback that drives the movement of the differentiation wave across the primordium. Only when the CE differentiation is midway the differentiation of the OCX region starts. (E) Schematic representation of the Hh feedforward signaling. In the retina precursor cells Hh binds to Ptc, blocking its repressing action on the pathway. As a consequence, CiA accumulates and activates the expression of target genes. Among these targets is Ptc itself. In the CE the nuclear factor *dachshund* (Dac) is required for full pathway activation (Bras-Pereira et al., 2016). Downstream of the signaling the photoreceptor differentiation pathway ("PR") is activated. With time, this pathway is expressed and the PR cells differentiate. As part of this differentiation program, PR cells activate the expression of Hh through a retina-specific enhancer, *bar3*. The transition from precursor cells to PRs happens as cells constrict transiently (MF). The direction of differentiation is marked with a red arrow (in B, D′). See text for references.

also about 50 μm (Garcia-Morales et al., 2019). Therefore, and with a cell diameter of 4–5 μm, the Hh signaling sources have a field of 10–15 cell wide ahead of them at the onset of differentiation.

We have measured the characteristic length of the Hh gradient (the length from the source at which the gradient reaches $1/e$ of its maximal amplitude and which gives an intuitive idea of the gradient's range), using a transgenic BAC that drives expression of a GFP-tagged Hh within its normal expression domains and at functional levels (Chen, Huang, Hatori, & Kornberg, 2017). The Hh's characteristic length is about constant in different organs (11–14 μm; Fig. 2). The range of action of the signaling pathway in receiving cells of the CE, monitored by the accumulation of the activator form of the Gli TF *ci*, CiA (Alexandre, Jacinto, & Ingham, 1996; Dominguez, Brunner, Hafen, & Basler, 1996), is about 30–40 μm (Fig. 1C). These ranges are definitely too short to explain a Hh-mediated differentiation of the CE, with a final width of more than 200 μm (Vollmer et al., 2016).

3. A Hh-driven positive feedback loop induces a differentiation wave across the developing CE

In *Drosophila*, the adult CE comprises about 16,000 cells, but some fly species (including common species such as house flies, hoverflies or horse flies) harbor eyes several times larger (in cell number). It is the vigorous proliferation of the eye's progenitor cells that drives the growth of the CE (reviewed in Amore & Casares, 2010; Casares & Almudi, 2016). Hh produced from the posterior center triggers the expression of *ato* (transforming the progenitors into retina precursors) and the ensuing differentiation of the first PR cells of the retina (Dominguez, 1999). These PRs (and some accessory cells) cluster in regular cellular arrays, the future ommatidia. Within each ommatidium, of the eight PRs cells (R1–8), all but R8 express Hh (Rogers et al., 2005). Therefore, with each new row of ommatidia that is added to the retina, the front of the Hh source moves forward, recruiting further progenitors into differentiating ommatidia, so that the moving differentiating wave sweeps across the primordium until no further progenitors are left and the eye reaches its final size (Fig. 1D and E; Fried et al., 2016; Treisman, 2013). Therefore, the positive feedback loop "Hh recruits PRs that produce more Hh" makes the differentiation wave-front move forward. At the wave-front, cells transiently constrict, generating a characteristic "furrow" (the so-called "morphogenetic furrow," or "MF," as it is

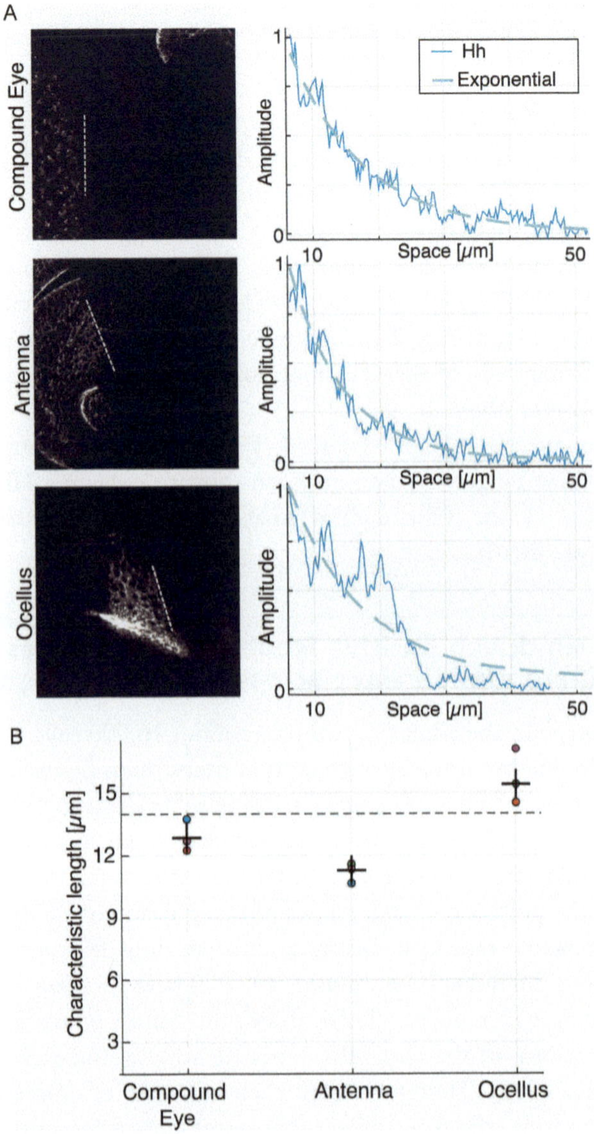

Fig. 2 Characteristic length of the Hh gradient. (A) Expression of GFP:Hh in different organs (left column) and its quantitative distribution (right). This distribution fits an exponential gradient. The characteristic length is similar in the three organs shown, including the compound eye (CE) and ocellus (OC) and about 12–15 μm.

associated to the initiation of cell differentiation and assembly into ommatidia). The regulatory mechanism enabling the positive feedback has been elucidated. A *hh* CE-specific enhancer (*hhbar3*) is activated in R1–7 PRs (Pauli, Seimiya, Blanco, & Gehring, 2005; Rogers et al., 2005) (by EGF signaling, with the Ets transcription factor Pointed (Pnt) (Rogers et al., 2005) and the Six2 transcription factor Sine-oculis (So) (Pauli et al., 2005) directly binding and activating this enhancer) (Fig. 1E). In mutants in which the *hhbar3* enhancer is deleted, only the posterior *hh* source contributes to retinogenesis and, in the absence of the self-propagating wave, the mutant eye is only a sixth of the normal one (Rogers et al., 2005).

A Hh-driven wave as a mechanism to pattern a large organ is also used during the patterning of the retina in vertebrates (Viets, Eldred, & Johnston, 2016). But in principle, there are alternative strategies for a morphogen to pattern a large organ. For example, in the case of the neural tube in amniotes, dorso-ventral (DV) patterning occurs in two phases: first, Shh (aided by anti-parallel signaling gradients of Wnt and BMP molecules) establishes a number of adjacent domains along the DV axis within its range of action, and then, once the pattern has been established, the organ grows and the pattern scales to sizes beyond Shh's range (reviewed in Kicheva & Briscoe, 2015). However, we note that these two patterning mechanisms differ substantially: while patterning and growth of the neural tube represent two functionally distinct phases in time, in the retina patterning and growth occur *simultaneously*. The patterning role of Hh in the neural tube is essentially spatial, while in the retina it is temporal. In this latter, the polarized expression of Hh links the paced recruitment of PRs to space, so that successive recruitment steps in time are also successive in space, generating a wave.

The Hh-driven wave not only allows for patterning a larger organ, but can also influence the final organ's size (Fried et al., 2016). For example, increasing the velocity of the wave (relative to the growth rate of the progenitors) may result in a smaller CE, as the wave would exhaust the progenitor pool faster. The regulation of the velocity of wave movement is controlled by Hh signaling (Fu & Baker, 2003; Greenwood & Struhl, 1999) and can be modulated by genes regulating the intensity of Hh signaling (Baker, Bhattacharya, & Firth, 2009; Bras-Pereira et al., 2016; Zhang et al., 2006). Also, the movement of the wave involves a number of processes (tissue furrowing, cell cycle synchronization, proneural gene activation, etc.), mostly mediated by Hh-regulated gene expression, that might suggest the need for a mechanism coordinating them all. However, this may not be necessary. The Hh-driven wave is built in a way that is "biochemically coupled." That is, by making the production of Hh dependent upon PR differentiation, which is itself Hh dependent, the velocity of differentiation is exactly the time that Hh-receiving cells take to differentiate and re-initiate Hh production. For example, in *dachshund* (*dac*) mutant cells Hh signaling is weakened (Bras-Pereira et al., 2016), and this results in the global slowing down of the differentiation wave, not in a desynchronization of the individual processes. This fact might allow the system to easily accommodate variations in biochemical rates, like those produced, for example, by temperature fluctuations, without affecting the process significantly.

4. Hh sets the stage for ocellus development

The three ocelli are the small eyes. After the OC complex region has been specified, a process in which Hh also plays a major role (Dominguez-Cejudo & Casares, 2015; Royet & Finkelstein, 1996), Hh is expressed in a small domain (Fig. 3A). From this domain, Hh induces the expression of retinal determination genes (Eya and its partner So/Six2) in two retina-competence domains: one posterior (p) and another anterior (a) which will give rise to one posterior ocellus (pOC) and half of the anterior ocellus (aOC), respectively (the left and right parts of the fly body derive from paired imaginal discs. The two half aOCs meet at the head midline to form the anterior single ocellus when the two discs fuse during metamorphosis) (Aguilar-Hidalgo, Casares, & Lemos, 2018; Aguilar-Hidalgo et al., 2013; Blanco et al., 2009). The aOC domain is smaller and its differentiation starts later than the pOC. Since both anterior and posterior OC depend on Hh for their development (and the aOC readily responds to increased Hh signaling by increasing dramatically its size (Fig. 3A and B)) signaling differences must exist, either in the quantity or the quality of the signal itself or in the "perception" of the signal between the anterior and posterior OCs that account for this different response. In fact, only the aOC expresses the Six3 transcription factor *optix* (Dominguez-Cejudo & Casares, 2015) indicating that, although very similar in morphology, the genetic makeup of both ocelli is distinct (Dominguez-Cejudo & Casares, 2015; Maynard-Smith & Sondhi, 1960). Still, it is not clear how this difference translates into differential Hh signaling.

5. The ocellus, the small eye, is perhaps still too large for Hh

Once the two ocellar domains have been genetically defined, Hh has the task of controlling the differentiation of these small retinas. The pOC is about 50 μm across (10–12 cells), and gives rise to a total of about 80 cells, of which 40 are photoreceptor neurons. In contrast with the CE, that comprises PR cells of several types, all OC PRs express rhodopsin 2 and

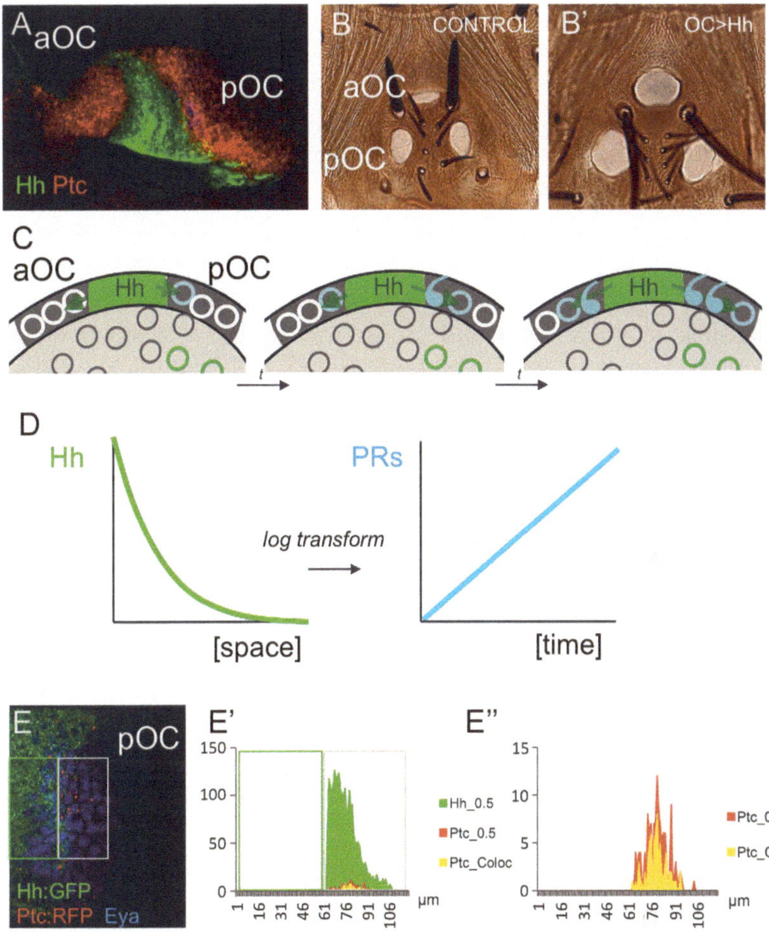

Fig. 3 Hh expression and the differentiation of the ocellus. (A) The ocellar region of the head primordium has one Hh-producing domain (green) that signals to two adjacent domains. These domains, which upregulate the expression of the Hh receptor Ptc will give rise to the posterior ocellus (pOC) and half of the anterior ocellus (aOC). The two head primordia fuse during metamorphosis to give rise to two pOC and one aOC, that results from the merging of the left and right halves (B). (B) View from the top of the ocellar region. The lenses are visible. Underneath these lenses lie the retinas. When Hh is overproduced in the OC primordia, the size of the resulting OCs is dramatically increased (B′, "OC > Hh"). (C) The differentiation of the ocellar retinas proceeds as a wave, with the differentiation of the aOC lagging behind the pOC. Upon receiving the Hh signal, OC primordium cells (white circles) activate the expression of proneural genes (cyan circles) to then differentiate (cyan comas). This differentiation proceeds as a wave, starting close to the Hh source (green) and moving away progressively. (D) A non-linearly decaying signal across the OC primordium is transformed into a differentiation wave of constant speed. Degree of colocalization of the fluorescently tagged forms of Hh (Hh:GFP, green) and Ptc (Ptc:RFP, red) in the pOC (co-stained for Eya, which marks the extent of the retina competent region, E). Most of Ptc is colocalized with Hh (represented in yellow) and the ratio of Ptc bound to Hh ("Ptc_Coloc," yellow) to total Ptc ("Ptc_0,5," red) is about constant across all the competence domain (E′,E″). In (E″) the Ptc and Ptc:Hh signal is shown.

therefore are likely of a single type (Mismer, Michael, Laverty, & Rubin, 1988). Ocellar PR differentiation has been described in detail and shown to proceed as a wave of constant speed (Garcia-Morales et al., 2019; Fig. 3C). The first PRs to differentiate are those close to the Hh source and then differentiation proceeds sequentially farther away. This is reminiscent of the CE differentiation wave. An obvious driving mechanism for this wave would be, again, that OC PRs expressed Hh as they differentiate, but this is not the case (Garcia-Morales et al., 2019). Another option is that the signaling gradient is capable of controlling the wave without self-propagation. However, the typical characteristic length of the Hh gradient at this time point is about 15 μm (Fig. 2), that is, three to four cells. This means that the cells closer to the source would be much more exposed to the signal than cells a bit farther apart. If the probability of a cell to differentiate is proportional to the level of (accumulated) Hh signaling (Dessaud et al., 2007), this would result in a non-linear pattern of differentiation—in contrast to what is observed (Fig. 3D and Garcia-Morales et al., 2019). In addition, with very sharp gradients, the effective range of action of the signaling gradient may be too short even to induce differentiation across the whole width of the ocellus. Are there ways for a Hh gradient to signal farther away?

6. Stretching and linearizing a gradient

At this point we will introduce the mathematical description of a gradient to explore mechanisms that may extend its reach. To simplify the equations, we will replace Hh, the ligand, by "u," and its receptor (mainly representing Ptc) by "v" (Box 1). In a typical morphogen gradient, the morphogen u is produced and secreted at a specific domain ("source") from which it spreads across the receiving cells (that express the morphogen receptor, v), via diffusion or other forms of passive or active transport (at a rate D_u). The morphogen also decays at a constant (k_u) rate. This scheme is often called synthesis-diffusion-degradation (SDD; Wartlick, Kicheva, & Gonzalez-Gaitan, 2009). Theoretical analysis of this scenario (reviewed in Box 1) shows that, at steady state, the gradient has an exponentially decaying profile in which its characteristic length ("λ") is proportional to (the square root of) the ratio between the effective diffusion D_u and degradation constant k_u of the morphogen (as in Eq. 9). In Fig. 2 we show that the experimentally determined Hh gradients in different organs of the developing fly fit well the exponential profile predicted by the SSD model.

BOX 1 Synthesis-diffusion-degradation theory of morphogen gradients

How morphogenetic molecules adjust to establish gradients of different length to drive patterning of tissues of different sizes is still not well understood. Here, we review the features that shape a morphogen gradient in the context of the synthesis-diffusion-degradation scenario (SDD) (Rogers & Schier, 2011; Wartlick et al., 2009), where a given molecule is produced at one specific region of the system, spreads and gets degraded at a constant rate. The Reaction-Diffusion partial differential equations that describe the dynamics of the system can be derived based on mass action kinetics with the general form:

$$\frac{\partial u(x,t)}{\partial t} = D_u \frac{\partial^2 u(x,t)}{\partial x^2} - k_u \cdot u(x,t) \tag{1}$$

The diffusion term is simply derived from Fick's second law, where D_u represents an *effective* diffusion constant. The degradation term is defined as a simple linear dependence on the morphogen concentration with rate k_u. Eq. (1) has a steady state solution $u_{ss}(x)$ where the second spatial derivative of u is proportional to itself.

$$\frac{\partial^2 u_{ss}(x)}{\partial x^2} = \frac{k_u}{D_u} u_{ss}(x) \tag{2}$$

Since exponential functions have the unique property of their derivatives being proportional to the function itself, a general solution for u takes the form:

$$u_{ss}(x) = A \cdot e^{-x/\lambda} + B \cdot e^{+x/\lambda} \tag{3}$$

The boundary conditions are used to determine the integration constants A and B. Since we are interested in a typical morphogen gradient that is being continuously produced at one edge of the system, the first boundary condition is defined as $u_{ss}(0) = u_0$.

$$u_{ss}(0) = A \cdot e^{-0/\lambda} + B \cdot e^{+0/\lambda} = A + B = u_0 \tag{4}$$

On the other hand, if the dimensions of the system are sufficiently large compared to the characteristic length λ of the exponential, we can define the concentration of the morphogen as zero at the opposite edge of the system ($u_{ss}(L) = 0$). Therefore,

$$u_{ss}(L) = A \cdot e^{-L/\lambda} + B \cdot e^{+L/\lambda} = 0 \tag{5}$$

Since the second term of Eq. (5) increases when x increases (i.e., we move away from the morphogen source), B has to be zero. This way, $A = u_0$ and the solution has the form a single exponential:

$$u_{ss}(x) = u_0 \cdot e^{-x/\lambda} \tag{6}$$

Continued

BOX 1 Synthesis-diffusion-degradation theory of morphogen gradients—cont'd

To study the explicit dependence of the characteristic length λ on the parameters of the reaction, we take the second derivative of Eq. (6):

$$\frac{\partial^2 u_{ss}(x)}{\partial^2 x} = \frac{1}{\lambda^2}\left(u_0 \cdot e^{-x/\lambda}\right) = \frac{u_{ss}}{\lambda^2} \tag{7}$$

This expression is then substituted into the steady state solution in Eq. (2)

$$\frac{u_{ss}(x)}{\lambda^2} = \frac{k_u}{D_u} u_{ss}(x) \tag{8}$$

Therefore, λ depends directly on the diffusion and degradation of the morphogen as

$$\lambda = \sqrt{\frac{D_u}{k_u}} \tag{9}$$

The previous equation represents the analytical solution to the synthesis-diffusion-degradation (SDD) interaction scheme, and shows that the distance that the morphogen travels results from the balance between its effective diffusion coefficient and its degradation rate.

To generate a Hh-driven differentiation wave, the simplest solution would be to couple cell differentiation to the increase in the characteristic length λ that the gradient experiences as it evolves toward the steady state. Using experimentally determined values for D_u and k_u (Fried et al., 2016) and solving Eq. (1) numerically, the time-evolution of the gradient shows that the dynamics of λ is fast (lines in Fig. 4A) and the steady state is reached in about 30 min (dots in Fig. 4A). This dynamics is much faster than the actual OC differentiation, which occurs in the order of days (Garcia-Morales et al., 2019), indicating that coupling between diffusion/degradation and cell differentiation is unlikely as a wave-generating mechanism.

Another strategy would be to modify directly some of the three major parameters of the SDD: the rates of morphogen synthesis, diffusion and degradation. The concentration of ligand calculated by SDD with measured D_u and k_u (Fried et al., 2016), and amplitude at the source "u_0" for Hh (Garcia-Morales et al., 2019), falls below 10% of its maximal value well before the 50 μm width of the ocellus. If we use this 10% as an arbitrary minimum for an effective Hh signaling, can changes in synthesis, diffusion or degradation during OC differentiation make the gradient reach farther?

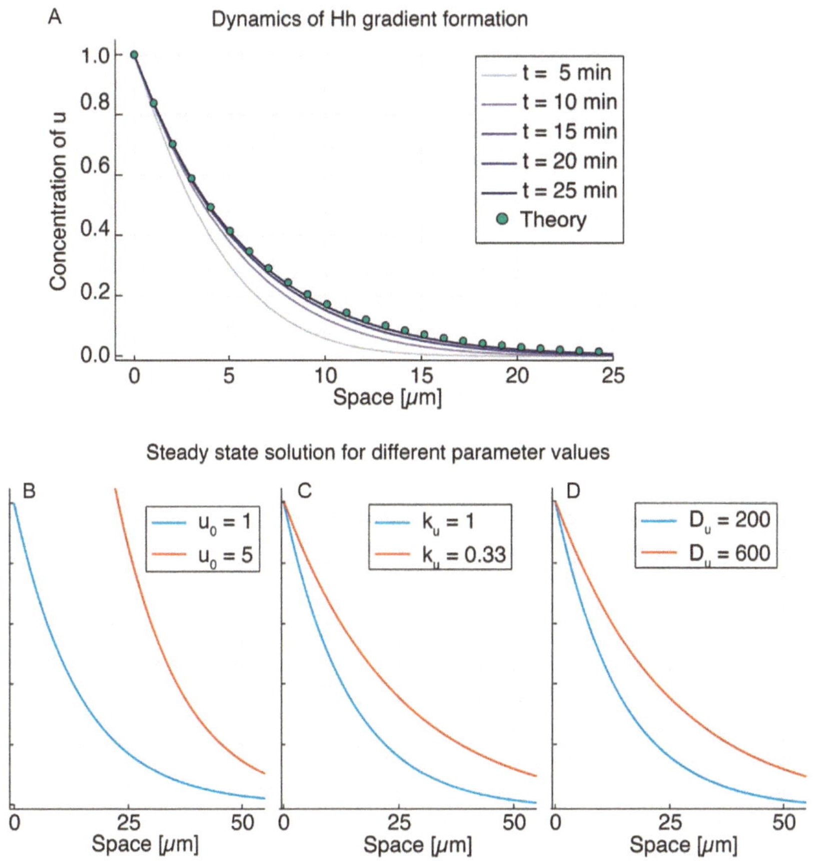

Fig. 4 Dynamics of Hh gradient formation. (A) Time evolution of the gradient profile in the SDD approach (reviewed in Box 1) using values of the effective diffusion ($D_u = 118\,\mu m^2 \times h^{-1}$) and effective degradation ($k_u = 3{,}6\,h^{-1}$) obtained from Fried et al. (2016). Each line corresponds to the profile every 5 min. The steady state profile predicted for these parameter values is plotted as dots. (B–D) Changes in the steady state profile when different parameters are modified to obtain an amount of *u* at the end of the region of at least a 10% value of the amplitude at the source: (B) rate of production, (C) effective diffusion or (D) effective degradation. Other parameters are chosen to fit the experimental gradients measured in Fig. 2.

In Fig. 4B we show that, to reach the end of the ocellus with concentrations of morphogen above 10%, the concentration of Hh at the source should increase at least five times. This is a large increase. And even if this was the strategy used in other organs, the measured trend in the ocellus is the opposite, with Hh maximal amplitude decreasing over developmental time (Garcia-Morales et al., 2019).

Modulation of the diffusion and/or degradation rates is yet another possibility. In fact, it has been shown that changes in the composition of the extracellular matrix proteoglycans, or in the morphogen degradation rates during development affect the signaling gradient of Hh, Wg/Wnt-1 or Dpp/BMP-2 (Akiyama et al., 2008; Neto, Aguilar-Hidalgo, & Casares, 2016; Rogers & Schier, 2011; Wartlick et al., 2009, 2011). We explored the impact of these variations (Fig. 4C and D): obtaining Hh concentrations above 10% maximal across the whole ocellus requires variations in k_u or D_u of at least three times. These are again large changes in rate constants.

So far, we have not made explicit a major biochemical step that shapes morphogen gradients: the interaction of the ligand with its receptor. This step has been shown to be of key relevance in the establishment of the Hh gradient, as the interaction with its receptor Ptc restricts Hh's mobility and availability (Briscoe, Chen, Jessell, & Struhl, 2001; Chen & Struhl, 1996). In Box 2, we expand the theoretical analysis of the SDD model by including the binding/unbinding of u and v and the clearance/inactivation of the ligand:receptor complex uv (by, for example, endocytosis), and shown schematically in Fig. 5A.

After making some basic assumptions, the system of equations can be reduced and solved analytically, and the resulting morphogen profile is again an exponentially decaying gradient (Eq. 24). Now, the characteristic length depends on the parameters describing the ligand:receptor interaction and, more importantly, on the relative amounts of receptors available for binding (Eq. 24). Fig. 5B shows the steady state solution of Eq. (23) for different receptor concentrations v_0, keeping all other variables unchanged. If the concentration of free receptor is reduced, the length of the gradient increases significantly (Fig. 5B).

For instance, a simple scenario of reduced free receptor is illustrated (Fig. 5C–E) using numerical simulations for the interactions between ligand and receptor in a spatial system. In the absence of saturation (Fig. 5C), the gradient is short and reaches a steady state in just 5 hours. However, when the amount of available receptor is limited by a strong and stable interaction with the ligand (Fig. 5D), the gradient moves further away from the source. This saturation effect is best observed in Fig. 5E, that shows that the majority of receptors are bound to u forming a complex uv, allowing the further spread of the ligand u.

To analyze whether this receptor saturation is seen *in vivo*, we measured the colocalization of Hh and Ptc in the ocelli, in larvae expressing fluorescently tagged Hh (GFP:Hh) and Ptc (RFP:Ptc) at endogenous levels

BOX 2 Synthesis-diffusion-degradation-clearance theory of morphogen gradients

The interaction of the diffusive ligand with its receptor can also affect the shape and reach of a morphogen gradient. This system can be also described in terms of partial differential equations based on mass action kinetics:

$$\frac{\partial u(x,t)}{\partial t} = D_u \frac{\partial^2 u(x,t)}{\partial x^2} - k_u \cdot u(x,t) + k_{off} \cdot uv(x,t) - k_{on} \cdot u(x,t) \cdot u(x,t) \quad (10)$$

$$\frac{\partial uv(x,t)}{\partial t} = D_{uv} \frac{\partial^2 uv(x,t)}{\partial x^2} - k_{uv} \cdot uv(x,t) - k_{off} \cdot uv(x,t) + k_{on} \cdot u(x,t) \cdot v(x,t) \quad (11)$$

$$\frac{\partial v(x,t)}{\partial t} = D_v \frac{\partial^2 v(x,t)}{\partial x^2} - k_v \cdot v(x,t) + k_{off} \cdot uv(x,t) - k_{on} \cdot u(x,t) \cdot v(x,t) \quad (12)$$

where u, v and uv refer to the concentration of ligand, receptor and ligand-receptor complex. Constants k_{on} and k_{off} correspond to the affinity and dissociation rates of the interaction between ligand and receptor. Similarly, D_u, D_v and D_{uv} are the diffusion constants, and k_u, k_v and k_{uv} are the clearance rate constants of u, v and uv. The condition for the conservation of mass of the morphogen is:

$$u_{total}(x,t) = u(x,t) + uv(x,t) \quad (13)$$

and the derivative of the last expression is simply:

$$\frac{\partial u_{total}(x,t)}{\partial t} = \frac{\partial u(x,t)}{\partial t} + \frac{\partial uv(x,t)}{\partial t} \quad (14)$$

The change of u_{total} is now written in terms of the change of the two forms of u:

$$\frac{\partial u_{total}(x,t)}{\partial t} = Du \frac{\partial^2 u(x,t)}{\partial x^2} - k_u \cdot u(x,t) + D_{uv} \frac{\partial^2 uv(x,t)}{\partial x^2} - k_{uv} \cdot uv(x,t) \quad (15)$$

As a first approximation, we will assume that the kinetics of binding and dissociation are fast (normally in the order of seconds) compared to the diffusion and degradation rates of the morphogen (normally in the order of minutes). This allows us to separate the two time scales and assume that there is a local equilibrium at all times between u and uv. In this condition, the ratio between the two forms of the ligand at equilibrium is simply:

$$uv = \frac{k_{on}}{k_{off}} u \cdot v = \frac{u \cdot v}{k_d} \quad (16)$$

where k_d is defined as the dissociation constant of the interaction between ligand and its receptor. Using this approximation into Eq. (14) we obtain:

$$\frac{\partial u_{total}(x,t)}{\partial t} = \frac{\partial u(x,t)}{\partial t} + \frac{1}{k_d} \frac{\partial (u(x,t) \cdot v(x,t))}{\partial t} \quad (17)$$

Continued

BOX 2 Synthesis-diffusion-degradation-clearance theory of morphogen gradients—cont'd

If we assume that only a small amount of the total receptors in the system is bound to a ligand, the concentration of v can be set as constant ($v(x,t) = v_0$), and therefore, it can be taken out of the derivative.

$$\frac{\partial u_{total}(x,t)}{\partial t} = \frac{k_d + v_0}{k_d}\frac{\partial u(x,t)}{\partial t} \tag{18}$$

Next, we assume that the diffusion of the receptor v and the ligand-receptor complex uv are much slower than the diffusion rate of the free ligand u (therefore, $D_v = D_{uv} = 0$). Taking this into account, and substituting Eq. (18) into Eq. (15), we obtain:

$$\frac{k_d + v_0}{k_d}\frac{\partial u(x,t)}{\partial t} = D_u\frac{\partial^2 u(x,t)}{\partial x^2} - k_u \cdot u(x,t) - k_{uv} \cdot \frac{u(x,t) \cdot v_0}{k_d} \tag{19}$$

and rearranging terms becomes:

$$\frac{\partial u(x,t)}{\partial t} = \frac{D_u \cdot k_d}{k_d + v_0}\frac{\partial^2 u(x,t)}{\partial x^2} - \frac{k_d}{k_d + v_0}\left(k_u + \frac{k_{uv} \cdot v_0}{k_d}\right) \cdot u(x,t) \tag{20}$$

Due to similitude with Eq. (1), we define the effective diffusion and degradation rate constants as:

$$D_{eff} = \frac{D_u \cdot k_d}{k_d + v_0} \tag{21}$$

$$k_{eff} = \frac{k_d}{k_d + v_0}\left(k_u + \frac{k_{uv} \cdot v_0}{k_d}\right) \tag{22}$$

Therefore, Eq. (20) can be rewritten simply as:

$$\frac{\partial u(x,t)}{\partial t} = D_{eff}\frac{\partial^2 u(x,t)}{\partial x^2} - k_{eff} \cdot u(x,t) \tag{23}$$

which can be solved analytically following the same strategy as Eq. (1) with the same solution (Eq. 6). The characteristic length of the exponential profile is given by Eq. (9), so substituting the expressions for effective diffusion and degradation rates (Eqs. 21 and 22), we obtain

$$\lambda = \sqrt{\frac{D_{eff}}{k_{eff}}} = \sqrt{\frac{D_u}{\left(k_u + \frac{k_{uv} \cdot v_0}{k_d}\right)}} \tag{24}$$

This equation shows that now the average distance traveled by a morphogen ligand increases when the concentration of its receptor decreases.

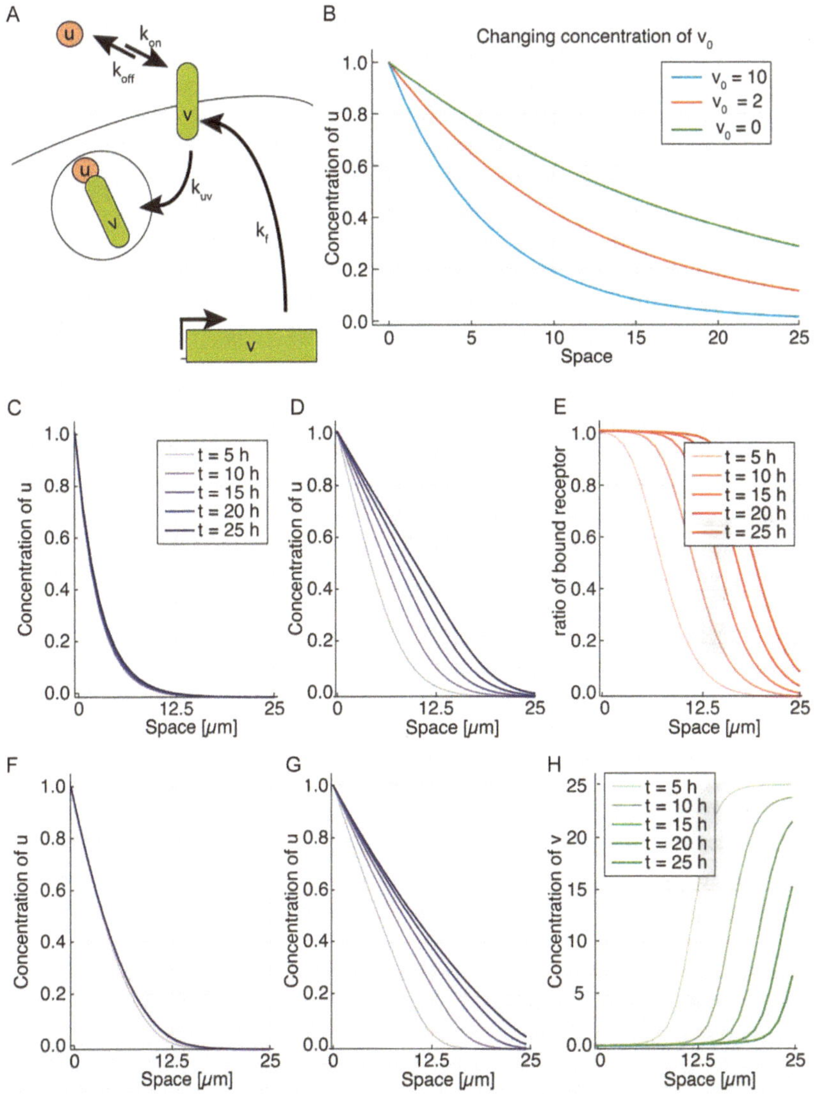

Fig. 5 Reduction in the concentration of receptor extends the characteristic length of the gradient. (A) Scheme of the interaction between ligand (*u*) and receptor (*v*) molecules, with the active complex *uv* being internalized and cleared. (B) Theoretical prediction of the changes in the gradient with the value of receptor v_0. (C–E) Comparison of gradient dynamics in conditions of (C) no receptor saturation and (D) receptor saturation. (E) Ratio of bound *uv* versus unbound *v* in conditions of receptor saturation. (F–H) Comparison of gradient dynamics in conditions of fast (F) and slow (G) rate of *v* production. (H) Spatial profile of free receptor *v* at different time points in conditions of slow rate of production.

(Chen et al., 2017). The results in Fig. 3E show that a large fraction of Ptc colocalizes with Hh. More importantly, this percentage is maintained all along the gradient (i.e., in both high and low Hh concentrations), suggesting that indeed the receptor Ptc is saturated throughout the tissue. Another important observation is that now the predicted dynamics of the gradient spreading in much slower and closer to the timing of ocellar differentiation (about 50 h).

A side effect of the receptor saturation is that the gradient changes from exponential to a more linear profile (Fig. 5D). This change is consistent with experimental observations in the Drosophila ocelli (Garcia-Morales et al., 2019), where the Hh profile becomes more linear as differentiation takes place.

The same effective reduction of available free receptor can be achieved if the rate of receptor transcription is much slower relative to the clearance rate of the bound receptor, which can be safely assumed to be the case. This scenario was simulated in Fig. 5F–H, where we compare the profile dynamics when the transcription and clearance rates are of the same scale (Fig. 5F) with when the production is slower (similar equilibrium concentration of v: Fig. 5G). In the latter, again the gradient extends farther and becomes more linear. The progressive reduction of free receptor is clearly shown in Fig. 5H as the ligand spreads in space. Comparison between the lines that illustrate the gradient at different time points (Fig. 5D and G), shows that the profile keeps advancing at a progressively decreasing speed. Therefore, the same 10% threshold for the morphogen concentration used above would be eventually reached if the simulation were allowed to run for a longer time.

7. When a negative feedback "log transforms" the gradient

It has recently been shown that, indeed, as the ocellar precursor cells differentiate as PRs, the Ptc receptor is transcriptionally repressed (Garcia-Morales et al., 2019). Cells closest to the source are the first to differentiate and to downregulate Ptc. As a consequence, these PRs cells no longer bind and consume Hh, and so the free morphogen travels further, triggering differentiation in adjacent cells, with the process propagating away from the source. To illustrate the effect of this feedback repression of the receptor v, we define a set of coupled partial differential equations for ligand u, receptor v and ligand:receptor complex uv, where the expression of v is modulated by the active uv complex following a Hill function.

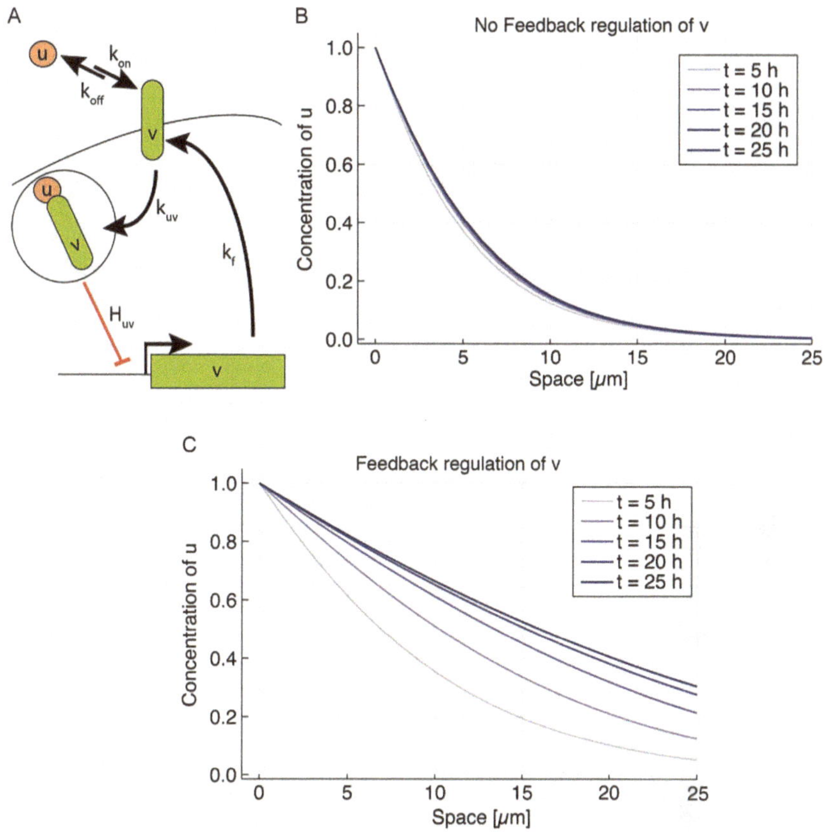

Fig. 6 Feedback loop in the concentration of receptor shapes the gradient dynamics. (A) Scheme of the interaction between ligand and receptor molecules, with the active complex *uv* being internalized and driving repression of the production of free receptor. (B) Numerical simulations of the gradient profile with no feedback interaction. (C) Numerical simulations of the system with the same parameter values, but including feedback repression of the receptor *v*.

Again, only free u is allowed to disperse and is degraded at a rate k_u. In addition, u binds v and the uv complex is cleared with a constant k_{uv}. A scheme showing these interactions is shown in Fig. 6A. Numerical solution of the equations either in the absence (Fig. 6B) or in the presence (Fig. 6C) of the negative feedback is shown for different time points.

The introduction of the feedback produces three major changes in the gradient dynamics: its characteristic length is longer (i.e., the gradient stretches farther); the steady state is reached more slowly due to the dynamic coupling between differentiation time (that represses v) and the gradient expansion; and the shape of the gradient is changed—no longer exponential but approaching linearity. A more complete two-dimensional model has been presented in Garcia-Morales et al. (2019).

8. A static morphogen source with a dynamic signaling works like a developmental metronome

By making the regulation of Hh depend on PR differentiation, the gradient experiences a dynamic change that occurs at the same rate as that of cell differentiation. Because the resulting dynamics linearizes the signaling output in space, Hh-induced PR differentiation occurs at a constant pace. In addition, colocalization studies suggest that the receptor Ptc is likely saturated (Fig. 3), which can also contribute to the linearization of the gradient, as shown in the previous sections (Fig. 5). This linearization is equivalent to a mathematical "log-transform." The morphogen gradient, this time, is not used to specify different cell types in space, as the OC retinas are composed by a single PR type all throughout, but to generate a wave of differentiation (Garcia-Morales et al., 2019). Rather than specifying space, Hh "marks a pace." It is important to note that even though the regulation of receptor expression occurs at the single-cell level, the "log-transform" operation emerges across a field of cells.

But how is this system expected to perform in the face of the inescapable biochemical noise? One potential problem might arise if the receptor Ptc had low occupancy, as this would increase binding noise (Lander, 2013). However, as mentioned above, Ptc is likely saturated along the whole signaling range, which would minimize this sensitivity. Ptc saturation, though, could in turn increase the sensitivity of the system to fluctuations in Hh production rate, unless signal measurements were carried out before the gradient reaches steady state (Lander, 2013). In fact, and as we mentioned above, the Hh gradient is not in steady state in the OC (Garcia-Morales et al., 2019). Then, the downregulation of the Ptc receptor as cells differentiate, generates an internal biochemical clock that, when linked to the dynamic Hh gradient, results in an increased resistance to fluctuations in Hh concentration (Garcia-Morales et al., 2019).

9. Intrinsic constraints to the variation of ocellar size imposed by the gradient

The patterning of the small eyes of flies is controlled by a dynamic (in space and time) gradient of Hh produced from a spatially fixed source. We have argued above that the reach of such a gradient is limited, unless some of its biochemical constants or the availability of the receptor change dramatically. Do these limitations set a maximum size for the ocelli? The range of sizes between different fly species is enormous, spanning two orders of magnitude in body length: from 0,5 mm of some Phorid flies to the 70 mm of *Gauromydas heros*. If all these flies used a Hh gradient to control the size and pattern of their ocelli similar to the one described in *Drosophila*, we would expect that the size of their ocelli would not scale perfectly with body size—that is, ocelli should be disproportionally smaller in large flies when compared to smaller species. This seems to be the case. In Fig. 7A the wing length (a correlate of body size) and the ocellar size

Fig. 7 Ocellar size in different species in relation to body size. (A) A large fly (top, *Episyrphus balteatus*) and a small fly (bottom, *Drosophila melanogaster*). The ratio of their wing length is indicated (*wl;* the wing length is a correlate of body size). Microscopic images of the adult ocelli of *Episyrphus* (A′) and *Drosophila* (A″) at the same magnification. The relative width of their posterior ocelli (*ocl*) is indicated. (B, C) Ocelli of the fly *Senotainia tricuspis* (NHMUK010579737) (Sarcophagidae) (B) and the wasp *Dusona bicoloripes* (NHMUK010579721) (Ichneumonidae) (C) at the same magnification, showing the much larger ocelli of the wasp.

(measured as the length of the major axis of the elliptical lens of the ocellus) of a small (*Drosophila melanogaster*) and a large fly (*Episyrphus balteatus*) are compared. While *Episyrphus* wings are four times longer, its ocelli are only ~1,6 times the width of *Drosophila*'s. Of course, this limit in ocellar size might be the result of natural selection: above a certain size, there might not be a selective advantage in having larger ocelli. However, very large ocelli are present in other insect groups, such as Odonata (dragonflies and damselflies) and Hymenoptera (wasps and bees) (Berry, Stange, & Warrant, 2007; Ribi, Warrant, & Zeil, 2011) (see also Fig. 7B and C). Therefore, it is possible that, indeed, ocellar size in flies is limited by the effective range of Hh action, as suggested previously (Aguilar-Hidalgo, Becerra-Alonso, Garcia-Morales, & Casares, 2016; Aguilar-Hidalgo et al., 2013). Ocelli larger than those of Diptera might need different mechanisms, although surely based on Hh, to be patterned.

10. Concluding remarks

In this review, and using as motivation the role played by Hh in the differentiation of the *Drosophila* eyes, we have explored how a morphogen gradient can stretch itself to pattern tissues of different size. In *Drosophila*, when the eye is larger than a certain size, new mechanisms come into place to overcome these limitations. Finally, we note that PR differentiation in both the small and the large eyes of flies proceeds at a constant pace, despite their using different regulatory mechanisms to achieve this linearity. In the case of the OC, perhaps the functional need of generating sufficiently large retinas resulted in the selection of subtle changes in the regulatory network that allowed the extension of the gradient reach without the need of more dramatic or pleiotropic modifications affecting, for example, the diffusivity or degradation of the morphogen itself. Here, we propose three different strategies: saturation, slow transcription (relative to diffusion) or downregulation of the receptor. In the particular case of the OC, we observe experimental evidence of the three, so maybe the system uses a combination of these strategies. As we have shown, this extension brings about immediately the linearization of the gradient which, in turn, can be used to induce differentiation at a constant pace. The CE, being larger, requires the introduction of at least one extra link within the PR regulatory network: the production of Hh itself by PRs. Still, a detailed study of Hh distribution and Ptc dynamics anterior to the differentiation wave in the CE is lacking, so it is not yet known whether Hh signaling in the CE is further modified using some of the strategies used in the OC.

In any case, why should PR differentiation proceed at a constant pace? Recent studies show how signal transduction through highly non-linear biochemical pathways often results in linear outputs (Goentoro & Kirschner, 2009; Nunns & Goentoro, 2018). This linearity in signal transduction increases the fidelity of the signaling. We suggest that linearity of patterning processes allows the uniform deployment of these processes throughout development, allowing an easier control. For example, the control of a process whose rate varies with developmental time may require also varying the intensity of the control mechanism as time passes (that is, *developmental time* becomes an additional variable), while a linear process may be controlled even when the perturbation has already happened. Also, a linear process is less prone to amplify or attenuate changes in the input signal. Finally, the coupling or coordination between developmental processes might be facilitated if these processes develop at constant speeds.

11. Materials and methods

Codes "NHMUK" are specimen codes of the Natural History Museum, London (UK) collection. The results presented in Figs. 1–3 were obtained using methods described in Garcia-Morales et al. (2019), except for colocalization studies of Hh:GFP and Ptc:GFP. For this analysis, eye-antennal discs of Hh:GFP/CyO; Ptc:RFP larvae (Chen et al., 2017) were dissected and stained with anti-GFP, anti-RFP and anti-Eya (this latter to label the OC-competent region) as in Garcia-Morales et al. (2019) and imaged under a Zeiss LSM880 Airyscan. Images were processed with Softworx Suite 2.0 (Applied Precision) and analyzed with Imaris (Bitplane). Punctate signal ("vesicles") in the red and green channels was selected with the function "slice." Only vesicles with diameters equal or larger than 0,5 μm were considered in order to minimize noise inclusion. The "spots" function was used to select the vesicles of the defined diameter and the "spots colocalize" function to identify those where the two signals overlapped. The three signals (green, red and yellow ("colocalized")) were represented in 3D with the "vantage plot" function and collapsed into 2D for data representation using bins of 0,05 μm. The data shown here is representative of six samples.

Acknowledgments and Funding

F.C. is funded through grants BFU2015-66040-P, PGC2018-093704-B-I00 and MDM-2016-0687, and D.G.M. through grants BFU2014-53299-P and RTI2018-096953-B-I00, all from Ministerio de Ciencia, Innovación y Universidades of Spain. A SYNTHESYS grant (GB-TAF 5341) allowed F.C. to visit the Natural History Museum (London, UK) to study and image the ocellar complex across a large sample of Diptera species (and some hymenoptera). F.C. thanks the collaboration of Daniel Whitmore and the help of Vladimir Blagoderov and Benjamin Price.

References

Aguilar-Hidalgo, D., Becerra-Alonso, D., Garcia-Morales, D., & Casares, F. (2016). Toward a study of gene regulatory constraints to morphological evolution of the Drosophila ocellar region. *Development Genes and Evolution*, *226*, 221–233.

Aguilar-Hidalgo, D., Casares, F., & Lemos, M. C. (2018). Patterning, dynamics and evolution in the ocellar complex of the fruit fly. In J. F. R. Archilla, F. Palmero, M. C. Lemos, B. Sanchez-Rey, & J. Casado-Pascual (Eds.), *Nonlinear systems. Nonlinear phenomena in biology, optics and condensed matter* (pp. 39–62). Cham, Switzerland: Springer International Publishing.

Aguilar-Hidalgo, D., Dominguez-Cejudo, M. A., Amore, G., Brockmann, A., Lemos, M. C., Cordoba, A., et al. (2013). A Hh-driven gene network controls specification, pattern and size of the Drosophila simple eyes. *Development*, *140*, 82–92.

Akiyama, T., Kamimura, K., Firkus, C., Takeo, S., Shimmi, O., & Nakato, H. (2008). Dally regulates Dpp morphogen gradient formation by stabilizing Dpp on the cell surface. *Developmental Biology*, *313*, 408–419.

Alexandre, C., Jacinto, A., & Ingham, P. W. (1996). Transcriptional activation of hedgehog target genes in Drosophila is mediated directly by the cubitus interruptus protein, a member of the GLI family of zinc finger DNA-binding proteins. *Genes & Development*, *10*, 2003–2013.

Amore, G., & Casares, F. (2010). Size matters: The contribution of cell proliferation to the progression of the specification Drosophila eye gene regulatory network. *Developmental Biology*, *344*, 569–577.

Baker, N. E., Bhattacharya, A., & Firth, L. C. (2009). Regulation of Hh signal transduction as Drosophila eye differentiation progresses. *Developmental Biology*, *335*, 356–366.

Berry, R. P., Stange, G., & Warrant, E. J. (2007). Form vision in the insect dorsal ocelli: An anatomical and optical analysis of the dragonfly median ocellus. *Vision Research*, *47*, 1394–1409.

Blanco, J., Pauli, T., Seimiya, M., Udolph, G., & Gehring, W. J. (2010). Genetic interactions of eyes absent, twin of eyeless and orthodenticle regulate sine oculis expression during ocellar development in Drosophila. *Developmental Biology*, *344*, 1088–1099.

Blanco, J., Seimiya, M., Pauli, T., Reichert, H., & Gehring, W. J. (2009). Wingless and Hedgehog signaling pathways regulate orthodenticle and eyes absent during ocelli development in Drosophila. *Developmental Biology*, *329*, 104–115.

Borod, E. R., & Heberlein, U. (1998). Mutual regulation of decapentaplegic and hedgehog during the initiation of differentiation in the Drosophila retina. *Developmental Biology*, *197*, 187–197.

Bras-Pereira, C., Bessa, J., & Casares, F. (2006). Odd-skipped genes specify the signaling center that triggers retinogenesis in Drosophila. *Development, 133*, 4145–4149.

Bras-Pereira, C., Potier, D., Jacobs, J., Aerts, S., Casares, F., & Janody, F. (2016). dachshund potentiates hedgehog signaling during Drosophila retinogenesis. *PLoS Genetics, 12*, e1006204.

Briscoe, J., Chen, Y., Jessell, T. M., & Struhl, G. (2001). A hedgehog-insensitive form of patched provides evidence for direct long-range morphogen activity of sonic hedgehog in the neural tube. *Molecular Cell, 7*, 1279–1291.

Brockmann, A., Dominguez-Cejudo, M. A., Amore, G., & Casares, F. (2011). Regulation of ocellar specification and size by twin of eyeless and homothorax. *Developmental Dynamics, 240*, 75–85.

Casares, F., & Almudi, I. (2016). Fast and furious 800. The retinal determination gene network in Drosophila. In J. Castelli-Gair Hombría & P. Bovolenta (Eds.), *Organogenetic gene networks* (pp. 95–124). Cham: Springer.

Chen, W., Huang, H., Hatori, R., & Kornberg, T. B. (2017). Essential basal cytonemes take up Hedgehog in the Drosophila wing imaginal disc. *Development, 144*, 3134–3144.

Chen, Y., & Struhl, G. (1996). Dual roles for patched in sequestering and transducing Hedgehog. *Cell, 87*, 553–563.

Dessaud, E., Yang, L. L., Hill, K., Cox, B., Ulloa, F., Ribeiro, A., et al. (2007). Interpretation of the sonic hedgehog morphogen gradient by a temporal adaptation mechanism. *Nature, 450*, 717–720.

Dominguez, M. (1999). Dual role for Hedgehog in the regulation of the proneural gene atonal during ommatidia development. *Development, 126*, 2345–2353.

Dominguez, M., Brunner, M., Hafen, E., & Basler, K. (1996). Sending and receiving the hedgehog signal: Control by the Drosophila Gli protein Cubitus interruptus. *Science, 272*, 1621–1625.

Dominguez, M., & Hafen, E. (1997). Hedgehog directly controls initiation and propagation of retinal differentiation in the Drosophila eye. *Genes & Development, 11*, 3254–3264.

Dominguez-Cejudo, M. A., & Casares, F. (2015). Anteroposterior patterning of Drosophila ocelli requires an anti-repressor mechanism within the hh pathway mediated by the Six3 gene Optix. *Development, 142*, 2801–2809.

Fried, P., Sanchez-Aragon, M., Aguilar-Hidalgo, D., Lehtinen, B., Casares, F., & Iber, D. (2016). A model of the spatio-temporal dynamics of Drosophila eye disc development. *PLoS Computational Biology, 12*, e1005052.

Friedrich, M. (2006). Ancient mechanisms of visual sense organ development based on comparison of the gene networks controlling larval eye, ocellus, and compound eye specification in Drosophila. *Arthropod Structure & Development, 35*, 357–378.

Fu, W., & Baker, N. E. (2003). Deciphering synergistic and redundant roles of Hedgehog, Decapentaplegic and Delta that drive the wave of differentiation in Drosophila eye development. *Development, 130*, 5229–5239.

Garcia-Morales, D., Navarro, T., Iannini, A., Pereira, P. S., Miguez, D. G., & Casares, F. (2019). Dynamic Hh signalling can generate temporal information during tissue patterning. *Development, 146*, dev176933.

Goentoro, L., & Kirschner, M. W. (2009). Evidence that fold-change, and not absolute level, of beta-catenin dictates Wnt signaling. *Molecular Cell, 36*, 872–884.

Greenwood, S., & Struhl, G. (1999). Progression of the morphogenetic furrow in the Drosophila eye: The roles of Hedgehog, Decapentaplegic and the Raf pathway. *Development, 126*, 5795–5808.

Haynie, J. L., & Bryant, P. J. (1986). Development of the eye-antenna imaginal disc and morphogenesis of the adult head in Drosophila melanogaster. *The Journal of Experimental Zoology, 237*, 293–308.

Kicheva, A., & Briscoe, J. (2015). Developmental pattern formation in phases. *Trends in Cell Biology*, *25*, 579–591.

Krapp, H. G. (2009). Ocelli. *Current Biology*, *19*, R435–R437.

Lander, A. D. (2013). How cells know where they are. *Science*, *339*, 923–927.

Ma, C., Zhou, Y., Beachy, P. A., & Moses, K. (1993). The segment polarity gene hedgehog is required for progression of the morphogenetic furrow in the developing Drosophila eye. *Cell*, *75*, 927–938.

Maynard-Smith, J., & Sondhi, K. C. (1960). The genetics of a pattern. *Genetics*, *45*, 1039–1050.

Mismer, D., Michael, W. M., Laverty, T. R., & Rubin, G. M. (1988). Analysis of the promoter of the Rh2 opsin gene in Drosophila melanogaster. *Genetics*, *120*, 173–180.

Neto, M., Aguilar-Hidalgo, D., & Casares, F. (2016). Increased avidity for Dpp/BMP2 maintains the proliferation of progenitors-like cells in the Drosophila eye. *Developmental Biology*, *418*, 98–107.

Neumann, C. J., & Nuesslein-Volhard, C. (2000). Patterning of the zebrafish retina by a wave of sonic hedgehog activity. *Science*, *289*, 2137–2139.

Nunns, H., & Goentoro, L. (2018). Signaling pathways as linear transmitters. *eLife*, 7, e33617.

Pauli, T., Seimiya, M., Blanco, J., & Gehring, W. J. (2005). Identification of functional sine oculis motifs in the autoregulatory element of its own gene, in the eyeless enhancer and in the signalling gene hedgehog. *Development*, *132*, 2771–2782.

Ribi, W., Warrant, E., & Zeil, J. (2011). The organization of honeybee ocelli: Regional specializations and rhabdom arrangements. *Arthropod Structure & Development*, *40*, 509–520.

Rogers, E. M., Brennan, C. A., Mortimer, N. T., Cook, S., Morris, A. R., & Moses, K. (2005). Pointed regulates an eye-specific transcriptional enhancer in the Drosophila hedgehog gene, which is required for the movement of the morphogenetic furrow. *Development*, *132*, 4833–4843.

Rogers, K. W., & Schier, A. F. (2011). Morphogen gradients: From generation to interpretation. *Annual Review of Cell and Developmental Biology*, *27*, 377–407.

Royet, J., & Finkelstein, R. (1996). hedgehog, wingless and orthodenticle specify adult head development in Drosophila. *Development*, *122*, 1849–1858.

Treisman, J. E. (2013). Retinal differentiation in Drosophila. *Wiley Interdisciplinary Reviews: Developmental Biology*, *2*, 545–557.

Viets, K., Eldred, K., & Johnston, R. J., Jr. (2016). Mechanisms of photoreceptor patterning in vertebrates and invertebrates. *Trends in Genetics*, *32*, 638–659.

Vollmer, J., Fried, P., Sanchez-Aragon, M., Lopes, C. S., Casares, F., & Iber, D. (2016). A quantitative analysis of growth control in the Drosophila eye disc. *Development*, *143*, 1482–1490.

Wartlick, O., Kicheva, A., & Gonzalez-Gaitan, M. (2009). Morphogen gradient formation. *Cold Spring Harbor Perspectives in Biology*, *1*, a001255.

Wartlick, O., Mumcu, P., Kicheva, A., Bittig, T., Seum, C., Julicher, F., et al. (2011). Dynamics of Dpp signaling and proliferation control. *Science*, *331*, 1154–1159.

Zhang, Q., Zhang, L., Wang, B., Ou, C. Y., Chien, C. T., & Jiang, J. (2006). A hedgehog-induced BTB protein modulates hedgehog signaling by degrading Ci/Gli transcription factor. *Developmental Cell*, *10*, 719–729.

CHAPTER ELEVEN

Genetic mechanisms controlling anterior expansion of the central nervous system

Behzad Yaghmaeian Salmani[a], **Stefan Thor**[b,*]
[a]Department of Cell and Molecular Biology, Karolinska Institute, Stockholm, Sweden
[b]School of Biomedical Sciences, University of Queensland, Saint Lucia, QLD, Australia
*Corresponding author: e-mail address: s.thor@uq.edu.au

Contents

Current Topics in Developmental Biology, Volume 137
ISSN 0070-2153
https://doi.org/10.1016/bs.ctdb.2019.11.005

Abstract

In bilaterally-symmetric animals (Bilateria), condensation of neurons and ganglia into a centralized nervous system (CNS) constitutes a salient feature. In most, if not all, Bilateria another prominent aspect is that the anterior regions of the CNS are typically larger than the posterior ones. Detailed studies in *Drosophila melanogaster* (*Drosophila*) have revealed that anterior expansion in this species stems from three major developmental features: the generation of more progenitors anteriorly, an extended phase of proliferation of anterior progenitors, and more proliferative daughter cells in anterior regions. These brain-specific features combine to generate a larger average lineage size and higher cell numbers in the brain, when compared to more posterior regions. Genetic studies reveal that these anterior-posterior (A-P) differences are controlled by the modulation of temporal programs, common to all progenitors, as well as by Hox homeotic genes, expressed in the nerve cord, and brain-specific factors. All of these regulatory features are gated by the action of the PRC2 epigenetic complex. Studies in mammals indicate that most, if not all of these anterior expansion principles and the underlying genetic programs are evolutionarily conserved. These findings further lend support for the recently proposed idea that the brain and nerve cord may have originated from different parts of the nervous system present in the Bilaterian ancestor. This brain-nerve cord "fusion" concept may help explain a number of the well-known fundamental differences in the biology of the brain, when compared to the nerve cord.

1. Introduction

The central nervous system (CNS) is a distinctive attribute of bilaterally-symmetric animals (Bilateria), and can be sub-divided, at an initial and general level, into the brain and nerve cord. One of the conspicuous, yet not extensively investigated, features of the CNS in the majority of species pertains to the obvious expansion of its anterior part (brain), when compared with the posterior one (nerve cord). Studies in *Drosophila melanogaster* (*Drosophila*) reveal that this evolutionarily conserved feature is driven in this species by three major developmental features: the generation of more progenitors anteriorly, an extended phase of proliferation of anterior progenitors, and more proliferative daughter cells in anterior regions. These brain-specific features combine to generate a larger average lineage size and higher cell numbers in the brain, when compared to the nerve cord. Studies point to that all three of these *Drosophila* anterior expansion features may be conserved into mammals.

Regarding the genetic control of these events, recent studies in *Drosophila* reveal that this A-P progenitor generation and proliferation gradient is governed by different developmental programs operating in the brain and the nerve cord. In the nerve cord, the Hox homeotic genes act in an anti-proliferative manner, while in the brain, brain-specific transcription factors (TFs) act in a pro-proliferative manner, as well as act to generate more progenitors. The selective expression of Hox and brain TFs is gated by the Polycomb Group epigenetic complex, in particular the PRC2 epigenetic complex. In the brain, PRC2 suppresses Hox expression, and promotes brain TF expression. Moreover, PRC2-Hox interplay helps create a gradient of expression of a core cassette of pan-progenitor TFs, which promote NB and daughter proliferation. A number of these regulatory features appear to be conserved into mammals.

Herein, we review these recent findings, and further discuss how they relate to recent models regarding a separate evolutionary origin of the brain and nerve cord.

2. Generation of CNS progenitors and proliferation modes: *Drosophila*

2.1 Progenitor (NB) generation

A powerful model system for addressing the molecular mechanisms underlying CNS formation is the fruit fly *Drosophila melanogaster* (*Drosophila*). From early- to mid-embryogenesis, a relatively small number (~1200) of neural stem cells, neuroblasts (NBs) delaminate from the neuroectoderm and via asymmetric divisions generate the *Drosophila* CNS (Birkholz, Rickert, Berger, Urbach, & Technau, 2013; Bossing, Udolph, Doe, & Technau, 1996; Schmid, Chiba, & Doe, 1999; Schmidt et al., 1997; Urbach, Jussen, & Technau, 2016; Urbach, Schnabel, & Technau, 2003; Wheeler, Stagg, & Crews, 2009; Younossi-Hartenstein, Nassif, Green, & Hartenstein, 1996). The CNS consists of 19 segments, which are categorized into the brain (B1–B3), suboesophageal area (S1–S3) and nerve cord (T1–T3 and A1–A10) segments (Birkholz et al., 2013; Urbach et al., 2016, 2003).

The NBs in the nerve cord delaminate from bilaterally symmetrical sheets of cells, the ventral neurogenic regions of the ectoderm. NBs form by the process of lateral inhibition, which singles out one cell per proneural cluster to delaminate as an NB, while the other cells remain in the periphery to become epidermal progenitor cells (Bhat, 1999; Skeath, 1999; Skeath & Doe, 1996; Skeath & Thor, 2003).

The developing brain originates from a bilaterally symmetrical area of head ectoderm, known as the procephalic neurogenic regions (Campos-Ortega, 1993; Younossi-Hartenstein et al., 1996). In the brain, the process of NB formation is more complex than that observed in the nerve cord, and can in some instances entail NB formation and delamination of adjacent cells (Urbach & Technau, 2003). As an effect of these altered processes, the most anterior brain segment, B1 (protocerebrum), demonstrates increased NB generation. Hence, the number of NBs in B1 is more than twofold higher, when compared to posterior segments: ~160 NBs in B1 compared to 28–70 NBs per segment for the rest of the posterior segments (B2–A10) (Alvarez & Diaz-Benjumea, 2018; Birkholz et al., 2013; Bossing et al., 1996; Schmid et al., 1999, 1997; Urbach et al., 2016, 2003; Walsh & Doe, 2017; Wheeler et al., 2009; Younossi-Hartenstein et al., 1996). Two factors contribute to the phenomenon of "super generation" of NBs in B1. First, group delamination of NBs from the proneural clusters of the procephalic neurogenic regions (Urbach & Technau, 2003), as opposed to the single NB delamination from the proneural clusters of the ventral neurogenic regions (Campos-Ortega, 1993). This is, in part, caused by reduced Notch signaling in the procephalic neuroectoderm (Stuttem & Campos-Ortega, 1991; Urbach & Technau, 2003). Second, expression of brain-specific gap genes *tailless* (*tll*) and *ocelliless* (*oc*; also known as *orthodenticle* (*otd*)) has also been linked to this phenomenon, and is important for NB generation, acting in part via activating *lethal of scute* expression in the procephalic neuroectoderm (Hirth et al., 1995; Younossi-Hartenstein et al., 1997).

2.2 Progenitor (NB) proliferation modes

Neurogenesis in *Drosophila* CNS is predicated upon asymmetric cell division of NBs (Doe, 2008; Knoblich, 2010). NBs undergo asymmetric divisions repeatedly generating two cells of different size, potential and thereby fate. The elaborate asymmetric division machinery ensures the asymmetric sequestration and therefore inheritance of different components by the two daughter cells; bestowing the progenitor identity upon the bigger, self-renewing daughter cell, while the smaller daughter cell will have less proliferative potential, and either divides terminally, to generate two neurons or glial cells, or directly differentiates (Fig. 1). The asymmetric inheritance of many of these components determine whether the recipient cell is bound for self-renewal or differentiation. There are two major apical complexes that segregate to the progenitor cell. (1) The Par protein complex

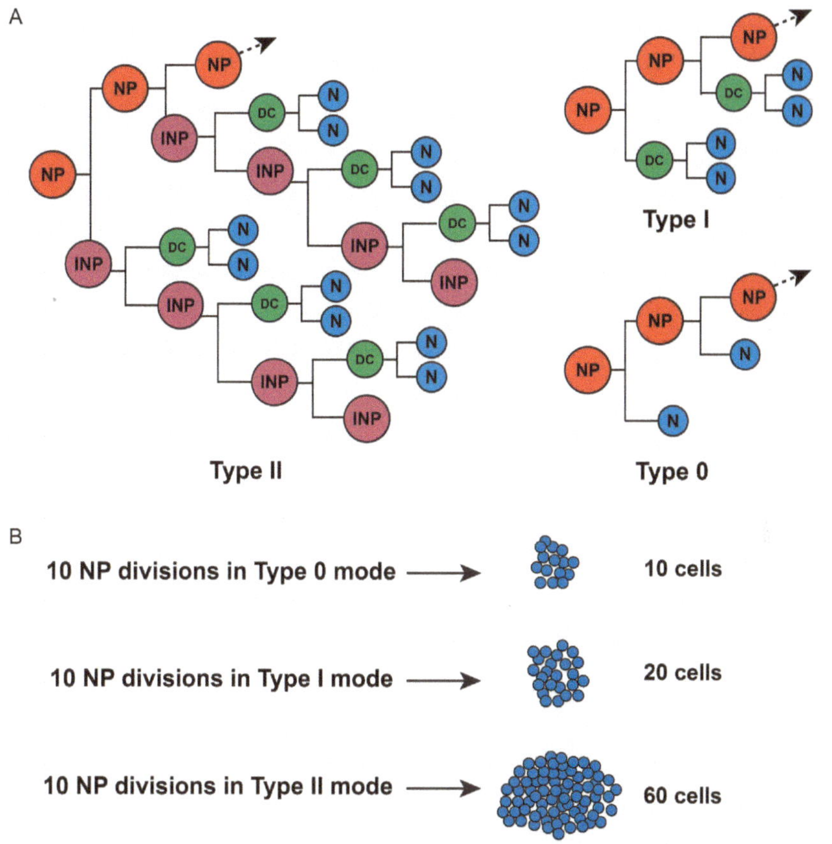

Fig. 1 Three modes of daughter proliferation in *Drosophila*. (A) Schematic of proliferation modes in the embryonic *Drosophila* CNS. (NP, neural progenitor; INP, intermediate neural progenitor; DC, daughter cell; N, neuron). (B) Cell number outcome of 10 rounds of NP divisions in the Type 0, Type I and Type II proliferation modes (based upon three divisions of each INP).

including Bazooka (Par3), aPKC, and Par6, which is mainly to ensure the exclusion of the basal components from the apical pole. Therefore, the differentiating factors of the basal complex such as Prospero and Numb will then segregate into the daughter cell. (2) The Gαi-Pins-Loco complex which is mainly responsible for the mitotic spindle formation and its alignment along the apico-basal axis and perpendicular to the cleavage plane (Doe, 2008; Knoblich, 2010; Sousa-Nunes & Somers, 2013).

The majority of NBs in the nerve cord begin lineage progression in the Type I mode, in which the NB buds off a daughter cell that divides once to generate two neurons or glial cells (Boone & Doe, 2008) (Fig. 1).

Later in nerve cord neurogenesis, many NBs switch to Type 0 mode, generating daughters that differentiate directly without any further division (Baumgardt et al., 2014; Baumgardt, Karlsson, Terriente, Diaz-Benjumea, & Thor, 2009; Karcavich & Doe, 2005; Monedero Cobeta, Yaghmaeian Salmani, & Thor, 2017). Within each nerve cord hemisegment, each of the ~30 NB lineages is unique in its timing of the Type I→0 switch, hence showing different window lengths in the Type I and 0 daughter proliferation modes. In addition, the final NB cell cycle exit also occurs after a programmed number of NB divisions. The timing of the switch and the NB exit is precisely regulated by the combinatorial activity of a hierarchy of transcription factors (see below).

Neurogenesis in the brain progresses along a modified program. Not only do most, if not all brain NBs stay in the Type I mode, they also have a longer proliferation window, when compared to the nerve cord (Yaghmaeian Salmani et al., 2018). Additionally, the anterior-most segment of the brain (B1) contains NBs with even more potent modes of proliferation: Type II NBs and mushroom body NBs (MBNB). Specifically, each B1 brain lobe contains eight Type II NBs and four MBNBs. Type II NBs generate daughter cells, known as the intermediate neural progenitor cells (INPs), which can divide multiple times, giving rise to daughter cells of their own, which in turn divide once to bud off either two neurons or glial cells (Alvarez & Diaz-Benjumea, 2018; Walsh & Doe, 2017) (Fig. 1). MBNBs generate 30–40 cells during embryonic neurogenesis and do not seem to bud off INPs (Kunz, Kraft, Technau, & Urbach, 2012). It is, therefore, likely that lineage progression of MBNBs is of Type I without switching to Type 0. However, MBNBs are perhaps the only type of NBs in the whole CNS that keep dividing throughout embryonic neurogenesis, as well as into larval stages and never enter quiescence. These three modes of "hyper-proliferation" in brain NBs; the extended, non-switching Type I, the Type II and the "never-stopping" MBNBs explains why the average lineage size is much larger in the brain when compared to nerve cord, by the end of embryogenesis (Yaghmaeian Salmani et al., 2018) (Fig. 1). The brain-specific "super generation" of NBs plus the "hyper-proliferative" modes of NBs result in the brain segments having many more cells than the posterior ones (Yaghmaeian Salmani et al., 2018).

The "hyper-proliferation" of brain NBs begs the question of how the key cell cycle genes are regulated during these events. Studies have shown that the pro-proliferation genes *Cyclin E* (*CycE*), *E2f1*, and *string* (*stg*; mammalian *Cdc25*), as well as the cell cycle inhibitor *dacapo* (*dap*; mammalian *Cdkn1a-c*), control these events, and are known to be important for the Type I → 0 switch and the precise NB exit (Baumgardt et al., 2014; Bivik et al., 2015). Studies reveal that all three pro-proliferative cell cycle proteins are indeed expressed at higher levels in mitotic brain NBs, when compared to the nerve cord (Yaghmaeian Salmani et al., 2018). Previous studies in the nerve cord demonstrated that NBs complete a cell cycle in about 40 min, while daughter cells complete it in about 100 min (Baumgardt et al., 2014; Hartenstein, Rudloff, & Campos-Ortega, 1987). Strikingly, DNA pulse labeling experiments revealed that while NB cell cycle length is the same across the A-P axis of the CNS, daughter cells cycle faster in the brain than in the nerve cord (Yaghmaeian Salmani et al., 2018).

3. Generation of CNS progenitors and proliferation modes: Mouse

3.1 Progenitor generation

In the mouse, we do not, as of yet, have knowledge of the precise number of CNS progenitors generated at different axial levels. Determining these numbers in the mouse presents a greater challenge than that in *Drosophila*, for reasons beyond mere tissue size and overall cell numbers. Specifically, in *Drosophila* all CNS progenitors (NBs), at all axial levels, clearly delaminate from the overlying neuroectoderm, and thereafter commence generating their lineages. By contrast, in the mouse, and by extension in most, if not all mammals, the majority of cells in the entire neuroectoderm can undergo a transition from neuroepithelial (NE) to radial glial cell (RGC; progenitor) state (Gotz & Huttner, 2005; Kriegstein, Noctor, & Martinez-Cerdeno, 2006; Llinares-Benadero & Borrell, 2019). However, this transition occurs during an extended developmental phase, and it is furthermore not clear if all neuroectodermal cells actually do undergo the NE-RGC transition. Further complicating the issue, and again different from *Drosophila*, is that the NE-RGC transition is different for different axial levels. For example, in the developing telencephalon, neuroepithelial cells undergo a clear NE-RGC transition, and then commence generating their lineages. In contrast, in the spinal cord progenitors retain neuroepithelial properties (Gotz & Huttner, 2005). Due to these complicating issues, we do not currently know the number of neural progenitors at different axial levels of the mouse CNS. But given the broader extent of the embryonic neuroectoderm in the anterior region

(Shimamura, Hartigan, Martinez, Puelles, & Rubenstein, 1995) it is tempting to speculate that there may indeed be more progenitors generated anteriorly. Nevertheless, akin to *Drosophila*, there is indeed an extended neurogenesis phase in the forebrain, when compared to the spinal cord (Caviness Jr., Takahashi, & Nowakowski, 1995; Huang et al., 2013; Kicheva et al., 2014; Yaghmaeian Salmani et al., 2018). Studies of the cell cycle speed also revealed, again similar to *Drosophila*, that progenitors in the mouse telencephalon had a faster cell cycle speed when compared to spinal cord progenitors (Yaghmaeian Salmani et al., 2018).

3.2 Progenitor proliferation modes

The proliferation modes of mammalian RGCs have been studied most intensively in the developing cortex. Similar to *Drosophila*, after transitioning from NE identity, RGCs typically divide asymmetrically, to self-renew and generate daughter cells with reduced "stemness." Daughter cells, in turn, either differentiate into neurons or divide one or several times (Gotz & Huttner, 2005; Kriegstein et al., 2006; Llinares-Benadero & Borrell, 2019). In mammals, the dividing daughter cells are most often referred to as basal progenitors (bPs), and they have been identified along the entire mouse CNS A-P axis (Haubensak, Attardo, Denk, & Huttner, 2004; Smart, 1972a, 1972b, 1973, 1976; Tarabykin, Stoykova, Usman, & Gruss, 2001; Wang, Bluske, Dickel, & Nakagawa, 2011). Interestingly, previous studies have revealed that the ratio of dividing bPs to RGCs was higher in the telencephalon, when compared to the hindbrain (Haubensak et al., 2004). RGCs will typically divide close to the lumen, while daughter cells divide further away from the lumen. Using distance to lumen as a proxy for RGC versus daughter cell division then, a recent study also found a higher ratio of dividing daughter cells in the telencephalon, when compared to the lumbo-sacral spinal cord (Yaghmaeian Salmani et al., 2018). While the mapping of dividing daughter cells along the mouse CNS A-P axis is incomplete, these studies lend support for the notion that there is a higher prevalence of proliferating daughter cells in the anterior mouse CNS.

Interestingly, there are indications that in more derived mammals, the clearly noticeable brain expansion, and the transition from the rodent-type lissencephalic cortex into the primate-type gyrencephalic cortex, may be, at least in part, caused by an increase of daughter cell proliferation (Fish, Dehay, Kennedy, & Huttner, 2008; Kriegstein et al., 2006;

Nonaka-Kinoshita et al., 2013). The increased prevalence of daughter cell proliferation in primates also entails the appearance of specialized bPs located in the outer subventricular zone (oRGCs) (Lui, Hansen, & Kriegstein, 2011; Paridaen & Huttner, 2014). However, the actual gyration of the cortex is generally believed not to be driven purely by increased daughter cell proliferation (Llinares-Benadero & Borrell, 2019).

4. Anterior-posterior determinants of graded proliferation: Programmed cell death (PCD)

4.1 *Drosophila*

Programmed cell death (PCD) plays a complex role during NB lineage progression. PCD has been shown to remove postmitotic cells, including NBs that have exited the cell cycle and postmitotic daughter cells (either immediately after they are born, or after differentiation) (Miguel-Aliaga & Thor, 2004, 2009). However, PCD in some cases acts as the actual trigger for stopping NB lineage progression: stop-by-PCD (Bivik et al., 2016; Novotny, Eiselt, & Urban, 2002). PCD of all types is more prevalent in the posterior nerve cord, thereby contributing in several ways to the final gradient of cell numbers along the nerve cord (Monedero Cobeta et al., 2017).

In the *Drosophila* brain, PCD was found to play an even greater role, but surprisingly, more, not less, PCD was observed in the brain (Yaghmaeian Salmani et al., 2018). Similar to the VNC, PCD reduced NB proliferation, and in addition, in contrast to the nerve cord, PCD also reduced daughter proliferation. In fact, there is no evidence that any brain NBs normally are removed by PCD, during or after neurogenesis.

4.2 Mouse

The contribution of PCD to mouse CNS development has not been rigorously analyzed, chiefly because of the lack of mutants with complete loss of PCD. Recently, however, triple mutants for the BCL-2 family genes *Bax*, *Bak* and *Bok* were described as completely lacking PCD in the developing embryo (Ke et al., 2018), and this opens up avenues for assessing the contribution of PCD to mouse CNS development.

5. Anterior-posterior determinants of graded proliferation: Hox genes

5.1 Hox gene expression patterns

5.1.1 *Drosophila*

In *Drosophila*, the early developmental decisions are chiefly made by maternally-deposited factors (proteins and mRNAs), such as Bicoid (Bcd), Caudal (Cad) and Nanos (Nos), which establish gradients of morphogens along the A-P axis to drive segmentation of the early embryo (Nusslein-Volhard, 1991; Sanson, 2001). These maternal factors exert control over a hierarchy of downstream factors including gap genes, pair-rule genes, segment polarity genes and Hox homeotic genes, which together sub-divide the embryo into segments along the A-P axis (Nusslein-Volhard, 1991; Sanson, 2001). Expression of gap genes, such as *hunchback* (*hb*) is under control of the maternally-deposited morphogens.

Hox homeotic genes are homeodomain-containing transcription factors that are highly conserved throughout evolution. They have been shown to play a pivotal role in A-P patterning, segment identification and limb development in both invertebrates and vertebrates alike (Philippidou & Dasen, 2013). It is hypothesized that Hox genes originated from the duplication of an ancient bilaterian Hox cluster (Garcia-Fernandez, 2005; Gehring, Kloter, & Suga, 2009; Holland, 2013).

In *Drosophila*, Hox genes are located on the third chromosome and organized into two clusters; the Antennapedia complex (ANT-C), including *labial* (*lab*), *proboscipedia* (*pb*), *Deformed* (*Dfd*), *Sex combs reduced* (*Scr*) and *Antennapedia* (*Antp*), and the Bithorax complex (BX-C), including *Ultrabithorax* (*Ubx*), *abdominal-A* (*abd-A*) and *Abdominal-B* (*Abd-B*). Hox genes are expressed in partly overlapping patterns along the A-P axis, with the anterior most part of the embryo, and later on the anterior CNS, completely void of detectable expression (Hirth, Hartmann, & Reichert, 1998; Yaghmaeian Salmani et al., 2018). The gap gene/early temporal factor *hb* plays an important early role in repressing Hox expression in anterior regions of the embryo (Lehmann & Nusslein-Volhard, 1987; White & Lehmann, 1986). This transcriptional repression is achieved via direct biding of Hb to Hox regulatory loci (Kehle et al., 1998; Zhang & Bienz, 1992). This repression is subsequently maintained by the Polycomb Complex (see below). The transcriptional control of Hox gene expression is also modulated by the order in which Hox genes are arranged in the genomic cluster. This mode of regulation is known as the spatial collinearity, whereby the Hox genes located at the 3′ end of the cluster are more anteriorly expressed while the genes at the 5′ end of the cluster are more posteriorly expressed (Garcia-Fernandez, 2005; Gehring et al., 2009; Holland, 2013).

5.1.2 Mouse

In the mouse, Hox homeotic genes are organized into 4 clusters: A, B, C and D as well as into 13 paralogy groups depending on their position within the four clusters (Philippidou & Dasen, 2013). In addition to spatial collinearity, studies have revealed the existence of temporal collinearity in Hox regulation in mammals, i.e., the Hox genes at the 3′ end of the cluster are expressed earlier in development than the ones toward the 5′ end (Parker & Krumlauf, 2017). However, it is noteworthy that there are many exceptions to these general models of Hox gene regulation.

In contrast to *Drosophila*, there is no single regulator, such as *hunchback*, that mediates anterior repression of Hox genes. Instead, there are several important factors/morphogens acting to control Hox gene expression along the A-P axis. Gradients of both Retinoic Acid (RA), peaking at caudal hindbrain and rostral spinal cord, and Fibroblast growth factor (FGF), peaking at the caudal region of the neural tube/spinal cord, play important roles in successive activation of different Hox genes and establishing roughly-defined domains along the A-P axis of the CNS. These two factors later interact with others to fine tune Hox expression domains (Philippidou & Dasen, 2013). Wnt signaling has also been shown to be pivotal in the regulation of Hox gene expression, especially for neural progenitors to assume caudal hindbrain and spinal cord identities, via adjusting the progenitors' responsiveness to primary morphogens such as RA and FGF (Philippidou & Dasen, 2013).

5.2 Hox genes suppress proliferation

5.2.1 Drosophila

As outlined above, Hox genes are expressed in overlapping domains along the A-P CNS axis. In addition to the well-studied spatial A-P pattern of Hox genes, they also are expressed in a restricted temporal manner in NBs. Specifically, during early NB formation and initial stages of lineage progression, although there is indeed Hox protein expression in neurons, there is little if any Hox protein expression inside NBs (Karlsson, Baumgardt, & Thor, 2010; Monedero Cobeta et al., 2017). Subsequently, Hox factors are turned on inside NBs, where they help trigger the Type I → 0 daughter proliferation switch and ultimately, NB exit. The dynamics of Hox protein expression in NBs is apparently controlled by two mechanisms. First, as exemplified in the case of Ubx, the binding of Elav protein, specifically expressed in neurons, to the Ubx mRNA positively regulates Ubx expression in neurons (Rogulja-Ortmann et al., 2014). Second, early NB factors (see below), which are expressed early in NBs, act in a repressive manner upon Hox genes (Bahrampour, Gunnar, Jonsson, Ekman, & Thor, 2017). Studies of Antp and the three BX-C genes have furthermore revealed that the posteriorly expressed Hox genes, such as Abd-B, act in a stronger anti-proliferative manner (Monedero Cobeta et al., 2017). Hence, the spatial (A-P) and temporal Hox expression pattern creates a proliferation gradient along the A-P axis of the nerve cord, with the Type I → 0 switch and NBs cell cycle exit occurring earlier in more posterior parts. This proliferation control of NBs and daughter cells combines to generate smaller average lineages and fewer cells in more posterior nerve cord segments when compared to anterior ones (Monedero Cobeta et al., 2017; Yaghmaeian Salmani et al., 2018).

5.2.2 Mouse

Similar to invertebrates, there is evidence that Hox genes act in an anti-proliferative manner, with the posterior Hox genes being more potent. Studies of *Hoxb13* mutants show an increase in the proliferation of posterior spinal cord (Economides, Zeltser, & Capecchi, 2003). More evidence in support of the connection between Hox genes and proliferation comes from studies in which proliferation is attenuated following the misexpression of the posterior Hox genes *Hoxb13* and *Hoxb9* in embryonic chick telencephalon (Yaghmaeian Salmani et al., 2018).

6. Anterior-posterior determinants of graded proliferation: Brain genes

6.1 *Drosophila*

In *Drosophila*, as outlined above, three rostral features contribute to the anterior CNS expansion: (1) super generation of NBs, (2) increased NB proliferation and (3) increased daughter cell proliferation. Recent studies reveal that all three features are driven by the action of brain-specific transcription factors; *tailless*, *otp/Rx/hbn*, *Doc1/2/3* and *earmuff* (Curt, Yaghmaeian Salmani, & Thor, 2019; Younossi-Hartenstein et al., 1997). Brain TF co-misexpression can give rise to brain-like proliferative behavior in the nerve cord and also reprogram the larval wing disks into brain NBs (Curt et al., 2019). These brain TFs are also important for Type II NBs generation and proliferation. Finally, *Rx* and *tll* have been shown to be expressed in MBNBs and are important for their proliferation and lineage progression but not NB generation (Kraft, Massey, Kolb, Walldorf, & Urbach, 2016; Kurusu et al., 2009). Hence, the anterior expansion of the *Drosophila* CNS is not only promoted by the absence of Hox expression, but also by the presence of brain TF expression (Fig. 2).

6.2 Mouse

In the mouse, the orthologous brain-specific TFs are: Nr2E1/Tlx (Tll); Otp, Rax and Arx (Otp); Tbx2/3/6 (Doc1/2/3); and FezF1/2 (Erm). Similar to *Drosophila*, the expression of these mouse TFs is also limited to the brain. Moreover, they are pivotal for mouse brain development, and evidence points to that these genes positively regulate proliferation (Eckler & Chen, 2014; Islam & Zhang, 2015; Kar et al., 2013; Kitamura et al., 2002; Murai et al., 2014; Price et al., 2009; Trowe et al., 2013; Wang & Lufkin, 2000). The regulation of these brain-specific TFs by the A-P patterning system has not been extensively analyzed. However, Nr2E1 has been found to be negatively regulated by RA (Yu et al., 2017).

7. Anterior-posterior determinants of graded proliferation: PcG

7.1 *Drosophila*

The absence of Hox homeotic gene expression in the brain is a common and evolutionarily conserved feature (Holland, 2013; Philippidou & Dasen, 2013). The Polycomb group complex, especially the PRC2 complex, is responsible for anterior repression of Hox gene expression (Struhl, 1983; Struhl & Akam, 1985). PRC2 catalyzes H3K27 trimethylation and thereby represses the target loci (Muller & Verrijzer, 2009). The *extra sex combs* (*esc*) gene encodes a core subunit of PRC2, and mutation of *esc* (mammalian *Eed*) causes the H3K27me3 mark to disappear completely in the *Drosophila* embryo (Yaghmaeian Salmani et al., 2018). This is accompanied by anterior

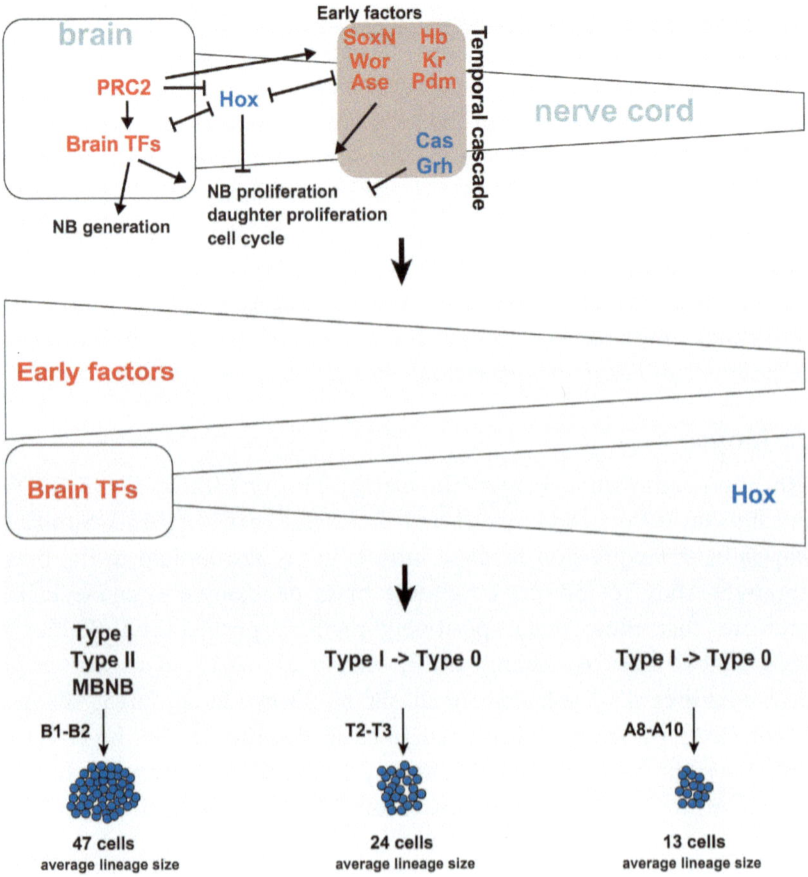

Fig. 2 Genetic program controlling anterior CNS expansion in *Drosophila*. Along the developing neuro-axis, a core cassette of early NB factors and temporal factors is expressed by most, if not all NBs (gray box). The expression of this core cassette is modified along the A-P axis by posterior Hox gene expression and anterior brain-specific TFs. These A-P differences are promoted by the gatekeeping activity of the PRC2 complex, whereby in the anterior brain, Hox genes are suppressed, brain TFs and early factors are highly expressed. Together with other regulatory interplay, such as the cross-repressive regulation of Hox genes upon each other, and the stronger anti-proliferative role of posterior Hox genes, this results in a gradient of Hox activity, and an opposing gradient of early NB factor activity. Brain TFs promote NB generation, and together with early factors promote NB and daughter cell proliferation, via driving the cell cycle. Hox genes and late temporal genes play opposite roles. The outcome of this regulatory interplay is the super generation of NBs in B1, and establishment of Type I, Type II and MBNB behavior in the brain, while the nerve cord displays fewer NB per segment, which display Type I→0 behavior. In the nerve cord, the gradient of Hox and early factor activity furthermore results in a gradient of the Type I→0 switch and NB cell cycle exit. The outcome of these alternate lineage behaviors is the generation of considerably larger average lineages in the brain, when compared to the nerve cord.

expression of Hox genes and reduced levels of brain TFs (Tll, Rx and Doc2) in brain NBs (Curt et al., 2019; Yaghmaeian Salmani et al., 2018). This results in reduced NB and daughter proliferation, reduced average lineage size and reduced overall brain cell numbers. Rescue experiments also lend support to the instructive interactions between these groups of genes in generating and maintaining the brain expansion. Specifically, transgenic expression of brain TFs can rescue the proliferation phenotype of *esc* mutants (Curt et al., 2019). Interestingly, the proliferation phenotype in *esc* mutants can also be fully rescued by *Abd-B* mutation, such that the NB and daughter cell proliferation in the brain is restored in *esc;Abd-B* double mutants (Bahrampour, Jonsson, & Thor, 2019).

7.2 Mouse

Similar to *Drosophila*, Hox genes are not expressed in the mouse telencephalon (Philippidou & Dasen, 2013). Mutation of PcG components causes Hox genes to be expressed anteriorly in the brain (Isono et al., 2005; Li et al., 2011; Suzuki et al., 2002; Wang, Mager, Schnedier, & Magnuson, 2002). In a CNS-specific *Eed* knockout (*Eed-cKO*) (component of PRC2) H3K27me3 disappears form the entire CNS around E11.5, followed by reduced proliferation of progenitors and daughter cells later in neurodevelopment (Yaghmaeian Salmani et al., 2018). This reduced proliferation was only observed in the brain not the spinal cord, which indicates the conserved repressive role of PRC2 toward Hox genes expression in the brain. *Eed-cKO* mutant mice are also microcephalic with no visible phenotypes in the spinal cord. In *Eed-cKO* telencephalon, there are more progenitors expressing $p27^{KIP1}$ (p27, Cdkn1b; *Drosophila* Dap), while levels of E2F3 decreases. The mouse embryonic telencephalon of *Eed-cKO* showed a premature decrease in Sox2+ progenitor area, attenuated proliferation, higher p27 expression and premature onset of PCD, which together can explain the microcephalic phenotype. In contrast, such effects were not observed in the *Eed-cKO* spinal cord (Yaghmaeian Salmani et al., 2018). Additionally, transcriptome analysis of the PRC2 mutant forebrains revealed clear posteriorization, with transcript detection of all 39 Hox genes, downregulation of brain-specific genes (*Dlx1/2/5*, *Tbr1*, *Nr2e1*, *Arx*, *Emx1*, *Foxg1*, *Fezf2*, *Tbr2*), and with several pro-proliferative genes being downregulated, whereas the cell cycle inhibitor genes Cdkn1a and Cdkn1c were upregulated (Yaghmaeian Salmani et al., 2018).

8. Temporal determinants of graded proliferation: Early and late factors

8.1 *Drosophila*

Drosophila embryonic NBs change their competence over time, evident by the generation of different types of neurons and glia at distinct stages of lineage progression (Kohwi & Doe, 2013; Li, Chen, & Desplan, 2013). Studies have revealed that these competence changes are controlled by a temporal gene expression cascade, where the sequential expression of the TFs Hb → Kr → Pdm1/2 → Cas → Grh act to dictate the change in competence of NBs over time (Kohwi & Doe, 2013; Li et al., 2013). Strikingly, studies reveal that this temporal cascade also controls the Type I → 0 daughter cell proliferation switch and the timing of NB cell cycle exit. Specifically, the early temporal factors Hb, Kr and Pdm appear to drive NB and daughter cell proliferation while the late factors Cas and Grh act in the opposite manner (Bahrampour et al., 2017, 2019; Baumgardt et al., 2014). These temporal genes play out these roles by controlling the key cell cycle factors Cyclin E (CycE), string (stg; Cdc25), E2f1, and dacapo (dap; mammalian p21CIP1/p27KIP1/p57Kip2) (Bahrampour et al., 2017, 2019; Baumgardt et al., 2014).

In addition to the temporal cascade, early NB factors, such as the Snail family (Worniou; Wor), SoxB family (SoxN) and the proneural-related factor Asense (Ase), act together with the three early temporal factors Hb, Kr and Pdm to positively regulate proliferation of NBs and daughter cells (Bahrampour et al., 2017). Strikingly, early factors are also expressed in an A-P gradient, with their levels being highest in the brain and lowest in posterior-most segments of the nerve cord (Bahrampour et al., 2019). In addition, both the early temporal factors Hb, Kr and Pdm, as well as the Wor, SoxN and Ase factors, are gradually downregulated in NBs and are replaced by late temporal factors. This precisely-controlled temporal transition is mainly responsible for NB cell cycle exit and Type I → 0 switch (Bahrampour et al., 2017; Baumgardt et al., 2014). Furthermore, combinatorial early factor misexpression can reprogram non-neurogenic tissue and trigger ectopic NB generation (Bahrampour et al., 2017).

In *esc* mutants, all the aforementioned early factors are also downregulated in brain NBs, indicating that the gate keeping role of PRC2 with regard to Hox genes and brain TFs also pertains to control of early

NB factor expression in the brain. Indeed, misexpression of a combination of early factors could rescue the proliferative reduction of *esc* mutants (Bahrampour et al., 2019).

8.2 Mouse

Genetic programs for generation and maintenance of neural stemness are evolutionarily conserved, as evident by the existence of *Drosophila* orthologues in mammals; *Sox 1–3* (SoxB family) are necessary for CNS development (Reiprich & Wegner, 2015); *Snail1* (*wor*) is needed for proliferation of mouse neural progenitors (Zander, Burns, Yang, Kaplan, & Miller, 2014) and also for reprograming into neural stem cells (Unternaehrer et al., 2014); *Ascl1* (*ase*) is vital for neurogenesis, and responsiveness to stimuli in order to exit quiescence and divide in adult mice (Andersen et al., 2014; Urban et al., 2016). While it is unclear if there is an A-P gradient of "neural stemness" factors in mammals, in a recent comparative transcriptome analysis, Sox1–3 expression was shown to be significantly higher in embryonic mouse telencephalon, when compared to the spinal cord (Yaghmaeian Salmani et al., 2018). In contract to *Drosophila*, in the mouse there is no evidence for a common temporal gene cascade, playing out at all axial levels of the CNS. Instead, different temporal gene cascades appear to be playing out at different levels.

In the developing hindbrain, a ventral progenitor domain sequentially generates three different cell types, i.e., motor neurons (MNs), serotonergic interneurons and oligodendrocytes. This temporal progression is gated by a combination of sequentially expressed transcription factors and TGFb signaling, acting to ensure the precise transition between the three competence stages (Deneris & Wyler, 2012; Dias, Alekseenko, Applequist, & Ericson, 2014; Jacob et al., 2007; Pattyn, Vallstedt, Dias, Samad, et al., 2003; Pattyn, Vallstedt, Dias, Sander, & Ericson, 2003; Vallstedt, Klos, & Ericson, 2005). Recent single-cell RNA analysis of the developing spinal cord has unraveled a whole program of temporal regulatory gene expression changes in progenitors (Delile et al., 2019), and future studies will reveal the extent to which these gene expression changes are functionally important.

The developing cortex is also subject to temporal changes in progenitors. The successive, inside-out mode of neurogenesis, together with early born versus late-born neurons having different morphology, maturation and migration processes, as well as different integration/destination sites in the layered cortex, point to the importance of temporal changes in cortical neurogenesis. As for the temporal changes in gene expression in progenitors, a number of studies demonstrate that progenitors sequentially express a number of transcription factors (reviewed in Holguera & Desplan, 2018; Kohwi & Doe, 2013; Llinares-Benadero & Borrell, 2019). However, the regulatory interplay between these factors, and their precise role in dictating temporal competence in cortical progenitors is not fully resolved. Finally, in spite of these insights into temporal patterning in the mouse CNS the connection between temporal cascades and progenitor and/or daughter cell proliferation is not well understood.

9. An evolutionary "fusion point" in the CNS?

The PcG epigenetic machinery, the temporal factors, the Hox homeotic genes, and the brain TFs discussed above are all conserved during evolution. PcG orthologous genes become apparent already in the budding yeast (Dumesic et al., 2015; Jamieson, Rountree, Lewis, Stajich, & Selker, 2013). The evolution of Hox homeotic genes has been extensively debated, but Hox homeotic genes appear to have emerged later than PcG, in the common ancestor of Cnidaria-bilateria (Garcia-Fernandez, 2005; Holland, 2013). The brain TFs discussed above, with strong driver roles in *Drosophila* embryonic brain development (Curt et al., 2019), i.e., *tailless* = *Nr2E1*, *otp/Rx/hbn* = *Otp/Arx/Rax* and *earmuff* = *FezF1/2*, are also well conserved and their appearance predates bilaterians (Arendt, Tosches, & Marlow, 2016; Tosches & Arendt, 2013). The temporal and pan-NB factors, SoxB, Snail and Ase are also well conserved. Thus, it appears that the appearance of PcG, many of the temporal and stem cell factors, the Hox genes and the brain TFs predates the emergence of the CNS, and thereby also the manifestation of anterior CNS expansion. Moreover, the complexity of the PcG-Hox-brain-TF system, e.g., regarding the number of genes, is to some extent paralleled by the increasing complexity of the CNS during evolution.

The evolution of the CNS remains under active investigation and thoughtful debate, but it is generally viewed to have emerged from dispersed nerve nets and ganglia present in the bilaterian ancestor (Arendt et al., 2016; Holland, 2003; Holland et al., 2013; Jekely, Paps, & Nielsen, 2015; Martin-Duran et al., 2017; Nielsen, 2012, 2015; Tosches & Arendt, 2013). Based upon a combination of gene expression and phylogenetic considerations it was recently proposed that the CNS may represent an evolutionary "fusion" of two separate nervous systems, present in the common ancestor; the apical and basal nervous systems (Arendt et al., 2016; Nielsen, 2012, 2015; Tosches & Arendt, 2013) (Fig. 3A). On this note, it is interesting to point out that in arthropods, e.g., *Drosophila*, there is initially separate development of the brain and nerve cord, and they only merge during subsequent stages of development (Hartenstein & Campos-Ortega, 1984) (Fig. 3D).

Focusing on the mouse, where would the "fusion point" be then? Three models are emerging. The first model, based upon gene expression, brain region/cell type function, and evolutionary considerations, placed the "fusion point" between the anterior and posterior hypothalamus (Tosches & Arendt, 2013) (Fig. 3B). Second, based solely upon Hox homeotic gene expression (Yaghmaeian Salmani et al., 2018), the "fusion point" could tentatively be placed more posteriorly. Specifically, with the expression of the anterior-most expressed Hox gene, *Hoxa1*, extending into the midbrain (Allen Brain Atlas), this would place the "fusion point" somewhat more posteriorly to that proposed by Tosches and Arendt, i.e., at the diencephalon-midbrain intersection (Fig. 3B–C). Both of these "fusion points" would also be supported by the expression of the mouse orthologues of the recently identified *Drosophila* brain proliferation driver TFs, i.e., *tailless*=*Nr2E1*, *otp/Rx/hbn*=*Otp/Arx/Rax* and *earmuff*=*FezF1/2* (Curt et al., 2019). Third, a recent publication, addressing the early embryonic cell origins and the epigenomic signature of the anterior versus posterior developing CNS, placed the "fusion point" much more posteriorly, at the junction of the hindbrain and spinal cord (Metzis et al., 2018) (Fig. 3B). This model conflicts with the notion of Hox genes as specifically belonging to the posteriorly originated CNS, since a number of Hox genes are expressed in the mouse hindbrain, and even into the midbrain (Fig. 3C). However, it should be noted that even in the *Drosophila* embryo, which develops with the brain (B1–B3 segments) and nerve cord initially separated (Hartenstein & Campos-Ortega, 1984), and hence develops with a de facto CNS fusion point, the most anteriorly expressed Hox genes *proboscipedia* and *labial* are indeed expressed into the B2 and B3 region of the brain (Hirth et al., 1998).

As previously pointed out (Tosches & Arendt, 2013), the "fusion point" may be difficult to pin-point with precision in the mammalian brain, due to intermingling of cell types and regions at the "fusion point." It is possible that continued comparative studies in a number of different species, including studies of progenitor numbers, gene expression analysis, as well as progenitor and daughter cell proliferation behaviors may help shed light on the location of the "fusion point."

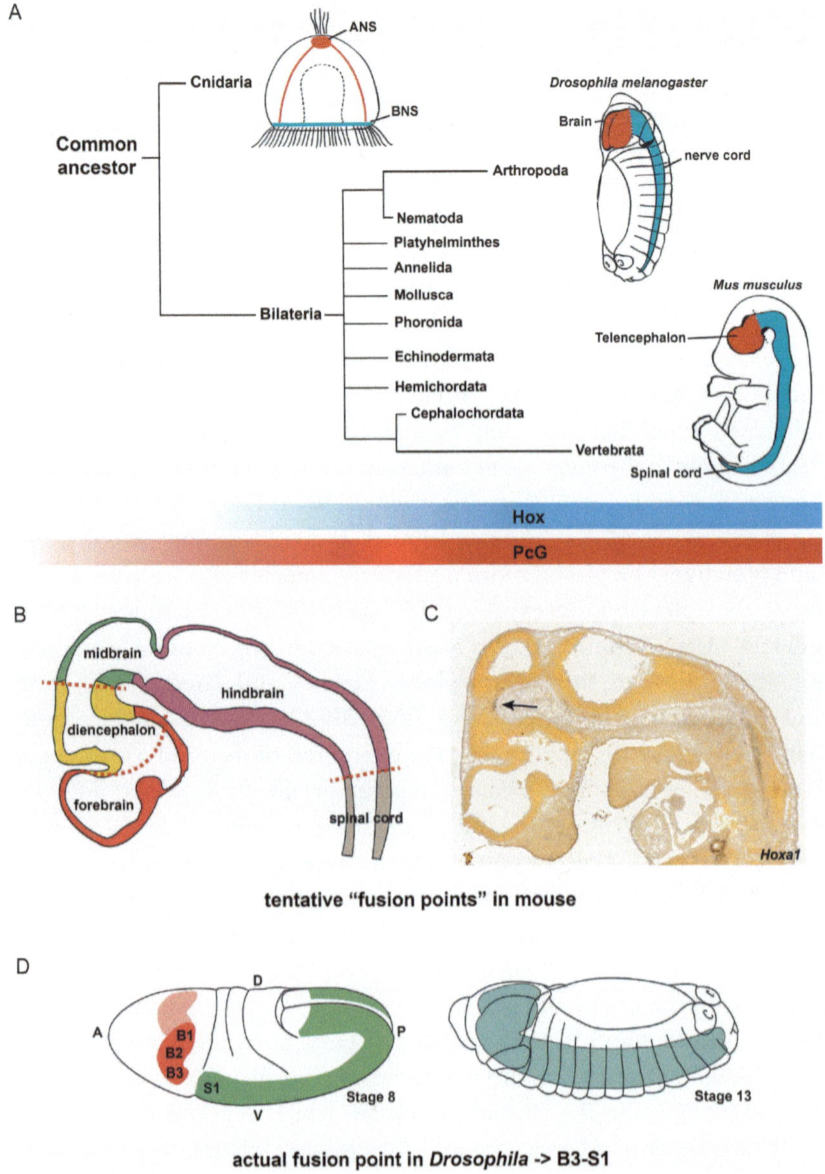

Fig. 3 The CNS; a fusion of the apical and basal nervous systems? (A) Recent proposals postulate that the Bilaterian CNS formed by merger of the apical and basal nervous systems (ANS and BNS, respectively), present in the Bilaterian-Cnidarian ancestor. Emergence of the PcG gene complex predates this merger, while Hox homeotic genes straddles these events. (B) Three models for the location of the remnant of the evolutionary "fusion point" of the apical and basal nervous systems in the developing mouse CNS, here depicted at E11.5. (C) Gene expression of *Hoxa1*, by in situ hybridization, shows expression in the midbrain-diencephalon boundary. (D) In *Drosophila*, the CNS actually forms by merger of the brain and nerve cord, with the brain 3 (B3) and subesophageal segment 1 (S1) as the fusion point. *Panel (A) image adapted, with the author's permission, from Yaghmaeian Salmani, B. (2018).* Genetic mechanisms regulating the spatio-temporal modulation of proliferation rate and mode in neural progenitors and daughter cells during CNS development. *Linkoping University Medical Dissertations No 1628, 1–63. Panel (C) image from Allen Brain Atlas; www.brain-map.org.*

10. Summary

The brain and nerve cord display striking differences in regional size, with brain expansion being a prominent feature. Based upon gross anatomy, this feature appears to be present in most, if not all Bilaterians.

Detailed studies in the *Drosophila* embryo have revealed the cellular mechanisms driving brain expansion, and found that three major developmental features underlie this phenomenon: the generation of more progenitors anteriorly, an extended phase of proliferation of anterior progenitors, and more proliferative daughter cells in anterior regions. These brain-specific features combine to generate two–three times larger average lineage sizes in the brain, when compared to the nerve cord. The combination of more progenitors and larger lineages in the brain results in greater cell numbers in the brain, when compared to the nerve cord.

In the mouse, the number of CNS progenitors generated at different axial levels has not been determined so far. It is however clear that neurogenesis spans a wider time-window in the brain, when compared to the spinal cord (Caviness Jr. et al., 1995; Huang et al., 2013; Kicheva et al., 2014; Yaghmaeian Salmani et al., 2018). Regarding the daughter proliferation profiles along the mouse CNS A-P axis this also remains to be rigorously and systematically addressed. However, observations indicate that there is a higher prevalence of dividing daughter cells in the brain than in the posterior regions of the CNS (Haubensak et al., 2004; Yaghmaeian Salmani et al., 2018). Hence, there are reasons to believe that similar principles as those unraveled in *Drosophila* also apply to the mouse and by extension most, if not all mammals.

There are likely to be a number of genetic mechanisms involved in driving the anterior expansion of the CNS. Here, we have focused on the role of the Hox genes, brain TFs and temporal genes, as well as the PcG system as a gate keeper of the expression of these genes, ensuring their differential expression along the CNS A-P axis. In *Drosophila*, a rather comprehensive model is emerging, where a core cassette of temporal factors, expressed in most of not all NBs, at all axial levels, control proliferation. Early factors drive NB and daughter cell proliferation (Type I and II modes) while late factors stop NBs and daughter cells (Type 0 mode). The activity of this core temporal cassette is modulated along the A-P axis by Hox gene expression posteriorly and brain TFs expression anteriorly. These A-P differences are gated by PRC2, which ensures that the brain stays free of Hox gene expression, and expresses brain TFs and high levels of early NB factors. In the mouse, orthologous genes exist to all these *Drosophila* genes, and in several cases (Hox, PRC2, Sox2, Snail1, brain TFs) appear to play related roles. But the vast increase in CNS cell numbers, of both progenitors and daughters, the challenges in lineage analysis, and the lack of markers that depict progenitors and different types of daughters at all axial levels have precluded the systematic analysis of these features in the mouse CNS. Hence, whether the mouse CNS develops along a similar logic as that now unraveled in *Drosophila* is still unclear.

In *Drosophila*, surprisingly, PCD does not drive anterior CNS expansion, rather it abrogates it (Monedero Cobeta et al., 2017; Yaghmaeian Salmani et al., 2018). The contribution of PCD to mouse CNS development has not yet been systematically addressed. However, the generation of mouse mutants completely lacking PCD during embryogenesis (Ke et al., 2018) opens up for assessing the contribution of PCD to mouse CNS development.

Subdividing the CNS into brain and nerve cord obviously represents an oversimplification, but helps in pointing out the most salient A-P expansion features. At a greater level of granularity, different parts of the fore-, mid-, and hindbrain, as well as different axial levels of the spinal cord, may display graded progenitor generation and proliferation behavior, similar to what has been observed in the *Drosophila* nerve cord (Monedero Cobeta et al., 2017).

There are several outstanding issues that would be important to address in the near future: Where is the "fusion point" in mammals between the apical and basal nervous systems? What is the proliferation behavior of daughter cells along the entire mammalian A-P CNS axis, and can one discern the "fusion point" based upon daughter cell proliferation behavior? What is the effect of PRC2 mutations on regions intermediate to the lumbo-sacral spinal cord and dorsal telencephalon? Are all 39 mammalian Hox homeotic genes truly anti-proliferative, or does this function primarily pertain to the posterior-most expressed Hox genes? What is the connection between PcG, Hox genes and brain TFs and the cell cycle machinery? Answers to these points may help unravel the cellular and genetic mechanisms underlying the anterior expansion of the CNS and the evolutionary origins of its A-P differences.

Author contributions

B.Y.S. and S.T. compiled the figures and wrote the manuscript.

Competing interests

No competing interests declared.

Funding

Financial support was provided by the University of Queensland, Australia, to S.T.

References

Alvarez, J. A., & Diaz-Benjumea, F. J. (2018). Origin and specification of type II neuroblasts in the Drosophila embryo. *Development*, *145*(7). https://doi.org/10.1242/dev.158394.

Andersen, J., Urban, N., Achimastou, A., Ito, A., Simic, M., Ullom, K., et al. (2014). A transcriptional mechanism integrating inputs from extracellular signals to activate hippocampal stem cells. *Neuron*, *83*(5), 1085–1097. https://doi.org/10.1016/j.neuron.2014.08.004.

Arendt, D., Tosches, M. A., & Marlow, H. (2016). From nerve net to nerve ring, nerve cord and brain–evolution of the nervous system. *Nature Reviews Neuroscience*, *17*(1), 61–72. https://doi.org/10.1038/nrn.2015.15.

Bahrampour, S., Gunnar, E., Jonsson, C., Ekman, H., & Thor, S. (2017). Neural lineage progression controlled by a temporal proliferation program. *Developmental Cell*, *43*(3), 332–348. e334. https://doi.org/10.1016/j.devcel.2017.10.004.

Bahrampour, S., Jonsson, C., & Thor, S. (2019). Brain expansion promoted by polycomb-mediated anterior enhancement of a neural stem cell proliferation program. *PLoS Biology*, *17*(2). e3000163. https://doi.org/10.1371/journal.pbio.3000163.

Baumgardt, M., Karlsson, D., Salmani, B. Y., Bivik, C., MacDonald, R. B., Gunnar, E., et al. (2014). Global programmed switch in neural daughter cell proliferation mode triggered by a temporal gene cascade. *Developmental Cell*, *30*(2), 192–208. https://doi.org/10.1016/j.devcel.2014.06.021.

Baumgardt, M., Karlsson, D., Terriente, J., Diaz-Benjumea, F. J., & Thor, S. (2009). Neuronal subtype specification within a lineage by opposing temporal feed-forward loops. *Cell*, *139*(5), 969–982.

Bhat, K. M. (1999). Segment polarity genes in neuroblast formation and identity specification during Drosophila neurogenesis. *BioEssays*, *21*(6), 472–485.

Birkholz, O., Rickert, C., Berger, C., Urbach, R., & Technau, G. M. (2013). Neuroblast pattern and identity in the Drosophila tail region and role of doublesex in the survival of sex-specific precursors. *Development*, *140*(8), 1830–1842. https://doi.org/10.1242/dev.090043.

Bivik, C., Bahrampour, S., Ulvklo, C., Nilsson, P., Angel, A., Fransson, F., et al. (2015). Novel genes involved in controlling specification of Drosophila FMRFamide neuropeptide cells. *Genetics*, *200*(4), 1229–1244. https://doi.org/10.1534/genetics.115.178483.

Bivik, C., MacDonald, R. B., Gunnar, E., Mazouni, K., Schweisguth, F., & Thor, S. (2016). Control of neural daughter cell proliferation by multi-level Notch/Su(H)/E(spl)-HLH signaling. *PLoS Genetics*, *12*(4). e1005984. https://doi.org/10.1371/journal.pgen.1005984.

Boone, J. Q., & Doe, C. Q. (2008). Identification of Drosophila type II neuroblast lineages containing transit amplifying ganglion mother cells. *Developmental Neurobiology*, *68*(9), 1185–1195.

Bossing, T., Udolph, G., Doe, C. Q., & Technau, G. M. (1996). The embryonic central nervous system lineages of Drosophila melanogaster. I. Neuroblast lineages derived from the ventral half of the neuroectoderm. *Developmental Biology*, *179*(1), 41–64.

Campos-Ortega, J. A. (1993). Mechanisms of early neurogenesis in Drosophila melanogaster. *Journal of Neurobiology*, *24*(10), 1305–1327.

Caviness, V. S., Jr., Takahashi, T., & Nowakowski, R. S. (1995). Numbers, time and neocortical neuronogenesis: a general developmental and evolutionary model. *Trends in Neurosciences*, *18*(9), 379–383.

Curt, J. R., Yaghmaeian Salmani, B., & Thor, S. (2019). Anterior CNS expansion driven by brain transcription factors. *eLife8*. https://doi.org/10.7554/eLife.45274.

Delile, J., Rayon, T., Melchionda, M., Edwards, A., Briscoe, J., & Sagner, A. (2019). Single cell transcriptomics reveals spatial and temporal dynamics of gene expression in the developing mouse spinal cord. *Development*, *146*(12), 1–14. https://doi.org/10.1242/dev.173807.

Deneris, E. S., & Wyler, S. C. (2012). Serotonergic transcriptional networks and potential importance to mental health. *Nature Neuroscience*, *15*(4), 519–527. https://doi.org/10.1038/nn.3039.

Dias, J. M., Alekseenko, Z., Applequist, J. M., & Ericson, J. (2014). Tgfbeta signaling regulates temporal neurogenesis and potency of neural stem cells in the CNS. *Neuron*, *84*(5), 927–939. https://doi.org/10.1016/j.neuron.2014.10.033.

Doe, C. Q. (2008). Neural stem cells: Balancing self-renewal with differentiation. *Development*, *135*(9), 1575–1587.

Dumesic, P. A., Homer, C. M., Moresco, J. J., Pack, L. R., Shanle, E. K., Coyle, S. M., et al. (2015). Product binding enforces the genomic specificity of a yeast polycomb repressive complex. *Cell*, *160*(1–2), 204–218. https://doi.org/10.1016/j.cell.2014.11.039.

Eckler, M. J., & Chen, B. (2014). Fez family transcription factors: Controlling neurogenesis and cell fate in the developing mammalian nervous system. *BioEssays*, *36*(8), 788–797. https://doi.org/10.1002/bies.201400039.

Economides, K. D., Zeltser, L., & Capecchi, M. R. (2003). Hoxb13 mutations cause overgrowth of caudal spinal cord and tail vertebrae. *Developmental Biology*, *256*(2), 317–330.

Fish, J. L., Dehay, C., Kennedy, H., & Huttner, W. B. (2008). Making bigger brains-the evolution of neural-progenitor-cell division. *Journal of Cell Science*, *121*(Pt. 17), 2783–2793. https://doi.org/10.1242/jcs.023465.

Garcia-Fernandez, J. (2005). The genesis and evolution of homeobox gene clusters. *Nature Reviews. Genetics*, *6*(12), 881–892. https://doi.org/10.1038/nrg1723.

Gehring, W. J., Kloter, U., & Suga, H. (2009). Evolution of the Hox gene complex from an evolutionary ground state. *Current Topics in Developmental Biology*, *88*, 35–61.

Gotz, M., & Huttner, W. B. (2005). The cell biology of neurogenesis. *Nature Reviews. Molecular Cell Biology*, *6*(10), 777–788. https://doi.org/10.1038/nrm1739.

Hartenstein, V., & Campos-Ortega, J. A. (1984). Early neurogenesis in wild-type *Drosophila melanogaster*. *Roux's Archives of Developmental Biology*, *193*, 308–325.

Hartenstein, V., Rudloff, E., & Campos-Ortega, J. A. (1987). The pattern of proliferation of the neuroblasts in the wild-type embryo of Drosophila melanogaster. *Roux's Archives of Developmental Biology*, *196*(8), 473–485.

Haubensak, W., Attardo, A., Denk, W., & Huttner, W. B. (2004). Neurons arise in the basal neuroepithelium of the early mammalian telencephalon: a major site of neurogenesis. *Proceedings of the National Academy of Sciences of the United States of America*, *101*(9), 3196–3201. https://doi.org/10.1073/pnas.0308600100.

Hirth, F., Hartmann, B., & Reichert, H. (1998). Homeotic gene action in embryonic brain development of Drosophila. *Development*, *125*(9), 1579–1589.

Hirth, F., Therianos, S., Loop, T., Gehring, W. J., Reichert, H., & Furukubo-Tokunaga, K. (1995). Developmental defects in brain segmentation caused by mutations of the homeobox genes orthodenticle and empty spiracles in Drosophila. *Neuron*, *15*(4), 769–778.

Holguera, I., & Desplan, C. (2018). Neuronal specification in space and time. *Science*, *362*(6411), 176–180. https://doi.org/10.1126/science.aas9435.

Holland, N. D. (2003). Early central nervous system evolution: An era of skin brains? *Nature Reviews Neuroscience*, *4*(8), 617–627. https://doi.org/10.1038/nrn1175.

Holland, P. W. (2013). Evolution of homeobox genes. *Wiley Interdisciplinary Reviews: Developmental Biology*, *2*(1), 31–45. https://doi.org/10.1002/wdev.78.

Holland, L. Z., Carvalho, J. E., Escriva, H., Laudet, V., Schubert, M., Shimeld, S. M., et al. (2013). Evolution of bilaterian central nervous systems: A single origin? *EvoDevo*, *4*(1), 27. https://doi.org/10.1186/2041-9139-4-27.

Huang, J., Chen, J., Wang, W., Wei, Y. Y., Cai, G. H., Tamamaki, N., et al. (2013). Birthdate study of GABAergic neurons in the lumbar spinal cord of the glutamic acid decarboxylase 67-green fluorescent protein knock-in mouse. *Frontiers in Neuroanatomy*, 7, 42. https://doi.org/10.3389/fnana.2013.00042.

Islam, M. M., & Zhang, C. L. (2015). TLX: A master regulator for neural stem cell maintenance and neurogenesis. *Biochimica et Biophysica Acta*, *1849*(2), 210–216. https://doi.org/10.1016/j.bbagrm.2014.06.001.

Isono, K., Fujimura, Y., Shinga, J., Yamaki, M., J, O. W., Takihara, Y., et al. (2005). Mammalian polyhomeotic homologues Phc2 and Phc1 act in synergy to mediate polycomb repression of Hox genes. *Molecular and Cellular Biology*, *25*(15), 6694–6706. https://doi.org/10.1128/MCB.25.15.6694-6706.2005.

Jacob, J., Ferri, A. L., Milton, C., Prin, F., Pla, P., Lin, W., et al. (2007). Transcriptional repression coordinates the temporal switch from motor to serotonergic neurogenesis. *Nature Neuroscience*, *10*(11), 1433–1439. https://doi.org/10.1038/nn1985.

Jamieson, K., Rountree, M. R., Lewis, Z. A., Stajich, J. E., & Selker, E. U. (2013). Regional control of histone H3 lysine 27 methylation in Neurospora. *Proceedings of the National Academy of Sciences of the United States of America*, *110*(15), 6027–6032. https://doi.org/10.1073/pnas.1303750110.

Jekely, G., Paps, J., & Nielsen, C. (2015). The phylogenetic position of ctenophores and the origin(s) of nervous systems. *EvoDevo*, *6*(1), 1–8. https://doi.org/10.1186/2041-9139-6-1.

Karcavich, R., & Doe, C. Q. (2005). Drosophila neuroblast 7-3 cell lineage: a model system for studying programmed cell death, Notch/Numb signaling, and sequential specification of ganglion mother cell identity. *The Journal of Comparative Neurology*, *481*(3), 240–251.

Karlsson, D., Baumgardt, M., & Thor, S. (2010). Segment-specific neuronal subtype specification by the integration of anteroposterior and temporal cues. *PLoS Biology*, *8*(5), e1000368.

Ke, F. F. S., Vanyai, H. K., Cowan, A. D., Delbridge, A. R. D., Whitehead, L., Grabow, S., et al. (2018). Embryogenesis and adult life in the absence of intrinsic apoptosis effectors BAX, BAK, and BOK. *Cell*, *173*(5). https://doi.org/10.1016/j.cell.2018.04.0361217–1230.e1217.

Kehle, J., Beuchle, D., Treuheit, S., Christen, B., Kennison, J. A., Bienz, M., et al. (1998). dMi-2, a hunchback-interacting protein that functions in polycomb repression. *Science*, *282*(5395), 1897–1900. https://doi.org/10.1126/science.282.5395.1897.

Kicheva, A., Bollenbach, T., Ribeiro, A., Valle, H. P., Lovell-Badge, R., Episkopou, V., et al. (2014). Coordination of progenitor specification and growth in mouse and chick spinal cord. *Science*, *345*(6204), 1254927. https://doi.org/10.1126/science.1254927.

Kitamura, K., Yanazawa, M., Sugiyama, N., Miura, H., Iizuka-Kogo, A., Kusaka, M., et al. (2002). Mutation of ARX causes abnormal development of forebrain and testes in mice and X-linked lissencephaly with abnormal genitalia in humans. *Nature Genetics*, *32*(3), 359–369. https://doi.org/10.1038/ng1009.

Knoblich, J. A. (2010). Asymmetric cell division: recent developments and their implications for tumour biology. *Nature Reviews. Molecular Cell Biology*, *11*(12), 849–860. https://doi.org/10.1038/nrm3010.

Kohwi, M., & Doe, C. Q. (2013). Temporal fate specification and neural progenitor competence during development. *Nature Reviews. Neuroscience*, *14*(12), 823–838.

Kraft, K. F., Massey, E. M., Kolb, D., Walldorf, U., & Urbach, R. (2016). Retinal homeobox promotes cell growth, proliferation and survival of mushroom body neuroblasts in the Drosophila brain. *Mechanisms of Development*, *142*, 50–61. https://doi.org/10.1016/j.mod.2016.07.003.

Kriegstein, A., Noctor, S., & Martinez-Cerdeno, V. (2006). Patterns of neural stem and progenitor cell division may underlie evolutionary cortical expansion. *Nature Reviews. Neuroscience*, 7(11), 883–890. https://doi.org/10.1038/nrn2008.

Kunz, T., Kraft, K. F., Technau, G. M., & Urbach, R. (2012). Origin of Drosophila mushroom body neuroblasts and generation of divergent embryonic lineages. *Development*, *139*(14), 2510–2522. https://doi.org/10.1242/dev.077883.

Kurusu, M., Maruyama, Y., Adachi, Y., Okabe, M., Suzuki, E., & Furukubo-Tokunaga, K. (2009). A conserved nuclear receptor, Tailless, is required for efficient proliferation and prolonged maintenance of mushroom body progenitors in the Drosophila brain. *Developmental Biology*, *326*(1), 224–236. https://doi.org/10.1016/j.ydbio.2008.11.013.

Lehmann, R., & Nusslein-Volhard, C. (1987). hunchback, a gene required for segmentation of an anterior and posterior region of the Drosophila embryo. *Developmental Biology*, *119*(2), 402–417. https://doi.org/10.1016/0012-1606(87)90045-5.

Li, X., Chen, Z., & Desplan, C. (2013). Temporal patterning of neural progenitors in Drosophila. *Current Topics in Developmental Biology*, *105*, 69–96. https://doi.org/10.1016/B978-0-12-396968-2.00003-8.

Li, X., Isono, K., Yamada, D., Endo, T. A., Endoh, M., Shinga, J., et al. (2011). Mammalian polycomb-like Pcl2/Mtf2 is a novel regulatory component of PRC2 that can differentially modulate polycomb activity both at the Hox gene cluster and at Cdkn2a genes. *Molecular and Cellular Biology*, *31*(2), 351–364. https://doi.org/10.1128/MCB.00259-10.

Llinares-Benadero, C., & Borrell, V. (2019). Deconstructing cortical folding: genetic, cellular and mechanical determinants. *Nature Reviews Neuroscience*, *20*(3), 161–176. https://doi.org/10.1038/s41583-018-0112-2.

Lu, F., Kar, D., Gruenig, N., Zhang, Z. W., Cousins, N., Rodgers, H. M., et al. (2013). Rax is a selector gene for mediobasal hypothalamic cell types. *The Journal of Neuroscience*, *33*(1), 259–272. https://doi.org/10.1523/JNEUROSCI.0913-12.2013.

Lui, J. H., Hansen, D. V., & Kriegstein, A. R. (2011). Development and evolution of the human neocortex. *Cell*, *146*(1), 18–36. https://doi.org/10.1016/j.cell.2011.06.030.

Martin-Duran, J. M., Pang, K., Borve, A., Le, H. S., Furu, A., Cannon, J. T., et al. (2017). Convergent evolution of bilaterian nerve cords. *Nature*, *553*, 45–50. https://doi.org/10.1038/nature25030.

Metzis, V., Steinhauser, S., Pakanavicius, E., Gouti, M., Stamataki, D., Ivanovitch, K., et al. (2018). Nervous system regionalization entails axial allocation before neural differentiation. *Cell*, *175*(4). https://doi.org/10.1016/j.cell.2018.09.0401105–1118.e1117.

Miguel-Aliaga, I., & Thor, S. (2004). Segment-specific prevention of pioneer neuron apoptosis by cell-autonomous, postmitotic Hox gene activity. *Development*, *131*(24), 6093–6105.

Miguel-Aliaga, I., & Thor, S. (2009). Programmed cell death in the nervous system—A programmed cell fate? *Current Opinion in Neurobiology*, *19*(2), 127–133. https://doi.org/10.1016/j.conb.2009.04.002.

Monedero Cobeta, I., Yaghmaeian Salmani, B. Y., & Thor, S. (2017). Anterior-posterior gradient in neural stem and daughter cell proliferation governed by spatial and temporal Hox control. *Current Biology*, *27*(8), 1161–1172. https://doi.org/10.1016/j.cub.2017.03.023.

Muller, J., & Verrijzer, P. (2009). Biochemical mechanisms of gene regulation by polycomb group protein complexes. *Current Opinion in Genetics & Development*, *19*(2), 150–158. https://doi.org/10.1016/j.gde.2009.03.001.

Murai, K., Qu, Q., Sun, G., Ye, P., Li, W., Asuelime, G., et al. (2014). Nuclear receptor TLX stimulates hippocampal neurogenesis and enhances learning and memory in a transgenic mouse model. *Proceedings of the National Academy of Sciences of the United States of America, 111*(25), 9115–9120. https://doi.org/10.1073/pnas.1406779111.

Nielsen, C. (2012). How to make a protostome. *Invertebrate Systematics, 26*(1), 25–40. https://doi.org/10.1071/IS11041.

Nielsen, C. (2015). Larval nervous systems: True larval and precocious adult. *Journal of Experimental Biology, 218*(4), 629–636. https://doi.org/10.1242/jeb.109603.

Nonaka-Kinoshita, M., Reillo, I., Artegiani, B., Martinez-Martinez, M. A., Nelson, M., Borrell, V., et al. (2013). Regulation of cerebral cortex size and folding by expansion of basal progenitors. *The EMBO Journal, 32*(13), 1817–1828. https://doi.org/10.1038/emboj.2013.96.

Novotny, T., Eiselt, R., & Urban, J. (2002). Hunchback is required for the specification of the early sublineage of neuroblast 7-3 in the Drosophila central nervous system. *Development, 129*(4), 1027–1036.

Nusslein-Volhard, C. (1991). Determination of the embryonic axes of Drosophila. *Development Supplement, 1*, 1–10.

Paridaen, J. T. M. L., & Huttner, W. B. (2014). Neurogenesis during development of the vertebrate central nervous system. *EMBO Reports, 15*(4), 351–364. https://doi.org/10.1002/embr.201438447.

Parker, H. J., & Krumlauf, R. (2017). Segmental arithmetic: summing up the Hox gene regulatory network for hindbrain development in chordates. *Wiley Interdisciplinary Reviews: Developmental Biology, 6*(6), 1–28. https://doi.org/10.1002/wdev.286.

Pattyn, A., Vallstedt, A., Dias, J. M., Samad, O. A., Krumlauf, R., Rijli, F. M., et al. (2003). Coordinated temporal and spatial control of motor neuron and serotonergic neuron generation from a common pool of CNS progenitors. *Genes & Development, 17*(6), 729–737.

Pattyn, A., Vallstedt, A., Dias, J. M., Sander, M., & Ericson, J. (2003). Complementary roles for Nkx6 and Nkx2 class proteins in the establishment of motoneuron identity in the hindbrain. *Development, 130*(17), 4149–4159.

Philippidou, P., & Dasen, J. S. (2013). Hox genes: choreographers in neural development, architects of circuit organization. *Neuron, 80*(1), 12–34. https://doi.org/10.1016/j.neuron.2013.09.020.

Price, M. G., Yoo, J. W., Burgess, D. L., Deng, F., Hrachovy, R. A., Frost, J. D., Jr., et al. (2009). A triplet repeat expansion genetic mouse model of infantile spasms syndrome, Arx(GCG)10+7, with interneuronopathy, spasms in infancy, persistent seizures, and adult cognitive and behavioral impairment. *The Journal of Neuroscience, 29*(27), 8752–8763. https://doi.org/10.1523/JNEUROSCI.0915-09.2009.

Reiprich, S., & Wegner, M. (2015). From CNS stem cells to neurons and glia: Sox for everyone. *Cell and Tissue Research, 359*(1), 111–124. https://doi.org/10.1007/s00441-014-1909-6.

Rogulja-Ortmann, A., Picao-Osorio, J., Villava, C., Patraquim, P., Lafuente, E., Aspden, J., et al. (2014). The RNA-binding protein ELAV regulates Hox RNA processing, expression and function within the Drosophila nervous system. *Development, 141*(10), 2046–2056. https://doi.org/10.1242/dev.101519.

Sanson, B. (2001). Generating patterns from fields of cells. Examples from Drosophila segmentation. *EMBO Reports, 2*(12), 1083–1088. https://doi.org/10.1093/embo-reports/kve255.

Schmid, A., Chiba, A., & Doe, C. Q. (1999). Clonal analysis of Drosophila embryonic neuroblasts: neural cell types, axon projections and muscle targets. *Development, 126*(21), 4653–4689.

Schmidt, H., Rickert, C., Bossing, T., Vef, O., Urban, J., & Technau, G. M. (1997). The embryonic central nervous system lineages of Drosophila melanogaster. II. Neuroblast lineages derived from the dorsal part of the neuroectoderm. *Developmental Biology, 189*(2), 186–204.

Shimamura, K., Hartigan, D. J., Martinez, S., Puelles, L., & Rubenstein, J. L. (1995). Longitudinal organization of the anterior neural plate and neural tube. *Development*, *121*(12), 3923–3933.

Skeath, J. B. (1999). At the nexus between pattern formation and cell-type specification: the generation of individual neuroblast fates in the Drosophila embryonic central nervous system. *BioEssays*, *21*(11), 922–931.

Skeath, J. B., & Doe, C. Q. (1996). The achaete-scute complex proneural genes contribute to neural precursor specification in the Drosophila CNS. *Current Biology*, *6*(9), 1146–1152.

Skeath, J. B., & Thor, S. (2003). Genetic control of Drosophila nerve cord development. *Current Opinion in Neurobiology*, *13*(1), 8–15.

Smart, I. H. (1972a). Proliferative characteristics of the ependymal layer during the early development of the mouse diencephalon, as revealed by recording the number, location, and plane of cleavage of mitotic figures. *Journal of Anatomy*, *113*(Pt. 1), 109–129.

Smart, I. H. (1972b). Proliferative characteristics of the ependymal layer during the early development of the spinal cord in the mouse. *Journal of Anatomy*, *111*(Pt. 3), 365–380.

Smart, I. H. (1973). Proliferative characteristics of the ependymal layer during the early development of the mouse neocortex: a pilot study based on recording the number, location and plane of cleavage of mitotic figures. *Journal of Anatomy*, *116*(Pt. 1), 67–91.

Smart, I. H. (1976). A pilot study of cell production by the ganglionic eminences of the developing mouse brain. *Journal of Anatomy*, *121*(Pt. 1), 71–84.

Sousa-Nunes, R., & Somers, W. G. (2013). Mechanisms of asymmetric progenitor divisions in the Drosophila central nervous system. *Advances in Experimental Medicine and Biology*, *786*, 79–102. https://doi.org/10.1007/978-94-007-6621-1_6.

Struhl, G. (1983). Role of the esc+ gene product in ensuring the selective expression of segment-specific homeotic genes in Drosophila. *Journal of Embryology and Experimental Morphology*, *76*, 297–331.

Struhl, G., & Akam, M. (1985). Altered distributions of Ultrabithorax transcripts in extra sex combs mutant embryos of Drosophila. *The EMBO Journal*, *4*(12), 3259–3264.

Stuttem, I., & Campos-Ortega, J. A. (1991). Cell commitment and cell interactions in the ectoderm of Drosophila melanogaster. *Development Supplement*, *2*, 39–46.

Suzuki, M., Mizutani-Koseki, Y., Fujimura, Y., Miyagishima, H., Kaneko, T., Takada, Y., et al. (2002). Involvement of the Polycomb-group gene Ring1B in the specification of the anterior-posterior axis in mice. *Development*, *129*(18), 4171–4183.

Tarabykin, V., Stoykova, A., Usman, N., & Gruss, P. (2001). Cortical upper layer neurons derive from the subventricular zone as indicated by Svet1 gene expression. *Development*, *128*(11), 1983–1993.

Tosches, M. A., & Arendt, D. (2013). The bilaterian forebrain: An evolutionary chimaera. *Current Opinion in Neurobiology*, *23*(6), 1080–1089. https://doi.org/10.1016/j.conb.2013.09.005.

Trowe, M. O., Zhao, L., Weiss, A. C., Christoffels, V., Epstein, D. J., & Kispert, A. (2013). Inhibition of Sox2-dependent activation of Shh in the ventral diencephalon by Tbx3 is required for formation of the neurohypophysis. *Development*, *140*(11), 2299–2309. https://doi.org/10.1242/dev.094524.

Unternaehrer, J. J., Zhao, R., Kim, K., Cesana, M., Powers, J. T., Ratanasirintrawoot, S., et al. (2014). The epithelial-mesenchymal transition factor SNAIL paradoxically enhances reprogramming. *Stem Cell Reports*, *3*, 691–698.

Urbach, R., Jussen, D., & Technau, G. M. (2016). Gene expression profiles uncover individual identities of gnathal neuroblasts and serial homologies in the embryonic CNS of Drosophila. *Development*, *143*(8), 1290–1301. https://doi.org/10.1242/dev.133546.

Urbach, R., Schnabel, R., & Technau, G. M. (2003). The pattern of neuroblast formation, mitotic domains and proneural gene expression during early brain development in Drosophila. *Development*, *130*(16), 3589–3606.

Urbach, R., & Technau, G. M. (2003). Early steps in building the insect brain: neuroblast formation and segmental patterning in the developing brain of different insect species. *Arthropod Structure & Development, 32*(1), 103–123.

Urban, N., van den Berg, D. L., Forget, A., Andersen, J., Demmers, J. A., Hunt, C., et al. (2016). Return to quiescence of mouse neural stem cells by degradation of a pro-activation protein. *Science, 353*(6296), 292–295. https://doi.org/10.1126/science.aaf4802.

Vallstedt, A., Klos, J. M., & Ericson, J. (2005). Multiple dorsoventral origins of oligodendrocyte generation in the spinal cord and hindbrain. *Neuron, 45*(1), 55–67.

Walsh, K. T., & Doe, C. Q. (2017). Drosophila embryonic type II neuroblasts: origin, temporal patterning, and contribution to the adult central complex. *Development, 144*(24), 4552–4562. https://doi.org/10.1242/dev.157826.

Wang, L., Bluske, K. K., Dickel, L. K., & Nakagawa, Y. (2011). Basal progenitor cells in the embryonic mouse thalamus—Their molecular characterization and the role of neurogenins and Pax6. *Neural Development, 6,* 35. https://doi.org/10.1186/1749-8104-6-35.

Wang, W., & Lufkin, T. (2000). The murine Otp homeobox gene plays an essential role in the specification of neuronal cell lineages in the developing hypothalamus. *Developmental Biology, 227*(2), 432–449. https://doi.org/10.1006/dbio.2000.9902.

Wang, J., Mager, J., Schnedier, E., & Magnuson, T. (2002). The mouse PcG gene eed is required for Hox gene repression and extraembryonic development. *Mammalian Genome, 13*(9), 493–503. https://doi.org/10.1007/s00335-002-2182-7.

Wheeler, S. R., Stagg, S. B., & Crews, S. T. (2009). MidExDB: A database of Drosophila CNS midline cell gene expression. *BMC Developmental Biology, 9,* 56.

White, R. A., & Lehmann, R. (1986). A gap gene, hunchback, regulates the spatial expression of Ultrabithorax. *Cell, 47*(2), 311–321. https://doi.org/10.1016/0092-8674(86)90453-8.

Yaghmaeian Salmani, B., Monedero Cobeta, I., Rakar, J., Bauer, S., Curt, J. R., Starkenberg, A., et al. (2018). Evolutionarily conserved anterior expansion of the central nervous system promoted by a common PcG-Hox program. *Development, 145*(7), 1–17. https://doi.org/10.1242/dev.160747.

Younossi-Hartenstein, A., Green, P., Liaw, G. J., Rudolph, K., Lengyel, J., & Hartenstein, V. (1997). Control of early neurogenesis of the Drosophila brain by the head gap genes tll, otd, ems, and btd. *Developmental Biology, 182*(2), 270–283. https://doi.org/10.1006/dbio.1996.8475.

Younossi-Hartenstein, A., Nassif, C., Green, P., & Hartenstein, V. (1996). Early neurogenesis of the Drosophila brain. *The Journal of Comparative Neurology, 370*(3), 313–32910.1002/(SICI)1096-9861(19960701)370:3<313::AID-CNE3>3.0.CO;2-7.

Yu, J., Guo, Q., Mu, J. B., Zhang, T., Li, R. K., & Xie, J. (2017). Nr2e1 downregulation is involved in excess retinoic acid-induced developmental abnormality in the mouse brain. *Biomedical and Environmental Sciences, 30*(3), 185–193. https://doi.org/10.3967/bes2017.026.

Zander, M. A., Burns, S. E., Yang, G., Kaplan, D. R., & Miller, F. D. (2014). Snail coordinately regulates downstream pathways to control multiple aspects of Mammalian neural precursor development. *Journal of Neuroscience, 34,* 5164–5175.

Zhang, C. C., & Bienz, M. (1992). Segmental determination in Drosophila conferred by hunchback (hb), a repressor of the homeotic gene Ultrabithorax (Ubx). *Proceedings of the National Academy of Sciences of the United States of America, 89*(16), 7511–7515. https://doi.org/10.1073/pnas.89.16.7511.

CHAPTER TWELVE

Temporal dynamics in the formation and interpretation of Nodal and BMP morphogen gradients

Andrew D. Economou, Caroline S. Hill*

Developmental Signalling Laboratory, The Francis Crick Institute, London, United Kingdom

*Corresponding author: e-mail address: caroline.hill@crick.ac.uk

Contents

Abstract

One of the most powerful ideas in developmental biology has been that of the morphogen gradient. In the classical view, a signaling molecule is produced at a local source from where it diffuses, resulting in graded levels across the tissue. This gradient provides positional information, with thresholds in the level of the morphogen determining the position of different cell fates. While experimental studies have uncovered numerous potential morphogens in biological systems, it is becoming increasingly apparent that one important feature, not captured in the simple model, is the role of time in both the formation and interpretation of morphogen gradients. We will focus on two members of the transforming growth factor-β family that are known to play a vital role as morphogens in early vertebrate development: the Nodals and the bone morphogenetic proteins (BMPs). Primarily drawing on the early zebrafish embryo, we will show how recent studies have demonstrated the importance of feedback and other interactions that evolve through time, in shaping morphogen gradients. We will further show how rather than simply reading out levels of a morphogen, the duration of ligand exposure can be a crucial determinant of how cells interpret morphogens, in particular through the unfolding of downstream transcriptional events and in their interactions with other pathways.

Current Topics in Developmental Biology, Volume 137
ISSN 0070-2153
https://doi.org/10.1016/bs.ctdb.2019.10.012

1. Introduction

Morphogen gradients have been one of the most influential ideas in developmental biology (Wolpert, 2011). They were initially proposed by Lewis Wolpert as a solution to the French Flag problem, which he formulated to focus his ideas on pattern formation. He proposed that a graded morphogen could act as a source of positional information across a tissue, whereby the concentration would encode a positional value. By interpreting the concentration of a morphogen relative to a threshold, cells in different parts of the embryo could differentiate into distinct fates.

One fundamental feature of developing embryos, which is excluded from this classical view of morphogens and has largely been neglected (although see Green, Howes, Symes, Cooke, & Smith, 1990; Pages & Kerridge, 2000), is how they continuously change over time. During development, patterns of gene expression are constantly changing in tissues that are reshaping through morphogenesis. This sets up the potential for feedback and feedforward interactions to shape a morphogen gradient and its effects on target genes through time. These interactions may be at the level of the signaling pathway downstream of the particular morphogen of interest, but they might also involve the induction or inhibition of other distinct signaling pathways that contribute to the regulation of the morphogen target genes. The importance of time is at odds with the idea of a static morphogen gradient, where fixed ligand concentrations give rise to distinct cell fates. This raises a number of questions: how do morphogen gradients develop through time? How does the timing of exposure to morphogens affect cell fate decisions?

In this review, we will address these questions by focusing on two members of the transforming growth factor-β (TGF-β) family that have been implicated as morphogens in early vertebrate embryos: Nodals and bone morphogenetic proteins (BMPs). Both are ancient signaling molecules which play central roles in embryonic development. Nodal signaling has a conserved role throughout the deuterostomes in the induction of mesoderm (prospective muscle and blood) and endoderm (prospective gut and other internal organs), as well as in left–right asymmetry (Schier, 2003), while BMP signaling is involved in many processes across the Metazoa, in particular a conserved role in dorsal–ventral (DV) patterning (Ramel & Hill, 2012). Indeed, it has been argued that intersecting gradients of Nodal and BMP alone are sufficient to organize the embryonic axes (Xu, Houssin, Ferri-Lagneau, Thisse, & Thisse, 2014).

As with all members of the TGF-β family, Nodals and BMPs signal through complexes of serine/threonine kinase receptors comprising two type I and two type II receptors (Heldin & Moustakas, 2016) (Fig. 1). In addition, in the case of Nodal signaling, the co-receptor Tdgf1 (also known

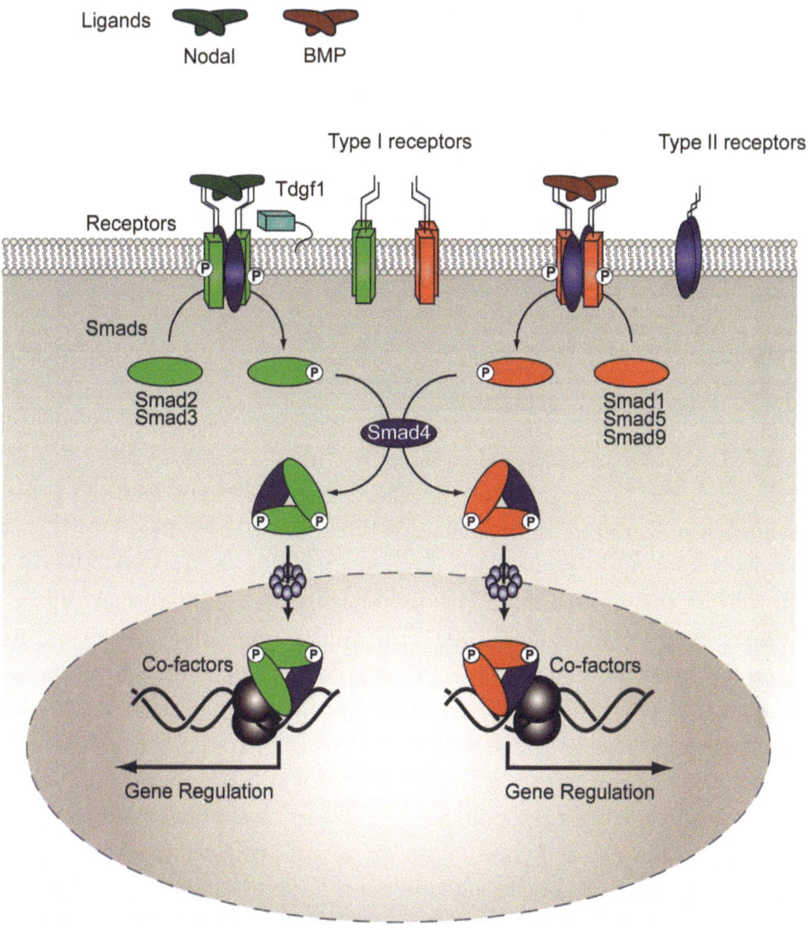

Fig. 1 Nodal and BMP signaling pathways. Schematic illustrating the main components of the Nodal and BMP signaling pathways. Ligands signal though a heterotetrameric complex of type I and type II receptors, which for Nodal also requires the co-receptor Tdgf1. Upon ligand binding constitutively active type II receptors phosphorylate type I receptors, resulting in the phosphorylation of receptor-regulated Smads—Smad2/3 for Nodal signaling, and Smad1/5/9 for BMP signaling. Phosphorylated Smads form trimeric complexes with Smad4 which accumulate in the nucleus where they regulate transcription. *Adapted from Schmierer, B., & Hill, C. S. (2007). TGFβ-SMAD signal transduction: molecular specificity and functional flexibility. Nature Reviews. Molecular Cell Biology, 8, 970–982.*

as Cripto, or One eyed pinhead (Oep)) is also required (Schier, 2003). Ligand binding results in the phosphorylation of the intracellular mediators of the pathway, the receptor-regulated Smads (R-Smads)—Smad2 or Smad3 for Nodal signaling; Smad1, Smad5 or Smad9 for BMP signaling—which complex with Smad4 and accumulate in the nucleus when the pathway is active (Miller & Hill, 2016) (Fig. 1). We will focus primarily on the zebrafish embryo, where many recent studies have shed light on the formation and interpretation of gradients of these two morphogens. We also include studies using *Drosophila* and *Xenopus*, which have informed the development of models particularly for BMP gradient formation. We will discuss the importance of timing and the involvement of additional cooperating signaling pathways in the establishment and interpretation of these morphogen gradients.

2. Gradient formation

2.1 Nodal

In the early zebrafish embryo, the two Nodal ligands Nodal-related 1 (Ndr1, also known as Squint) (Feldman et al., 1998) and Nodal-related 2 (Ndr2, also known as Cyclops) (Rebagliati, Toyama, Haffter, & Dawid, 1998) are required for the induction of the mesodermal and endodermal cell fates (Feldman et al., 1998; Schier, 2003). Starting at sphere stage (4 h post fertilization (hpf)) when the blastoderm has the form of a cap of cells on top of a ball of yolk, the Nodal ligands are initially produced in an extra-embryonic tissue known as the yolk syncytial layer (YSL) (Erter, Solnica-Krezel, & Wright, 1998; Feldman et al., 1998). In the subsequent 90 min, as the embryo progresses through epiboly (the process by which the blastoderm thins and spreads over the yolk—Fig. 2A), it was proposed that these ligands produced a gradient of Nodal signaling at the margin of the embryo where the mesendodermal cell fates are induced (Chen & Schier, 2001). These cells subsequently involute during gastrulation (from 5.7 hpf—see van Boxtel et al., 2015 for stages and timings).

It was initially postulated that Ndr1 acts as a classical morphogen in inducing these cell fates (Chen & Schier, 2001). By injecting *ndr1* mRNA at the animal pole, these authors were able to reproduce the patterns of gene expression at the margin around the clones. Importantly, the induction of Nodal target genes occurred in an apparent dose-dependent manner, with low sensitivity targets such as *goosecoid* (*gsc*) requiring high levels of *ndr1*

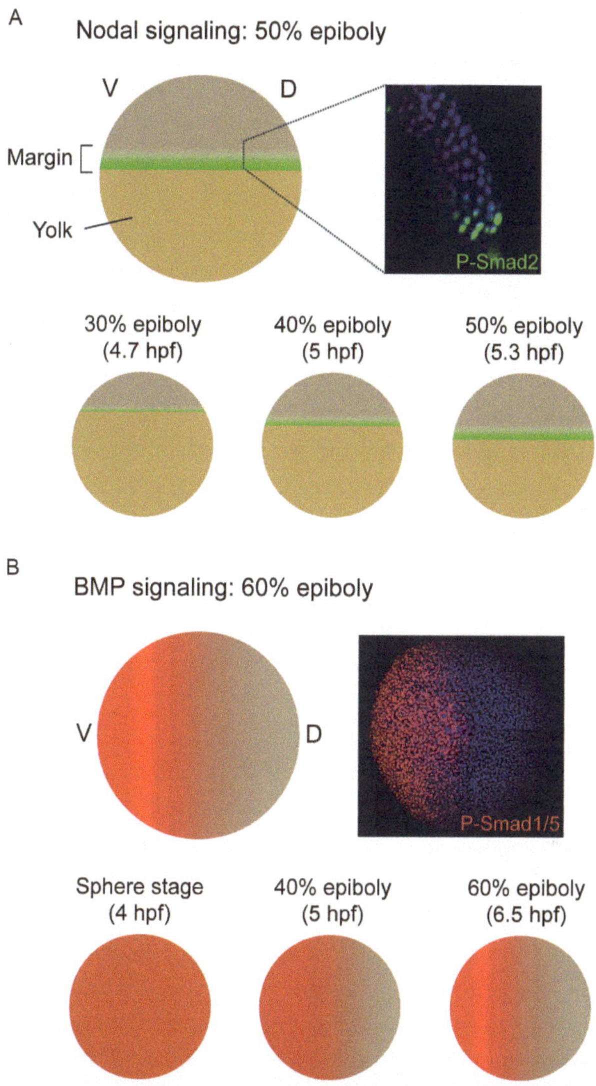

Fig. 2 Nodal and BMP gradients in the early zebrafish embryo. (A) A schematic showing the gradient of Nodal signaling at the margin in a 50% epiboly embryo extending over the first five cell tiers, as seen by immunofluorescence for P-Smad2 (green), shown in cross section. Nuclei are stained with DAPI (blue). This gradient forms gradually, as the signal spreads from the YSL between 30 and 50% epiboly, as shown in schematics of embryos viewed laterally. The timings of these stages are also given in hours post fertilization (hpf). (B) A schematic showing the gradient of BMP signaling along the DV axis of a 60% epiboly embryo seen from the animal pole, as visualized by immunofluorescence for P-Smad1/5 (red), shown as a maximum projection. Nuclei are stained with DAPI (blue). The gradient forms in reverse, starting with uniform low levels of signaling across the embryo at sphere stage, before clearing dorsally and intensifying with time. All schematics shown with dorsal (D) to the right and ventral (V) to the left.

mRNA for their induction, and high sensitivity targets such as *tbxta* (also known as *no-tail, ta* or *brachyury*), *noto* (also known as *floating head*) and *bhikhari* (*bik*) requiring lower levels. Moreover, Ndr1 acted at long range, with the high sensitivity targets being induced up to 8 cell tiers away, while importantly, the low sensitivity targets were only induced in 2 cell tiers surrounding the clones. These results, taken together with further transplantation experiments, led the authors to argue that Ndr1 acted as a morphogen which diffused through tissue to set up a gradient and induce cell fates at different ranges, in accordance with a classical French Flag model (Fig. 3A).

This view of Nodal acting as a classical morphogen does not consider that the Nodal ligands Ndr1 and Ndr2 are themselves targets of Nodal signaling (Whitman, 2001), as are the secreted antagonists of Nodal signaling, Lefty1 and Lefty2 (Meno et al., 1999). Consequently, an evolving gradient of Nodal signaling would be subject to both positive and negative feedback. A number of studies have considered potential consequences of these feedbacks on gradient formation, with Chen and Schier initially proposing that they are in fact crucial for the establishment of the domain of Nodal signaling (Chen & Schier, 2002). It was proposed that the observed behaviors of Nodal and Lefty1/2 are consistent with a Turing-type reaction–diffusion system (Chen & Schier, 2002; Meinhardt & Gierer, 2000; Saijoh et al., 2000). In a reaction–diffusion system the activator should induce its own expression and that of the antagonist, which should also function at a longer range than the activator. Chen and Schier argued that in addition to the well-established regulation of the Nodals and Leftys by Nodal signaling, a crucial feature of such a system is that Lefty should act to diminish the range of Nodal signaling by antagonizing Nodal signaling (Fig. 3B). Indeed, by knocking down the Leftys they showed an expansion in the expression domains of Nodal targets *tbxta*, *bik* and *gsc*. This could occur independently of Nodal autoregulation, as domains of Nodal target gene expression around clones of Ndr1 at the animal pole expanded upon Lefty knockdown, even in *ndr1*$^{-/-}$;*ndr2*$^{-/-}$ double mutants where autoregulation of Nodal was not possible. Another key feature of reaction–diffusion systems is that the antagonist is more diffusive than the activator. This was demonstrated using overexpressed GFP-tagged Nodal and Lefty proteins, further supporting a reaction–diffusion model (Muller et al., 2012).

More recently however, an alternative model for Nodal gradient formation was proposed (van Boxtel et al., 2015). In this model, feedback through the Lefty proteins was shown to be essential for establishing the Nodal morphogen gradient, but in addition, the timing of inhibition through the Leftys

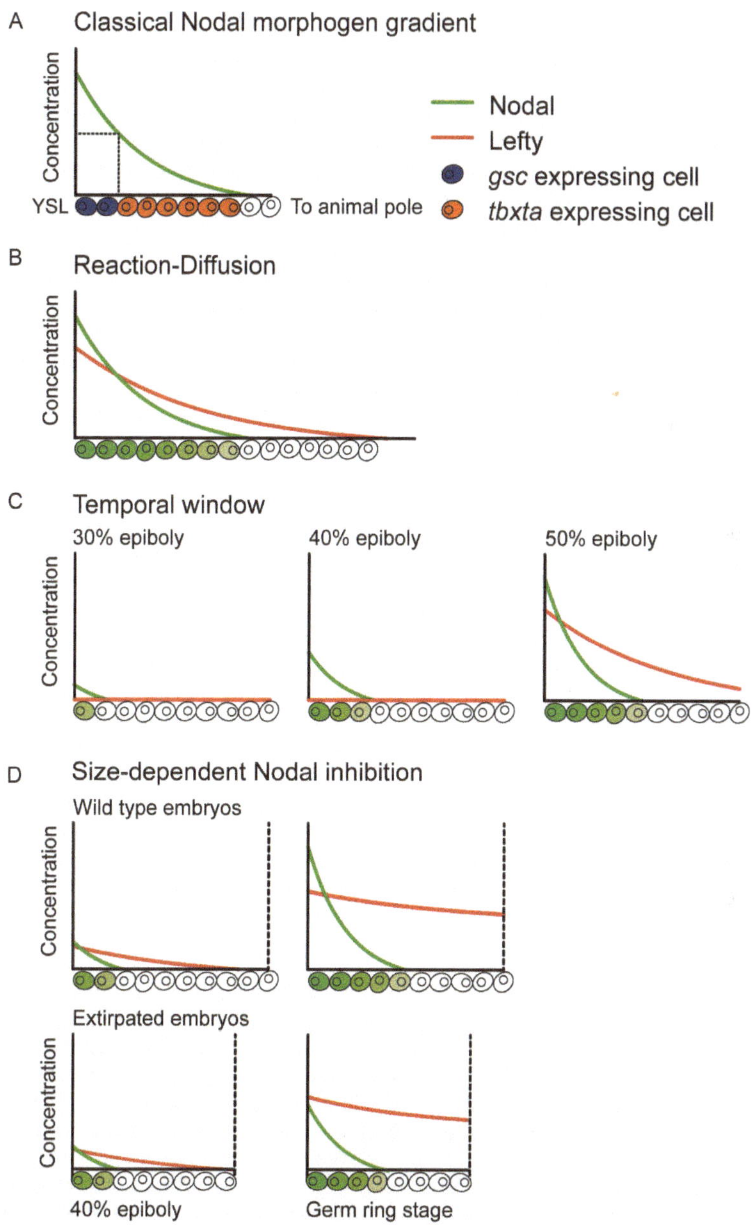

Fig. 3 Models of Nodal morphogen gradient formation. (A) Schematic of Nodal as a morphogen in the classical sense, as illustrated by the induction of cells to express *goosecoid* (*gsc*) or *tbxta*. A gradient of Nodal extends from the yolk syncytial layer (YSL) over approximately eight cell tiers toward the animal pole. Above a threshold signaling concentration (dashed black line) cells are induced to express *gsc*, below *tbxta*.

(Continued)

was a crucial feature. Using a sensitive transcriptional reporter of Nodal signaling alongside immunofluorescence for phosphorylated Smad2 (P-Smad2, a direct readout of the pathway), this study provided the first visualization of the domain of Nodal signaling at the zebrafish margin (Fig. 2A). Surprisingly, it was demonstrated that the range of Nodal signaling is much shorter than previously thought, reaching only around 5 cell tiers from the YSL (the initial source of Nodal ligands), and therefore, not able to account for expression of supposed long-range Nodal targets such as such as *tbxta* and *noto*. Nodal, in fact, was shown to regulate these genes at long-range through the Fgf signaling pathway, with the ligands Fgf3 and Fgf8a themselves being transcriptional targets of Nodal signaling (see below). Importantly, the Nodal ligands are expressed in the same domain as the pathway is active, indicating that unlike Fgf, Nodal does not signal at a long range, but rather spreads from the YSL by a relay, with cells signaling to their immediate neighbors and inducing ligand expression. The gradient of Nodal signaling, as visualized by P-Smad2 staining, is therefore not a readout of ligand diffusing from a source, but rather a temporal gradient, with cells closest to the YSL having been exposed to ligand for longest. Van Boxtel et al. also demonstrated that the length of this gradient is determined by a delay in the translation of *lefty* mRNA, which is regulated by the miRNA *miR-430*. Thus, before *lefty* mRNA is translated, the domain of Nodal signaling can grow, but upon Lefty translation at around 50% epiboly, the spread of Nodal between cell tiers is stopped (van Boxtel et al., 2015) (Fig. 3C). A more recent study has made use of *lft1*$^{-/-}$;*lft2*$^{-/-}$ double mutants and

Fig. 3—Cont'd (B) A reaction–diffusion model for Nodal gradient formation. Nodal signaling induces its faster diffusing inhibitor Lefty, which inhibits further spread of Nodal. (C) Temporal window model of Nodal morphogen gradient formation. The profile of Nodal signaling and Lefty at successive time points showing Nodal signaling spreading by a relay from the YSL—note that Nodal signaling does not diffuse over a long range. Initially, Lefty translation is suppressed (see text for details) before suppression is overcome by 50% epiboly and the spread of Nodal signaling is inhibited. (D) Schematic illustrating size-dependent inhibition of Nodal signaling. Nodal initially spreads from the YSL as in the temporal window model, and produces a fast diffusing Lefty. The spread of Lefty is bounded (dashed line) resulting in an increase in Lefty levels as the embryonic tissue saturates. When a threshold level of Lefty is reached, the spread of Nodal is inhibited. In smaller extirpated embryos (below), while the initial spread of Nodal is the same, levels of Lefty increase more rapidly (compare Lefty profiles), resulting in an earlier inhibition of the spread of Nodal, and a resulting pattern scaling. 40% epiboly corresponds to 1 h post extirpation, and germ ring stage to 2 h post extirpation.

shown that there is a small elevation of P-Smad2 at dome stage in these embryos relative to wild type (WT) embryos, which is then much more dramatic at 50% epiboly (Rogers et al., 2017). This indicates that there must be a low level of Lefty translation before 50% epiboly in WT embryos, which has a slight dampening effect on the growth of the Nodal domain. However, growth of the domain is then terminated at 50% epiboly when levels of the Leftys are maximal.

By considering how the Nodal gradient scales with embryo size, others have achieved further insights into Nodal gradient establishment. When zebrafish embryos are extirpated at 4 hpf (sphere stage), such that around 30% of their cells are removed, surprisingly they still develop normally, giving rise to smaller, but correctly proportioned, juveniles (Almuedo-Castillo et al., 2018). Looking at markers of mesoderm and ectoderm, Almuedo-Castillo and colleagues showed that whereas initially the mesodermal domain was the same size between normal and extirpated embryos and the prospective ectoderm was smaller in extirpated embryos, the two domains subsequently scaled in the smaller embryos. Through a computational screen designed to capture the interactions between Nodal and Lefty which would allow scaling, these authors recovered a model where the fast diffusivity of Lefty was crucial. In this model, graded Nodal signaling expands toward the animal pole, as do levels of Lefty, which due to its fast diffusion, saturates the tissue. Consequently, local Lefty levels increase more rapidly in smaller embryos, reaching levels at which they prevent the growth of the Nodal domain more rapidly, resulting in an earlier arrest of Nodal domain growth. As a result, the Nodal gradient, and therefore the mesoderm, scales to the smaller size of the extirpated embryo (Fig. 3D). Interestingly, the authors found that although the diffusivity of Lefty was crucial for scaling, it was not required in WT embryos, which developed normally even when the only Lefty produced in the embryo was at the margin and membrane associated, and thus poorly diffusible (Almuedo-Castillo et al., 2018).

While the above studies have proposed a key role for the Lefty proteins in determining the range of the Nodal signaling gradient, it has recently been argued that the Lefty proteins are not required for Nodal function per se, but rather the negative feedbacks they support confer robustness to Nodal function (Rogers et al., 2017). Using the *lft1*$^{-/-}$;*lft2*$^{-/-}$ double mutants which lack any Lefty protein, Rogers and colleagues demonstrated that the phenotype of these mutants can be rescued by reducing the dose of Nodal signaling by introducing *ndr1*$^{-/-}$ or *ndr2*$^{-/-}$ mutations, or by treating with a low dose of the Nodal receptor inhibitor SB-505124 (DaCosta Byfield, Major, Laping, & Roberts, 2004), even after Nodal signaling has initiated. Interestingly, Lefty production does not need to be coupled to Nodal, as the phenotype can be rescued by a Lefty-producing clone situated at the animal pole. This has led to the idea that the role of Lefty is to provide negative feedback and buffer any fluctuations in Nodal signaling. Indeed, the embryos rescued by drug treatment are more susceptible to perturbations of Nodal signaling by injecting *lft1* mRNA or *CA-smad2* mRNA (which encodes constitutively active Smad2) than WT embryos (Rogers et al., 2017).

In addition to the feedback through the Leftys, it has also recently been shown that the Nodal ligands Ndr1 and Ndr2 do not act as morphogens alone as homodimers, as had previously been thought. Rather, three recent studies demonstrated that they require another ligand of the TGF-β family, Gdf3 (also known as Vg1), with which they form heterodimers. Using either TALENs or CRISPR/Cas9, Gdf3 mutant zebrafish lines were generated (Bisgrove, Su, & Yost, 2017; Montague & Schier, 2017; Pelliccia, Jindal, & Burdine, 2017). Whereas the zygotic mutant did not show a phenotype, maternal and maternal–zygotic (MZ) mutants showed strong phenotypes resembling the loss of Nodal signaling seen in MZ*oep* mutants or $ndr1^{-/-};ndr2^{-/-}$ double mutants. The studies went on to show that maternal Gdf3 is required for Nodal function, and that the graded domain of Nodal signaling in the embryonic margin is likely through heterodimers of spatially-restricted Nodal ligands with a ubiquitously distributed Gdf3. It has been argued that this makes the embryo more responsive to low levels of Nodal. Interestingly, BMP signaling at these same stages also functions exclusively through Bmp2-Bmp7 heterodimers (Little & Mullins, 2009).

Ultimately, it is likely that no single mechanism acts alone to explain the growth of the Nodal gradient. In the future, it will be important to consider how these different mechanisms could potentially interact, and what predictions this could make for the growth of the gradient. It is also important to address how these Nodal-based systems could interact with other signaling pathways responsible for patterning other domains in the early embryo. For example, Almuedo-Castillo described a scaling system based on Nodal and Lefty (Almuedo-Castillo et al., 2018). However, it is not just the mesoderm that scales; other axes must also scale to produce a miniature embryo. It is interesting to speculate how the Nodal–Lefty based scaling mechanism interacts with potential mechanisms that scale the other axes of the embryo, such as those proposed for BMP signaling (see below).

Taking all these studies together, it is clear that the establishment of a Nodal morphogen gradient is not simply the result of diffusion of ligands from the YSL to form a concentration gradient as would be suggested in a classical morphogen model. Rather, the gradient of Nodal signaling appears to be a temporal gradient resulting from the gradual spread of the signaling domain in the margin, requiring feedbacks from the Lefty proteins and dimerization with Gdf3. These interactions further confer properties such as scalability, robustness and responsiveness to the Nodal gradient.

2.2 BMP

In the early zebrafish embryo, the BMPs are required for DV patterning, with mutants of the ligand Bmp2b (*swirl*) (Kishimoto, Lee, Zon, Hammerschmidt, & Schulte-Merker, 1997), Bmp7 (*snailhouse*) (Dick et al., 2000), the receptor Alk8 (*lost-a-fin*) (equivalent of mammalian ACVR1/ALK2) (Mintzer et al., 2001) and the intracellular transducer Smad5 (*somitabun*) (Hild et al., 1999) all showing dorsalized phenotypes (Mullins et al., 1996), where dorsal cell fates, such as neural and somatic mesoderm, are expanded at the expense of ventral cell fates such as blood and tail. Like Nodal, BMP signaling forms a gradient, but in this case, running from ventral to dorsal across the embryo during blastula stages (Tucker, Mintzer, & Mullins, 2008). Therefore, BMP has been argued to act as a morphogen in the allocation of ventral to dorsal cell fates (Fig. 2B).

As with Nodal signaling, the zebrafish gradient of BMP signaling does not appear to form by diffusion of ligand from a localized source. By visualizing the activity of the pathway through immunofluorescence for phosphorylated Smad5 (P-Smad5), it was evident that there is initially a uniform low level of signaling throughout the early blastula embryo (4 hpf), which progressively clears dorsally and intensifies in the ventral-most regions (Tucker et al., 2008). This produces a steep gradient which persists through gastrulation. Ramel and Hill further demonstrated that this same pattern of uniform expression followed by dorsal clearance is seen for the transcription of the BMP ligands *bmp2b*, *bmp7a* and *bmp4* (Ramel & Hill, 2013). The coincidence of the signaling domain with the expression domains of the ligands at 40% epiboly (5 hpf) suggested that diffusion does not play a major role in the establishment of the morphogen gradient, but rather that it is dependent on the spatial regulation of transcription. Ramel and Hill further demonstrated that dorsal clearance of a uniform transcript, which is the basis for the distribution of the ligands, is initially dependent on the dorsal expression of the transcription repressor Dharma (Koos & Ho, 1999), followed by inhibitory dorsal Fgf signaling (Furthauer, Van Celst, Thisse, & Thisse, 2004), and finally by the dorsal expression of the inhibitor of BMP signaling Chordin (Chd) (Ramel & Hill, 2013; Schulte-Merker, Lee, McMahon, & Hammerschmidt, 1997) (Fig. 4).

Interestingly, this mode of gradient formation, where graded signaling is a result of graded ligand expression, is in marked contrast to BMP gradient formation in the *Drosophila* wing disc, which has been a very influential system for thinking about morphogen gradients. Here the BMP ligands Decapentaplegic (Dpp) and Glass bottom boat (Gbb) are essential for growth and patterning (Raftery & Umulis, 2012). While Gbb is uniformly expressed throughout the disc, Dpp is expressed in a localized medial stripe. Unlike the situation in the fish, graded BMP signaling, as read out by phosphorylation of Mad (the *Drosophila* ortholog of Smad1/5/9) extends across the disc from this localized stripe. In this sense, Dpp in the wing disc apparently resembles a classic morphogen gradient. Two major models of Dpp dispersal have been proposed: a receptor-mediated transcytosis model, where Dpp is transported through cells via endocytic, intracellular trafficking (Entchev, Schwabedissen, & Gonzalez-Gaitan, 2000); and a hindered extracellular diffusion model, where Dpp spreads by diffusion which is inhibited by binding to extracellular matrix components such as heparin sulfate proteoglycans (Schwank et al., 2011). More recent data favors the hindered diffusion model (Muller, Rogers, Yu, Brand, & Schier, 2013).

BMP signaling also plays a role in DV patterning in the early *Drosophila* embryo, but here yet another mechanism of Dpp dispersal is at play (O'Connor, Umulis, Othmer, & Blair, 2006). However, in this system the activity gradient runs from dorsal to ventral, the reverse to what is seen in the zebrafish and other vertebrates (De Robertis, 2008). In the embryo, the BMP signaling gradient forms a sharp dorsal peak, as visualized by P-Mad, whereas neither of the two BMP ligands expressed in the early embryo show such a pattern, with Dpp expressed in a broad dorsal domain, and another BMP ligand, Screw (Scw) expressed uniformly throughout the embryo. Eldar and colleagues addressed the formation of this gradient, in particular focusing on interactions between the BMP ligands, Short gastrulation (Sog), and Tolloid (Tld) (Eldar et al., 2002). Sog is the *Drosophila* Chd ortholog, which is expressed ventrally (again the reverse of what is seen in the zebrafish), and Tld is the protease that cleaves Sog. These authors asked how robust the shape of the gradient is to changes in gene dose and showed that heterozygotes of *scw, sog* and *tld* all exhibited WT BMP activity

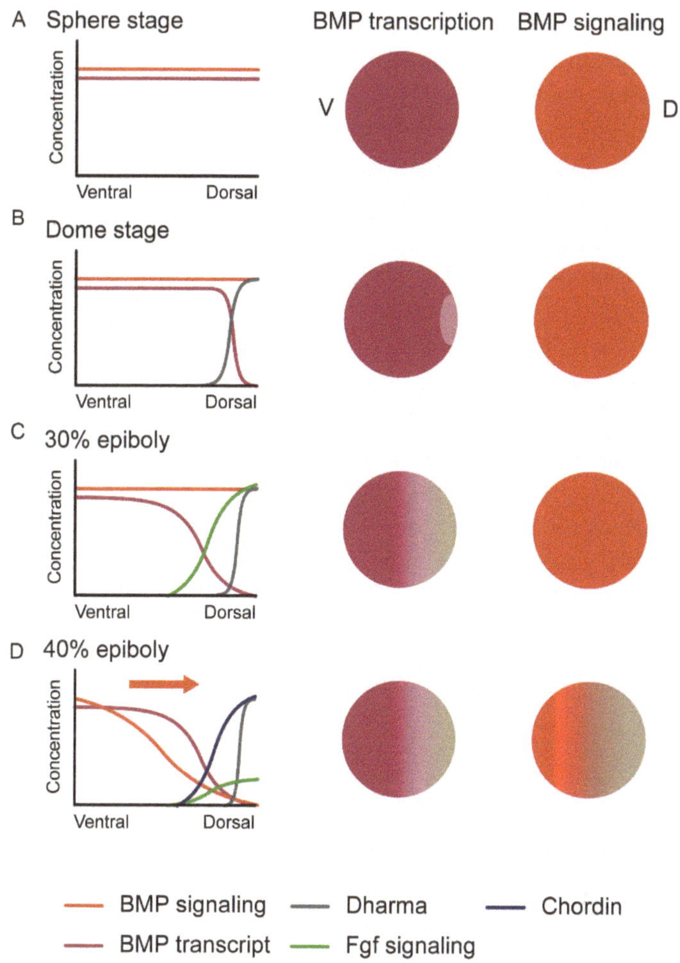

Fig. 4 Regulation of the BMP signaling domain through time. (A) Profiles and animal pole views of embryos showing uniform BMP transcription and signaling at sphere stage. (B) At dome stage, dorsal expression of the transcriptional repressor Dharma inhibits dorsal-most transcription of BMP ligands, but signaling remains uniform. (C) Dorsal Fgf signaling domain at 30% epiboly further represses BMP transcription, leading to a larger clearance of BMP transcript dorsally, with signaling still uniform. (D) At 40% epiboly, dorsal expression of BMP antagonist Chd results in a dorsal clearance of BMP ligands, resulting in a signaling gradient. Dorsal depletion of BMP ligand by Chd produces a local sink, leading to ventral-to-dorsal diffusion of BMP (red arrow) as described by a source–sink model (Zinski et al., 2017). For simplicity, only shapes of profiles are shown, with variations in BMP signaling and transcript levels between stages (as described in text and Fig. 2) not shown. In all profiles and embryo schematics, dorsal (D) to right, ventral (V) to left.

gradients, although haploinsufficiency was seen for *dpp*. The authors considered the diffusion of the BMP ligands, Sog and the BMP–Sog complex, as well as the cleavage of Sog by Tld, both when complexed to BMP and when free. They performed a computational screen for networks that showed robustness to changes in dose for Sog, Tld and BMP ligands. Rather than a source–sink model where the ventral inhibitor Sog depletes a dorsally-produced BMP ligand, they identified a shuttling model (Eldar et al., 2002). In this model, the cleavage of Sog by Tld is facilitated by complexing with BMP, and Sog–BMP is broadly diffusible, while the diffusion of BMP alone is restricted. By diffusing dorsally in complex with BMP, Sog concentrates BMPs to the dorsal side, where Sog is cleaved, resulting in a sharp dorsal peak of BMP activity, even in the presence of broad ligand expression. This model has subsequently been validated experimentally (see for example Sawala, Sutcliffe, & Ashe, 2012; Shimmi, Umulis, Othmer, & O'Connor, 2005). Thus, this mechanism for DV patterning contrasts significantly with how the BMP gradient is formed in the wing disc, and highlights the range of biological mechanisms that *Drosophila* use to pattern tissues, even when utilizing the same morphogen.

A potential role for shuttling in the establishment of the vertebrate DV BMP gradient has also been proposed, in the context of embryonic scaling (Ben-Zvi, Shilo, Fainsod, & Barkai, 2008). These authors considered the ability of the dorsal halves of bisected *Xenopus* embryos to produce well-proportioned tadpoles, as well as the ability of grafted Spemann organizers to induce well-proportioned second axes. Accounting for the interactions between the BMPs (Bmp2/4/7) and another BMP ligand Admp, which is dorsally localized, as well as interactions with Chd and the Tld orthologue Xlr, a computational screen was performed for parameter sets where the pattern could scale in a dorsal half embryo. This screen identified shuttling in all networks capable of robust scaling. The presence of shuttling was then inferred from experiments where they injected Myc-tagged Bmp4 into the dorsal side of the embryo, and observed that Bmp4 spread to the ventral pole only in the presence of Chd (Ben-Zvi et al., 2008). However, others have disputed the ability of dorsal half embryos to scale, further arguing against the agreement of the model predictions with known BMP and Admp phenotypes (Francois, Vonica, Brivanlou, & Siggia, 2009). It was also asserted that shuttling is not required for axis duplication, rather, they proposed that a reaction–diffusion model could recapitulate the observed patterns (Francois et al., 2009).

Given the diversity of models for DV BMP gradient formation from studies in *Drosophila* and *Xenopus*, two recent studies reconsidered the shaping of the DV BMP activity gradient in zebrafish (Pomreinke et al., 2017; Zinski et al., 2017). Zinski et al. considered four possible models for gradient formation: the transcriptional gradient (as described above); a shuttling model as in *Drosophila* DV patterning; a counter gradient model where diffusion of Chd from the dorsal side established the gradient in BMP activity; and a source–sink model where dorsally-expressed Chd depletes BMP ligands, thus establishing a BMP diffusion gradient from ventral to dorsal (Zinski et al., 2017). These models were formalized mathematically, and 10^6 parameter sets were screened to identify which model fit the observed BMP gradient measured by quantifying P-Smad1/5/9 immunofluorescence, as well as the observed expression domains of BMP, Chd and Noggin in WT embryos. Through further analysis of Chd mutant embryos, these authors found surprisingly that most of the parameter sets that fit the data were source–sink or counter gradient models, with little support for the shuttling or transcriptional models. As the source–sink and counter gradient models predict different BMP diffusivity (high in source–sink, low in counter gradient), they measured BMP diffusion by FRAP on Venus-tagged Bmp2b, and found diffusion rates consistent with their source–sink model (Zinski et al., 2017) (Fig. 4D).

In a similar study, Pomreinke et al. considered the source–sink, transcriptional and shuttling models for establishment of a BMP signaling gradient, as well as a long-range accumulation and feedback model (Inomata, Shibata, Haraguchi, & Sasai, 2013), where the gradient depends on differential stability of BMP and Chd rather than diffusion, and the self-regulating reaction–diffusion model referred to above (Francois et al., 2009), where the gradient of BMP can form noisy initial conditions as part of a self-organizing system (Pomreinke et al., 2017). A series of constraints were identified that the different models place on the biophysical properties of BMP and Chd, in particular on their diffusion and degradation. Through quantifying the evolution of the BMP activity gradient, using immunofluorescence for P-Smad1/5/9 in both WT and Chd mutants, and measuring the stability and diffusion of superfolder GFP (sfGFP)-tagged BMP and Chd, the authors showed that the proteins had similar diffusion and degradation rates. Chd did not affect the diffusion of BMP, nor was it important for peak levels of P-Smad1/5/9 activity as would be predicted in a shuttling model. As with the Zinski study, they argued that their measured properties were most consistent with the source–sink model (Fig. 4D).

Therefore, taken together, these studies suggest that the BMP gradient in zebrafish DV patterning evolves through time in a manner highly dependent on the spatial pattern of transcription of the ligand, but also through interaction with other DV-localized proteins in the early embryo. While in the wing disc the model of BMP gradient formation may show elements of a classical model of a morphogen diffusing from a local source to set up a gradient, this does not capture the variety of ways morphogen gradients are set up in biological systems. Also, as with the Nodal signaling gradient, these studies on BMP signaling in early zebrafish embryos have largely focused on which of a set of distinct models can best describe the formation of a BMP signaling gradient. Again, it is likely that no single model can capture the entire dynamics. For example, while the studies of Zinski and Promienke argue that a source–sink model plays a role in shaping the gradient of BMP signaling once Chd is expressed, this is not inconsistent with an earlier establishment of a signaling gradient through the refinement of an initial uniform ligand transcriptional domain to a graded ligand transcriptional domain. What is also interesting is how the patterning networks consisting of the evolutionarily-conserved components of BMP, Chd and Tld pattern the embryo in different ways in vertebrates compared to *Drosophila*.

3. Gradient interpretation

The above studies have demonstrated that Nodal and BMP gradients form through time in the context of various feedbacks and other inputs. We will now consider how information from these gradients is transformed into transcriptional profiles. Studies of the interpretation of Nodal and BMP signals have suggested that rather than simply responding to a concentration threshold, time is also crucial to the interpretation of these morphogens gradients. Using timed inhibition of the Nodal pathway by small molecule receptor inhibitors, it was shown that cell fate specification depends on the duration of exposure to Nodal ligands (Hagos & Dougan, 2007). By inhibiting Nodal signaling at progressively later time points, it was observed that cell fates specified closer to the embryonic margin require a longer exposure to Nodal signaling. These authors further argued that cell fates are determined by the overall dose of Nodal signaling received, since in *squint* mutants (lacking Ndr1), where the dose of Nodal signaling is reduced, all cell fates are delayed.

Following on from the idea that different durations of Nodal signaling are important for induction of different targets, Dubrulle et al. proposed that the kinetics of transcription are predictive of gene expression patterns (Dubrulle et al., 2015). Using transplanted cells of a GFP-Smad2 embryo as a sensor for Nodal signaling, they showed that absolute levels of signaling (as measured by the Smad2 nuclear: cytoplasmic ratio) did not correlate with the transcriptional output. What was important was the duration of exposure to signal. They found that expression of a long-range gene such as *tbxta* commenced after 1 h exposure to Nodal, while a short-range gene such as *gsc* required 2 h of signaling. Then, by fitting experimental measures of P-Smad2 levels and transcript levels to a mathematical model of Nodal signaling, they argued that this delay could be explained by differences in the transcription rates of *tbxta* and *gsc*, and that such differences in the kinetics of transcription can explain the different ranges of expression of the different target genes. Therefore, at least for some genes, the responses to the Nodal gradient could be influenced by the rates of transcription in response to a growing gradient of Nodal.

While many studies have considered direct transcriptional responses to morphogens, it is important to remember that responses to morphogens occur in the context of an overall developmental program, as part of a series of signaling, transcriptional and morphogenetic events. As targets of morphogens could themselves potentially affect subsequent events, another crucial requirement for an extended duration of signaling is the time required for a series of such feedforward interactions to unfold. This is clearly seen in the induction of both mesodermal and endodermal fates in cells at the embryonic margin by Nodal. Prospective endodermal cells first express *sox32* and are found among the cells closest to the YSL (Ober, Field, & Stainier, 2003). These cells show the highest levels of P-Smad2 and have experienced the longest duration of signaling (see further below), while mesodermal markers such as *tbxta* and *noto* extend into regions where cells have lower levels of P-Smad2, or no P-Smad2 at all. It is becoming increasingly apparent that these targets are not simple readouts of either the concentration or duration of Nodal signaling.

With the advent of optogenetics it is now possible to more precisely modulate the duration of signaling. The kinase domains of the Nodal receptors, Acvr1b and Acvr2b, have been fused to light-oxygen-voltage domains which dimerise with blue light, thereby activating Nodal signaling, the duration of which can be thus controlled by the length of the light pulse (Sako et al., 2016). In the context of MZ*oep* embryos, which lack the co-receptor of the pathway and are therefore deficient in endogenous Nodal signaling, altering Nodal signaling duration had differential effects on mesoderm and endoderm induction. Increasing windows of signaling from high stage (early blastula) to shield stage (mid-gastrulation) led to an increasing induction of the mesodermal markers *gsc* and *tbxta*, as well as the endodermal marker *sox17*—which is itself induced by Sox32. Interestingly, 2 h of exposure to Nodal signaling followed by a further hour without, led to a stronger induction of *sox17* expression, than exposure to Nodal signaling for 3 h continuously, which was shown to be due to Gsc inhibiting the induction of *sox17*. When signaling is sustained, the resulting increased levels of Gsc leads to suppression of *sox17*. While demonstrating the importance of the transcriptional events downstream of Nodal in modulating its function, and the significance of signaling duration, it should be noted that *gsc* is only endogenously expressed in the prospective prechordal plate mesoderm—a small portion of anterior mesoderm specified in a dorsal region of the embryo—while *sox32* and subsequently *sox17* are induced all around the margin. The leaves open the question of how general this particular mechanism of regulating endoderm induction actually is.

The importance of sequential transcriptional events has also been clearly demonstrated in other recent studies of how the Nodal gradient separates the endoderm and mesoderm lineages (van Boxtel et al., 2015; van Boxtel, Economou, Heliot, & Hill, 2018) (Fig. 5). As discussed above, the mesodermal markers, *tbxta* and *noto* are in fact targets of Fgf signaling, which is itself downstream of Nodal, with the ligands Fgf8a and Fgf3 being direct transcriptional targets of the Nodal signaling pathway (Mathieu et al., 2004). While demonstrating that mesodermal targets downstream of Nodal are not necessarily induced directly, this also set up a potential paradox, as it has long been known that Fgf signaling inhibits endoderm induction (Mizoguchi, Izawa, Kuroiwa, & Kikuchi, 2006; Poulain, Furthauer, Thisse, Thisse, & Lepage, 2006). Further work by van Boxtel et al. resolved this. Erk MAP kinase signaling downstream of Fgf, which is responsible for the inhibitory effect of Fgf signaling on endoderm formation, was shown to be suppressed in the cells closest to the YSL by the P-Erk inhibitor Dusp4, which is itself a transcriptional target of Nodal signaling (van Boxtel et al., 2018). Moreover, expression of *dusp4* and therefore the timing of Fgf inhibition is also the result of a series of feedforward interactions, resulting from a coherent feedforward, or self-enabling mechanism. As well as being a direct target of Nodal signaling (Brown et al., 2008), *dusp4* transcription requires the transcription factor Mixl1 (Nelson et al., 2017), which is itself a target of Nodal signaling (Nelson et al., 2014). Therefore, the requirement for an extended duration of signaling for the induction of *sox32* expression

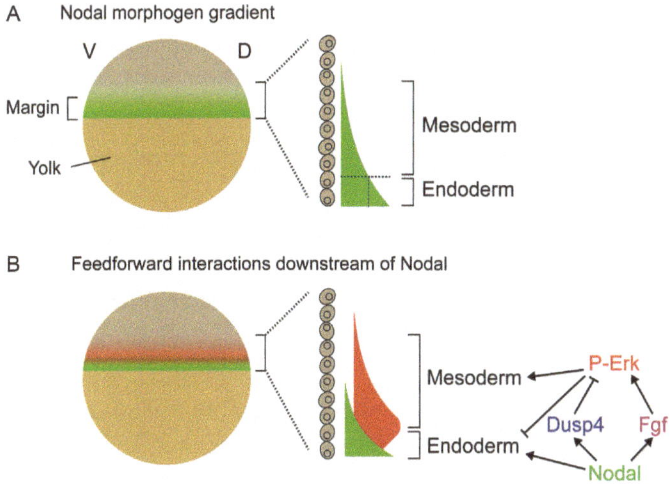

Fig. 5 Models of mesoderm and endoderm separation downstream of Nodal. (A) Schematic of a 50% epiboly embryo depicting the separation of the mesoderm from the endoderm through a Nodal morphogen gradient, as in the classical morphogen model. Graded Nodal signaling (green) extends approximately eight cell tiers from the YSL. Marginal-most cells, experiencing the highest levels of Nodal signaling take an endodermal fate, while cells further from the margin experiencing lower levels of Nodal take a mesodermal fate. (B) Schematic of 50% epiboly embryo depicting the separation of the mesoderm from the endoderm through a series of feedforward interactions downstream of Nodal. A smaller domain of graded Nodal signaling (green) extend approximately five cell tiers from the embryonic margin, while a larger domain of Fgf signaling through P-Erk (red) extends to around eight cell tiers. Profiles of Nodal and Fgf signaling also show a suppression of Fgf signaling at the margin. Network depicts feedforward interactions downstream of Nodal, whereby Fgf ligands are induced and act at long-range to promote P-Erk, while Dusp4 is induced locally and suppresses P-Erk. The resultant shaping of the domain of P-Erk leads to endoderm induction in a marginal domain where it is promoted by Nodal signaling and inhibitory Fgf signaling is suppressed by Dusp4, and mesoderm induction through Fgf further from the margin. Note that Nodal signaling may also provide a direct input into transcription of mesodermal genes closer to the YSL.

and the endodermal cell fate is in part due to the time required for successive expression of *mixl1*, then *dusp4*, in order to repress Fgf signaling. Rather than resulting from a simple read out of the levels or duration of Nodal signaling, it is becoming increasingly apparent that the separation of mesodermal and endodermal cell fates is the result of a cascade of transcriptional and signaling events initiated by the spread of Nodal signaling in the margin (Fig. 5).

The importance of signaling duration is also seen in the interpretation of the DV gradient of BMP signaling. While the BMP gradient forms during the blastula stages and is present throughout the gastrula stages, it is evident that the positional information encoded by this gradient is not read out simultaneously, but rather that there is a temporal requirement in its interpretation (Tucker et al., 2008). Using a heat shock-inducible Chd to rapidly turn off BMP signaling at different times, these authors demonstrated that inhibiting BMP signaling at progressively later time points led to the dorsalization of more posterior tissues, with more anterior tissues being unaffected. This is not due to the requirement for a longer duration of signaling to specify posterior tissues, but instead appears to be due to a later requirement for BMP signaling. Combining an MZ-*lost-a-fin* background which is mutant for the type I BMP receptor Alk8 and is therefore deficient in BMP signaling, with a heat shock Alk8 to rescue BMP signaling, revealed that switching BMP signaling on at late time points could rescue posterior structures (Tucker et al., 2008). Therefore, a late temporal window rather than a long duration of signaling is required for the DV patterning of the posterior embryo.

4. Molecular sensing of ligand concentration and signaling duration

From the above discussion it is evident that morphogen gradients are not purely static ligand concentration gradients across fields of cells that impart positional information via the exact levels of ligand experienced by a particular cell. Duration of signaling experienced by cells at different positions across a tissue is also important, as is the induction or inhibition of other signaling pathways that can influence the transcriptional regulation of morphogen target genes. In this section we will therefore consider the molecular mechanisms whereby cells interpret ligand concentrations and signaling duration for the BMP and Nodal pathways, as well as how cells additionally integrate information from other signaling pathways.

At the heart of the French Flag model is the idea that cells transform ligand concentrations into specific gene expression patterns, which in turn dictate cell fate specification across the morphogen gradient (Wolpert, 2011). The simplest mechanism whereby small differences in signal strength could be translated into all or none changes in gene expression would be via enhancers in target genes having different affinities for transcription factors activated directly by the signaling pathway (Driever, Thoma, & Nusslein-Volhard, 1989; Rogers & Schier, 2011). Note that this view also relies on the assumption that there is a direct correspondence between ligand concentration in the extracellular space and numbers of receptors activated, but this is very difficult to measure in vivo, and it is not clear over what range of ligand concentrations this might occur.

For Nodal and BMP signaling in zebrafish embryos, the transcription factors in question would be the activated Smad complexes, in particular P-Smad2/3–Smad4 complexes downstream of Nodal and P-Smad1/5/9–Smad4 complexes downstream of the BMPs. While there is no direct information about Smad complex binding affinities to enhancers of zebrafish Nodal and BMP target genes, there is precedent for this type of mechanism from other systems. In *Xenopus* embryos the mesodermal genes *gsc* and *xbra* (*Xenopus* equivalent of *tbxta*) can be induced by different concentrations of Activin (which mimics Nodal signaling) and also by different concentrations of exogenous Smad2, with *gsc* requiring higher concentrations of ligand and Smad2 than *xbra* (Dyson & Gurdon, 1998; Shimizu & Gurdon, 1999). Importantly, P-Smad2–Smad4 complexes do not actually bind DNA directly themselves, but are recruited to DNA via other transcription factors, such as FOXH1 and Mixer (also called Mixl1 in zebrafish) (Hill, 2016), and the Smad2 interaction motifs in these transcription factors have been characterized. FOXH1 and Mixer share a common motif, denoted the SIM (Smad interaction motif), but FOXH1 has an additional motif, denoted the FM (FOXH1 motif), which results in FOXH1 having a higher affinity for activated Smad2–Smad4 complexes than Mixer (Randall et al., 2004). Thus, a precedent exists for differential binding affinity of P-Smad2–Smad4 complexes to DNA. However, the functional consequences of this for Nodal gradient interpretation are not known.

In the case of BMP gradients, most of our information about how different ligand concentrations are interpreted at the level of gene expression comes from studies in *Drosophila*. For example, in the context of the early *Drosophila* embryo, the gene *Race* is activated only at the highest Dpp concentrations on the dorsal side of the embryo (Wharton, Basu, & Ashe, 2004). This would be consistent with the Smad binding sites in its enhancer having a relatively low affinity for P-Mad–Medea complexes (*Drosophila* P-Smad1–Smad4 complexes). Indeed, increasing the affinity of the Smad binding sites in the *Race* enhancer broadened the expression pattern of a linked reporter gene, giving this gene a new threshold response (Wharton et al., 2004). Thus transcription factor affinity for DNA can play a role in reading out ligand concentration and hence signal strength.

Our discussion above, however, also highlights the essential role of time in generating morphogen gradients, which raises the question of how cells monitor signaling duration. For Nodal signaling, the levels of P-Smad2 have been shown in zebrafish to be directly proportional to the time of ligand exposure (van Boxtel et al., 2015), and recent insights into the receptor trafficking behavior of Activin receptors (which are shared with Nodal) suggest a possible mechanism (Miller, Schmierer, & Hill, 2019). Activin binds its receptors relatively weakly and although signaling receptors constantly internalize to signal from endosomes, the pool of receptors at the cell surface is constantly replenished and does not deplete. Thus, over time, activated signaling receptors accumulate in endosomes. The continuous shuttling of the receptor-regulated Smads, Smad2 and Smad3 between the cytoplasm and nucleus then provides a mechanism to ensure that the levels of activated Smad complexes in the nucleus reflects the levels of activated signaling receptors at any given time (Schmierer & Hill, 2007; Schmierer, Tournier, Bates, & Hill, 2008). As a result, the levels of phosphorylated Smad2 and Smad3 in the nucleus are proportional to the time that cells are exposed to ligand. A similar mechanism is likely also operating for BMP signaling (Miller et al., 2019). While prolonged signaling would presumably result in a steady state, where this relationship would break down, this is not reached in rapid processes such as Nodal gradient formation in zebrafish embryos, which only takes about 1.3h. Therefore, for the Nodal and BMP signaling pathways, cells are able to monitor signal duration, and transform this information into levels of activated nuclear Smad complexes. But how is this temporal information interpreted? The importance of feedforward and feedback loops discussed in the previous section indicates that temporal information is encoded in sequences of events that must be completed for specific genes to be activated by a given signal (Fig. 5).

Another major principle that is emerging from the study of morphogen gradients in zebrafish is that these gradients do not function alone, but do so in combination with other signaling pathways that either may be initiated downstream of the morphogen, as the Fgf, Wnt and Notch pathways are downstream of Nodal (Kikuchi et al., 2004; Mathieu et al., 2004; Narayanan & Lekven, 2012; Schier & Talbot, 2005), or may be regulated independently. The activity of these cooperating pathways can be integrated with the morphogen signal at the enhancers of target genes. A recent example of this would be the co-occupancy of the *Gsc, Foxa2* and *Eomes* enhancers by Wnt3-activated Tcf3 and Nodal-activated Smad2/3 in mouse embryonic stem cells (Wang et al., 2017). Another is the observation that in zebrafish embryos *tbxta* expression at the margin is activated by a combination of Nodal, Fgf, Wnt and BMP signaling (Harvey, Tumpel, Dubrulle, Schier, & Smith, 2010; van Boxtel et al., 2015). An example of a downstream pathway acting negatively is the effect of the Erk MAP kinase pathway downstream of Fgf on *sox32* and *sox17* induction. It is has been shown that Erk-induced phosphorylation of Sox32 inhibits its ability to induce *sox17* (Poulain et al., 2006), although, it is not yet understood how Erk inhibits Nodal-induced *sox32* expression. Possible mechanisms could include a negatively acting Erk-induced phosphorylation of a transcriptional activator of *sox32*, or activation of a repressor (direct or indirect) of *sox32* transcription downstream of the Erk pathway.

5. Outlook

Wolpert proposed graded morphogens as a source of positional information in developing embryos. While this idea has provided an extremely useful framework for thinking about patterning, it is becoming increasingly apparent through studies such as those in the early zebrafish embryo, that such a model is overly simplistic. As discussed here, morphogen gradients grow through time in the context of various feedbacks and interactions with a changing environment of other molecules. Not only are these interactions vital in shaping the gradients, but in doing so, they also potentially confer properties such as robustness to embryonic development. Moreover, differences in the duration of signaling that distinct populations of cells are exposed to while morphogen gradients form, play a crucial role in their interpretation, with responses often resulting from a series of transcriptional and signaling events downstream of the morphogen. While we have focused on Nodal and BMP signaling, it is important to remember that these signals do not pattern the embryo in isolation, either from each other or other morphogens. As we have discussed, there are interactions with Fgf signaling, and there are almost certainly interactions with other pathways such as Wnt and Notch signaling, which are known to be active in the early zebrafish embryo. In the future, a full understanding of how morphogens pattern developing embryos will come not from focusing on these pathways in isolation, but with consideration of how interactions between them evolve through time.

Acknowledgments

We would like to thank Scott Wilcockson, Daniel Miller and Kristina Stapornwongkul for very helpful comments on the manuscript. We thank Isobel Bradley for the P-Smad1/5 immunostaining shown in Fig. 2. Work in the Hill lab is funded by the Francis Crick Institute, which receives its core funding from Cancer Research UK (FC001095), the UK Medical Research Council (FC001095), and the Wellcome Trust (FC001095).

References

Almuedo-Castillo, M., Blassle, A., Morsdorf, D., Marcon, L., Soh, G. H., Rogers, K. W., et al. (2018). Scale-invariant patterning by size-dependent inhibition of Nodal signalling. *Nature Cell Biology, 20*, 1032–1042.

Ben-Zvi, D., Shilo, B. Z., Fainsod, A., & Barkai, N. (2008). Scaling of the BMP activation gradient in *Xenopus embryos. Nature, 453*, 1205–1211.

Bisgrove, B. W., Su, Y. C., & Yost, H. J. (2017). Maternal Gdf3 is an obligatory cofactor in Nodal signaling for embryonic axis formation in zebrafish. *eLife, 6*, e28534.

Brown, J. L., Snir, M., Noushmehr, H., Kirby, M., Hong, S. K., Elkahloun, A. G., et al. (2008). Transcriptional profiling of endogenous germ layer precursor cells identifies dusp4 as an essential gene in zebrafish endoderm specification. *Proceedings of the National Academy of Sciences of the United States of America, 105*, 12337–12342.

Chen, Y., & Schier, A. F. (2001). The zebrafish Nodal signal Squint functions as a morphogen. *Nature, 411*, 607–610.

Chen, Y., & Schier, A. F. (2002). Lefty proteins are long-range inhibitors of squint-mediated nodal signaling. *Current Biology, 12*, 2124–2128.

DaCosta Byfield, S., Major, C., Laping, N. J., & Roberts, A. B. (2004). SB-505124 is a selective inhibitor of transforming growth factor-β type I receptors ALK4, ALK5, and ALK7. *Molecular Pharmacology, 65*, 744–752.

De Robertis, E. M. (2008). Evo-devo: Variations on ancestral themes. *Cell, 132*, 185–195.

Dick, A., Hild, M., Bauer, H., Imai, Y., Maifeld, H., Schier, A. F., et al. (2000). Essential role of Bmp7 (snailhouse) and its prodomain in dorsoventral patterning of the zebrafish embryo. *Development, 127*, 343–354.

Driever, W., Thoma, G., & Nusslein-Volhard, C. (1989). Determination of spatial domains of zygotic gene expression in the Drosophila embryo by the affinity of binding sites for the bicoid morphogen. *Nature, 340*, 363–367.

Dubrulle, J., Jordan, B. M., Akhmetova, L., Farrell, J. A., Kim, S. H., Solnica-Krezel, L., et al. (2015). Response to Nodal morphogen gradient is determined by the kinetics of target gene induction. *eLife, 4*, e05042.

Dyson, S., & Gurdon, J. B. (1998). The interpretation of position in a morphogen gradient as revealed by occupancy of activin receptors. *Cell, 93*, 557–568.

Eldar, A., Dorfman, R., Weiss, D., Ashe, H., Shilo, B. Z., & Barkai, N. (2002). Robustness of the BMP morphogen gradient in Drosophila embryonic patterning. *Nature, 419*, 304–308.

Entchev, E. V., Schwabedissen, A., & Gonzalez-Gaitan, M. (2000). Gradient formation of the TGF-β homolog Dpp. *Cell, 103*, 981–991.

Erter, C. E., Solnica-Krezel, L., & Wright, C. V. (1998). Zebrafish nodal-related 2 encodes an early mesendodermal inducer signaling from the extraembryonic yolk syncytial layer. *Developmental Biology, 204*, 361–372.

Feldman, B., Gates, M. A., Egan, E. S., Dougan, S. T., Rennebeck, G., Sirotkin, H. I., et al. (1998). Zebrafish organizer development and germ-layer formation require nodal-related signals. *Nature, 395*, 181–185.

Francois, P., Vonica, A., Brivanlou, A. H., & Siggia, E. D. (2009). Scaling of BMP gradients in *Xenopus embryos. Nature, 461*, E1, discussion E2.

Furthauer, M., Van Celst, J., Thisse, C., & Thisse, B. (2004). Fgf signalling controls the dorsoventral patterning of the zebrafish embryo. *Development, 131*, 2853–2864.

Green, J. B., Howes, G., Symes, K., Cooke, J., & Smith, J. C. (1990). The biological effects of XTC-MIF: Quantitative comparison with Xenopus bFGF. *Development, 108*, 173–183.

Hagos, E. G., & Dougan, S. T. (2007). Time-dependent patterning of the mesoderm and endoderm by Nodal signals in zebrafish. *BMC Developmental Biology*, 7, 22.

Harvey, S. A., Tumpel, S., Dubrulle, J., Schier, A. F., & Smith, J. C. (2010). No tail integrates two modes of mesoderm induction. *Development, 137*, 1127–1135.

Heldin, C. H., & Moustakas, A. (2016). Signaling receptors for TGF-β family members. *Cold Spring Harbor Perspectives in Biology*, *8*, a022053.
Hild, M., Dick, A., Rauch, G. J., Meier, A., Bouwmeester, T., Haffter, P., et al. (1999). The smad5 mutation somitabun blocks Bmp2b signaling during early dorsoventral patterning of the zebrafish embryo. *Development*, *126*, 2149–2159.
Hill, C. S. (2016). Transcriptional control by the SMADs. *Cold Spring Harbor Perspectives in Biology*, *8*, a022079.
Inomata, H., Shibata, T., Haraguchi, T., & Sasai, Y. (2013). Scaling of dorsal-ventral patterning by embryo size-dependent degradation of Spemann's organizer signals. *Cell*, *153*, 1296–1311.
Kikuchi, Y., Verkade, H., Reiter, J. F., Kim, C. H., Chitnis, A. B., Kuroiwa, A., et al. (2004). Notch signaling can regulate endoderm formation in zebrafish. *Developmental Dynamics*, *229*, 756–762.
Kishimoto, Y., Lee, K. H., Zon, L., Hammerschmidt, M., & Schulte-Merker, S. (1997). The molecular nature of zebrafish swirl: BMP2 function is essential during early dorsoventral patterning. *Development*, *124*, 4457–4466.
Koos, D. S., & Ho, R. K. (1999). The nieuwkoid/dharma homeobox gene is essential for bmp2b repression in the zebrafish pregastrula. *Developmental Biology*, *215*, 190–207.
Little, S. C., & Mullins, M. C. (2009). Bone morphogenetic protein heterodimers assemble heteromeric type I receptor complexes to pattern the dorsoventral axis. *Nature Cell Biology*, *11*, 637–643.
Mathieu, J., Griffin, K., Herbomel, P., Dickmeis, T., Strahle, U., Kimelman, D., et al. (2004). Nodal and Fgf pathways interact through a positive regulatory loop and synergize to maintain mesodermal cell populations. *Development*, *131*, 629–641.
Meinhardt, H., & Gierer, A. (2000). Pattern formation by local self-activation and lateral inhibition. *BioEssays*, *22*, 753–760.
Meno, C., Gritsman, K., Ohishi, S., Ohfuji, Y., Heckscher, E., Mochida, K., et al. (1999). Mouse Lefty2 and zebrafish antivin are feedback inhibitors of nodal signaling during vertebrate gastrulation. *Molecular Cell*, *4*, 287–298.
Miller, D. S. J., & Hill, C. S. (2016). TGF-β superfamily signalling. In R. A. Bradshaw & P. D. Stahl (Eds.), *Encyclopedia of cell biology* (pp. 37–50). Elsevier.
Miller, D. S. J., Schmierer, B., & Hill, C. S. (2019). TGF-β family ligands exhibit distinct signalling dynamics that are driven by receptor localisation. *Journal of Cell Science*, *132*, jcs234039.
Mintzer, K. A., Lee, M. A., Runke, G., Trout, J., Whitman, M., & Mullins, M. C. (2001). Lost-a-fin encodes a type I BMP receptor, Alk8, acting maternally and zygotically in dorsoventral pattern formation. *Development*, *128*, 859–869.
Mizoguchi, T., Izawa, T., Kuroiwa, A., & Kikuchi, Y. (2006). Fgf signaling negatively regulates Nodal-dependent endoderm induction in zebrafish. *Developmental Biology*, *300*, 612–622.
Montague, T. G., & Schier, A. F. (2017). Vg1-Nodal heterodimers are the endogenous inducers of mesendoderm. *eLife*, *6*, e28183.
Muller, P., Rogers, K. W., Jordan, B. M., Lee, J. S., Robson, D., Ramanathan, S., et al. (2012). Differential diffusivity of Nodal and Lefty underlies a reaction-diffusion patterning system. *Science*, *336*, 721–724.
Muller, P., Rogers, K. W., Yu, S. R., Brand, M., & Schier, A. F. (2013). Morphogen transport. *Development*, *140*, 1621–1638.
Mullins, M. C., Hammerschmidt, M., Kane, D. A., Odenthal, J., Brand, M., van Eeden, F. J., et al. (1996). Genes establishing dorsoventral pattern formation in the zebrafish embryo: the ventral specifying genes. *Development*, *123*, 81–93.
Narayanan, A., & Lekven, A. C. (2012). Biphasic wnt8a expression is achieved through interactions of multiple regulatory inputs. *Developmental Dynamics*, *241*, 1062–1075.

Nelson, A. C., Cutty, S. J., Gasiunas, S. N., Deplae, I., Stemple, D. L., & Wardle, F. C. (2017). In vivo regulation of the Zebrafish endoderm progenitor niche by T-Box transcription factors. *Cell Reports, 19*, 2782–2795.

Nelson, A. C., Cutty, S. J., Niini, M., Stemple, D. L., Flicek, P., Houart, C., et al. (2014). Global identification of Smad2 and Eomesodermin targets in zebrafish identifies a conserved transcriptional network in mesendoderm and a novel role for Eomesodermin in repression of ectodermal gene expression. *BMC Biology, 12*, 81.

Ober, E. A., Field, H. A., & Stainier, D. Y. (2003). From endoderm formation to liver and pancreas development in zebrafish. *Mechanisms of Development, 120*, 5–18.

O'Connor, M. B., Umulis, D., Othmer, H. G., & Blair, S. S. (2006). Shaping BMP morphogen gradients in the Drosophila embryo and pupal wing. *Development, 133*, 183–193.

Pages, F., & Kerridge, S. (2000). Morphogen gradients. A question of time or concentration? *Trends in Genetics, 16*, 40–44.

Pelliccia, J. L., Jindal, G. A., & Burdine, R. D. (2017). Gdf3 is required for robust Nodal signaling during germ layer formation and left-right patterning. *eLife, 6*, e28635.

Pomreinke, A. P., Soh, G. H., Rogers, K. W., Bergmann, J. K., Blassle, A. J., & Muller, P. (2017). Dynamics of BMP signaling and distribution during zebrafish dorsal-ventral patterning. *eLife, 6*, e25861.

Poulain, M., Furthauer, M., Thisse, B., Thisse, C., & Lepage, T. (2006). Zebrafish endoderm formation is regulated by combinatorial Nodal, FGF and BMP signalling. *Development, 133*, 2189–2200.

Raftery, L. A., & Umulis, D. M. (2012). Regulation of BMP activity and range in Drosophila wing development. *Current Opinion in Cell Biology, 24*, 158–165.

Ramel, M. C., & Hill, C. S. (2012). Spatial regulation of BMP activity. *FEBS Letters, 586*, 1929–1941.

Ramel, M. C., & Hill, C. S. (2013). The ventral to dorsal BMP activity gradient in the early zebrafish embryo is determined by graded expression of BMP ligands. *Developmental Biology, 378*, 170–182.

Randall, R. A., Howell, M., Page, C. S., Daly, A., Bates, P. A., & Hill, C. S. (2004). Recognition of phosphorylated-Smad2-containing complexes by a novel Smad interaction motif. *Molecular and Cellular Biology, 24*, 1106–1121.

Rebagliati, M. R., Toyama, R., Haffter, P., & Dawid, I. B. (1998). Cyclops encodes a nodal-related factor involved in midline signaling. *Proceedings of the National Academy of Sciences of the United States of America, 95*, 9932–9937.

Rogers, K. W., Lord, N. D., Gagnon, J. A., Pauli, A., Zimmerman, S., Aksel, D. C., et al. (2017). Nodal patterning without Lefty inhibitory feedback is functional but fragile. *eLife, 6*, e28785.

Rogers, K. W., & Schier, A. F. (2011). Morphogen gradients: From generation to interpretation. *Annual Review of Cell and Developmental Biology, 27*, 377–407.

Saijoh, Y., Adachi, H., Sakuma, R., Yeo, C. Y., Yashiro, K., Watanabe, M., et al. (2000). Left-right asymmetric expression of lefty2 and nodal is induced by a signaling pathway that includes the transcription factor FAST2. *Molecular Cell, 5*, 35–47.

Sako, K., Pradhan, S. J., Barone, V., Ingles-Prieto, A., Muller, P., Ruprecht, V., et al. (2016). Optogenetic control of Nodal signaling reveals a temporal pattern of Nodal signaling regulating cell fate specification during gastrulation. *Cell Reports, 16*, 866–877.

Sawala, A., Sutcliffe, C., & Ashe, H. L. (2012). Multistep molecular mechanism for bone morphogenetic protein extracellular transport in the Drosophila embryo. *Proceedings of the National Academy of Sciences of the United States of America, 109*, 11222–11227.

Schier, A. F. (2003). Nodal signaling in vertebrate development. *Annual Review of Cell and Developmental Biology, 19*, 589–621.

Schier, A. F., & Talbot, W. S. (2005). Molecular genetics of axis formation in zebrafish. *Annual Review of Genetics, 39*, 561–613.

Schmierer, B., & Hill, C. S. (2007). TGFβ-SMAD signal transduction: molecular specificity and functional flexibility. *Nature Reviews Molecular Cell Biology*, *8*, 970–982.
Schmierer, B., Tournier, A. L., Bates, P. A., & Hill, C. S. (2008). Mathematical modeling identifies Smad nucleocytoplasmic shuttling as a dynamic signal-interpreting system. *Proceedings of the National Academy of Sciences of the United States of America*, *105*, 6608–6613.
Schulte-Merker, S., Lee, K. J., McMahon, A. P., & Hammerschmidt, M. (1997). The zebrafish organizer requires chordino. *Nature*, *387*, 862–863.
Schwank, G., Dalessi, S., Yang, S. F., Yagi, R., de Lachapelle, A. M., Affolter, M., et al. (2011). Formation of the long range Dpp morphogen gradient. *PLoS Biology*, *9*, e1001111.
Shimizu, K., & Gurdon, J. B. (1999). A quantitative analysis of signal transduction from activin receptor to nucleus and its relevance to morphogen gradient interpretation. *Proceedings of the National Academy of Sciences of the United States of America*, *96*, 6791–6796.
Shimmi, O., Umulis, D., Othmer, H., & O'Connor, M. B. (2005). Facilitated transport of a Dpp/Scw heterodimer by Sog/Tsg leads to robust patterning of the Drosophila blastoderm embryo. *Cell*, *120*, 873–886.
Tucker, J. A., Mintzer, K. A., & Mullins, M. C. (2008). The BMP signaling gradient patterns dorsoventral tissues in a temporally progressive manner along the anteroposterior axis. *Developmental Cell*, *14*, 108–119.
van Boxtel, A. L., Chesebro, J. E., Heliot, C., Ramel, M. C., Stone, R. K., & Hill, C. S. (2015). A temporal window for signal activation dictates the dimensions of a Nodal signaling domain. *Developmental Cell*, *35*, 175–185.
van Boxtel, A. L., Economou, A. D., Heliot, C., & Hill, C. S. (2018). Long-range signaling activation and local inhibition separate the mesoderm and endoderm lineages. *Developmental Cell*, *44*, 179–191. e175.
Wang, Q., Zou, Y., Nowotschin, S., Kim, S. Y., Li, Q. V., Soh, C. L., et al. (2017). The p53 family coordinates Wnt and Nodal inputs in mesendodermal differentiation of embryonic stem cells. *Cell Stem Cell*, *20*, 70–86.
Wharton, S. J., Basu, S. P., & Ashe, H. L. (2004). Smad affinity can direct distinct readouts of the embryonic extracellular Dpp gradient in Drosophila. *Current Biology*, *14*, 1550–1558.
Whitman, M. (2001). Nodal signaling in early vertebrate embryos: themes and variations. *Developmental Cell*, *1*, 605–617.
Wolpert, L. (2011). Positional information and patterning revisited. *Journal of Theoretical Biology*, *269*, 359–365.
Xu, P. F., Houssin, N., Ferri-Lagneau, K. F., Thisse, B., & Thisse, C. (2014). Construction of a vertebrate embryo from two opposing morphogen gradients. *Science*, *344*, 87–89.
Zinski, J., Bu, Y., Wang, X., Dou, W., Umulis, D., & Mullins, M. C. (2017). Systems biology derived source-sink mechanism of BMP gradient formation. *eLife*, *6*, e22199.

CHAPTER THIRTEEN

Signaling regulation during gastrulation: Insights from mouse embryos and *in vitro* systems

Sophie M. Morgani[a,b,*], **Anna-Katerina Hadjantonakis**[a,*]
[a]Developmental Biology Program, Sloan Kettering Institute, Memorial Sloan Kettering Cancer Center, New York, NY, United States
[b]Wellcome Trust-Medical Research Council Cambridge Stem Cell Institute, University of Cambridge, Jeffrey Cheah Biomedical Centre Cambridge Biomedical Campus, Cambridge, United Kingdom
[*]Corresponding authors: e-mail address: morganis@mskcc.org; hadj@mskcc.org

Contents

Abstract

Gastrulation is the process whereby cells exit pluripotency and concomitantly acquire and pattern distinct cell fates. This is driven by the convergence of WNT, BMP, Nodal and FGF signals, which are tightly spatially and temporally controlled, resulting in regional and stage-specific signaling environments. The combination, level and duration of signals that a cell is exposed to, according its position within the embryo and the developmental time window, dictates the fate it will adopt. The key pathways driving gastrulation exhibit complex interactions, which are difficult to disentangle *in vivo* due to the complexity of manipulating multiple signals in parallel with high spatiotemporal resolution. Thus, our current understanding of the signaling dynamics regulating gastrulation is limited. *In vitro* stem cell models have been established, which undergo

Current Topics in Developmental Biology, Volume 137
ISSN 0070-2153
https://doi.org/10.1016/bs.ctdb.2019.11.011

organized cellular differentiation and patterning. These provide amenable, simplified, deconstructed and scalable models of gastrulation. While the foundation of our understanding of gastrulation stems from experiments in embryos, *in vitro* systems are now beginning to reveal the intricate details of signaling regulation. Here we discuss the current state of knowledge of the role, regulation and dynamic interaction of signaling pathways that drive mouse gastrulation.

1. The morphogenetic cell behaviors associated with gastrulation establish the blueprint of an organism

Gastrulation is the process whereby pluripotent cells of the embryo differentiate into the three definitive germ layers, the ectoderm, endoderm and mesoderm, in a spatially organized manner to establish the blueprint of the adult organism. The specification and patterning of cell fates is instructed by the distinct combination of signals present in different regions of the embryo.

Prior to the onset of gastrulation, the embryo comprises the pluripotent epiblast that will give rise to the embryo-proper, and the extraembryonic supporting tissues, including the visceral endoderm (VE) and the extraembryonic ectoderm (ExE), that will give rise to the yolk sac and fetal portion of the placenta, respectively (Fig. 1A). At embryonic day (E) 6.25, a combination of Wingless-INT (WNT), Bone Morphogenetic Protein (BMP), Nodal and Fibroblast Growth Factor (FGF) signals instructs proximal posterior epiblast cells to undergo an epithelial-mesenchymal transition (EMT). This involves a reduction in cell-cell contacts and an increase in migratory behavior that facilitates morphogenetic movements (Fig. 1B).

Cells that undergo an EMT then leave the epiblast, within a region known as the primitive streak (PS), and migrate proximally into the extraembryonic domain or laterally and anteriorly between the epiblast and VE (Fig. 1C). As gastrulation proceeds, the PS extends distally to span the length of the posterior epiblast. The first cells exiting the proximal epiblast through the posterior PS generate extraembryonic mesoderm (ExM), primordial germ cells (PGCs) and cranial and cardiac mesoderm followed by lateral and paraxial mesoderm. Cells subsequently exiting the distal epiblast through the anterior PS generate definitive endoderm (DE) and axial mesoderm (AxM). Cells that remain within the epiblast form neurectoderm (Kinder et al., 1999; Parameswaran & Tam, 1995). Experiments in which cells were isolated from one part of the epiblast and grafted to another location,

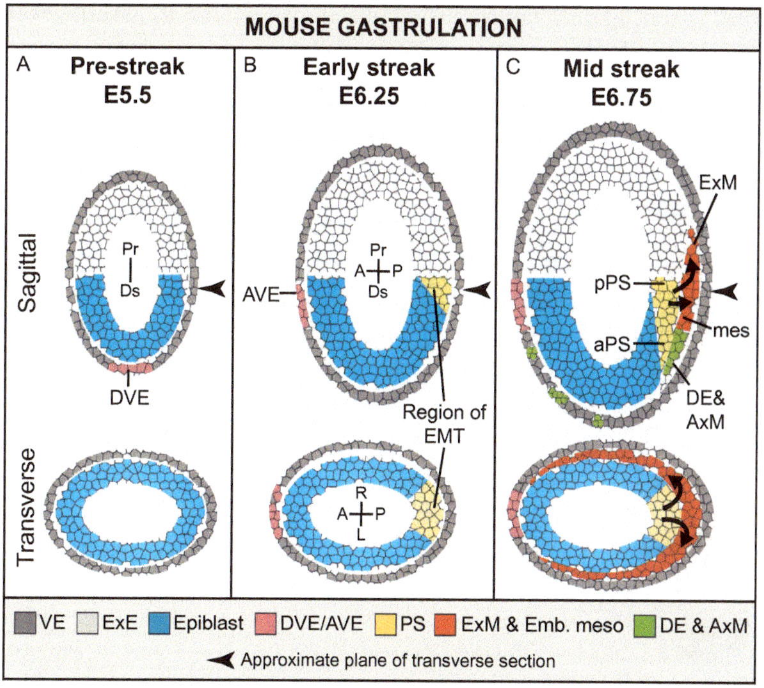

Fig. 1 Schematic diagram depicting mouse gastrulation. Shown are schematic diagrams of sagittal and transverse sections through mouse embryos during gastrulation. Arrowheads mark the plane of the transverse section. (A) Prior to the onset of gastrulation, at embryonic day (E) 5.5, the mouse embryo comprises three epithelial layers, the embryonic epiblast that will give rise to the embryo proper, the visceral endoderm (VE) that will give rise to the yolk sac and the extraembryonic ectoderm (ExE) that will give rise to the placenta. At the distal tip of the embryo, there is a thickening of the VE layer that acts as a specializing signaling center in the embryo, known as the distal visceral endoderm (DVE). (B) By the onset of gastrulation, around E6.25, the DVE has migrated anteriorly and is now known as the anterior visceral endoderm (AVE). Cells within the proximal posterior of the epiblast begin to undergo an epithelial-mesenchymal transition (EMT) and delaminate from the epithelial layer in a region known as the primitive streak (PS). Cells leaving the PS region first from the posterior PS (pPS) migrate proximally to form the extraembryonic *meso*derm (ExM) or anteriorly and laterally between the epiblast and VE to form the embryonic mesoderm (Emb. meso). (C) As gastrulation proceeds, the PS extends distally and at E6.75, the PS extends almost the whole length of the posterior epiblast. Cells traversing the PS more distally at later time points through the anterior PS (aPS) will give rise to definitive endoderm (DE) and axial mesoderm (AxM). DE cells then intercalate into the outer VE cell layer. Arrows indicate the direction of cell migration. Pr, proximal; Ds, distal; A, anterior; P, posterior.

revealed that cell fate is not spatially pre-determined but is instead governed by the local environment (Tam & Zhou, 1996), i.e., the timing and position that a cell traverses the PS (Kinder et al., 1999).

2. Signaling interactions during gastrulation

Gastrulation is driven by embryonic and extraembryonic signaling centers, sources of ligands or inhibitors that exert spatial and temporal control over signaling activity. Little is known about how signaling factors travel within the mammalian embryo but experiments in other organisms suggest that WNT, BMP, Nodal and FGF act as morphogens, diffusing from their source to establish concentration gradients that direct cell fate.

At the onset of gastrulation, localized ligand production and complex signaling pathway interactions establish a signaling hub within the proximal posterior epiblast (Fig. 2A). Initially, uncleaved pro-NODAL, produced by the epiblast, induces the expression of *Bmp4* within the adjacent ExE (Ben-Haim et al., 2006; Winnier et al., 1995). The ExE is also a source of other BMP ligands including *Bmp8b* (Ying et al., 2000), *8a*, *1* and 7 (Pijuan-Sala et al., 2019), while the VE produces *Bmp2* (Ying & Zhao, 2001). Additionally, pro-NODAL induces the expression of its own convertase enzymes in the ExE (Ben-Haim et al., 2006). These enzymes convert pro-NODAL it to its active NODAL form in the adjacent proximal epiblast, which then augments its own expression through an autoregulatory loop (Norris et al., 2002; Saijoh et al., 2000). *Wnt3* is expressed by posterior VE cells overlying the site of prospective PS formation (Rivera-Perez & Magnuson, 2005). *Wnt3* stimulates its own expression within the epiblast as well as that of *Nodal* (Ben-Haim et al., 2006; Norris et al., 2002; Tortelote et al., 2013; Yoon et al., 2015). BMP signaling also induces *Wnt3* expression (Ben-Haim et al., 2006; Miura, Singh, & Mishina, 2010). Together, these interactions enhance WNT, BMP and Nodal signaling activity within the proximal posterior epiblast.

Mechanisms are also in place to actively confine WNT, BMP and Nodal signaling responses to the posterior of the embryo, namely the expression of inhibitors by the anterior visceral endoderm (AVE), a signaling center overlying the anterior epiblast. Secreted CER1 and LEFTY1 inhibit BMP and Nodal signaling while DKK1 inhibits WNT activity (Belo et al., 1997; Meno et al., 1996; Glinka et al., 1998; Kawano & Kypta, 2003; Kemp et al., 2005) within the anterior epiblast. While FGF is critical for

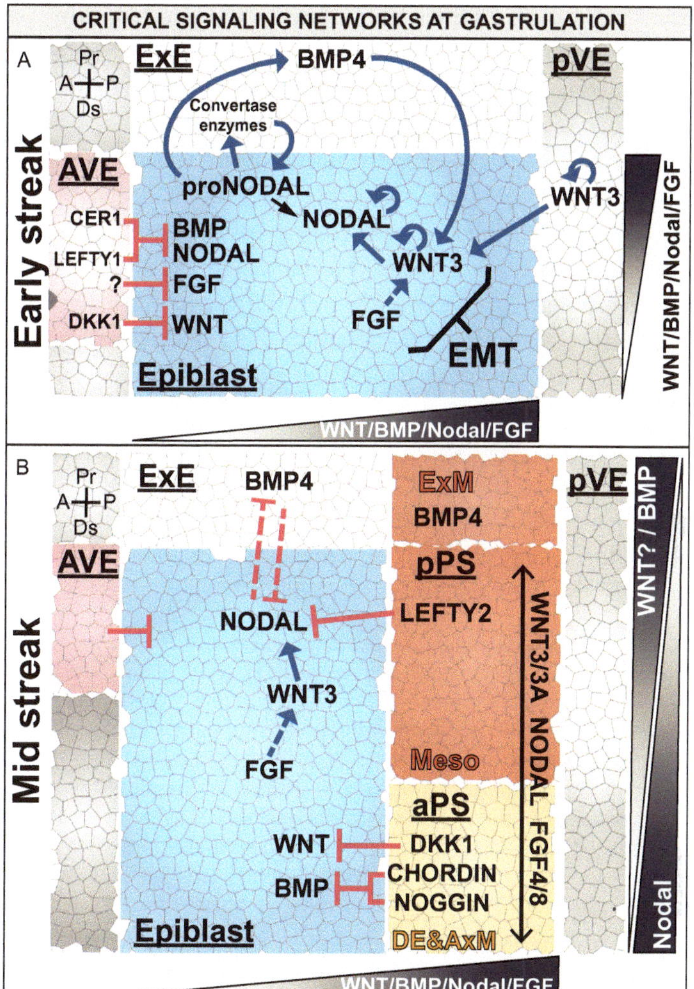

Fig. 2 Critical signaling networks at gastrulation. Schematic diagram summarizing the complex feedback loops between WNT, BMP, Nodal and FGF signaling that regulate gastrulation. Blue arrows denote positive interactions. Pink lines denote negative interactions. Dashed lines indicate interactions that are suggested but still unclear. (A) At the onset of gastrulation, uncleaved pro-NODAL, in the epiblast, induces the expression of *Bmp4* and its own convertase enzymes in the extraembryonic ectoderm (ExE). The convertase enzymes cleave pro-NODAL to active NODAL, which induces its own expression through an autoregulatory loop. *Wnt3* is produced by posterior visceral endoderm (pVE) cells at the embryonic-extraembryonic border. *Wnt3* stimulates its own expression as well as that of *Nodal*. BMP signaling also induces *Wnt3* expression. In addition, signaling is confined to the posterior by inhibitors secreted from the anterior visceral endoderm (AVE). CERL1 and LEFTY1 inhibit BMP and Nodal and DKK1 inhibits WNT signaling. FGF ligands are also expressed within the proximal posterior region and throughout the PS.

(Continued)

gastrulation, it is less clear how it interacts with the other signaling pathways and how its expression is restricted to the posterior of the embryo.

During gastrulation, the embryo substantially changes size and shape. These extensive morphological transformations shift the position of cells and signals such that the signaling landscape of the embryo is continuously evolving. As development proceeds, the PS extends distally, as does the expression domain of *Wnt3*, *Wnt3a* (Kemp et al., 2005; Liu et al., 1999; Takada et al., 1994) and *Nodal* (Norris & Robertson, 1999) ligands (Fig. 2B). However, *Bmp4* is expressed only within the ExE. Therefore, cells traversing the PS at later time points within the distal region of the embryo are furthest from the source of BMP4 and show reduced signaling activity compared to cells within the proximal region of the embryo (Morgani et al., 2018a). Additionally, BMP and Nodal signaling pathways cross inhibit one another, hence the extension of the PS relieves Nodal inhibition by BMP resulting in elevated Nodal signaling within the anterior PS (Chhabra et al., 2019; Heemskerk et al., 2019; Senft et al., 2019). Cells that have traversed the PS also express signaling inhibitors, which reinforce discrete

Fig. 2—Cont'd While FGF is critical for gastrulation, there is less information on its interactions with this signaling network. However, genetic experiments suggest that it may be upstream of WNT signaling. Additionally, although FGF signaling activity is restricted to the posterior of the embryo, it is unclear whether there are inhibitors expressed within the anterior that mediate this restriction. Therefore, the proximal posterior epiblast exhibits high WNT, BMP, Nodal and FGF activity. The combination of these signals stimulates epiblast cells to undergo an epithelial-mesenchymal transition (EMT) in order to exit the epiblast through the primitive streak (PS). (B) As gastrulation proceeds, the embryo grows in size, the PS extends distally and new cell types arise. Cells within the PS express WNT3, WNT3A, FGF4, FGF8 and NODAL ligands. Cells that traverse the posterior PS (pPS) express LEFTY2, a NODAL antagonist. Furthermore, *in vitro* data suggests that Nodal and BMP signaling pathways inhibit one another. Thus, NODAL activity is restricted posteriorly. As BMP4 is expressed proximally, by the ExE and extraembryonic mesoderm (ExM), the expansion of the embryo moves the distal cells further from the source of BMP. Additionally, cells within the anterior PS (aPS) secrete CHORDIN, NOGGIN and DKK1 restricting BMP and WNT signaling distally. Therefore, the proximal posterior of the embryo (pPS) is high in BMP and low in Nodal signaling activity, an environment that promotes embryonic mesoderm (Mes) and extraembryonic mesoderm (ExM) fates. Conversely, the distal embryo (aPS) is low in BMP and high in Nodal signaling activity, promoting definitive endoderm (DE) and axial mesoderm (AxM) fates. The expression of WNT inhibitors in the aPS suggests that there may also be a proximal-distal WNT gradient although this has not been carefully characterized. Similarly, it is unclear whether there is a gradient of FGF signaling activity across the PS. During these later gastrulation stages, the interactions in panel A likely remain however, we removed these details from this panel for simplicity. pPS, posterior PS; Pr, proximal; Ds, distal; A, anterior; P, posterior.

proximal and distal signaling environments that promote different cell fates. The nascent mesoderm cells within the proximal, posterior PS express *Lefty2* (Meno et al., 1997; Peng et al., 2019), a Nodal pathway inhibitor, while cells at the anterior PS express the WNT antagonist *Dkk1* and the BMP antagonists *Chordin* and *Noggin* (Klingensmith et al., 1999; McMahon et al., 1998; Pijuan-Sala et al., 2019). Thus, BMP signaling activity is high proximally and Nodal signaling is high distally. These observations also suggest that WNT signaling may be higher in the posterior compared to anterior PS, although this has not been clearly demonstrated. It is also unknown whether there is an FGF signaling gradient across the PS.

While we know what cells express signaling pathway components, it is not always clear which cells respond to these signals. This will be discussed in subsequent sections.

3. Technical hurdles to studying gastrulation

Genetic experiments have identified the critical signals driving gastrulation, but we know little about how they are regulated in time and space. Are signals stable or oscillatory? How do signaling molecules travel? How do neighboring cells exposed to the same environment specify distinct fates? The limitations to addressing such questions are largely technical.

Gastrulation occurs after the conceptus has implanted into the uterus and, while embryos can be extracted, *ex utero* culture is challenging and requires culture medium with a high percentage of rat serum, which is variable and contains undefined signaling factors that may not be present at this time *in vivo*. Additionally, the degree that the three-dimensional embryo is permeable to various cytokines and small molecule inhibitors is unclear, hence it may not be possible to use these tools as a means to simply manipulate signaling with temporal resolution. Furthermore, the drastic change in size and shape of the gastrulating embryo poses problems for imaging these events. Recent technological advances have enabled the visualization of development *ex vivo* from early gastrulation at E6.5 to the onset of organogenesis at E8.5 (McDole et al., 2018). Nevertheless, imaging the entire embryo at sufficient temporal and spatial resolution to capture individual cell behaviors remains challenging.

Single cell sequencing (Nowotschin et al., 2019; Pijuan-Sala et al., 2019) and spatial transcriptomics (Peng et al., 2016, 2019) of gastrulating embryos show when and where signaling components are expressed at the mRNA level. However, protein level information, in particular that of the post-translational modifications that govern signaling activity, is lacking. Although high-throughput single cell proteomics is currently not a possibility (Marx, 2019), recent studies have started to probe the regulation of signaling at the protein level *in vivo* for example by looking at receptor localization (Zhang et al., 2019).

The response of cells to signaling factors may be assessed at a number of levels such as analyzing the phosphorylation of signaling pathway components, for example, pSMAD1/5/9 downstream of BMP signaling, pSMAD2/3 downstream of Nodal signaling or pERK downstream of FGF signaling. However, immunostaining for phosphorylated proteins is notoriously difficult and thus, for the majority of the critical signaling pathways, it has not been possible to generate consistent single-cell resolution data at gastrulation stages of development. Alternatively, fluorescent reporters may be used to visualize signaling activity in real time. Those that are currently available to monitor signaling *in vivo* (Covert et al., 2015; Ferrer-Vaquer et al., 2010; Morgani et al., 2018b; Nowotschin et al., 2013a; Regot et al., 2014; Serup et al., 2012) utilize distinct fluorescent protein tags and report on different tiers of the pathway. For example, the BMP response element (BRE) reporter (Javier et al., 2012; Serup et al., 2012) indicates promoter activity while the *Spry4* FGF reporter (Morgani et al., 2018b) reports on downstream target expression. Thus, it is difficult to compare the signaling dynamics of various pathways. Furthermore, read outs of signaling based on a particular promoter sequence (Javier et al., 2012) or downstream target may not reveal all sites of pathway activity. Such studies are also constrained by the number of spectrally distinct reporters that can be generated and coordinately visualized, impeding the interrogation of pathways in parallel. Even so, combining several reporters necessitates lengthy exposure times, increasing phototoxicity. Additionally, as fluorescent protein tags are often stable over long periods of time, they may reveal initial signaling activation, but not rapid dynamics. More recently developed signaling reporters directly read out downstream kinase activity *via* FRET biosensors (Kamioka et al., 2013) or phosphorylation-induced nuclear-cytoplasmic shuttling of a fluorescent tag (Regot et al., 2014). These tools will be valuable to expose signaling dynamics but have not yet been analyzed in detail in mammalian systems *in vivo*. Visualizing signaling molecules directly (Venkiteswaran et al., 2013) would also provide key information about how ligands move and how concentration gradients are established over time, but tagging small secreted molecules can alter their function and mobility. As such, the investigation of signaling dynamics *in vivo* is a daunting task.

4. *In vitro* systems represent simplified gastrulation models

Due to the complexity of the embryo and technical barriers to studying these processes *in vivo*, simplified, readily manipulated systems are required to investigate the detailed mechanisms regulating signaling during organized cell fate specification. Pluripotent stem cells (PSCs) are the *in vitro* counterpart of the epiblast that can self-renew, providing an abundance of material for biochemical experiments and screens, and are readily genetically and chemically manipulated. Like the epiblast, PSCs differentiate when exposed to defined signals (Gadue et al., 2006; Turner et al., 2014). However, they specify cell fates in a disorganized manner that is inconsistent with germ layer patterning during gastrulation. Recently, several *in vitro* systems have been established that elicit organized germ layer specification, representing deconstructed, simplified models of gastrulation with which to probe complex signaling dynamics. These systems recapitulate different aspects of gastrulation, hence leveraging the advantages of each model will generate valuable insights into early development. Here we describe a number of the most commonly used *in vitro* gastrulation models.

4.1 Micropattern differentiation

Standard PSC cultures comprise colonies of a range of sizes and morphologies. This heterogeneity undoubtedly influences the response of cells to signals and causes variable, disorganized differentiation. Colony morphology may be standardized through the use of 'micropatterned' surfaces that limit cell binding to extracellular matrix (ECM)-coated regions of defined size and shape (Peerani et al., 2009). Mouse and human PSCs that are differentiated in 500 μm to 1 mm diameter circular colonies self-organize cell fates in concentric rings, patterning epiblast, PS and germ layer derivatives (Morgani et al., 2018a; Warmflash et al., 2014) in a temporal sequence that is comparable to the gastrulating mouse embryo (Morgani et al., 2018a). While colonies are exposed homogeneously to signaling factors, cells display radially dependent responses. Initially, all cells demonstrate signaling activity but differences in receptor localization and exposure to endogenous inhibitors gradually restrict the response to outer cells (Warmflash et al., 2014) (see Section 5.2). This formation of discrete signaling domains, also observed in the embryo, underpins patterning.

The spatial restriction of signaling within micropatterned colonies requires a confluent epithelium. In an epithelial layer, BMP signaling receptors are basolaterally localized preventing ligand binding (Etoc et al., 2016) (see Section 5.2). At low cell density, or if adherens junctions are disrupted, a signaling response is observed throughout the colony and consequently patterning is altered. Parallels might be drawn with gastrulating mouse embryos which, although expressing signaling receptors throughout the epiblast, only robustly respond to the surrounding signals once they undergo EMT and exit the epithelium (Ferrer-Vaquer et al., 2010; Morgani et al., 2018a, 2018b). Therefore, receptor localization may likewise regulate signaling *in vivo* (Zhang et al., 2019).

In the embryo, the extraembryonic AVE establishes anterior-posterior polarity by secreting signaling inhibitors at the anterior of the embryo (Fig. 2A). However, the PSC micropatterned colonies do not contain an AVE hence specify cell fates with radial symmetry (Fig. 3A). However, the incorporation of microfluidic-induced signaling gradients can mimic the role of the AVE and impart anterior-posterior polarity (Manfrin et al., 2019). The gastrulating mouse embryo also has proximal-distal polarity (Fig. 1). While providing different signaling factors can promote proximal or distal fates within the micropattern system (Morgani et al., 2018a), conditions have not been identified that support the full repertoire of proximal and distal fates in a single colony. It is thus possible that the relatively two-dimensional colonies of defined diameter can only establish a limited range of signaling niches. In keeping with this hypothesis, reducing the colony diameter eradicates the central cell fates (Manfrin et al., 2019; Morgani et al., 2018a; Warmflash et al., 2014), which can be rescued by reducing the ligand concentration (Tewary et al., 2017). Hence, at a defined ligand concentration, a minimum colony diameter may be required to set up a sufficient signaling gradient to specify all cell fates and therefore larger colonies might expand the signaling range and cell fate potential.

The uniformity of micropatterned colonies facilitates direct quantitative comparisons of large numbers of colonies to generate robust information about cell fate specification and patterning (Ostblom et al., 2019; Tewary et al., 2017). Furthermore, the flattened morphology of colonies renders them amenable to live imaging and signaling manipulation with small molecules. In addition, as both mouse (Morgani et al., 2018a) and human (Warmflash et al., 2014) micropattern differentiation protocols have been established, this model can be used as a platform for cross-species comparison.

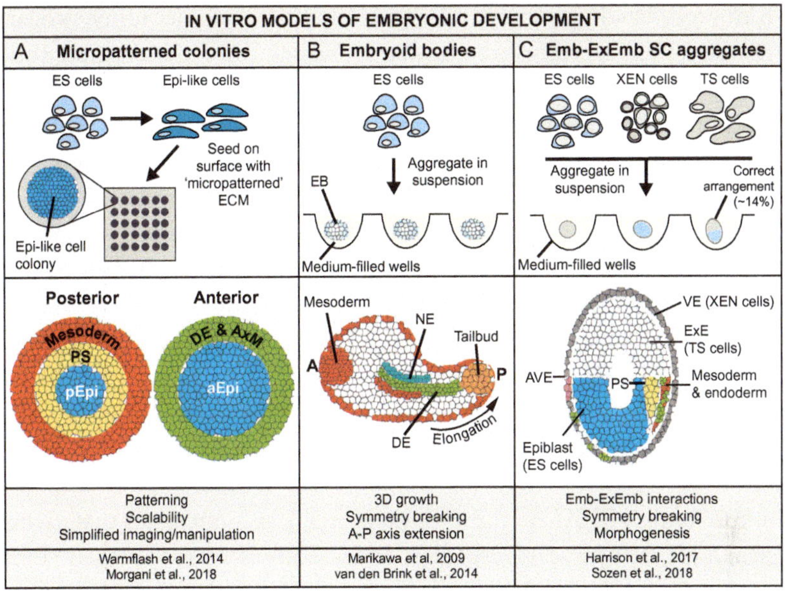

Fig. 3 Schematic diagram depicting *in vitro* gastrulation models. A number of simplified *in vitro* pluripotent stem cell (PSC) -based models have been established that recapitulate aspects of *in vivo* gastrulation. Shown are simplified schematic diagrams of some of these systems. Upper panels indicate the starting conditions for each system. Lower panels indicate the end point of differentiation. Advantages of each system are listed below. (A) For micropattern-based mouse PSC differentiation, embryonic stem cells (ES cells) are converted to epiblast like cells (Epi-like cells). Epi-like cells are seeded onto surfaces that are 'micropatterned' with extracellular matrix (ECM) within circular regions of defined diameter to facilitate the formation of uniform circular PSC colonies. When PSCs are exposed to various combinations of gastrulation-promoting signals, they pattern embryonic germ layers with radial symmetry. In the presence of FGF, ACTIVIN, WNT and BMP signaling, PSCs pattern cell fates formed in the posterior of the embryo: a central posterior epiblast (pEpi) region followed by an intermediate primitive streak (PS)—like region and an outer layer of cells comprised of later germ layer derivatives including embryonic and extraembryonic mesoderm. Alternatively, culturing cells in the presence of FGF, ACTIVIN and WNT (without BMP) gives rise to distinct cell fates found in the anterior of the embryo, such as anterior epiblast (aEpi), definitive endoderm (DE) and axial mesoderm (AxM). The spatial organization of cell fates facilitates the investigation of interactions between neighboring cell types. Micropatterned surfaces contain many colonies hence this system is scalable. The relatively two-dimensional morphology of colonies means that they can be easily imaged and manipulated with small molecules. (B) Embryoid bodies (EBs) are three-dimensional cell clusters formed by aggregating ES cells in suspension. EBs exhibit spontaneous symmetry breaking forming a posterior-like domain at one pole that expresses markers of the tailbud. During prolonged culture, EBs undergo posterior elongation and differentiate into various germ layer derivatives. Tube-like DE structures are observed in the posterior-center of these structures, flanked ventrally by neurectoderm (NE) cells and laterally by mesoderm.

(Continued)

Although flat-disc micropatterned colonies are topologically similar to gastrulating human embryos, they are distinct from cup-shaped rodent embryos. Micropatterned colonies are also of fixed dimension while gastrulating embryos rapidly change size and shape. In addition, while cells within micropatterned colonies undergo EMT, they show minimal directional movement (Chhabra et al., 2019). Hence, it is unclear whether this system can recapitulate all aspects of *in vivo* signaling regulation. Therefore, alternative physically unrestricted models, or the liberation of colonies from micropatterned surfaces, may be required to understand the evolution of the signaling landscape as the embryo grows and undergoes tissue rearrangements.

4.2 Embryoid bodies

Embryoid bodies (EBs) are three-dimensional PSC aggregates that give rise to germ layer derivatives (Desbaillets et al., 2000; Kurosawa, 2007). Under homogeneous conditions EBs establish an anterior-posterior axis by spontaneously initiating posterior marker expression at one pole (Marikawa et al., 2009; ten Berge et al., 2008; van den Brink et al., 2014). The extended culture of EBs under defined conditions additionally induces presumptive dorsal-ventral and left-right axes (Beccari et al., 2018; Turner et al., 2017). Furthermore, robust EB elongation and gene expression signatures associated with anterior-posterior axis extension is observed in the majority

Fig. 3—Cont'd Mesoderm is also observed at the proposed anterior domain as well as around the periphery of the aggregate. This system is useful to study signaling regulation in three-dimensional space in a morphologically dynamic structure and, in particular, axial elongation. (C) PSCs, representative of the embryonic epiblast (ES cells), can be aggregated with extraembryonic endoderm (XEN) cells representative of the visceral endoderm (VE), and trophoblast stem (TS) cells representative of the extraembryonic ectoderm (ExE). When these cells are aggregated in suspension, approximately 14% of aggregates arrange the cell types in a manner similar to the gastrulating mouse embryo, with ES cells adjacent to TS cells and surrounded by XEN cells. Aggregates establish an anterior-posterior axis through the asymmetric expression of PS markers, as in the gastrulating embryo. These aggregates also generate DE-like cells which show evidence of intercalation into the outer XEN cell layer, similar to the intercalation of DE with VE *in vivo*. Furthermore, some aggregates display anterior visceral endoderm (AVE) gene expression. However, it is currently unclear how different cell populations including PS/mesoderm *versus* DE, and AVE *versus* PS are positioned in relation to one another. Embryonic-extraembryonic stem cell (Emb-ExEmb SC) aggregates are a valuable tool to probe embryonic-extraembryonic interactions and morphogenetic events such as cavitation.

of aggregates (Marikawa et al., 2009; van den Brink et al., 2014) (Fig. 3B). These elongating EB structures have been termed 'gastruloids' (van den Brink et al., 2014).

The origin of EB patterning is unclear as extraembryonic tissues, which provide localized signals regulating axis formation *in vivo*, are not present within this system (Ding et al., 1998; Kimura-Yoshida et al., 2005; Migeotte et al., 2010; Nowotschin et al., 2013b). This suggests that the extraembryonic lineages are not be absolutely required for axis formation but, as axis formation occurs with varying frequency *in vitro*, they may be necessary to ensure developmental robustness (Marikawa et al., 2009; van den Brink et al., 2014; Simunovic et al., 2019). Furthermore, inhibitors produced by the extraembryonic AVE are required to generate the anterior-most neural fates (Ang et al., 1994), hence these are not present in EBs (Beccari et al., 2018; van den Brink et al., 2014; Turner et al., 2017) and likely not in micropatterns. In contrast to EBs, spontaneous symmetry breaking is not observed in micropatterned colonies without an exogenous asymmetric signal (Manfrin et al., 2019). This could either indicate either that this is not possible in two-dimensions, unlikely as most mammalian embryos are disc-shaped (Sheng, 2015), or that further modifications to the experimental conditions, such as colony diameter or ligand concentration, could permit these events.

While EBs exhibit a temporal gene expression pattern that is consistent with *in vivo* development, the cells that are specified do not recapitulate the morphogenetic movements that occur during gastrulation and thus, cells are not arranged into ordered germ layers or tissues (Beccari et al., 2018). For example, during gastrulation epiblast cells undergo EMT and subsequently migrate between the epiblast and the outer extraembryonic VE layer. As EBs do not contain VE cells and are grown in suspension, a fraction of cells undergoing EMT instead dissociate into the surrounding medium (van den Brink et al., 2014; Simunovic et al., 2019). However, EBs may also be generated inside a solid matrix (Bedzhov & Zernicka-Goetz, 2014; Simunovic et al., 2019), which conceivably could support further morphogenesis.

4.3 PSC-extraembryonic stem cell aggregates

Micropatterned colonies and EBs are comprised solely of PSCs hence do not permit the interrogation of the critical embryonic-extraembryonic interactions that drive gastrulation. When PSCs are cultured in suspension together with the *in vitro* counterparts of the ExE (trophoblast stem (TS) cells) and of the VE (extraembryonic endoderm (XEN) cells), they can form three-dimensional aggregates that bear remarkable resemblance to gastrulation-stage embryos (Harrison et al., 2017; Sozen et al., 2018) (Fig. 3C). The efficiency of the formation of these structures is low, however approximately half of the correctly assembled aggregates asymmetrically induce a PS-like population (Sozen et al., 2018).

Unlike micropatterned colonies and EBs, mixed cell aggregates undergo gastrulation-like cell movements whereby DE cells are specified and intercalate into the outer XEN layer (Kwon, Viotti, & Hadjantonakis, 2008; Sozen et al., 2018). Additional analysis is needed to determine whether mesoderm cells exhibit collective, directional migration between the PSC (epiblast) and XEN (VE) layers as *in vivo*. These aggregates therefore represent a unique system to study cell fate specification and morphogenetic events.

The morphological similarity of this three-dimensional system to the embryo makes it a powerful tool to investigate developmental interactions *in vitro*. However, they also evoke the same technical issues as working with embryos, particularly with regards to imaging and permeability to exogenous factors. Hence, both three-dimensional and simplified two-dimensional models will play valuable roles in understanding these processes.

5. Critical signaling pathways at gastrulation

WNT, BMP, Nodal and FGF signals cooperate to initiate gastrulation and germ layer specification. Disruption of any of these pathways results in cessation of gastrulation and early embryonic lethality. Here we discuss the current state of knowledge of the roles and regulation of these signals during gastrulation, gleaned from mouse embryos and PSCs.

5.1 WNT signaling

WNT ligands are secreted glycolipoproteins that activate canonical and non-canonical signaling responses. Here we focus primarily on the canonical WNT pathway. To establish signaling gradients, WNT ligands can diffuse in the extracellular space freely (Kicheva et al., 2007) or in association with lipid vesicles (Greco, Hannus, & Eaton, 2001; Panakova et al., 2005) as well as being transported by membrane extensions (Stanganello et al., 2015). Perhaps surprisingly, when artificially tethered to the cell membrane, WNT ligands can successfully pattern the Drosophila wing disc

(Alexandre, Baena-Lopez, & Vincent, 2014), indicating that diffusion is not required for WNT-mediated patterning in all contexts. Nevertheless, it remains unclear how WNT ligands travel within the mammalian embryo.

Under steady state conditions, the WNT transcriptional effector β-CATENIN is targeted for degradation by an AXIN-APC-GSK3-CK1 destruction complex. Binding of WNT to its receptors, Frizzled (FZ) and LRP5/6, leads to AXIN recruitment and breakdown of the destruction complex. This allows cytoplasmic β-CATENIN to accumulate and translocate to the nucleus where it regulates gene expression in combination with co-factors such as TCF/LEF (MacDonald, Tamai, & He, 2009). WNT signaling is regulated by antagonists, including SFRP and DKK1. In certain contexts, *Dkk1* expression is induced by WNT (Lewis et al., 2008), but in other contexts WNT signaling restricts *Dkk1* expression (Miura et al., 2010). These distinct effects may be WNT ligand specific (Lewis et al., 2008). When *Wnt3* and *Wnt3a* epiblast expression is reduced, the expression domain of *Dkk1* is expanded posteriorly, (Miura et al., 2010) suggesting that during gastrulation WNT ligands repress *Dkk1*. *Dkk1* expression within the VE is also required for correct migration of the AVE to the anterior of the embryo (Miura et al., 2010; Kimura-Yoshida et al., 2005).

Abrogation of the WNT pathway results in severe gastrulation phenotypes. Mutant embryos do not initiate gastrulation (Barrow et al., 2007; Biechele, Cox, & Rossant, 2011; Fu et al., 2009; Huelsken et al., 2000; Hsieh et al., 2003; Kelly, Pinson, & Skarnes, 2004; Liu et al., 1999; Tortelote et al., 2013) and likewise, WNT inhibition *in vitro* blocks germ layer differentiation (Chhabra et al., 2019; Marikawa et al., 2009; Martyn, Brivanlou, & Siggia, 2019; Martyn et al., 2018; Morgani et al., 2018a; Simunovic et al., 2019). In wild type embryos, cells that do not traverse the PS and remain within the epiblast give rise to neural cell types. However, specification of the anterior neural fates requires signals from the PS-derived anterior endoderm and mesoderm (Ang et al., 1994; Camus et al., 2000). Therefore, in WNT mutant embryos where a PS is absent, the epiblast remains pluripotent (Biechele et al., 2011; Kelly et al., 2004; Liu et al., 1999). Conversely, ectopic activation of WNT signaling or loss of WNT antagonists stimulates premature EMT, expansion of the PS and posterior fates and axis duplication, both *in vitro* and *in vivo* (Kemler et al., 2004; Martyn et al., 2019; Metzis et al., 2018; Morgani et al., 2018a; Popperl et al., 1997; Simunovic et al., 2019; Turner et al., 2017; Zeng et al., 1997). Hence, WNT is necessary and sufficient for PS formation. Furthermore, different levels of WNT along the anterior-posterior epiblast axis drive lineage-specific chromatin organization required to specify later anterior and posterior neural fates (Metzis et al., 2018).

From the onset of gastrulation *Wnt3* is expressed throughout the PS and later, at the mid-streak stage, *Wnt3a* is expressed within the same region (Kemp et al., 2005; Takada et al., 1994). WNT signaling is regulated by antagonists, including SFRP and DKK1, which are produced by the AVE and restrict WNT activity to the posterior of the embryo (Ding et al., 1998; Finley, Tennessen, & Shawlot, 2003; Glinka et al., 1998; Kawano & Kypta, 2003; Kemp et al., 2005; Pijuan-Sala et al., 2019; Yoon et al., 2015). Signaling reporters show that WNT signaling activity essentially recapitulates the expression pattern of WNT ligands with cells in the PS, embryonic and extraembryonic mesoderm exhibiting a response (Ferrer-Vaquer et al., 2010; Sundararajan et al., 2012). The transcription factor *Lef1*, which mediates the nuclear response to WNT, is also a target of this pathway (Kengaku et al., 1998). Thus, from around E7.0 *Lef1* expression is enriched within the posterior epiblast (Peng et al., 2019), which may reinforce WNT signaling activity within this region. *Wnt3* expression is induced by proximal BMP signaling (Ben-Haim et al., 2006) and WNT inhibitors are also secreted by anterior PS derivatives (Glinka et al., 1998; Kawano & Kypta, 2003; Kemp et al., 2005; Pijuan-Sala et al., 2019). Together, this may institute a proximal-distal WNT gradient in the embryo. While this has not been quantitatively assessed, WNT activity does appear increasingly heterogeneous in the anterior PS (Sundararajan et al., 2012).

Prior to gastrulation, at E5.5, *Wnt3* is expressed at the embryonic-extraembryonic border of the posterior VE (Rivera-Perez & Magnuson, 2005), representing the first known anterior-posterior asymmetry in the mouse embryo. Consistently, WNT signaling activity is also elevated within cells of the posterior VE during gastrulation (Ferrer-Vaquer et al., 2010). The basis of this early *Wnt3* asymmetry is unknown. However, WNT signaling is influenced by physical properties including ECM stiffness, three-dimensional culture and mechanical forces (Brunet et al., 2013; Du et al., 2016; Marikawa et al., 2009; Rotherham & El Haj, 2015). Hence, changes in cell density caused by the rapid proliferation of epiblast cells prior to gastrulation (Snow, 1977) and changes in tissue morphology could feed into this pathway.

WNT3 expressed by the VE induces its own expression in the underlying epiblast and PS (Tortelote et al., 2013; Yoon et al., 2015). Loss of *Wnt3* in either the epiblast or VE delays PS formation (Tortelote et al., 2013; Yoon et al., 2015) hence, a threshold of WNT may be required to trigger downstream events. In the absence of *Wnt3* from the VE, after this initial delay gastrulation proceeds normally (Yoon et al., 2015) but loss of epiblast *Wnt3* or a reduction of WNT signaling activity leads to an accumulation of cells at the PS, loss of paraxial mesoderm and embryonic lethality (Barrow et al., 2007; Kelly et al., 2004; Tortelote et al., 2013), mirroring the phenotype of FGF pathway mutants (see Section 5.4) (Ciruna et al., 1997; Ciruna & Rossant, 2001; Deng et al., 1994; Guo & Li, 2007; Sun et al., 1999; Yamaguchi et al., 1994). As FGF signaling is unaffected in WNT mutants (Kelly et al., 2004), WNT may be downstream of FGF or independently regulate related processes.

Wnt3a is expressed from approximately E7.0 (Pijuan-Sala et al., 2019; Sundararajan et al., 2012; Takada et al., 1994). In *Wnt3a*$^{-/-}$ embryos and EBs, expression of the target gene *Brachyury* is lost and axial elongation is perturbed (Marikawa et al., 2009; Takada et al., 1994; Yamaguchi et al., 1999). Hence, while *Wnt3* and *Wnt3a* are expressed in overlapping domains, they are either not functionally redundant or cells are sensitive to WNT dose and duration. In certain *in vitro* contexts, WNT signaling is adaptive, i.e., sustained ligand exposure produces only a transient response (Massey et al., 2019). *In vitro*, when, BMP and/or ACTIVIN are added to cells alongside WNT3A, WNT signaling adaptation is no longer observed and instead, cells show a sustained downstream response (Massey et al., 2019). Thus, BMP and Nodal signaling within the PS may be required to maintain the WNT response. It is also interesting to speculate whether WNT3A could overcome adaptation to WNT3.

Gastrulation involves both cell fate specification and morphogenesis. β-CATENIN is the transcriptional effector of the WNT pathway but also interacts with cadherins at cell junctions where it regulates adhesion and therefore migration and morphogenesis (Gottardi & Gumbiner, 2004). A simple model proposes that the downregulation of *E*-CADHERIN during the gastrulation EMT increases WNT signaling capacity by releasing junctional β-CATENIN (Ciruna & Rossant, 2001; Howard et al., 2011). Consistently, cells that undergo EMT display elevated WNT activity (Ferrer-Vaquer et al., 2010). However, EMT is associated with a switch from E to N -CADHERIN, both of which bind β-CATENIN, hence the regulatory mechanisms must be more complex. Additionally, counter-intuitive to this model, when cadherins are lost, WNT signaling is no longer elevated (Hendriksen et al., 2008; Howard et al., 2011; Punovuori et al., 2019; van de Wetering et al., 2001). Therefore, the initial cadherin interaction may prime β-CATENIN for signaling, supported by the finding that membrane and cytoplasmic β-CATENIN are molecularly distinct (Gottardi & Gumbiner, 2004). In other contexts, it has been shown that WNT signaling induces EMT rather than *vice versa* (Zhan, Rindtorff, & Boutros, 2017). The TCF/LEF transcription factors downstream of WNT, suppress *E-cadherin* expression directly (Huber et al., 1996; Jamora et al., 2003) as well as indirectly through the induction of negative regulators including *Twist* and *Slug* (Howe et al., 2003; Vallin et al., 2001). Thus, WNT signaling causes a reduction in *E-cadherin* expression and a corresponding reduction in E-CADHERIN protein can further elevate WNT signaling activity. However, these interactions remain to be verified in the mouse embryo.

During gastrulation, the PS and correlated WNT signaling domain extends distally. Analogously, during micropattern differentiation *in vitro*, WNT signaling activity progresses from the edge to the center of colonies (Chhabra et al., 2019; Martyn et al., 2019). This 'wave' of WNT signaling is independent of cell movement (Chhabra et al., 2019) and could arise by *Wnt3* transcriptional autoregulatory mechanisms (Tortelote et al., 2013) or alternatively *via* an "EMT cascade" (Fig. 4). Cells at the outer colony edge, corresponding to those in the proximal posterior epiblast, have a cell contact on only one surface and thus reduced *E*-CADHERIN compared to their neighbors. Low E-CADHERIN may promote low junctional and high cytoplasmic β-CATENIN, elevated WNT signaling and subsequent EMT. By virtue of a cell undergoing EMT, it reduces adhesion with its immediate neighbor, propagating this process (Martyn et al., 2019). Gastrulation may encompass a combination of these interactions whereby WNT signaling initially induces EMT, which subsequently leads to further elevation of WNT activity. Remarkably, if WNT secretion is blocked after the initiation of differentiation, the level of signaling is reduced but the wave dynamics remain the same (Chhabra et al., 2019). This could indicate that the ligand is stable over time and distance, or that the nuclear localization of β-CATENIN (used as a signaling read-out) is induced by other factors such as BMP (Massey et al., 2019) (see Section 5.2). As no live imaging of WNT activity during gastrulation has been performed, it is unknown whether similar dynamics are apparent in the embryo.

5.2 BMP signaling

BMPs are secreted factors that form part of the TGF-β superfamily of proteins. There is evidence both that BMP diffuses freely in the extracellular

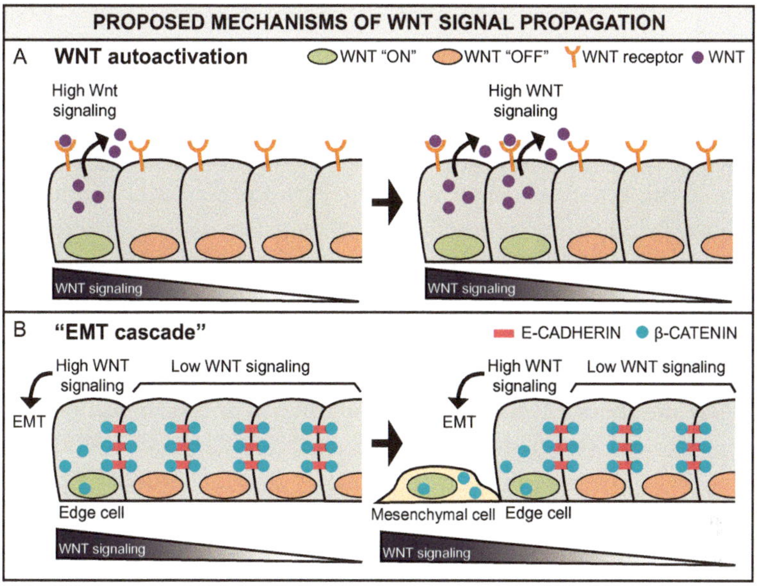

Fig. 4 Potential mechanisms for WNT signaling propagation. During gastrulation, the WNT signaling domain expands in a proximal-distal direction, correlating with primitive streak (PS) expansion. Similarly, *in vitro* in micropatterned colonies, WNT signaling activity expands from the colony edge to the center over time. The mechanism regulating this 'wave' of WNT activity is unknown, but it appears to be independent of cell movement and could involve WNT autoregulation (A) or an "EMT cascade" (B). (A) WNT induces its own expression. Therefore, a cell with high WNT signaling activity will produce and secrete WNT ligand that will act on its neighboring cell to activate WNT signaling and propagate this process. (B) The WNT transcriptional effector β-CATENIN can regulate adhesion by binding at adherens junction complexes with *E*-CADHERIN. It has been proposed that cells at the edge of an epithelium, for example at the edge of a micropatterned colony or within the proximal epiblast *in vivo*, have reduced E-CADHERIN compared to their neighbors as they only have a single cell-cell junction. As such, less β-CATENIN is sequestered at the membrane and more is free in the cytoplasm to mediate WNT signaling. Hence, WNT signaling activity is higher in edge cells, stimulating an EMT. The EMT causes the outer cell to delaminate from the epithelium, effectively producing a new outer cell and propagating this process.

space as well as moves through tissues by transcytosis (Entchev, Schwabedissen, & Gonzalez-Gaitan, 2000; Kicheva et al., 2007; Pomreinke et al., 2017). BMPs bind and activate type II receptors (BMPR-II, ACTR-IIA, ACTR-IIB), which in turn phosphorylate and activate associated type I receptors (BMPR-IA/ALK3, BMPR-IB/ALK6, ACTR-I/ALK2) (Wrana et al., 1994). Type I receptors phosphorylate receptor-regulated (R-) Smads, SMAD1, SMAD5 and SMAD8/9, mediating their dimerization with a common mediator (Co-) Smad, SMAD4. *In vitro*, Smads continuously shuttle between the nucleus and cytoplasm (Heemskerk et al., 2019; Nicolas et al., 2004; Xiao et al., 2001). It is unclear whether Smads are similarly dynamic *in vivo* but preventing shuttling does not affect mouse development or homeostasis (Biondi et al., 2007) indicating that this is not critical for signaling. Pathway activation reduces Smad nuclear export allowing accumulation in the nucleus and regulation of target gene expression. Unlike WNT and Activin/Nodal signaling (see Sections 5.1 and 5.3), the transcriptional response downstream of BMP shows little adaptation (Heemskerk et al., 2019; Nemashkalo et al., 2017; Yoney et al., 2018). The BMP pathway is regulated at a number of levels including *via* cofactors, inhibitory Smads (I-Smads) that competitively bind type I receptors, and secreted factors such as NOGGIN, CHORDIN and CERL1 that bind and inhibit BMP (Belo et al., 1997, 2000; Miyazawa & Miyazono, 2017; Piccolo et al., 1996; Xiao et al., 2003; Zimmerman, DeJesusEscobar, & Harland, 1996).

Disruption of BMP signaling, for example, by mutating *Bmp2*, *Bmp4*, *Bmp8b*, *Bmpr1a*, *AcvrI* or *Bmpr2*, results in embryonic lethality at gastrulation, with mutant embryos exhibiting precocious neural differentiation and a reduction or absence of posterior cell fates such as mesoderm and PGCs (Beppu et al., 2000; Gu et al., 1999; Lawson et al., 1999; Mishina et al., 1995; Xiao et al., 2003; Ying et al., 2000; Zhang & Bradley, 1996).

BMP signaling induces *Wnt3* expression and WNT signaling activity (Ben-Haim et al., 2006; Miura et al., 2010; van den Brink et al., 2014). BMP signaling within the VE is additionally required to induce *Dkk1* expression required for AVE migration (Miura et al., 2010), and hence to establish anterior-posterior polarity. Unexpectedly, BMP can also stimulate nuclear localization of β-CATENIN independent of WNT activity (Massey et al., 2019). If BMP is inhibited at any point during micropattern differentiation, WNT activity is reduced and the expansion of the WNT signaling domain is arrested (Chhabra et al., 2019). Hence, sustained BMP signaling is required to maintain WNT signaling and likely PS elongation. The interaction between BMP and WNT confounds mutant phenotypes as direct *versus* indirect effects are unclear. *In vitro* systems allow the disruption of pathways in isolation by supplying each factor exogenously in order to tease apart their individual roles. BMP is required for the specification of posterior mesoderm fates *in vitro* (Beccari et al., 2018; Morgani et al., 2018a; Tewary et al., 2017; Turner et al., 2017) but, in the absence of BMP, anterior PS derivatives can still be generated if WNT and/or ACTIVIN/NODAL are added exogenously (Morgani et al., 2018a; ten Berge et al., 2008). Hence, the primary role of BMP is to induce posterior fates in cells exiting the PS.

The dose of BMP that cells are exposed to affects the rate of signaling response, the size of the signaling domain and the cell fates specified. During micropattern differentiation, increasing the concentration of BMP results in a more rapid activation of the BMP response (Manfrin et al., 2019). Furthermore, the signaling domain is expanded and cell fates that are typically specified only within regions closest to the source are now also specified at greater distances (Manfrin et al., 2019). This suggests that increasing the ligand concentration at the source increases the distance that the morphogen travels. In addition to signaling level, the length of signal exposure regulates cell fate (Chhabra et al., 2019; Tewary et al., 2017; Zhang et al., 2008). *In vitro* data suggests that continued BMP signaling is required to drive cell fate specification (Chhabra et al., 2019). However, prolonging BMP signaling activity by disrupting receptor endocytosis results in severe morphological defects and embryonic lethality (Aoyama et al., 2012). Nevertheless, we know little about the dynamics of BMP signaling during gastrulation. The generation of novel fluorescent reporter tools and improved imaging techniques may shed light on this in the future.

The ExE is a source of BMP ligands (Winnier et al., 1995; Ying et al., 2000; Ying & Zhao, 2001) and the AVE secretes the antagonist CER1 (Belo et al., 1997, 2000), establishing high BMP signaling in the proximal posterior epiblast (Fig. 3). At the onset of gastrulation, cells within the proximal epiblast and VE, adjacent to the ExE, exhibit BMP signaling activity (Di-Gregorio et al., 2007; Hayashi et al., 2002; Javier et al., 2012). While the distal epiblast at this stage does not show a significant BMP signaling response, if distal epiblast explants are treated with BMP4, they respond to BMP (Okamura, Hayashi, & Matsui, 2005), suggesting that the difference in signaling response is based on a limited diffusion distance of BMP from the ExE. Cells of the ExE do not respond to BMP signaling. As ExE cells express similar levels of BMP receptors, downstream effectors and inhibitors of the BMP pathway (Pijuan-Sala et al., 2019), signaling response may be regulated at the protein level.

As gastrulation proceeds, the embryo grows in size and the PS extends distally. Cells within the distal epiblast that traverse the anterior PS are further away from the ExE source of BMP than cells that traverse the posterior PS at earlier stages of development. Consistently, cells within the posterior but not anterior PS respond to BMP (Morgani et al., 2018a). However, in contrast to at earlier stages, distal epiblast explants no longer respond to BMP (Okamura et al., 2005) indicating that, as well as limitations to the distance of BMP diffusion, additional mechanisms are in place to control BMP response. Nevertheless, it should be noted these explant experiments used BMP-mediated PGC specification as an indirect readout of signaling activity.

One possible mode of regulation is the expression of downstream effector Smads as loss of individual effectors, such as *Smad5*, is sufficient to reduce the BMP response (Monteiro et al., 2008). As *Smad1* and *Smad5* are expressed at lower levels in the distal compared to the proximal epiblast (Peng et al., 2019; Okamura et al., 2005), this could contribute to the reduced BMP response observed. The BMP antagonists NOGGIN and CHORDIN are also expressed by derivatives of the anterior PS (Klingensmith et al., 1999; McMahon et al., 1998) hence could regulate the distal signaling activity. However, it is unknown whether BMP signaling activity is altered in *Chordin/Noggin* double mutants (Bachiller et al., 2000). Thus, it remains an open question whether distal BMP signaling is actively restricted or if this stems purely from BMP diffusion limitations.

In vitro BMP signaling gradients in micropatterned colonies result in highest signaling activity in cells closest to and lowest in cells farthest from the ligand source (Manfrin et al., 2019). However, the outermost cells in the colony, including those furthest from the source, demonstrate signaling activity suggesting that edge cells are more responsive than central cells. This peripheral BMP activity is suppressed by an opposing inhibitor gradient (Manfrin et al., 2019), akin to that supplied by the AVE. It is unclear whether edge cells, perhaps corresponding to the most proximal epiblast cells, absolutely require inhibitors to prevent signaling or whether, at a certain distance from the source, they are no longer receptive to the signal.

In the human micropattern system, expression of the BMP inhibitor *NOGGIN* is induced by exposure to BMP and is required for the edge-restriction of signaling activity and consequent pattern formation (Etoc et al., 2016; Tewary et al., 2017). NOGGIN is proposed to be lost from the colony periphery thereby specifically increasing the receptivity of edge cells to BMP while maintaining inhibition in central cells (Etoc et al., 2016). It is uncertain how this inhibitor loss translates to the embryo, but it conceivably could represent loss of inhibitor from the proximal epiblast-ExE interface. It is also not clear how these findings relate to the mouse as expression of *Noggin* is not observed in the mouse epiblast prior to gastrulation (McMahon et al., 1998; Pijuan-Sala et al., 2019) and is not induced by BMP in mouse PSCs (Etoc et al., 2016). The same role may be played by alternative inhibitors that are expressed earlier, such as *Follistatin* or *Smad7* (Pijuan-Sala et al., 2019). While loss of these particular inhibitors individually does not yield gastrulation phenotypes (Chen et al., 2009; Lee et al., 2010), it is possible that several factors work together to regulate signaling. Alternatively, this initial widespread signaling activity and subsequent restriction observed in micropatterns may not occur *in vivo* and inhibitors may only be required at later stages to stabilize already established signaling gradients.

An additional component to signaling regulation is receptor localization. Tight junctions between epithelial cells, such as the epiblast, physically separate the apical and basolateral membrane domains. Hence receptors can be asymmetrically distributed, restricting signal detection to one surface (Etoc et al., 2016; Hobert & Carlin, 1995; Saitoh et al., 2013). BMP receptors are basolaterally-localized in the mouse epiblast and micropatterned colonies (Etoc et al., 2016; Zhang et al., 2019) and hence bind BMP at their VE-facing, but not apical cavity-facing, surface. Receptor localization constrains signaling activity as loss of tight junctions, unrestricted basolateral ligand exposure (Etoc et al., 2016) or apical mis-expression of BMPRs produces an ectopic BMP response (Zhang et al., 2019). In contrast to free ligand diffusion within the wide cavity, the narrow epiblast-VE interstitial space may limit the distance that a ligand can travel before being captured by a receptor, and hence influence the proximal-distal signaling gradient (Zhang et al., 2019). The thin basement membrane between micropatterned colonies and the underlying surface may similarly control ligand movement (Zhang et al., 2019). The influence of these properties on signaling gradients could be tested by altering the height of the basement membrane.

BMP receptors (*Bmpr1a* and *Bmpr2*) are expressed in both the epiblast and mesoderm cells at similar levels during gastrulation (Pijuan-Sala et al., 2019). However, strong signaling activity, indicated by nuclear localized phosphorylated SMAD1/5/9, is observed only in cells that undergo an EMT and migrate through the PS (Morgani et al., 2018a). This could be related to receptor localization as cells undergoing an EMT expose more of their basolateral receptor-bearing surface. Another possibility is that co-factor or inhibitor expression is altered during differentiation. For example, BMP endothelial cell precursor-derived regulator (*Bmper*) (Heinke et al., 2008; Serpe et al., 2008), which enhances BMP signaling activity, is upregulated in nascent mesoderm and recent spatial transcriptomic analysis shows that the inhibitor *Fst* is enriched in cells of the posterior epiblast compared to the mesoderm. Thus, there are many unanswered questions regarding the spatiotemporal regulation of BMP signaling.

5.3 Activin/Nodal signaling

NODAL and ACTIVIN are part of the TGF-β superfamily of signaling proteins. In chick and *Xenopus*, NODAL and NODAL inhibitor homologs rapidly diffuse across large distances (up to 500 μm) to establish signaling gradients (Sakuma et al., 2002; Williams et al., 2004). NODAL and ACTIVIN bind to and activate the type II receptors ACTRIIA/ACVR2A and ACTRIIB/ACVR2B, which in turn activate the type I receptor ACTRIB/ACVR1B (Reissmann et al., 2001). NODAL additionally requires the EGF-CFC co-receptor CRIPTO (Yan et al., 2002; Yeo & Whitman, 2001). ACTIVIN is not required for gastrulation (Matzuk, Kumar, & Bradley, 1995) but, due to the low activity of recombinant NODAL, is often used as a substitute *in vitro*, hence we occasionally refer to Activin/Nodal signaling. Activated receptors phosphorylate and activate the receptor (R-) Smads, SMAD2 and SMAD3, which form complexes with the Co-Smad, SMAD4. SMAD4 and SMAD2 shuttle between the nucleus and cytoplasm even in the absence of signaling (Schmierer & Hill, 2005). When signaling is active, heteromeric SMAD2/3-SMAD4 complexes accumulate in the nucleus where they regulate target gene expression (Kumar et al., 2001; Massague, Seoane, & Wotton, 2005) with cofactors such as FOXH1 (Hill, 2016; Labbe et al., 1998).

While the expression pattern of *Smad2* and *Smad3* differ, they exhibit a degree of functional redundancy. *Smad2* is expressed throughout the gastrulating mouse embryo (Pijuan-Sala et al., 2019; Tremblay et al., 2000). *Smad3* is expressed within the ExE and, at late gastrulation stages from around E7.75, within the epiblast (Pijuan-Sala et al., 2019). In contrast to *Smad2*, *Smad3* is not expressed within the VE (Tremblay et al., 2000). *Smad3* mutants are viable and fertile (Datto et al., 1999; Yang et al., 1999; Zhu et al., 1998), indicating that SMAD2 alone can transduce sufficient Nodal signaling. As *Smad3* is not expressed until later in development, and thus cannot compensate for loss of *Smad2* at early stages, *Smad2*$^{-/-}$ mice are embryonic lethal (Dunn et al., 2004; Heyer et al., 1999; Nomura & Li, 1998; Waldrip et al., 1998; Weinstein et al., 1998). However, *Smad3* expression from the *Smad2* locus rescues these defects (Dunn et al., 2005) and, while embryos lacking epiblast *Smad2* are relatively normal, removal of *Smad3* in this sensitized context results in the loss of anterior PS derivatives (Dunn et al., 2004; Vincent et al., 2003) indicating that these factors play overlapping roles.

In vitro, Activin/Nodal signaling is highly adaptive (Heemskerk et al., 2019; Nemashkalo et al., 2017; Sorre et al., 2014; Warmflash et al., 2012;

Yoney et al., 2018). In the continuous presence of a defined amount of ligand, SMAD4 nuclear localization and target gene expression is transiently elevated before returning to a low baseline level (Heemskerk et al., 2019; Warmflash et al., 2012; Yoney et al., 2018). Adaptation enables cells to respond both to ligand concentration and signal duration. The Activin/Nodal pathway is also sensitive to the speed of ligand delivery with high rates of delivery stimulating a maximal transcriptional response (Heemskerk et al., 2019; Sorre et al., 2014). Still, not all target genes exhibit adaptive behavior for example, *Nodal*, *Wnt3*, *Lefty* and *Cerl1* are expressed stably in response to ACTIVIN/NODAL (Heemskerk et al., 2019; Yoney et al., 2018). The non-adaptive expression of *Nodal* and *Wnt3* would result in gradual ligand accumulation, perhaps to sufficient levels to robustly induce the expression of downstream targets and differentiation. This kind of regulation could guard against premature differentiation resulting from transient signaling fluctuations.

The mechanisms that govern adaptive *versus* non-adaptive gene responses are unclear but could be mediated by distinct SMAD binding affinities (Dubrulle et al., 2015). Genes with high affinity SMAD binding sites require low levels of nuclear SMAD for expression and hence could be activated even at baseline signaling levels. In contrast, genes with low affinity binding sites require high nuclear SMAD and will not be expressed after adaptation. During Activin/Nodal adaptation, the transient expression of mesoderm and endoderm-associated target genes is not sufficient to drive differentiation (Yoney et al., 2018). However, the adaptive response can be overcome by WNT signaling (Heemskerk et al., 2019; Yoney et al., 2018). As WNT and Nodal downstream pathway components, β-CATENIN and SMAD2/3, cooperate to regulate key target genes, the presence of both factors may be necessary for stable expression (Funa et al., 2015; Wang et al., 2017).

Nodal is expressed both within the epiblast and the VE (Varlet, Collignon, & Robertson, 1997). When Nodal signaling is disrupted, embryos are morphologically disorganized with an undefined embryonic-extraembryonic boundary and an aberrantly folded epiblast (Ding et al., 1998; Dunn et al., 2004; Gu et al., 1999; Nomura & Li, 1998; Sirard et al., 1998; Waldrip et al., 1998). The cause of these morphological abnormalities is unknown, although Nodal mutant epiblasts have more cells hence this could be a result of over-proliferation. Furthermore, the AVE of mutant embryos, which normally blocks posteriorizing WNT, BMP and Activin/Nodal activity within the anterior epiblast, is not correctly specified or localized (Brennan et al., 2001; Ding et al., 1998; Heyer et al., 1999; Kumar et al., 2015; Norris et al., 2002; Waldrip et al., 1998; Yamamoto et al., 2004). Hence, AVE-deficient *Smad2*$^{-/-}$ embryos ectopically express posterior markers throughout the epiblast (Brennan et al., 2001; Waldrip et al., 1998). This expression is transient indicating that Nodal activity is necessary to maintain posterior gene expression. Yet, in *Nodal*$^{-/-}$ embryos, although the AVE is also not formed, posterior markers are absent (Brennan et al., 2001; Waldrip et al., 1998). This discrepancy may be explained by the *Smad2*-independent action pro-NODAL to induce posteriorizing *Bmp4* expression (Ben-Haim et al., 2006; Brennan et al., 2001; Brennan, Norris, & Robertson, 2002). WNT mutants exhibit a similar lack of posterior markers (Biechele et al., 2011; Kelly et al., 2004; Liu et al., 1999) (see Section 5.1). However, in WNT mutants, the epiblast remains pluripotent whereas in Nodal mutants it differentiates (Brennan et al., 2001; Ding et al., 1998).

When Nodal signaling is specifically disrupted within the epiblast, severe mutants lack a PS, embryonic mesoderm and definitive endoderm (Aragon et al., 2019; Dunn et al., 2004; Iannaccone et al., 1992; Nomura & Li, 1998). This is consistent with reports that Activin/Nodal signaling is needed prior to gastrulation to make cells competent for differentiation (Mulas, Kalkan, & Smith, 2017). However, as pro-NODAL regulates *Bmp4*, which in turn regulates *Wnt3*, it is difficult to decipher direct *versus* indirect Nodal functions. In fact, chimera experiments with *Smad2*$^{-/-}$ cells suggest that there is only a cell autonomous requirement for Nodal signaling in the DE (Tremblay et al., 2000), although comparable experiments with *Smad2/3* double knockout cells are required to rule out the possibility that *Smad3* rescues Nodal signaling in other cell types. Consistently, in less severe mutant embryos, teratomas and *in vitro* assays, Nodal mutants form a PS, mesoderm and ExM but fail to extend the PS distally and specify anterior fates (Conlon et al., 1994; Ding et al., 1998; Heyer et al., 1999; Nomura & Li, 1998; Norris et al., 2002; Waldrip et al., 1998). Furthermore, if Activin/Nodal signaling is blocked in the presence of exogenous BMP or WNT, ExM and mesoderm, respectively, are formed but DE is absent (Martyn et al., 2018; Tewary et al., 2017; Turner et al., 2017; Warmflash et al., 2014; Yoney et al., 2018). Nodal signaling mutants also exhibit an enlarged allantois (Dunn et al., 2004), an ExM structure, suggesting that Nodal acts as a rheostat, with low doses promoting posterior and high doses promoting anterior fates (Costello et al., 2011). This is in keeping with data showing that Nodal signaling inhibits the BMP pathway and ExM formation *in vitro* (Heemskerk et al., 2019) and *in vivo* (Senft et al., 2019) and *vice versa* (Chhabra et al., 2019). Thus, as the embryo grows and the PS extends, the distal cells of the embryo will be furthest from the Nodal inhibitory proximal source of BMP and thus will have high Nodal signaling activity. *In vivo*, these BMP-Nodal pathway interactions reinforce the anterior and posterior PS signaling domains (Fig. 2).

The dose-dependent cell fate responses to NODAL suggest that target genes sense signaling levels. One proposed mechanism is the dose sensitive de-repression of genes. In the absence of NODAL, various SMAD2/3 targets are occupied by the corepressor SNON (Stroschein et al., 1999). Upon activation of Nodal signaling, SMAD2/3 are phosphorylated and form complexes with SNON that are subsequently degraded, mediated by the E3 ubiquitin ligase, ARKADIA (Levy et al., 2007; Nagano et al., 2007). This mechanism of de-repression requires high levels of NODAL (Carthy, Ioannou, & Episkopou, 2018) and thus, SNON-bound target genes are NODAL dose sensitive.

Prior to gastrulation, *Nodal* and its coreceptor *Cripto* are expressed throughout the epiblast but are subsequently restricted to the proximal posterior region (Conlon et al., 1994; Ding et al., 1998; Varlet et al., 1997; Zhou et al., 1993) by the AVE-secreted inhibitors, CER1 and LEFTY1 (Belo et al., 1997; Meno et al., 1996). In addition, the antagonist *Lefty2* is expressed by the nascent mesoderm (Meno et al., 1997; Pijuan-Sala et al., 2019) but is absent from the DE (Meno et al., 1997; Pijuan-Sala et al., 2019), suggesting that it restricts Nodal activity to the anterior PS. Loss of these inhibitors results in an expansion of the anterior PS and, in some cases, axis duplication (Meno et al., 1999; Perea-Gomez et al., 2002). Similarly, *in vitro*, loss of *CER1* and *LEFTY1* causes expansion of the PS domain within micropatterned colonies (Warmflash et al., 2014). These phenotypes are rescued by loss of one allele of *Nodal* (Meno et al., 1999; Perea-Gomez et al., 2002), again highlighting the importance of ligand dose in cell fate specification.

While inhibitors restrict the *Nodal* expression domain *in vivo*, *in vitro* experiments show that they are not necessary for the initial *Nodal* asymmetry (Turner et al., 2017). Approximately twice as many PSC-extraembryonic stem cell aggregates display *Nodal* asymmetry than have an AVE-like domain (Sozen et al., 2018), suggesting that AVE-produced inhibitors do not drive this asymmetry. Furthermore, EB symmetry breaking occurs in the absence of *CER1* and *LEFTY1* (Simunovic et al., 2019). Hence, similar to WNT signaling (see Section 5.1), inhibitors may maintain rather than establish asymmetric expression domains. Asymmetric *Nodal* expression is mediated by auto-activated *Nodal* enhancers (Norris et al., 2002; Saijoh et al., 2000) but conceivably is reinforced by the induction of *Nodal* expression by the localized source of WNT3 on the posterior side of the embryo (Ben-Haim et al., 2006).

Over time, as the PS extends, *Nodal* expression expands distally and later becomes restricted to the node where it regulates left-right patterning (Brennan et al., 2001). Intriguingly, in Nodal pathway mutants, although anterior PS markers are observed, the PS is truncated (Song et al., 1999). Therefore, Nodal signaling may regulate PS elongation. Like BMP signaling, in micropatterned colonies the Activin/Nodal response is initially elevated at the periphery, due to differential receptor localization and endogenous inhibitor secretion (Etoc et al., 2016) (see Section 5.2). However, unlike BMP activity that remains at the colony edge, Activin/Nodal signaling spreads inwards over time in response to BMP or WNT (Chhabra et al., 2019; Warmflash et al., 2014). This Activin/Nodal wave rapidly becomes independent of WNT and BMP but continued WNT signaling is required for mesoderm cell fate specification (Chhabra et al., 2019). Nodal signaling propagation is likely based on *Nodal* autoactivation as, *in vivo*, disruption of the autoregulatory enhancer blocks *Nodal* distal expansion (Brennan et al., 2001; Norris et al., 2002).

Together these data show that Nodal signaling is required for the formation of specific cell types during gastrulation. However, it is still unclear which cells respond to NODAL and the signaling dynamics. This is largely due to the fact that immunostaining for the downstream effectors of Activin/Nodal signaling, phosphorylated SMAD2 and SMAD3, has proven difficult *in vivo* and, while Nodal signaling reporters exist (Granier et al., 2011), they have not been leveraged to study these stages of development in detail.

5.4 FGF signaling

FGF ligands are secreted into the extracellular space and, in gastrulating zebrafish embryos, move by free diffusion (Yu et al., 2009). FGFs bind to dimerized FGF receptor tyrosine kinases (FGFRs) (Sarabipour & Hristova, 2016), an interaction stabilized by heparan sulfate proteoglycan (Pye & Gallagher, 1999; Spivakkroizman et al., 1995). Heparan sulfate proteoglycan within the ECM can also bind FGF, reducing its diffusion rate (Yu et al., 2009) and thus, the ECM composition likely plays a role in establishing signal gradients within the embryo. FGF-bound receptors cross-phosphorylate one another and phosphorylate signaling components within the cytoplasm leading to the activation of downstream pathways including MAPK, PI3K, PLCγ and JAK/STAT (Ornitz & Itoh, 2015). FGF signaling is heavily regulated by negative feedback mechanisms (Neben et al., 2019). The critical role of FGF at gastrulation involves the MAPK pathway (Oki, Kitajima, & Meno, 2010) but, as comparable phenotypes are observed in PI3K pathway mutants (Bloomekatz et al., 2012), FGF may function *via* several downstream pathways simultaneously. *In vitro* gastrulation models could untangle the complex networks downstream of FGF but, to date, there has been little investigation into the FGF pathway using these tools.

Upon binding of FGF ligands to their receptor, the receptor is internalized and therefore target cells act as an FGF sink. In zebrafish, this sink activity regulates the FGF concentration gradient whereby increased receptor internalization establishes a steeper ligand gradient and *vice versa* (Yu et al., 2009).

During gastrulation, FGF regulates the movement of cells away from the PS. When FGF signaling is attenuated, the first cells to traverse the PS migrate proximally and specify ExM derivatives and PGCs (Ciruna et al., 1997; Deng et al., 1994; Sun et al., 1999; Yamaguchi et al., 1994) but, over time, a 'traffic jam' of cells accumulates within the posterior epiblast causing patterning defects (Deng et al., 1994; Garcia-Garcia & Anderson, 2003; Guo & Li, 2007; Meyers, Lewandoski, & Martin, 1998; Sun et al., 1999; Yamaguchi et al., 1994). The impaired cell migration is triggered by a gradual reduction in the expression of *Snail*, a factor that promotes the down-regulation of the epithelial adhesion molecule *E*-CADHERIN (Carver et al., 2001; Ciruna & Rossant, 2001). Hence, in the absence of FGF signaling, cells remain tightly attached to one another, do not undergo EMT and show reduced mesoderm and endoderm specification (Deng et al., 1994; Garcia-Garcia & Anderson, 2003; Guo & Li, 2007; Meyers et al., 1998; Sun et al., 1999; Yamaguchi et al., 1994).

The phenotype of FGF signaling mutants is similar to that of WNT pathway mutants (see Section 5.1) indicating that these pathways may interact. While FGF target genes are expressed normally in WNT mutants (Kelly et al., 2004), WNT target genes are disrupted in *Fgfr1*$^{-/-}$ embryos (Ciruna & Rossant, 2001). Hence, WNT is likely downstream of FGF. The WNT transcriptional effector, β-CATENIN forms complexes at cell junctions with *E*-CADHERIN hence the persistence of E-CADHERIN in FGF mutants could sequester β-CATENIN, reducing WNT signaling capacity (Ciruna & Rossant, 2001). Consistently, blocking E-CADHERIN function in FGF mutants rescues WNT target gene expression (Ciruna & Rossant, 2001). Nevertheless, the interaction between FGF and WNT is not clear as contrary data shows that FGF signaling mutants exhibit WNT activity and express target genes normally (Deng et al., 1994; Garcia-Garcia & Anderson, 2003; Yamaguchi et al., 1994).

In addition to controlling cell migration, FGF regulates neural differentiation and/or patterning. FGF mutants exhibit a posterior expansion of anterior neural markers (Sun et al., 1999) and ectopic neural tube formation (Ciruna et al., 1997; Deng et al., 1997). Therefore, in wild type embryos, FGF may restrict neural differentiation during gastrulation. Although the suppression of BMP and Nodal activity by the AVE is thought to be required for neural fates (see Sections 5.2 and 5.3), the expansion of neural expression in FGF mutants occurs even though *Bmp4* and *Nodal* are expressed within the posterior of the embryo (Sun et al., 1999). Hence, the ability of these signals to suppress neural differentiation may involve FGF.

During gastrulation, several FGF ligands, including *Fgf3*, *Fgf4* and *Fgf8*, are expressed at low levels within the VE and epiblast but are significantly elevated in cells of the PS and nascent mesoderm (Crossley & Martin, 1995; Haub & Goldfarb, 1991; Hebert, Boyle, & Martin, 1991; Maruoka et al., 1998; Niswander & Martin, 1992; Wilkinson et al., 1988). *Fgf5* is also expressed at this time within the epiblast but at lower levels within the nascent mesoderm (Khoa et al., 2016; Pijuan-Sala et al., 2019). However, *Fgf8* and *Fgfr1* are the only ligand-receptor combination shown to be required for gastrulation (Ciruna et al., 1997; Deng et al., 1994; Sun et al., 1999; Yamaguchi et al., 1994). While multiple FGF ligands are expressed at gastrulation, *Fgf8* single mutants exhibit gastrulation defects, presumably as the expression of other FGF ligands, such as *Fgf4*, are also lost in this context (Sun et al., 1999). *Fgf4* mutants are embryonic lethal prior to gastrulation (Feldman et al., 1995) hence FGF4 may also play a role at gastrulation that could be investigated using tissue-specific genetic ablation. *Fgfr1* is expressed throughout the epiblast, PS, mesoderm and ExE. *Fgfr1* mutants exhibit gastrulation defects with cells failing to exit the primitive streak and subsequently acquire lineage identities (Orr-Urtreger et al., 1991; Pijuan-Sala et al., 2019; Yamaguchi, Conlon, & Rossant, 1992). In contrast, *Fgfr2* is expressed most highly in the ExE and only at low levels in epiblast-derived cell types suggesting that it does not play a critical role within the epiblast during gastrulation. A downstream target of FGF/MAPK signaling, *Spry4*, is expressed within the VE and epiblast and at higher levels within the PS and mesoderm (Morgani et al., 2018b).

Therefore, as with WNT signaling, cells that produce or are close to the source of FGF ligands are those that show the strongest signaling response. As the *Spry4* reporter is a downstream target of FGF/MAPK, it does not show the signaling dynamics and may not capture all sites of FGF signaling activity. Thus, further analysis using dynamic biosensors, such as the ERK-KTR reporter that indicates the kinase activity of an FGF/MAPK pathway component, is necessary to reveal this information (Regot et al., 2014).

While cells undergoing an EMT within the posterior of the embryo exhibit elevated FGF signaling activity, neural cells within the anterior of the embryo, which also make an *E*-CADHERIN to N-CADHERIN switch later in gastrulation, show a reduced FGF signaling response (Punovuori et al., 2019). However, it is unknown how cadherins differentially affect signaling activity within distinct regions of the embryo.

FGF is perhaps the least well studied of the signaling pathways required for gastrulation and it is still unclear how FGF ligands become enriched at the PS and how this pathway interacts with other signals during gastrulation.

6. Conclusions

Although *in vivo* genetic experiments established the critical signaling pathways driving cell fate specification at gastrulation, technical issues have hampered detailed mechanistic insights into the dynamic interactions of signaling pathways in the vicinity of the PS. WNT, BMP, Nodal and FGF pathway mutants exhibit early, embryonic lethal, gastrulation defects. As these pathways cross-regulate one another at numerous levels, it is difficult to tease apart the individual roles of each pathway.

In addition, while WNT, BMP, Nodal and FGF are the primary drivers of gastrulation, many questions remain about how they interact with other pathways, such as Hippo (Beyer et al., 2013; Estaras, Benner, & Jones, 2015) and Notch (Souilhol et al., 2015), to regulate lineage specification. *In vitro* gastrulation models can be imaged over substantial periods of time and individual signaling pathways manipulated with temporal accuracy. Thus, they provide simplified systems to probe these processes and have already started to generate novel insights. However, as these PSC systems do not recapitulate all aspects of *in vivo* gastrulation, continuous comparisons to the embryo are necessary to validate *in vitro* findings.

References

Alexandre, C., Baena-Lopez, A., & Vincent, J. P. (2014). Patterning and growth control by membrane-tethered Wingless. *Nature*, *505*(7482), 180–185.

Ang, S. L., et al. (1994). Positive and negative signals from mesoderm regulate the expression of mouse Otx2 in ectoderm explants. *Development*, *120*(10), 2979–2989.

Aoyama, M., et al. (2012). Spatial restriction of bone morphogenetic protein signaling in mouse gastrula through the mVam2-dependent endocytic pathway. *Developmental Cell*, *22*(6), 1163–1175.

Aragon, E., et al. (2019). Structural basis for distinct roles of SMAD2 and SMAD3 in FOXH1 pioneer-directed TGF-beta signaling. *Genes & Development*, *33*, 1506–1524.

Bachiller, D., et al. (2000). The organizer factors Chordin and Noggin are required for mouse forebrain development. *Nature*, *403*(6770), 658–661.

Barrow, J. R., et al. (2007). Wnt3 signaling in the epiblast is required for proper orientation of the anteroposterior axis. *Developmental Biology*, *312*(1), 312–320.

Beccari, L., et al. (2018). Multi-axial self-organization properties of mouse embryonic stem cells into gastruloids. *Nature*, *562*(7726), 272–276.

Bedzhov, I., & Zernicka-Goetz, M. (2014). Self-organizing properties of mouse pluripotent cells initiate morphogenesis upon implantation. *Cell*, *156*(5), 1032–1044.

Belo, J. A., et al. (1997). Cerberus-like is a secreted factor with neuralizing activity expressed in the anterior primitive endoderm of the mouse gastrula. *Mechanisms of Development*, *68*(1–2), 45–57.

Belo, J. A., et al. (2000). Cerberus-like is a secreted BMP and nodal antagonist not essential for mouse development. *Genesis*, *26*(4), 265–270.

Ben-Haim, N., et al. (2006). The nodal precursor acting via activin receptors induces mesoderm by maintaining a source of its convertases and BMP4. *Developmental Cell*, *11*(3), 313–323.

Beppu, H., et al. (2000). BMP type II receptor is required for gastrulation and early development of mouse embryos. *Developmental Biology*, *221*(1), 249–258.

Beyer, T. A., et al. (2013). Switch enhancers interpret TGF-beta and Hippo signaling to control cell fate in human embryonic stem cells. *Cell Reports*, *5*(6), 1611–1624.

Biechele, S., Cox, B. J., & Rossant, J. (2011). Porcupine homolog is required for canonical Wnt signaling and gastrulation in mouse embryos. *Developmental Biology*, *355*(2), 275–285.

Biondi, C. A., et al. (2007). Mice develop normally in the absence of Smad4 nucleocytoplasmic shuttling. *The Biochemical Journal*, *404*(2), 235–245.

Bloomekatz, J., et al. (2012). Pten regulates collective cell migration during specification of the anterior-posterior axis of the mouse embryo. *Developmental Biology*, *364*(2), 192–201.

Brennan, J., Norris, D. P., & Robertson, E. J. (2002). Nodal activity in the node governs left-right asymmetry. *Genes & Development*, *16*(18), 2339–2344.

Brennan, J., et al. (2001). Nodal signalling in the epiblast patterns the early mouse embryo. *Nature*, *411*(6840), 965–969.

Brunet, T., et al. (2013). Evolutionary conservation of early mesoderm specification by mechanotransduction in Bilateria. *Nature Communications*, *4*, 2821.

Camus, A., et al. (2000). The morphogenetic role of midline mesendoderm and ectoderm in the development of the forebrain and the midbrain of the mouse embryo. *Development*, *127*(9), 1799–1813.

Carthy, J. M., Ioannou, M., & Episkopou, V. (2018). Arkadia degrades SNON to activate level-specific NODAL responses. *bioRxiv*, 487371.

Carver, E. A., et al. (2001). The mouse snail gene encodes a key regulator of the epithelial-mesenchymal transition. *Molecular and Cellular Biology*, *21*(23), 8184–8188.

Chen, Q., et al. (2009). Smad7 is required for the development and function of the heart. *Journal of Biological Chemistry*, *284*(1), 292–300.

Chhabra, S., et al. (2019). Dissecting the dynamics of signaling events in the BMP, WNT, and NODAL cascade during self-organized fate patterning in human gastruloids. *PLoS Biology*, *17*(10), e3000498.
Ciruna, B., & Rossant, J. (2001). FGF signaling regulates mesoderm cell fate specification and morphogenetic movement at the primitive streak. *Developmental Cell*, *1*(1), 37–49.
Ciruna, B. G., et al. (1997). Chimeric analysis of fibroblast growth factor receptor-1 (Fgfr1) function: A role for FGFR1 in morphogenetic movement through the primitive streak. *Development*, *124*(14), 2829–2841.
Conlon, F. L., et al. (1994). A primary requirement for nodal in the formation and maintenance of the primitive streak in the mouse. *Development*, *120*(7), 1919–1928.
Costello, I., et al. (2011). The T-box transcription factor eomesodermin acts upstream of Mesp1 to specify cardiac mesoderm during mouse gastrulation. *Nature Cell Biology*, *13*(9), 1084–U108.
Covert, M. W., et al. (2015). High-sensitivity measurements of multiple kinase activities in live single cells. *Molecular Biology of the Cell*, *26*, 1724–1734.
Crossley, P. H., & Martin, G. R. (1995). The mouse Fgf8 gene encodes a family of polypeptides and is expressed in regions that direct outgrowth and patterning in the developing embryo. *Development*, *121*(2), 439–451.
Datto, M. B., et al. (1999). Targeted disruption of Smad3 reveals an essential role in transforming growth factor beta-mediated signal transduction. *Molecular and Cellular Biology*, *19*(4), 2495–2504.
Deng, C. X., et al. (1994). Murine Fgfr-1 Is required for early postimplantation growth and axial organization. *Genes & Development*, *8*(24), 3045–3057.
Deng, C. X., et al. (1997). Fibroblast growth factor receptor-1 (FGFR-1) is essential for normal neural tube and limb development. *Developmental Biology*, *185*(1), 42–54.
Desbaillets, I., et al. (2000). Embryoid bodies: An in vitro model of mouse embryogenesis. *Experimental Physiology*, *85*(6), 645–651.
Di-Gregorio, A., et al. (2007). BMP signalling inhibits premature neural differentiation in the mouse embryo. *Development*, *134*(18), 3359–3369.
Ding, J. X., et al. (1998). Cripto is required for correct orientation of the anterior-posterior axis in the mouse embryo. *Nature*, *395*(6703), 702–707.
Du, J., et al. (2016). Extracellular matrix stiffness dictates Wnt expression through integrin pathway. *Scientific Reports*, *6*, 20395.
Dubrulle, J., et al. (2015). Response to Nodal morphogen gradient is determined by the kinetics of target gene induction. *eLife*, *4*, e05042.
Dunn, N. R., et al. (2004). Combinatorial activities of Smad2 and Smad3 regulate mesoderm formation and patterning in the mouse embryo. *Development*, *131*(8), 1717–1728.
Dunn, N. R., et al. (2005). Mice exclusively expressing the short isoform of Smad2 develop normally and are viable and fertile. *Genes & Development*, *19*(1), 152–163.
Entchev, E. V., Schwabedissen, A., & Gonzalez-Gaitan, M. (2000). Gradient formation of the TGF-beta homolog Dpp. *Cell*, *103*(6), 981–991.
Estaras, C., Benner, C., & Jones, K. A. (2015). SMADs and YAP compete to control elongation of beta-catenin: LEF-1-recruited RNAPII during hESC differentiation. *Molecular Cell*, *58*(5), 780–793.
Etoc, F., et al. (2016). A Balance between secreted inhibitors and edge sensing controls gastruloid self-organization. *Developmental Cell*, *39*(3), 302–315.
Feldman, B., et al. (1995). Requirement of FGF-4 for postimplantation mouse development. *Science*, *267*(5195), 246–249.
Ferrer-Vaquer, A., et al. (2010). A sensitive and bright single-cell resolution live imaging reporter of Wnt/ss-catenin signaling in the mouse. *BMC Developmental Biology*, *10*, 121.
Finley, K. R., Tennessen, J., & Shawlot, W. (2003). The mouse Secreted frizzled-related protein 5 gene is expressed in the anterior visceral endoderm and foregut endoderm during early post-implantation development. *Gene Expression Patterns*, *3*(5), 681–684.

Fu, J., et al. (2009). Reciprocal regulation of Wnt and Gpr177/mouse Wntless is required for embryonic axis formation. *Proceedings of the National Academy of Sciences of the United States of America*, *106*(44), 18598–18603.
Funa, N. S., et al. (2015). Beta-catenin regulates primitive streak induction through collaborative interactions with SMAD2/SMAD3 and OCT4. *Cell Stem Cell*, *16*(6), 639–652.
Gadue, P., et al. (2006). Wnt and TGF-beta signaling are required for the induction of an in vitro model of primitive streak formation using embryonic stem cells. *Proceedings of the National Academy of Sciences of the United States of America*, *103*(45), 16806–16811.
Garcia-Garcia, M. J., & Anderson, K. V. (2003). Essential role of glycosaminoglycans in Fgf signaling during mouse gastrulation. *Cell*, *114*(6), 727–737.
Glinka, A., et al. (1998). Dickkopf-1 is a member of a new family of secreted proteins and functions in head induction. *Nature*, *391*(6665), 357–362.
Gottardi, C. J., & Gumbiner, B. M. (2004). Distinct molecular forms of beta-catenin are targeted to adhesive or transcriptional complexes. *Journal of Cell Biology*, *167*(2), 339–349.
Granier, C., et al. (2011). Nodal cis-regulatory elements reveal epiblast and primitive endoderm heterogeneity in the peri-implantation mouse embryo. *Developmental Biology*, *349*(2), 350–362.
Greco, V., Hannus, M., & Eaton, S. (2001). Argosomes: A potential vehicle for the spread of morphogens through epithelia. *Cell*, *106*(5), 633–645.
Gu, Z. Y., et al. (1999). The type I serine threonine kinase receptor ActRIA (ALK2) is required for gastrulation of the mouse embryo. *Development*, *126*(11), 2551–2561.
Guo, Q. X., & Li, J. Y. H. (2007). Distinct functions of the major Fgf8 spliceform, Fgf8b, before and during mouse gastrulation. *Development*, *134*(12), 2251–2260.
Harrison, S. E., et al. (2017). Assembly of embryonic and extraembryonic stem cells to mimic embryogenesis in vitro. *Science*, *356*(6334), eaal1810.
Haub, O., & Goldfarb, M. (1991). Expression of the fibroblast growth factor-5 gene in the mouse embryo. *Development*, *112*(2), 397–406.
Hayashi, K., et al. (2002). SMAD1 signaling is critical for initial commitment of germ cell lineage from mouse epiblast. *Mechanisms of Development*, *118*(1–2), 99–109.
Hebert, J. M., Boyle, M., & Martin, G. R. (1991). mRNA localization studies suggest that murine FGF-5 plays a role in gastrulation. *Development*, *112*(2), 407–415.
Heemskerk, I., et al. (2019). Rapid changes in morphogen concentration control self-organized patterning in human embryonic stem cells. *eLife*, *8*, e40526.
Heinke, J., et al. (2008). BMPER is an endothelial cell regulator and controls bone morphogenetic protein-4-dependent angiogenesis. *Circulation Research*, *103*(8), 804–U65.
Hendriksen, J., et al. (2008). Plasma membrane recruitment of dephosphorylated beta-catenin upon activation of the Wnt pathway. *Journal of Cell Science*, *121*(11), 1793–1802.
Heyer, J., et al. (1999). Postgastrulation Smad2-deficient embryos show defects in embryo turning and anterior morphogenesis. *Proceedings of the National Academy of Sciences of the United States of America*, *96*(22), 12595–12600.
Hill, C. S. (2016). Transcriptional control by the SMADs. *Cold Spring Harbor Perspectives in Biology*, *8*(10), a022079.
Hobert, M., & Carlin, C. (1995). Cytoplasmic juxtamembrane domain of the human Egf receptor is required for basolateral localization in Mdck cells. *Journal of Cellular Physiology*, *162*(3), 434–446.
Howard, S., et al. (2011). A positive role of cadherin in Wnt/beta-catenin signalling during epithelial-mesenchymal transition. *PLoS One*, *6*(8), e23899.
Howe, L. R., et al. (2003). Twist is up-regulated in response to Wnt1 and inhibits mouse mammary cell differentiation. *Cancer Research*, *63*(8), 1906–1913.
Hsieh, J. C., et al. (2003). Mesd encodes an LRP5/6 chaperone essential for specification of mouse embryonic polarity. *Cell*, *112*(3), 355–367.

Huber, O., et al. (1996). Nuclear localization of beta-catenin by interaction with transcription factor LEF-1. *Mechanisms of Development, 59*(1), 3–10.
Huelsken, J., et al. (2000). Requirement for beta-catenin in anterior-posterior axis formation in mice. *Journal of Cell Biology, 148*(3), 567–578.
Iannaccone, P. M., et al. (1992). Insertional mutation of a gene involved in growth-regulation of the early mouse embryo. *Developmental Dynamics, 194*(3), 198–208.
Jamora, C., et al. (2003). Links between signal transduction, transcription and adhesion in epithelial bud development (vol 422, pg 317, 2003). *Nature, 424*(6951), 974.
Javier, A. L., et al. (2012). Bmp indicator mice reveal dynamic regulation of transcriptional response. *PLoS One*, 7(9), e42566.
Kamioka, Y., et al. (2013). Live imaging of transgenic mice expressing FRET biosensors. *Conference Proceedings: Annual International Conference of the IEEE Engineering in Medicine and Biology Society, 2013*, 125–128.
Kawano, Y., & Kypta, R. (2003). Secreted antagonists of the Wnt signalling pathway. *Journal of Cell Science, 116*(13), 2627–2634.
Kelly, O. G., Pinson, K. I., & Skarnes, W. C. (2004). The Wnt co-receptors Lrp5 and Lrp6 are essential for gastrulation in mice. *Development, 131*(12), 2803–2815.
Kemler, R., et al. (2004). Stabilization of beta-catenin in the mouse zygote leads to premature epithelial-mesenchymal transition in the epiblast. *Development, 131*(23), 5817–5824.
Kemp, C., et al. (2005). Expression of all Wnt genes and their secreted antagonists during mouse blastocyst and postimplantation development. *Developmental Dynamics, 233*(3), 1064–1075.
Kengaku, M., et al. (1998). Distinct WNT pathways regulating AER formation and dorsoventral polarity in the chick limb bud. *Science, 280*(5367), 1274–1277.
Khoa, L. T. P., et al. (2016). Visualization of the epiblast and visceral endodermal cells using Fgf5-P2A-venus BAC transgenic mice and epiblast stem cells. *PLoS One, 11*(7), e0159246.
Kicheva, A., et al. (2007). Kinetics of morphogen gradient formation. *Science, 315*(5811), 521–525.
Kimura-Yoshida, C., et al. (2005). Canonical Wnt signaling and its antagonist regulate anterior-posterior axis polarization by guiding cell migration in mouse visceral endoderm. *Developmental Cell, 9*(5), 639–650.
Kinder, S. J., et al. (1999). The orderly allocation of mesodermal cells to the extraembryonic structures and the anteroposterior axis during gastrulation of the mouse embryo. *Development, 126*(21), 4691–4701.
Klingensmith, J., et al. (1999). Neural induction and patterning in the mouse in the absence of the node and its derivatives. *Developmental Biology, 216*(2), 535–549.
Kumar, A., et al. (2001). Nodal signaling uses activin and transforming growth factor-beta receptor-regulated Smads. *Journal of Biological Chemistry, 276*(1), 656–661.
Kumar, A., et al. (2015). Nodal signaling from the visceral endoderm is required to maintain Nodal gene expression in the epiblast and drive DVE/AVE migration. *Developmental Biology, 400*(1), 1–9.
Kurosawa, H. (2007). Methods for inducing embryoid body formation: In vitro differentiation system of embryonic stem cells. *Journal of Bioscience and Bioengineering, 103*(5), 389–398.
Kwon, G. S., Viotti, M., & Hadjantonakis, A. K. (2008). The endoderm of the mouse embryo arises by dynamic widespread intercalation of embryonic and extraembryonic lineages. *Developmental Cell, 15*(4), 509–520.
Labbe, E., et al. (1998). Smad2 and Smad3 positively and negatively regulate TGF beta-dependent transcription through the forkhead DNA-binding protein FAST2. *Molecular Cell, 2*(1), 109–120.

Lawson, K. A., et al. (1999). Bmp4 is required for the generation of primordial germ cells in the mouse embryo. *Genes & Development, 13*(4), 424–436.

Lee, S. J., et al. (2010). Regulation of muscle mass by follistatin and activins. *Molecular Endocrinology, 24*(10), 1998–2008.

Levy, L., et al. (2007). Arkadia activates Smad3/Smad4-dependent transcription by triggering signal-induced SnoN degradation. *Molecular and Cellular Biology, 27*(17), 6068–6083.

Lewis, S. L., et al. (2008). Dkk1 and Wnt3 interact to control head morphogenesis in the mouse. *Development, 135*(10), 1791–1801.

Liu, P., et al. (1999). Requirement for Wnt3 in vertebrate axis formation. *Nature Genetics, 22*(4), 361–365.

MacDonald, B. T., Tamai, K., & He, X. (2009). Wnt/beta-catenin signaling: Components, mechanisms, and diseases. *Developmental Cell, 17*(1), 9–26.

Manfrin, A., et al. (2019). Engineered signaling centers for the spatially controlled patterning of human pluripotent stem cells. *Nature Methods, 16*(7), 640–648.

Marikawa, Y., et al. (2009). Aggregated P19 mouse embryonal carcinoma cells as a simple in vitro model to study the molecular regulations of mesoderm formation and axial elongation morphogenesis. *Genesis, 47*(2), 93–106.

Martyn, I., Brivanlou, A. H., & Siggia, E. D. (2019). A wave of WNT signaling balanced by secreted inhibitors controls primitive streak formation in micropattern colonies of human embryonic stem cells. *Development, 146*(6), dev172791.

Martyn, I., et al. (2018). Self-organization of a human organizer by combined Wnt and Nodal signalling. *Nature, 558*(7708), 132–135.

Maruoka, Y., et al. (1998). Comparison of the expression of three highly related genes, Fgf8, Fgf17 and Fgf18, in the mouse embryo. *Mechanisms of Development, 74*(1–2), 175–177.

Marx, V. (2019). A dream of single-cell proteomics. *Nature Methods, 16*(9), 809–812.

Massague, J., Seoane, J., & Wotton, D. (2005). Smad transcription factors. *Genes & Development, 19*(23), 2783–2810.

Massey, J., et al. (2019). Synergy with TGFbeta ligands switches WNT pathway dynamics from transient to sustained during human pluripotent cell differentiation. *Proceedings of the National Academy of Sciences of the United States of America, 116*(11), 4989–4998.

Matzuk, M. M., Kumar, T. R., & Bradley, A. (1995). Different phenotypes for mice deficient in either activins or activin receptor-type-II. *Nature, 374*(6520), 356–360.

McDole, K., et al. (2018). In toto imaging and reconstruction of post-implantation mouse development at the single-cell level. *Cell, 175*(3), 859–876.e33.

McMahon, J. A., et al. (1998). Noggin-mediated antagonism of BMP signaling is required for growth and patterning of the neural tube and somite. *Genes & Development, 12*(10), 1438–1452.

Meno, C., et al. (1996). Left-right asymmetric expression of the TGF beta-family member lefty in mouse embryos. *Nature, 381*(6578), 151–155.

Meno, C., et al. (1997). Two closely-related left-right asymmetrically expressed genes, lefty-1 and lefty-2: Their distinct expression domains, chromosomal linkage and direct neuralizing activity in Xenopus embryos. *Genes to Cells, 2*(8), 513–524.

Meno, C., et al. (1999). Mouse Lefty2 and zebrafish antivin are feedback inhibitors of nodal signaling during vertebrate gastrulation. *Molecular Cell, 4*(3), 287–298.

Metzis, V., et al. (2018). Nervous system regionalization entails axial allocation before neural differentiation. *Cell, 175*(4), 1105–1118.e17.

Meyers, E. N., Lewandoski, M., & Martin, G. R. (1998). An Fgf8 mutant allelic series generated by Cre- and Flp-mediated recombination. *Nature Genetics, 18*(2), 136–141.

Migeotte, I., et al. (2010). Rac1-dependent collective cell migration is required for specification of the anterior-posterior body axis of the mouse. *PLoS Biology, 8*(8), e1000442.

Mishina, Y., et al. (1995). Bmpr encodes a type I bone morphogenetic protein receptor that is essential for gastrulation during mouse embryogenesis. *Genes & Development*, *9*(24), 3027–3037.

Miura, S., Singh, A. P., & Mishina, Y. (2010). Bmpr1a is required for proper migration of the AVE through regulation of Dkk1 expression in the pre-streak mouse embryo. *Developmental Biology*, *341*(1), 246–254.

Miyazawa, K., & Miyazono, K. (2017). Regulation of TGF-beta family signaling by inhibitory Smads. *Cold Spring Harbor Perspectives in Biology*, *9*(3).

Monteiro, R. M., et al. (2008). Real time monitoring of BMP Smads transcriptional activity during mouse development. *Genesis*, *46*(7), 335–346.

Morgani, S. M., et al. (2018a). Micropattern differentiation of mouse pluripotent stem cells recapitulates embryo regionalized cell fate patterning. *eLife*, 7, e32839.

Morgani, S. M., et al. (2018b). A Sprouty4 reporter to monitor FGF/ERK signaling activity in ESCs and mice. *Developmental Biology*, *441*(1), 104–126.

Mulas, C., Kalkan, T., & Smith, A. (2017). Nodal secures pluripotency upon embryonic stem cell progression from the ground state. *Stem Cell Reports*, *9*(1), 77–91.

Nagano, Y., et al. (2007). Arkadia induces degradation of SnoN and c-Ski to enhance transforming growth factor-beta signaling. *The Journal of Biological Chemistry*, *282*(28), 20492–20501.

Neben, C. L., et al. (2019). Feedback regulation of RTK signaling in development. *Developmental Biology*, *447*(1), 71–89.

Nemashkalo, A., et al. (2017). Morphogen and community effects determine cell fates in response to BMP4 signaling in human embryonic stem cells. *Development*, *144*(17), 3042–3053.

Nicolas, F. J., et al. (2004). Analysis of Smad nucleocytoplasmic shuttling in living cells. *Journal of Cell Science*, *117*(18), 4113–4125.

Niswander, L., & Martin, G. R. (1992). Fgf-4 expression during gastrulation, myogenesis, limb and tooth development in the mouse. *Development*, *114*(3), 755–768.

Nomura, M., & Li, E. (1998). Smad2 role in mesoderm formation, left-right patterning and craniofacial development. *Nature*, *393*(6687), 786–790.

Norris, D. P., & Robertson, E. J. (1999). Asymmetric and node-specific nodal expression patterns are controlled by two distinct cis-acting regulatory elements. *Genes & Development*, *13*(12), 1575–1588.

Norris, D. P., et al. (2002). The Foxh1-dependent autoregulatory enhancer controls the level of Nodal signals in the mouse embryo. *Development*, *129*(14), 3455–3468.

Nowotschin, S., et al. (2013a). A bright single-cell resolution live imaging reporter of Notch signaling in the mouse. *BMC Developmental Biology*, *13*, 15.

Nowotschin, S., et al. (2013b). The T-box transcription factor Eomesodermin is essential for AVE induction in the mouse embryo. *Genes & Development*, *27*(9), 997–1002.

Nowotschin, S., et al. (2019). The emergent landscape of the mouse gut endoderm at single-cell resolution. *Nature*, *569*(7756), 361–367.

Okamura, D., Hayashi, K., & Matsui, Y. (2005). Mouse epiblasts change responsiveness to BMP4 signal required for PGC formation through functions of extraembryonic ectoderm. *Molecular Reproduction and Development*, *70*(1), 20–29.

Oki, S., Kitajima, K., & Meno, C. (2010). Dissecting the role of Fgf signaling during gastrulation and left-right axis formation in mouse embryos using chemical inhibitors. *Developmental Dynamics*, *239*(6), 1768–1778.

Ornitz, D. M., & Itoh, N. (2015). The fibroblast growth factor signaling pathway. *Wiley Interdisciplinary Reviews: Developmental Biology*, *4*(3), 215–266.

Orr-Urtreger, A., et al. (1991). Developmental expression of two murine fibroblast growth factor receptors, flg and bek. *Development*, *113*(4), 1419–1434.

Ostblom, J., et al. (2019). Context-explorer: Analysis of spatially organized protein expression in high-throughput screens. *PLoS Computational Biology*, *15*(1), e1006384.

Panakova, D., et al. (2005). Lipoprotein particles are required for Hedgehog and Wingless signalling. *Nature*, *435*(7038), 58–65.

Parameswaran, M., & Tam, P. P. L. (1995). Regionalization of cell fate and morphogenetic movement of the mesoderm during mouse gastrulation. *Developmental Genetics*, *17*(1), 16–28.

Peerani, R., et al. (2009). Patterning mouse and human embryonic stem cells using microcontact printing. *Methods in Molecular Biology*, *482*, 21–33.

Peng, G. D., et al. (2016). Spatial transcriptome for the molecular annotation of lineage fates and cell identity in mid-gastrula mouse embryo. *Developmental Cell*, *36*(6), 681–697.

Peng, G. D., et al. (2019). Molecular architecture of lineage allocation and tissue organization in early mouse embryo. *Nature*, *572*(7770), 528–532.

Perea-Gomez, A., et al. (2002). Nodal antagonists in the anterior visceral endoderm prevent the formation of multiple primitive streaks. *Developmental Cell*, *3*(5), 745–756.

Piccolo, S., et al. (1996). Dorsoventral patterning in xenopus: Inhibition of ventral signals by direct binding of Chordin to BMP-4. *Cell*, *86*(4), 589–598.

Pijuan-Sala, B., et al. (2019). A single-cell molecular map of mouse gastrulation and early organogenesis. *Nature*, *566*(7745), 490–495.

Pomreinke, A. P., et al. (2017). Dynamics of BMP signaling and distribution during zebrafish dorsal ventral patterning. *eLife*, *6*.

Popperl, H., et al. (1997). Misexpression of Cwnt8C in the mouse induces an ectopic embryonic axis and causes a truncation of the anterior neuroectoderm. *Development*, *124*(15), 2997–3005.

Punovuori, K., et al. (2019). N-cadherin stabilises neural identity by dampening anti-neural signals. *Development*, *146*(21), Article id: dev183269.

Pye, D. A., & Gallagher, J. T. (1999). Monomer complexes of basic fibroblast growth factor and heparan sulfate oligosaccharides are the minimal functional unit for cell activation. *The Journal of Biological Chemistry*, *274*(19), 13456–13461.

Regot, S., et al. (2014). High-sensitivity measurements of multiple kinase activities in live single cells. *Cell*, *157*(7), 1724–1734.

Reissmann, E., et al. (2001). The orphan receptor ALK7 and the Activin receptor ALK4 mediate signaling by Nodal proteins during vertebrate development. *Genes & Development*, *15*(15), 2010–2022.

Rivera-Perez, J. A., & Magnuson, T. (2005). Primitive streak formation in mice is preceded by localized activation of Brachyury and Wnt3. *Developmental Biology*, *288*(2), 363–371.

Rotherham, M., & El Haj, A. J. (2015). Remote activation of the Wnt/beta-catenin signalling pathway using functionalised magnetic particles. *PLoS One*, *10*(3), e0121761.

Saijoh, Y., et al. (2000). Left-right asymmetric expression of lefty2 and nodal is induced by a signaling pathway that includes the transcription factor FAST2. *Molecular Cell*, *5*(1), 35–47.

Saitoh, M., et al. (2013). Basolateral BMP signaling in polarized epithelial cells. *PLoS One*, *8*(5), e62659.

Sakuma, R., et al. (2002). Inhibition of Nodal signalling by Lefty mediated through interaction with common receptors and efficient diffusion. *Genes to Cells*, 7(4), 401–412.

Sarabipour, S., & Hristova, K. (2016). Mechanism of FGF receptor dimerization and activation. *Nature Communications*, 7, 10262.

Schmierer, B., & Hill, C. S. (2005). Kinetic analysis of Smad nucleocytoplasmic shuttling reveals a mechanism for transforming growth factor beta-dependent nuclear accumulation of Smads. *Molecular and Cellular Biology*, *25*(22), 9845–9858.

Senft, A. D., et al. (2019). Genetic dissection of Nodal and Bmp signalling requirements during primordial germ cell development in mouse. *Nature Communications*, *10*(1), 1089.
Serpe, M., et al. (2008). The BMP-binding protein Crossveinless 2 is a short-range, concentration-dependent, biphasic modulator of BMP signaling in Drosophila. *Developmental Cell*, *14*(6), 940–953.
Serup, P., et al. (2012). Partial promoter substitutions generating transcriptional sentinels of diverse signaling pathways in embryonic stem cells and mice. *Disease Models & Mechanisms*, *5*(6), 956–966.
Sheng, G. (2015). Epiblast morphogenesis before gastrulation. *Developmental Biology*, *401*(1), 17–24.
Simunovic, M., et al. (2019). A 3D model of a human epiblast reveals BMP4-driven symmetry breaking. *Nature Cell Biology*, *21*(7), 900–910.
Sirard, C., et al. (1998). The tumor suppressor gene Smad4/Dpc4 is required for gastrulation and later for anterior development of the mouse embryo. *Genes & Development*, *12*(1), 107–119.
Snow, M. H. L. (1977). Gastrulation in mouse—Growth and regionalization of epiblast. *Journal of Embryology and Experimental Morphology*, *42*, 293–303.
Song, J. W., et al. (1999). The type II activin receptors are essential for egg cylinder growth, gastrulation, and rostral head development in mice. *Developmental Biology*, *213*(1), 157–169.
Sorre, B., et al. (2014). Encoding of temporal signals by the TGF-beta pathway and implications for embryonic patterning. *Developmental Cell*, *30*(3), 334–342.
Souilhol, C., et al. (2015). NOTCH activation interferes with cell fate specification in the gastrulating mouse embryo. *Development*, *142*(21), 3649–3660.
Sozen, B., et al. (2018). Self-assembly of embryonic and two extra-embryonic stem cell types into gastrulating embryo-like structures (vol 20, 979, 2018). *Nature Cell Biology*, *20*(10), 1229.
Spivakkroizman, T., et al. (1995). How do heparin and heparan-sulfate activate Fgf-mitogenic activity. *Trends in Glycoscience and Glycotechnology*, 7(37), 447–449.
Stanganello, E., et al. (2015). Filopodia-based Wnt transport during vertebrate tissue patterning. *Nature Communications*, *6*, 5846.
Stroschein, S. L., et al. (1999). Negative feedback regulation of TGF-beta signaling by the SnoN oncoprotein. *Science*, *286*(5440), 771–774.
Sun, X., et al. (1999). Targeted disruption of Fgf8 causes failure of cell migration in the gastrulating mouse embryo. *Genes & Development*, *13*(14), 1834–1846.
Sundararajan, S., et al. (2012). A fast and sensitive alternative for beta-galactosidase detection in mouse embryos. *Development*, *139*(23), 4484–4490.
Takada, S., et al. (1994). Wnt-3a regulates somite and tailbud formation in the mouse embryo. *Genes & Development*, *8*(2), 174–189.
Tam, P. P. L., & Zhou, S. X. (1996). The allocation of epiblast cells to ectodermal and germ-line lineages is influenced by the position of the cells in the gastrulating mouse embryo. *Developmental Biology*, *178*(1), 124–132.
ten Berge, D., et al. (2008). Wnt signaling mediates self-organization and axis formation in embryoid bodies. *Cell Stem Cell*, *3*(5), 508–518.
Tewary, M., et al. (2017). A stepwise model of reaction-diffusion and positional information governs self-organized human peri-gastrulation-like patterning. *Development*, *144*(23), 4298–4312.
Tortelote, G. G., et al. (2013). Wnt3 function in the epiblast is required for the maintenance but not the initiation of gastrulation in mice. *Developmental Biology*, *374*(1), 164–173.
Tremblay, K. D., et al. (2000). Formation of the definitive endoderm in mouse is a Smad2-dependent process. *Development*, *127*(14), 3079–3090.

Turner, D. A., et al. (2014). Brachyury cooperates with Wnt/beta-catenin signalling to elicit primitive-streak-like behaviour in differentiating mouse embryonic stem cells. *BMC Biology*, *12*, 63.
Turner, D. A., et al. (2017). Anteroposterior polarity and elongation in the absence of extra-embryonic tissues and of spatially localised signalling in gastruloids: Mammalian embryonic organoids. *Development*, *144*(21), 3894–3906.
Vallin, J., et al. (2001). Cloning and characterization of three Xenopus Slug promoters reveal direct regulation by Lef/beta-catenin signaling. *Journal of Biological Chemistry*, *276*(32), 30350–30358.
van de Wetering, M., et al. (2001). Mutant E-cadherin breast cancer cells do not display constitutive Wnt signaling. *Cancer Research*, *61*(1), 278–284.
van den Brink, S. C., et al. (2014). Symmetry breaking, germ layer specification and axial organisation in aggregates of mouse embryonic stem cells. *Development*, *141*(22), 4231–4242.
Varlet, I., Collignon, J., & Robertson, E. J. (1997). Nodal expression in the primitive endoderm is required for specification of the anterior axis during mouse gastrulation. *Development*, *124*(5), 1033–1044.
Venkiteswaran, G., et al. (2013). Generation and dynamics of an endogenous, self-generated signaling gradient across a migrating tissue. *Cell*, *155*(3), 674–687.
Vincent, S. D., et al. (2003). Cell fate decisions within the mouse organizer are governed by graded Nodal signals. *Genes & Development*, *17*(13), 1646–1662.
Waldrip, W. R., et al. (1998). Smad2 signaling in extraembryonic tissues determines anterior-posterior polarity of the early mouse embryo. *Cell*, *92*(6), 797–808.
Wang, Q., et al. (2017). The p53 family coordinates Wnt and nodal inputs in mesendodermal differentiation of embryonic stem cells. *Cell Stem Cell*, *20*(1), 70–86.
Warmflash, A., et al. (2012). Dynamics of TGF-beta signaling reveal adaptive and pulsatile behaviors reflected in the nuclear localization of transcription factor Smad4. *Proceedings of the National Academy of Sciences of the United States of America*, *109*(28), E1947–E1956.
Warmflash, A., et al. (2014). A method to recapitulate early embryonic spatial patterning in human embryonic stem cells. *Nature Methods*, *11*(8), 847–854.
Weinstein, M., et al. (1998). Failure of egg cylinder elongation and mesoderm induction in mouse embryos lacking the tumor suppressor smad2. *Proceedings of the National Academy of Sciences of the United States of America*, *95*(16), 9378–9383.
Wilkinson, D. G., et al. (1988). Expression of the Fgf-related proto-oncogene int-2 during gastrulation and neurulation in the mouse. *EMBO Journal*, 7(3), 691–695.
Williams, P. H., et al. (2004). Visualizing long-range movement of the morphogen Xnr2 in the Xenopus embryo (vol 14, pg 1916, 2004). *Current Biology*, *14*(24), 2312.
Winnier, G., et al. (1995). Bone morphogenetic protein-4 is required for mesoderm formation and patterning in the mouse. *Genes & Development*, *9*(17), 2105–2116.
Wrana, J. L., et al. (1994). Mechanism of activation of the Tgf-Beta receptor. *Nature*, *370*(6488), 341–347.
Xiao, Z., et al. (2001). Nucleocytoplasmic shuttling of Smad1 conferred by its nuclear localization and nuclear export signals. *Journal of Biological Chemistry*, *276*(42), 39404–39410.
Xiao, C. C., et al. (2003). Ecsit is required for Bmp signaling and mesoderm formation during mouse embryogenesis. *Genes & Development*, *17*(23), 2933–2949.
Yamaguchi, T. P., Conlon, R. A., & Rossant, J. (1992). Expression of the fibroblast growth-factor receptor Fgfr-1/Flg during gastrulation and segmentation in the mouse embryo. *Developmental Biology*, *152*(1), 75–88.
Yamaguchi, T. P., et al. (1994). Fgfr-1 is required for embryonic growth and mesodermal patterning during mouse gastrulation. *Genes & Development*, *8*(24), 3032–3044.
Yamaguchi, T. P., et al. (1999). T (Brachyury) is a direct target of Wnt3a during paraxial mesoderm specification. *Genes & Development*, *13*(24), 3185–3190.

Yamamoto, M., et al. (2004). Nodal antagonists regulate formation of the anteroposterior axis of the mouse embryo. *Nature, 428*(6981), 387–392.
Yan, Y. T., et al. (2002). Dual roles of Cripto as a ligand and coreceptor in the nodal signaling pathway. *Developmental Biology, 247*(2), 455.
Yang, X., et al. (1999). Targeted disruption of SMAD3 results in impaired mucosal immunity and diminished T cell responsiveness to TGF-beta. *The EMBO Journal, 18*(5), 1280–1291.
Yeo, C. Y., & Whitman, M. (2001). Nodal signals to Smads through Cripto-dependent and Cripto-independent mechanisms. *Molecular Cell, 7*(5), 949–957.
Ying, Y., & Zhao, G. Q. (2001). Cooperation of endoderm-derived BMP2 and extraembryonic ectoderm-derived BMP4 in primordial germ cell generation in the mouse. *Developmental Biology, 232*(2), 484–492.
Ying, Y., et al. (2000). Requirement of Bmp8b for the generation of primordial germ cells in the mouse. *Molecular Endocrinology, 14*(7), 1053–1063.
Yoney, A., et al. (2018). WNT signaling memory is required for ACTIVIN to function as a morphogen in humam gastruloids. *eLife, 7*.
Yoon, Y., et al. (2015). Extra-embryonic Wnt3 regulates the establishment of the primitive streak in mice. *Developmental Biology, 403*(1), 80–88.
Yu, S. R., et al. (2009). Fgf8 morphogen gradient forms by a source-sink mechanism with freely diffusing molecules. *Nature, 461*(7263), 533–U100.
Zeng, L., et al. (1997). The mouse fused locus encodes Axin, an inhibitor of the Wnt signaling pathway that regulates embryonic axis formation. *Cell, 90*(1), 181–192.
Zhan, T., Rindtorff, N., & Boutros, M. (2017). Wnt signaling in cancer. *Oncogene, 36*(11), 1461–1473.
Zhang, H. B., & Bradley, A. (1996). Mice deficient for BMP2 are nonviable and have defects in amnion chorion and cardiac development. *Development, 122*(10), 2977–2986.
Zhang, P., et al. (2008). Short-term BMP-4 treatment initiates mesoderm induction in human embryonic stem cells. *Blood, 111*(4), 1933–1941.
Zhang, Z. C., et al. (2019). Mouse embryo geometry drives formation of robust signaling gradients through receptor localization. *Nature Communications, 10*, 4516.
Zhou, X. L., et al. (1993). Nodal is a novel Tgf-beta-like gene expressed in the mouse node during gastrulation. *Nature, 361*(6412), 543–547.
Zhu, Y., et al. (1998). Smad3 mutant mice develop metastatic colorectal cancer. *Cell, 94*(6), 703–714.
Zimmerman, L. B., DeJesusEscobar, J. M., & Harland, R. M. (1996). The Spemann organizer signal noggin binds and inactivates bone morphogenetic protein 4. *Cell, 86*(4), 599–606.

CHAPTER FOURTEEN

Just passing through: The auxin gradient of the root meristem

Bruno Guillotin, Kenneth D. Birnbaum*
New York University, The Department of Biology, The Center for Genomics and Systems Biology, New York, NY, United States
*Corresponding author: e-mail address: kdb5@nyu.edu

Contents

Abstract

The root meristem—one of the plant's centers of continuous growth—is a conveyer belt in which cells of different identities are pushed through gradients along the root's longitudinal axis. An auxin gradient has long been implicated in controlling the progression of cell states in the root meristem. Recent work has shown that a PLETHORA (PLT) protein transcription factor gradient, which is under a delayed auxin response, has a dose-dependent effect on the differentiation state of cells. The direct effect of auxin

Current Topics in Developmental Biology, Volume 137
ISSN 0070-2153
https://doi.org/10.1016/bs.ctdb.2019.12.001

concentration on differential transcriptional outputs remains unclear. Genomic and other analyses of regulatory sequences show that auxin responses are likely controlled by combinatorial inputs from transcription factors outside the core auxin signaling pathway. The passage through the meristem exposes cells to varying positional signals that could help them interpret auxin inputs independent of gradient effects. One open question is whether cells process information from the changes in the gradient over time as they move through the auxin gradient.

1. Introduction

1.1 Proximo-distal maturation gradient of the root

The growing tissue of the Root Apical Meristem (RAM) continuously produces cells on a homeostatic anatomy, like a road crew adding new sections to a multilane highway (Fig. 1). In the RAM, cells are born around a structure known as the quiescent center, QC (Dolan et al., 1993; Haecker, 2004; van den Berg, Willemsen, Hendriks, Weisbeek, & Scheres, 1997). While plant cells do not migrate, the cylinder-like root contains files of cells that are displaced from the tip by the birth, expansion, and division of new cells behind them. Meanwhile, the gradients that form in the meristem are stationary with respect to the growing tip (Motte, Vanneste, & Beeckman, 2019), analogous to a highly organized road crew moving along as the highway gets built. Thus, behind the root cap, cells in the meristem progress in an assembly-line fashion, such that, in theory, a morphogen gradient along this length-wise (proximo-distal) axis of the root could stage each step of cellular maturation for a cell in a given file (Fig. 1). We consider the prospects of such a maturation-instructive morphogen here—a substance whose concentration instructs the progressive differentiation states of cells.

Below the QC, a separate set of divisions gives rise to the columella cells of the cap. The QC has been shown to be the source of signals that maintain columella stem cells (Pi et al., 2015; Sarkar et al., 2007), but the maturation gradient in that axis is extremely short so here we focus on the gradients above. This means that the root really has two conveyer belts of cell production, with the very slowly dividing QC remaining stationary in the middle (Cruz-Ramirez et al., 2013; Rahni & Birnbaum, 2018). Thus, as will be discussed below, the QC (and surrounding cells) appear to serve as a focal point of signals that help polarize the gradients above the cap.

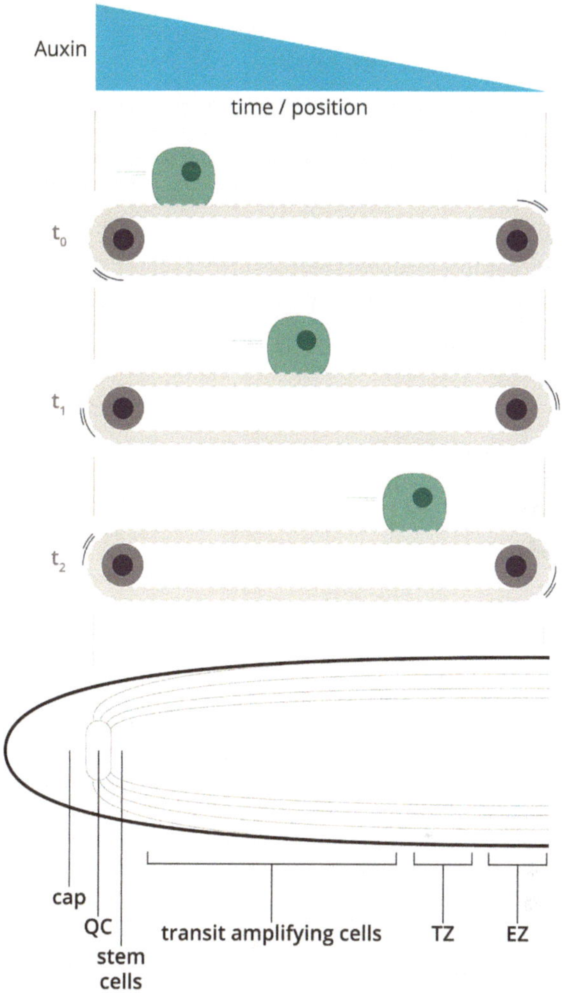

Fig. 1 Cells in the plant meristem pass through the auxin gradient of the root. At top, for most of the radial files of the root, there is a decreasing gradient of auxin within the division zone as cells mature. At middle, the division and expansion of cells create a conveyer-belt like mechanism whereby cells are pushed through the meristem over time, passing through each level of the gradient. At bottom, the key landmarks in the root meristem on the maturation axis are indicated, with the QC surrounded by stem cells, followed by a zone of transit amplification, a transition zone (TZ), and then an elongation zone (EZ) at the start of differentiation.

2. Signals across cell walls

2.1 Plasmodesmata

To understand the gradients that form in the roots, it's useful to consider the plant's common modes of signaling—some of which are unique to plants. One such feature of plant signaling is the gap-junction-like structure known as plasmodesmata, which forms channels through cell walls connecting the cytoplasm of neighboring cells (Sager & Lee, 2018). Remarkably, transcription factors use these conduits to move from cell to cell, acting as direct signaling agents. For example, the transcription factor SHORTROOT (SHR) moves at the protein level from the vascular and surrounding pericycle cells into the adjacent ground tissue to mediate a tissue-forming division and help specify the endodermal cell identity (Nakajima, Sena, Nawy, & Benfey, 2001). Constructs that block plasmodesmata prevented the movement of SHR (Vatén et al., 2011). Many other plant transcription factors and microRNAs have now been shown to move from cell to cell (Daum, Medzihradszky, Suzaki, & Lohmann, 2014; Mähönen et al., 2014; Miyashima et al., 2019; Pi et al., 2015; Skopelitis et al., 2018; Yadav et al., 2011), presumably mediated by passage through the plasmodesmata.

Transport through plasmodesmata—known as symplastic signaling—appears to have some size exclusion limit, although there may also be active mechanisms that control movement (Sager & Lee, 2018). For example, small RNA diffusion was also shown to be directional—that is, polarized in one direction (Skopelitis et al., 2018). There is also some evidence that metabolites travel through plasmodesmata (Benitez-Alfonso, Faulkner, Ritzenthaler, & Maule, 2010; Sivaguru et al., 2000; Vatén et al., 2011). Thus, the plasmodesmata appear to mediate symplastic gradients that allow a certain level of diffusion for transcription factors, much like the syncytium of the Drosophila embryo (Daniels, Rikhy, Renz, Dobrowsky, & Lippincott-Schwartz, 2012).

Plants, like animals, also signal through peptides and cognate membrane bound receptor-like kinases (Olsson et al., 2019). Although it's feasible that peptides could diffuse through cell walls, there are several reports that receptors and their ligands interact at the site of plasmodesmata (Faulkner, 2013; Stahl et al., 2013; Vaddepalli et al., 2014), although more research is needed to determine how plasmodesmata mediate peptide signaling.

2.2 Auxin

Auxin is not the only plant hormone that affects root development (Kamiya, 2010), but auxin takes center stage, not only for its ubiquity in plant development, but also because of both its necessity and sufficiency for root development. Seminal experiments showed that treating callus—a blastula-like mass of regenerative cells—with high auxin and low cytokinin leads to the regeneration of a fully functional root (Skoog & Miller, 1957). Mutations in *Auxin Response Factor 5* (*ARF5* or *MONOPTEROS, MP*)—in a family of transcription factors that mediate direct auxin responses—led to rootless embryos (Hardtke & Berleth, 1998), while blocking auxin signaling in roots leads to severe defects in the root meristem (Overvoorde, Fukaki, & Beeckman, 2010; Sato & Yamamoto, 2008).

Auxin is a small metabolite derived from tryptophan in just a few enzymatic steps (Zhao, 2012). The bulk of auxin synthesis is in the shoot and its long-distance transport to the root relies, in part, on the transport system of the phloem (Robert & Friml, 2009; Weijers, Nemhauser, & Yang, 2018). However, auxin's localization within organs is precisely fluxed and positioned by what amounts to its own private circulatory system. Auxin appears to be able to enter cells by passive mechanisms, although auxin influx carriers also have essential roles in development (van Berkel, de Boer, Scheres, & ten Tusscher, 2013). Once inside the cell, auxin is unable to passively diffuse through the membrane. Thus, much of auxin flux is mediated by efflux carriers, known as PINs, that are polarly localized in the cell, such that auxin is shuttled along in one cell and out the other in specific directions (Adamowski & Friml, 2015). This planar polarity among cells shifts at strategic points fluxing auxin through organs like traffic patterns through city streets (Benkova et al., 2003; Friml et al., 2003). In the root, distinct PINs flux auxin down the vasculature toward the tip, then outwards toward the outer cap and epidermis, then upwards in the epidermis toward the shoot, and finally back into the vascular flux in what has been likened to a reverse fountain (Grieneisen, Xu, Marée, Hogeweg, & Scheres, 2007; Swarup & Bennett, 2003).

Downstream, auxin signaling is mediated by an E3-ligase receptor complex that leads to the degradation of a set of repressors called Aux/IAAs (Dharmasiri, Dharmasiri, & Estelle, 2005; Kepinski & Leyser, 2005), the core of the widely used auxin degron (Pierre-Jerome, Jang, Havens, Nemhauser, & Klavins, 2014). The repressors bind to a subset of ARF family members (Ulmasov, Hagen, & Guilfoyle, 1997;

Ulmasov, Murfett, Hagen, & Guilfoyle, 1997), which mediate a large set of transcriptional responses (Bargmann et al., 2013; Lewis et al., 2013). There are more than 20 members each in both ARF and Aux/IAA gene families in Arabidopsis with similarly high gene family expansion in other flowering plants, allowing for combinations of the core machinery with different responses to auxin (Leyser, 2018; Roosjen, Paque, & Weijers, 2018). Some of the direct targets of positively regulating ARFs are, in fact, their own repressors, the Aux/IAAs (Leyser, 2018). Thus, there is a strong feedback built into core response, with auxin leading to the degradation of Aux/IAAs but also inducing their expression. In addition, while genes in the ARF5 clade function as activators, genes in another ARF clade appear to act as repressors and may compete for the ARF5 clade binding sites (Roosjen et al., 2018; Wang & Estelle, 2014). The negative feedback and repressive modes of auxin signaling are increasingly at the forefront of auxin signaling research, as we will cover below.

3. The auxin gradient of the root

3.1 Setting up the gradient

Several lines of evidence have established the existence of an auxin gradient within the root meristem with a high concentration at the stem cell niche that gradually decreases in the differentiation zone of the root (Petersson et al., 2009; Sabatini et al., 1999; Santuari et al., 2011). In the last several years, new auxin sensors, which are based on the Aux/IAA receptor rather than transcriptional responses mediated by ARF binding sites, have painted a more complex picture of the auxin gradient. In addition to a diminishing gradient, auxin appears to have a second peak after the transition from cell division as cells elongate and differentiate (Brunoud et al., 2012). The epidermal layers appear to have lower auxin concentration near the stem cell niche, with the auxin peak building up in cells positioned several cell lengths toward the shoot. Thus, the auxin gradient is bimodal over the entire root, with most cells experiencing a decreasing gradient within the meristem before the onset of differentiation with cell elongation.

3.2 Positional inputs to the auxin gradient

Computational models suggest that the polar auxin transport system described above could generate a local maximum at the stem cell niche (Grieneisen et al., 2007). A recent extension of this model includes the opposing effects of the plant hormone cytokinin on auxin at the end of the meristem (Di Mambro et al., 2017). The model shows that cytokinin determines the position of the auxin minimum and the transition zone by both downregulating PIN transporters and inactivating auxin (Di Mambro et al., 2017). In a validation of the model, manipulation of a *GRETCHEN HAGEN3* gene, which deactivates auxin and was shown to be downstream of cytokinin, moved the position of the transition zone and extended the meristem (Di Mambro et al., 2017). This mechanism sheds light on how the familiar antagonism between auxin and cytokinin helps shape the root auxin gradient.

Furthermore, recent work shows how specific cell files can modulate the auxin gradient locally to "customize" their own developmental timing. The work showed that two phloem-specific plasma membrane-localized proteins interact with PINs to mediate auxin flux, resulting in a steeper gradient within this vascular cell type (Marhava et al., 2018). When the cell-specific gradient was perturbed, cellular maturation was altered (Marhava et al., 2018). The phloem work showed that cell-type specific mechanisms could alter the auxin gradient and speed the maturation of one cell file—analogous to a lane of the highway being constructed early. It is plausible there is a need to speed the maturation of a cell type that delivers sucrose to an organ that is an energy sink.

Both of these studies draw attention to the way local signals reshape the auxin gradient to influence maturation, either across many cell types, as in the case of cytokinin inputs, or within a specific cell type, as in phloem.

3.3 Local auxin synthesis influences the root maximum

While auxin flux via PINs is of central importance to auxin localization, a body of work now shows that local auxin biosynthesis also has a key role in the establishment of the RAM auxin gradient (Brumos et al., 2018; He et al., 2011; Stepanova et al., 2008; Yang et al., 2014). For example, mutants in the auxin biosynthesis gene *WEAK ETHYLENE INSENSITIVE 8/Tryptophan aminotransferase-related protein 2*, which is localized to the QC, columella and surrounding cells, fail to maintain an auxin maximum centered around the QC (Brumos et al., 2018; Stepanova et al., 2008). This work firmly established the need to account for local auxin synthesis in the root when considering auxin's ability to self-organize its own flux and localization. That is, again, positional signals outside the core auxin signaling machinery may help the auxin transport system establish or maintain fluxes.

4. PLETHORAs mediate auxin but form their own gradient

4.1 Downstream auxin responses

To date, auxin's clearest role in regulating progressive stages of differentiation in the root comes from its rather indirect effects on members of the *PLETHORA* (*PLT*) transcription factor family—*PLT1*, *PLT2*, *PLT3* and *BABY BOOM* (*BBM*, also known as *PLT4*). Several lines of evidence put the PLTs downstream of auxin. First, several *PLT*s are no longer expressed in the embryo in *ARF5* mutants (Aida et al., 2004). Still, in the primary root, it was shown that auxin reporters respond almost immediately to auxin treatment, while *PLT* reporters required 72 h of treatment to show dramatic up regulation (Mähönen et al., 2014). Thus, the *PLT*s appear to be a slow response to auxin, which led to a model where PLTs regulated auxin's role in long-term developmental processes while rapid auxin responses mediated faster processes like auxin-mediated gravitropism (Mähönen et al., 2014).

4.2 Gene level regulation of PLTs

Interestingly, a menagerie of signaling mechanisms that are not directly tied to auxin appear to shape the PLT protein gradient. At the transcriptional level, *PLT*s are induced in the zone of highest auxin concentration within the stem cell niche and some surrounding daughter cells (Mähönen et al., 2014). However, above the niche, the *PLT*s are repressed by *Growth Regulating Factors* (*GRF*s; Kim, Choi, & Kende, 2003). In Arabidopsis roots, *GRF1, 2,* and *3* are expressed above the QC (Fig. 2; Rodriguez et al., 2015) and are co-activated by *GRF Interacting Factors, GIF*s (Fig. 2; Debernardi et al., 2014; Ercoli et al., 2018; Horiguchi, Kim, & Tsukaya, 2005; Kim & Kende, 2004; Nelissen et al., 2015). Thus, the *GRF*s keep *PLT* mRNA expression within a narrow zone near the QC and stem cells.

Meanwhile, *PLT* expression appears to be shielded from its negative regulators in the niche by microRNA396 family, which directly targets 7 out of 9 *GRF*s for transcriptional degradation (Rodriguez et al., 2015). For example, it was shown that miRNA resistant *GRF*s are ectopically expressed in the QC resulting in a smaller meristem size and development defects (Ercoli et al., 2018; Rodriguez et al., 2015). Finally, this local expression of *PLT* mRNAs appears to be on a feedforward loop, as high levels of PLT proteins directly promote miR396s expression

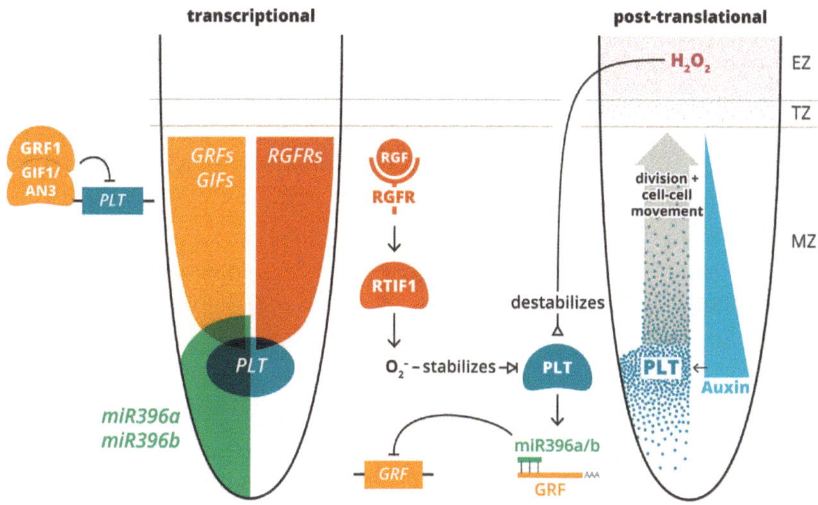

Fig. 2 The PLETHORA gradient is shaped at both the transcriptional and posttranscriptional level starting with an auxin gradient. On the right, the PLT gradient is induced transcriptionally in the region of the auxin maximum near the stem cell niche. The GRFs and GIFs mediate PLT transcriptional down regulation in the transit amplifying zone, while miR396 protects PLT expression levels around the stem cell niche. PLTs form a gradient through both cell–cell movement and through protein carry over and dilution from cell division. In addition, an RGF peptide gradient recognized by RGF receptors (RGFRs) helps stabilize the PLT transcription factor gradient, through RTIF1 and reactive oxygen species.

and protect *PLT* transcription in the stem cell niche from *GRF* activity (Rodriguez et al., 2015). Thus, rather than a gradient at this point, the transcriptional regulatory circuit appears to concentrate the *PLT* expression in a narrow zone of high concentration near the QC (Fig. 2).

4.3 Protein level regulation of PLTs

In addition to the mechanisms that restrict PLT mRNA expression, protein-level regulation shapes the PLT gradient. First, PLT proteins appear quite stable, allowing them to a "ride" a wave of cell divisions in the transit amplifying zone of the meristem, where dilution helps form a decreasing gradient (Mähönen et al., 2014). Second, in a very plant-centric signaling mode, the PLTs also appear to be able to move from cell to cell, most likely through the plasmodesmata, to spread out toward the transition zone. For example, one litmus test for plasmodesmata-mediated mobility is attaching a triple GFP protein to the candidate transcription factor (Pi et al., 2015). These bulky fusions appear to exceed the plasmodesmatal size exclusion limit. Aiding in phenotypic interpretation, they can typically rescue cell-autonomous but not cell nonautonomous defects in their own mutant backgrounds (Pi et al., 2015; Schlereth et al., 2010). In the PLT longitudinal gradient, a nonmobile PLT2-3xYFP exhibited a shorter gradient than a mobile version when cell division was halted with a chemical block, showing that protein movement as well as protein persistence and division dilution were important in generating the gradient (Mähönen et al., 2014).

Critically, the PLT protein gradient is strongly shaped by members of the *CLAVATA-LIKE* (*CLE*) peptide family, known as *Root meristem Growth Factors* (*RGFs*). The small peptide subfamily has nine members, more than half of which are expressed around the QC, stem cell niche, and cap area (Matsuzaki, Ogawa-Ohnishi, Mori, & Matsubayashi, 2010). The proteins diffuse locally out of the QC region with the effect of stabilizing PLT proteins (Matsuzaki et al., 2010). The gradient of RGFs is critical for PLT function and meristem maintenance, as a triple mutant in the family, *rgf123*, causes a short-root phenotype characterized by a decrease in meristematic cell number. Exogenous treatment with the RGF peptides drastically increases the PLT gradient length and meristem size (Shinohara, Mori, Yasue, Sumida, & Matsubayashi, 2016; Yamada, Han, & Benfey, 2019). Furthermore, the increase in PLT protein is visible within only a few hours, implying a fairly direct action on PLTs (Shinohara et al., 2016; Yamada et al., 2019).

Recently, several key components of the pathway downstream of RGFs were discovered, including the *RGF* receptors, *RGFRs* (Fig. 2, Shinohara et al., 2016) and a downstream mediator *RGF1 INDUCIBLE TRANSCRIPTION FACTOR 1* (*RITF1*), which then signals to the PLTs via reactive oxygen species (Yamada et al., 2019). This entire circuit appears to be independent of auxin since auxin treatment does not affect the *RGFs* (Matsuzaki et al., 2010).

4.4 PLTs show dose dependent effects

The information conveyed by the PLT gradient has largely been assessed by loss- and gain-of-function experiments. For example, overexpression of PLT2 increases the potency of distal cells to regenerate after wounding, associating high levels of PLTs with cellular potency and youth

(Marhava et al., 2019). At the next level, moderate PLT concentrations appear to maintain cell division in the transit amplifying zone of the meristem, as increasing *PLT* expression with 35S:PLT2 increases the root meristem size (Galinha et al., 2007). Finally, low concentrations of PLTs correlate with the shift from cell division to elongation—the transition zone that marks meristem size. Indeed, decreasing the numbers of functional PLTs in higher level mutants led to a decrease in meristem size (Galinha et al., 2007). Thus, PLT levels correlate with maturation landmarks in the root both in stereotypical development and upon perturbations.

How the PLT gradient is perceived by cells remains an open question. An in-depth study of downstream PLT regulation showed that many PLT targets were expressed in discrete zones within the meristem (Santuari et al., 2016). This could be the result of threshold effects stemming from PLT dosage. However, it was not clear if PLT dosage or combinatorial regulation with other spatially discrete factors generated the zoned expression (Santuari et al., 2016). A PLT binding site has been identified and it was enriched under PLT ChIP-seq peaks in the root studies (Santuari et al., 2016). However, no analysis so far has uncovered a relationship between binding site affinity or arrangement and spatial expression along the PLT gradient. This may be a prospect for future work with more information on PLT binding affinities.

In addition, another level of feedback complexity is evident from the list of PLT target genes. One of the direct targets of PLT proteins was the auxin biosynthetic gene, *YUCCA3* (Santuari et al., 2016), providing evidence, along with other studies, that the PLTs provide feedback to influence auxin levels and transport in the root (Aida et al., 2004; Santuari et al., 2016).

Thus, the PLTs offer one of the clearest examples of a gradient in plants that can instruct cellular maturation. The variety of positional signals that go into shaping the gradient raises an important question on how the system assembles. A hallmark of plant development is the capacity for *de-novo* meristem assembly both in regeneration and postembryonic development like lateral root formation. For example, the entire meristem and its maturation gradient can reform over a few days after severe injury (Sena, Wang, Liu, Hofhuis, & Birnbaum, 2009). Is there an order to the assembly of the positional signals that shape meristem gradients? For example, is there is a signaling center that emerges first to set up positional signals, or, could physical properties like distal tip vs internal proximal positions provide cues that provide a proximo-distal coordinate system at once?

5. Direct outputs of the auxin gradient

5.1 Auxin has effects independent of PLTs

Downstream analyses of the auxin and PLT pathways showed a significant overlap in their targets (Santuari et al., 2016). Still, global studies of auxin responsive genes suggest a much wider set of genes affected by auxin than are regulated by the PLTs (Bargmann et al., 2013; Lewis et al., 2013; Santuari et al., 2016). In addition, perturbations to the auxin gradient can alter the position of the transition zone relatively rapidly, suggesting a role for auxin signaling independent of the slowly induced PLTs (Dello Ioio et al., 2008; Di Mambro et al., 2017). Is there a role for the auxin gradient in cellular maturation aside from its effect on the PLTs?

The question goes to the heart of one of the central issues in auxin biology. How does this single hormone, which has a role in virtually every developmental process, exert so many different effects?

In terms of a dose response to ARFs, there is also evidence that ARF-binding motif variants, copy number, and arrangement in regulatory regions can determine the sensitivity to auxin signaling (Boer et al., 2014; Liao et al., 2015; Pierre-Jerome et al., 2014; Pierre-Jerome, Moss, Lanctot, Hageman, & Nemhauser, 2016). However, evidence has yet to emerge of binding motif patterns that could correlate with local expression along the gradient of the root.

5.2 Prepatterned responses

Overall, auxin appears to be more of a trigger than a signal whose level conveys specific information to cells (Benkova, Ivanchenko, Friml, Shishkova, & Dubrovsky, 2009; Leyser, 2018; Stewart & Nemhauser, 2010). Auxin has been likened to an orchestra conductor instructing cells to "carry out" their preprogrammed response (Leyser, 2006). It has also been compared to currency, informing cells of the "wealth" of auxin concentration in their vicinity (Stewart & Nemhauser, 2010), implying a role for concentration.

The differential competence of cells to respond to auxin is illustrated in an experiment in which roots were treated with auxin, and cell types differentially labeled with fluorescent reporters were separated by cell sorting to measure their global response (Bargmann et al., 2013). The analysis showed that hundreds of genes respond differently to the same treatment in different cell types (Bargmann et al., 2013). This showed the extent to which intrinsic cell state could determine a cell's response to auxin.

Mechanistically, generating differential auxin responses along the gradient might be controlled by differences in the auxin response machinery itself, and there is evidence that different ARFs and/or AUX/IAA repressors allow cells to respond differently to auxin inputs (Bargmann & Birnbaum, 2009; O'Malley et al., 2016; Pierre-Jerome et al., 2014; Vernoux et al., 2011; Weijers et al., 2018; Zemlyanskaya, Wiebe, Omelyanchuk, Levitsky, & Mironova, 2016). For example, in the embryo, ARF9 is expressed in specific root-precursor cells, mediating a response that cannot be rescued by other ARFs (Rademacher et al., 2012). Such a change in the auxin response machinery along the meristem gradient could explain the differentiation effects of auxin along the gradient. However, it is not clear that such differences in the core auxin machinery could explain the differential responses to auxin along the maturation axis of the root meristem. For example, in a recent report in the yeast system, activating ARFs did not show specific preferences in binding motifs, suggesting their first-order interactions with regulatory sequences cannot determine differences in a cell's intrinsic response to auxin (Lanctot, Taylor-Teeples, Oki, & Nemhauser, 2019).

5.3 Combinatorial inputs

Transcription factors outside of the core auxin machinery that are expressed in zones along the meristem could help auxin regulate downstream genes in discrete zones. For example, ARFs have been shown to interact with other transcription factors (Roosjen et al., 2018; Shin et al., 2007; Varaud et al., 2011). In global analyses, genes responding at different times to auxin treatments showed enrichments of both ARF-responsive motifs and binding sites for other transcription factors such as basic helix-loop-helix proteins (Cherenkov et al., 2017; Mironova, Omelyanchuk, Wiebe, & Levitsky, 2014). These studies strongly suggest that differences in auxin responses are likely the result of combinatorial regulation that includes transcription factors outside the ARF gene family.

Combinatorial control of auxin responses could explain differential responses along the auxin gradient without the need for auxin concentration to carry any information other than off or on signaling. It's not necessarily an argument against gradient control, as the current efforts in the field may be progressing more quickly on the combinatorial aspects of auxin's transcriptional regulation. Nonetheless, concentration effects for auxin do not appear obvious. For example, we treated cells with varying levels of auxin in culture and then cell sorted specific cell types to measure their global response to auxin dose. The cells showed only quantitative responses of largely the same target genes rather than qualitatively different responses at different auxin levels that are indicative of different maturation states or cell identities (unpublished results). Of course, many caveats in such an experiment exist, including the need for tissue context in regulating a cell's primary response to auxin.

5.4 Repressive lockdown states

Some of the information contained in the auxin gradient may be encoded in its disappearance. Low auxin states are known to carry information for plant cells (Dubrovsky et al., 2011; Sorefan et al., 2009; Wang, Kohlen, Rossmann, Vernoux, & Theres, 2014), such as when low auxin marks the site where valve margin separation will form in the seed-bearing organ, the silique (Sorefan et al., 2009). Mechanistically, ARF-Aux/IAA complexes were shown to recruit *TOPLESS*, a Groucho-like homolog, that interacts with histone deacetylases in low auxin states to form a repressive complex (Szemenyei, Hannon, & Long, 2008)—a flip switch-type lockdown of auxin targets. In turn, *ARF5* can recruit SWI/SNF chromatin remodelers to make chromatin accessible upon auxin treatment (Wu et al., 2015). In this way, different auxin environments can regulate differential chromatin accessibility and, potentially, a cell's competence to respond to other signals. Going from a high to a low auxin environment could conceivably prime a specific response. Still, it's still not clear if such a progression of chromatin states occurs along the root axis.

Alternative to combinatorial or repression "flip switch" models, auxin levels may be read out directly by a mechanism yet to be identified. Finally, it remains possible that auxin levels may not have a primary role in staging the maturation of cells outside of their effects on PLTs.

5.5 Prospects

Most of our understanding of the role of auxin in staging root maturation has emerged through a more detailed understanding of the PLTs. Extensive work has shown how the PLT gradient is set up by an auxin maximum at the root tip. The strong foundational work on the PLTs has made it perhaps one of the best characterized morphogen candidates in plants. Perturbations strongly suggest that a gradient of PLTs carries information at different levels. This comes largely through the genetic manipulation of PLT levels that show corresponding shifts in the meristem size.

A key question remains whether PLT levels directly affect downstream gene regulation. It is possible that the multiple effects of PLTs arise from combinatorial regulation with other transcription factors that are localized to different parts of the meristem. The eventual dissipation of the gradient could then serve as a switch to allow other genetic programs to take over. With direct targets now in hand (Santuari et al., 2016), there will be new opportunities to gain a mechanistic understanding of how PLT levels are interpreted at the promoter/regulatory sequence level. Thus, the PLTs provide an important tool in understanding how mobile transcription factors—a primary signaling mode in plants—can act as morphogens.

The role of an auxin gradient remains a more difficult question. Two important properties of auxin will likely be important considerations in analyzing a gradient role in the root. The so-called "auxin code" represents the concept that a prepatterned, intrinsic cellular state dictates a cell's response to an auxin signal. Progress has been made to show that differences in the core auxin machinery can prime a specific response to auxin. Increasingly, evidence for intrinsic cellular mechanisms points to combinatorial regulation with transcription factors outside the auxin machinery. In the root meristem, cells at different stages of maturity respond differently to the same auxin input (Bargmann et al., 2013), at least in part because of inputs from cytokinin (Di Mambro et al., 2017). Positional signals appear to set up differences in the auxin response code that allow cells near the stem cell niche to respond differently than those near the transition zone. Could auxin levels provide another level of fine tuning to a combinatorial code? A strong reliance on combinatorial inputs along the gradient does not leave a clear role for different levels of the gradient to carry specific information for cells.

Still, auxin is notoriously self-organizing, for example, influencing its own transport polarity (Leyser, 2018). The conveyer belt-like nature of the meristem means that, cells not only experience each level of the auxin gradient, but they do so in a particular order. Others have raised the possibility that a prior auxin state could influence a future auxin state (Pierre-Jerome, Moss, & Nemhauser, 2013). Extending that idea to the root meristem, cells may interpret the sequence of auxin inputs rather than the immediate level of auxin. It would be a tangled web: auxin would set up the prepattern that shapes its response to the next auxin level. Still, such a scenario brings the auxin gradient back into a focus as a primary organizing feature of the meristem.

Auxin's feedback and feedforward behaviors have been difficult to disentangle. The yeast system has been powerful in assessing the function of the core response machinery (Pierre-Jerome et al., 2014). However, the role of the gradient in the root will require more context from the meristematic environment. One potential tool lies in the plant's high capacity to self-organize its meristems. The root regeneration protocols mentioned above provide "organoid" systems in which the order of meristem assembly can be tracked (Skoog & Miller, 1957).

In addition, advances in fine-scale analyses at the molecular and imaging level have the potential to dissect cellular responses at progressive stages of the gradient. For example, single-cell RNA-seq studies have ordered cells of the root by both cell type and developmental trajectories represented in pseudo-time (Denyer et al., 2019; Jean-Baptiste et al., 2019; Mironova & Xu, 2019; Ryu, Huang, Kang, & Schiefelbein, 2019; Wang, Ryu, Barron, & Schiefelbein, 2019; Zhang, Xu, Shang, & Wang, 2019). This tool offers a comparative analysis of different cell types moving through a similar, if not a uniform, gradient for each cellular position in the meristem. In addition, the root is highly amenable to live imaging, which should allow analyses of auxin levels and responses that could be mapped back to high-content RNA-seq profiles of single cells. Such fine-scale temporal analysis of meristematic cells could provide models of how cells transition from one state to another during maturation.

For testing the function of auxin feedback mechanisms, loss- and gain-of-function mutants have been useful, but auxin's involvement in every step of plant development make perturbations highly pleiotropic, complicating interpretations of phenotype. We anticipate gene-editing techniques that allow localized, conditional perturbations will help parse out auxin's potential sequential roles in meristem maturation. These tools hold promise to provide a perspective on auxin signaling that has been difficult to assess: how do cells respond to dynamic auxin environments over time?

References

Adamowski, M., & Friml, J. (2015). PIN-dependent auxin transport: Action, regulation, and evolution. *The Plant Cell*, *27*, 20–32.

Aida, M., Beis, D., Heidstra, R., Willemsen, V., Blilou, I., Galinha, C., et al. (2004). The PLETHORA genes mediate patterning of the Arabidopsis root stem cell niche. *Cell*, *119*, 109–120.

Bargmann, B. O., & Birnbaum, K. D. (2009). Positive fluorescent selection permits precise, rapid, and in-depth overexpression analysis in plant protoplasts. *Plant Physiology*, *149*, 1231–1239.

Bargmann, B. O., Vanneste, S., Krouk, G., Nawy, T., Efroni, I., Shani, E., et al. (2013). A map of cell type-specific auxin responses. *Molecular Systems Biology*, *9*, 688.

Benitez-Alfonso, Y., Faulkner, C., Ritzenthaler, C., & Maule, A. J. (2010). Plasmodesmata: Gateways to local and systemic virus infection. *Molecular Plant-Microbe Interactions*, *23*, 1403–1412.

Benkova, E., Ivanchenko, M. G., Friml, J., Shishkova, S., & Dubrovsky, J. G. (2009). A morphogenetic trigger: Is there an emerging concept in plant developmental biology? *Trends in Plant Science, 14*, 189–193.

Benkova, E., Michniewicz, M., Sauer, M., Teichmann, T., Seifertova, D., Jurgens, G., et al. (2003). Local, efflux-dependent auxin gradients as a common module for plant organ formation. *Cell, 115*, 591–602.

Boer, D. R., Freire-Rios, A., van den Berg, W. A., Saaki, T., Manfield, I. W., Kepinski, S., et al. (2014). Structural basis for DNA binding specificity by the auxin-dependent ARF transcription factors. *Cell, 156*, 577–589.

Brumos, J., Robles, L. M., Yun, J., Vu, T. C., Jackson, S., Alonso, J. M., et al. (2018). Local auxin biosynthesis is a key regulator of plant development. *Developmental Cell, 47*, 306–318.e305.

Brunoud, G., Wells, D. M., Oliva, M., Larrieu, A., Mirabet, V., Burrow, A. H., et al. (2012). A novel sensor to map auxin response and distribution at high spatio-temporal resolution. *Nature, 482*, 103–106.

Cherenkov, P., Novikova, D., Omelyanchuk, N., Levitsky, V., Grosse, I., Weijers, D., et al. (2017). Diversity of cis-regulatory elements associated with auxin response in *Arabidopsis thaliana*. *Journal of Experimental Botany, 69*, 329–339.

Cruz-Ramirez, A., Diaz-Trivino, S., Wachsman, G., Du, Y., Arteaga-Vazquez, M., Zhang, H., et al. (2013). A SCARECROW-RETINOBLASTOMA protein network controls protective quiescence in the Arabidopsis root stem cell organizer. *PLoS Biology, 11*, e1001724.

Daniels, B. R., Rikhy, R., Renz, M., Dobrowsky, T. M., & Lippincott-Schwartz, J. (2012). Multiscale diffusion in the mitotic *Drosophila melanogaster* syncytial blastoderm. *Proceedings of the National Academy of Sciences of the United States of America, 109*, 8588–8593.

Daum, G., Medzihradszky, A., Suzaki, T., & Lohmann, J. U. (2014). A mechanistic framework for noncell autonomous stem cell induction in Arabidopsis. *Proceedings of the National Academy of Sciences of the United States of America, 111*, 14619–14624.

Debernardi, J. M., Mecchia, M. A., Vercruyssen, L., Smaczniak, C., Kaufmann, K., Inze, D., et al. (2014). Post-transcriptional control of GRF transcription factors by microRNA miR396 and GIF co-activator affects leaf size and longevity. *The Plant Journal, 79*, 413–426.

Dello Ioio, R., Nakamura, K., Moubayindin, L., Perilli, S., Taniguchi, M., Morita, M. T., et al. (2008). A genetic framework for genetic control of cell division and differentiation in the root meristem. *Science, 322*, 1543–1547.

Denyer, T., Ma, X., Klesen, S., Scacchi, E., Nieselt, K., & Timmermans, M. C. P. (2019). Spatiotemporal developmental trajectories in the Arabidopsis root revealed using high-throughput single-cell RNA sequencing. *Developmental Cell, 48*, 840–852 e845.

Dharmasiri, N., Dharmasiri, S., & Estelle, M. (2005). The F-box protein TIR1 is an auxin receptor. *Nature, 435*, 441–445.

Di Mambro, R., De Ruvo, M., Pacifici, E., Salvi, E., Sozzani, R., Benfey, P. N., et al. (2017). Auxin minimum triggers the developmental switch from cell division to cell differentiation in the Arabidopsis root. *Proceedings of the National Academy of Sciences of the United States of America, 114*, E7641–E7649.

Dolan, L., Janmaat, K., Willemsen, V., Linstead, P., Poethig, S., Roberts, K., et al. (1993). Cellular organisation of the *Arabidopsis thaliana* root. *Development, 119*, 71–84.

Dubrovsky, J. G., Napsucialy-Mendivil, S., Duclercq, J., Cheng, Y., Shishkova, S., Ivanchenko, M. G., et al. (2011). Auxin minimum defines a developmental window for lateral root initiation. *The New Phytologist, 191*, 970–983.

Ercoli, M. F., Ferela, A., Debernardi, J. M., Perrone, A. P., Rodriguez, R. E., & Palatnik, J. F. (2018). GIF transcriptional coregulators control root meristem homeostasis. *The Plant Cell, 30*, 347–359.

Faulkner, C. (2013). Receptor-mediated signaling at plasmodesmata. *Frontiers in Plant Science, 4*, 1–6.

Friml, J., Vieten, A., Sauer, M., Weijers, D., Schwarz, H., Hamann, T., et al. (2003). Efflux-dependent auxin gradients establish the apical-basal axis of Arabidopsis. *Nature, 426*, 147–153.

Galinha, C., Hofhuis, H., Luijten, M., Willemsen, V., Blilou, I., Heidstra, R., et al. (2007). PLETHORA proteins as dose-dependent master regulators of Arabidopsis root development. *Nature, 449*, 1053–1057.

Grieneisen, V. A., Xu, J., Marée, A. F. M., Hogeweg, P., & Scheres, B. (2007). Auxin transport is sufficient to generate a maximum and gradient guiding root growth. *Nature, 449*, 1008–1013.

Haecker, A. (2004). Expression dynamics of WOX genes mark cell fate decisions during early embryonic patterning in *Arabidopsis thaliana. Development, 131*, 657–668.

Hardtke, C. S., & Berleth, T. (1998). The Arabidopsis gene MONOPTEROS encodes a transcription factor mediating embryo axis formation and vascular development. *The EMBO Journal, 17*, 1405–1411.

He, W., Brumos, J., Li, H., Ji, Y., Ke, M., Gong, X., et al. (2011). A small-molecule screen identifies l-kynurenine as a competitive inhibitor of TAA1/TAR activity in ethylene-directed auxin biosynthesis and root growth in Arabidopsis. *The Plant Cell, 23*, 3944–3960.

Horiguchi, G., Kim, G.-T., & Tsukaya, H. (2005). The transcription factor AtGRF5 and the transcription coactivator AN3 regulate cell proliferation in leaf primordia of *Arabidopsis thaliana. The Plant Journal, 43*, 68–78.

Jean-Baptiste, K., McFaline-Figueroa, J. L., Alexandre, C. M., Dorrity, M. W., Saunders, L., Bubb, K. L., et al. (2019). Dynamics of gene expression in single root cells of *Arabidopsis thaliana. Plant Cell, 31*, 993–1011.

Kamiya, Y. (2010). Plant hormones: Versatile regulators of plant growth and development. *Annual Review of Plant Biology, 61*, 301–307.

Kepinski, S., & Leyser, O. (2005). The Arabidopsis F-box protein TIR1 is an auxin receptor. *Nature, 435*, 446–451.

Kim, J. H., Choi, D., & Kende, H. (2003). The AtGRF family of putative transcription factors is involved in leaf and cotyledon growth in Arabidopsis. *The Plant Journal: For Cell and Molecular Biology, 36*, 94–104.

Kim, J. H., & Kende, H. (2004). A transcriptional coactivator, AtGIF1, is involved in regulating leaf growth and morphology in Arabidopsis. *Proceedings of the National Academy of Sciences of the United States of America, 101*, 13374–13379.

Lanctot, A., Taylor-Teeples, M., Oki, E. A., & Nemhauser, J. L. (2019). *Specificity in auxin responses is not explained by the promoter preferences of activator ARFs* (p. 843391). bioRxiv.

Lewis, D. R., Olex, A. L., Lundy, S. R., Turkett, W. H., Fetrow, J. S., & Muday, G. K. (2013). A kinetic analysis of the auxin transcriptome reveals cell wall remodeling proteins that modulate lateral root development in Arabidopsis. *The Plant Cell, 25*, 3329–3346.

Leyser, O. (2006). Dynamic integration of auxin transport and signalling. *Current Biology, 16*, 424–433.

Leyser, O. (2018). Auxin signaling. *Plant Physiology, 176*, 465–479.

Liao, C. Y., Smet, W., Brunoud, G., Yoshida, S., Vernoux, T., & Weijers, D. (2015). Reporters for sensitive and quantitative measurement of auxin response. *Nature Methods, 12*, 207–210. 202 p following 210.

Mähönen, A. P., Tusscher, K. T., Siligato, R., Smetana, O., Díaz-Triviño, S., Salojärvi, J., et al. (2014). PLETHORA gradient formation mechanism separates auxin responses. *Nature, 515*, 125–129.

Marhava, P., Bassukas, L., Zourelidou, M., Kolb, M., Moret, B., Fastner, A., et al. (2018). A molecular rheostat adjusts auxin flux to promote root protophloem differentiation. *Nature, 558*, 297–300.

Marhava, P., Hoermayer, L., Yoshida, S., Marhavý, P., Benková, E., & Friml, J. (2019). Re-activation of stem cell pathways for pattern restoration in plant wound healing. *Cell*, *177*, 957–969.
Matsuzaki, Y., Ogawa-Ohnishi, M., Mori, A., & Matsubayashi, Y. (2010). Secreted peptide signals required for maintenance of root stem cell niche in Arabidopsis. *Science*, *329*, 1065–1067.
Mironova, V. V., Omelyanchuk, N. A., Wiebe, D. S., & Levitsky, V. G. (2014). Computational analysis of auxin responsive elements in the *Arabidopsis thaliana* L. genome. *BMC Genomics*, *15*(Suppl. 12), S4.
Mironova, V., & Xu, J. (2019). A single-cell view of tissue regeneration in plants. *Current Opinion in Plant Biology*, *52*, 149–154.
Miyashima, S., Roszak, P., Sevilem, I., Toyokura, K., Blob, B., Heo, J. O., et al. (2019). Mobile PEAR transcription factors integrate positional cues to prime cambial growth. *Nature*, *565*, 490–494.
Motte, H., Vanneste, S., & Beeckman, T. (2019). Molecular and environmental regulation of root development. *Annual Review of Plant Biology*, *70*, 465–488.
Nakajima, K., Sena, G., Nawy, T., & Benfey, P. N. (2001). Intercellular movement of the putative transcription factor SHR in root patterning. *Nature*, *413*, 307–311.
Nelissen, H., Eeckhout, D., Demuynck, K., Persiau, G., Walton, A., van Bel, M., et al. (2015). Dynamic changes in ANGUSTIFOLIA3 complex composition reveal a growth regulatory mechanism in the maize leaf. *The Plant Cell*, *27*, 1605–1619.
Olsson, V., Joos, L., Zhu, S., Gevaert, K., Butenko, M. A., & De Smet, I. (2019). Look closely, the beautiful may be small: Precursor-derived peptides in plants. *Annual Review of Plant Biology*, *70*, 153–186.
O'Malley, R. C., Huang, S. C., Song, L., Lewsey, M. G., Bartlett, A., Nery, J. R., et al. (2016). Cistrome and epicistrome features shape the regulatory DNA landscape. *Cell*, *166*, 1598.
Overvoorde, P., Fukaki, H., & Beeckman, T. (2010). Auxin control of root development. *Cold Spring Harbor Perspectives in Biology*, *2*, a001537.
Petersson, S. V., Johansson, A. I., Kowalczyk, M., Makoveychuk, A., Wang, J. Y., Moritz, T., et al. (2009). An auxin gradient and maximum in the Arabidopsis root apex shown by high-resolution cell-specific analysis of IAA distribution and synthesis. *The Plant Cell*, *21*, 1659–1668.
Pi, L., Aichinger, E., van der Graaff, E., Llavata-Peris, C. I., Weijers, D., Hennig, L., et al. (2015). Organizer-derived WOX5 signal maintains root columella stem cells through chromatin-mediated repression of CDF4 expression. *Developmental Cell*, *33*, 576–588.
Pierre-Jerome, E., Jang, S. S., Havens, K. A., Nemhauser, J. L., & Klavins, E. (2014). Recapitulation of the forward nuclear auxin response pathway in yeast. *Proceedings of the National Academy of Sciences of the United States of America*, *111*, 9407–9412.
Pierre-Jerome, E., Moss, B. L., Lanctot, A., Hageman, A., & Nemhauser, J. L. (2016). Functional analysis of molecular interactions in synthetic auxin response circuits. *Proceedings of the National Academy of Sciences of the United States of America*, *113*, 11354–11359.
Pierre-Jerome, E., Moss, B. L., & Nemhauser, J. L. (2013). Tuning the auxin transcriptional response. *Journal of Experimental Botany*, *64*, 2557–2563.
Rademacher, E. H., Lokerse, A. S., Schlereth, A., Llavata-Peris, C. I., Bayer, M., Kientz, M., et al. (2012). Different auxin response machineries control distinct cell fates in the early plant embryo. *Developmental Cell*, *22*, 211–222.
Rahni, R., & Birnbaum, K. D. (2018). *Week-long imaging of cell divisions in the Arabidopsis root meristem* (p. 268102). bioRxiv. Provisionally accepted at Plant Methods.
Robert, H. S., & Friml, J. (2009). Auxin and other signals on the move in plants. *Nature Chemical Biology*, *5*, 325–332.

Rodriguez, R. E., Ercoli, M. F., Debernardi, J. M., Breakfield, N. W., Mecchia, M. A., Sabatini, M., et al. (2015). MicroRNA miR396 regulates the switch between stem cells and transit-amplifying cells in Arabidopsis roots. *The Plant Cell*, *27*, 3354–3366.

Roosjen, M., Paque, S., & Weijers, D. (2018). Auxin response factors: Output control in auxin biology. *Journal of Experimental Botany*, *69*, 179–188.

Ryu, K. H., Huang, L., Kang, H. M., & Schiefelbein, J. (2019). Single-cell RNA sequencing resolves molecular relationships among individual plant cells. *Plant Physiology*, *179*, 1444–1456.

Sabatini, S., Beis, D., Wolkenfelt, H., Murfett, J., Guilfoyle, T., Malamy, J., et al. (1999). An auxin-dependent distal organizer of pattern and polarity in the Arabidopsis root. *Cell*, *99*, 463–472.

Sager, R. E., & Lee, J.-Y. (2018). Plasmodesmata at a glance. *Journal of Cell Science*, *131*, jcs209346-jcs209346.

Santuari, L., Sanchez-Perez, G. F., Luijten, M., Rutjens, B., Terpstra, I., Berke, L., et al. (2016). The PLETHORA gene regulatory network guides growth and cell differentiation in Arabidopsis roots. *The Plant Cell*, *28*, 2937–2951.

Santuari, L., Scacchi, E., Rodriguez-Villalon, A., Salinas, P., Dohmann, E. M. N., Brunoud, G., et al. (2011). Positional information by differential endocytosis splits auxin response to drive Arabidopsis root meristem growth. *Current Biology*, *21*, 1918–1923.

Sarkar, A. K., Luijten, M., Miyashima, S., Lenhard, M., Hashimoto, T., Nakajima, K., et al. (2007). Conserved factors regulate signalling in *Arabidopsis thaliana* shoot and root stem cell organizers. *Nature*, *446*, 811–814.

Sato, A., & Yamamoto, K. T. (2008). Overexpression of the non-canonical Aux/IAA genes causes auxin-related aberrant phenotypes in Arabidopsis. *Physiologia Plantarum*, *133*, 397–405.

Schlereth, A., Moller, B., Liu, W., Kientz, M., Flipse, J., Rademacher, E. H., et al. (2010). MONOPTEROS controls embryonic root initiation by regulating a mobile transcription factor. *Nature*, *464*, 913–916.

Sena, G., Wang, X., Liu, H. Y., Hofhuis, H., & Birnbaum, K. D. (2009). Organ regeneration does not require a functional stem cell niche in plants. *Nature*, *457*, 1150–1153.

Shin, R., Burch, A. Y., Huppert, K. A., Tiwari, S. B., Murphy, A. S., Guilfoyle, T. J., et al. (2007). The Arabidopsis transcription factor MYB77 modulates auxin signal transduction. *Plant Cell*, *19*, 2440–2453.

Shinohara, H., Mori, A., Yasue, N., Sumida, K., & Matsubayashi, Y. (2016). Identification of three LRR-RKs involved in perception of root meristem growth factor in Arabidopsis. *Proceedings of the National Academy of Sciences of the United States of America*, *113*, 3897–3902.

Sivaguru, M., Fujiwara, T., Šamaj, J., Baluška, F., Yang, Z., Osawa, H., et al. (2000). Aluminum-induced 1→3-β-d-glucan inhibits cell-to-cell trafficking of molecules through plasmodesmata. A new mechanism of aluminum toxicity in plants. *Plant Physiology*, *124*, 991–1006.

Skoog, F., & Miller, C. O. (1957). Chemical regulation of growth and organ formation in plant tissues cultured in vitro. *Symposia of the Society for Experimental Biology*, *54*, 118–130.

Skopelitis, D. S., Hill, K., Klesen, S., Marco, C. F., von Born, P., Chitwood, D. H., et al. (2018). Gating of miRNA movement at defined cell-cell interfaces governs their impact as positional signals. *Nature Communications*, *9*, 1–14.

Sorefan, K., Girin, T., Liljegren, S. J., Ljung, K., Robles, P., Galvan-Ampudia, C. S., et al. (2009). A regulated auxin minimum is required for seed dispersal in Arabidopsis. *Nature*, *459*, 583–586.

Stahl, Y., Grabowski, S., Bleckmann, A., Kühnemuth, R., Weidtkamp-Peters, S., Pinto, K. G., et al. (2013). Moderation of Arabidopsis root stemness by CLAVATA1 and ARABIDOPSIS CRINKLY4 receptor kinase complexes. *Current Biology*, *23*, 362–371.

Stepanova, A. N., Robertson-Hoyt, J., Yun, J., Benavente, L. M., Xie, D.-Y., Doležal, K., et al. (2008). TAA1-mediated auxin biosynthesis is essential for hormone crosstalk and plant development. *Cell, 133*, 177–191.

Stewart, J. L., & Nemhauser, J. L. (2010). Do trees grow on money? Auxin as the currency of the cellular economy. *Cold Spring Harbor Perspectives in Biology, 2*, 1–14.

Swarup, R., & Bennett, M. (2003). Auxin transport: The fountain of life in plants? *Developmental Cell, 5*, 824–826.

Szemenyei, H., Hannon, M., & Long, J. A. (2008). TOPLESS mediates auxin-dependent transcriptional repression during Arabidopsis embryogenesis. *Science, 319*, 1384–1386.

Ulmasov, T., Hagen, G., & Guilfoyle, T. J. (1997). ARF1, a transcription factor that binds to auxin response elements. *Science, 276*, 1865–1868.

Ulmasov, T., Murfett, J., Hagen, G., & Guilfoyle, T. J. (1997). Aux/IAA proteins repress expression of reporter genes containing natural and highly active synthetic auxin response elements. *Plant Cell, 9*, 1963–1971.

Vaddepalli, P., Herrmann, A., Fulton, L., Oelschner, M., Hillmer, S., Stratil, T. F., et al. (2014). The C2-domain protein QUIRKY and the receptor-like kinase STRUBBELIG localize to plasmodesmata and mediate tissue morphogenesis in *Arabidopsis thaliana*. *Development, 141*, 4139–4148.

van Berkel, K., de Boer, R. J., Scheres, B., & ten Tusscher, K. (2013). Polar auxin transport: Models and mechanisms. *Development, 140*, 2253–2268.

van den Berg, C., Willemsen, V., Hendriks, G., Weisbeek, P., & Scheres, B. (1997). Short-range control of cell differentiation in the Arabidopsis root meristem. *Nature, 390*, 287–289.

Varaud, E., Brioudes, F., Szecsi, J., Leroux, J., Brown, S., Perrot-Rechenmann, C., et al. (2011). AUXIN RESPONSE FACTOR8 regulates Arabidopsis petal growth by interacting with the bHLH transcription factor BIGPETALp. *Plant Cell, 23*, 973–983.

Vatén, A., Dettmer, J., Wu, S., Stierhof, Y. D., Miyashima, S., Yadav, S. R., et al. (2011). Callose biosynthesis regulates symplastic trafficking during root development. *Developmental Cell, 21*, 1144–1155.

Vernoux, T., Brunoud, G., Farcot, E., Morin, V., Van den Daele, H., Legrand, J., et al. (2011). The auxin signalling network translates dynamic input into robust patterning at the shoot apex. *Molecular Systems Biology, 7*, 508.

Wang, R., & Estelle, M. (2014). Diversity and specificity: Auxin perception and signaling through the TIR1/AFB pathway. *Current Opinion in Plant Biology, 21*, 51–58.

Wang, Q., Kohlen, W., Rossmann, S., Vernoux, T., & Theres, K. (2014). Auxin depletion from the leaf axil conditions competence for axillary meristem formation in Arabidopsis and tomato. *Plant Cell, 26*, 2068–2079.

Wang, W., Ryu, K. H., Barron, C., & Schiefelbein, J. (2019). Root epidermal cell patterning is modulated by a critical residue in the WEREWOLF transcription factor. *Plant Physiology, 181*, 1239–1256.

Weijers, D., Nemhauser, J., & Yang, Z. (2018). Auxin: Small molecule, big impact. *Journal of Experimental Botany, 69*, 133–136.

Wu, M. F., Yamaguchi, N., Xiao, J., Bargmann, B., Estelle, M., Sang, Y., et al. (2015). Auxin-regulated chromatin switch directs acquisition of flower primordium founder fate. *eLife, 4*, e09269.

Yadav, R. K., Perales, M., Gruel, J., Girke, T., Jonsson, H., & Reddy, G. V. (2011). WUSCHEL protein movement mediates stem cell homeostasis in the Arabidopsis shoot apex. *Genes & Development, 25*, 2025–2030.

Yamada, M., Han, X., & Benfey, P. N. (2019). RGF1 controls root meristem size through ROS signalling. *Nature*. https://www.nature.com/articles/s41586-019-1819-6.pdf.

Yang, Z.-B., Geng, X., He, C., Zhang, F., Wang, R., Horst, W. J., et al. (2014). TAA1-regulated local auxin biosynthesis in the root-apex transition zone mediates the aluminum-induced inhibition of root growth in Arabidopsis. *The Plant Cell, 26*, 2889–2904.

Zemlyanskaya, E. V., Wiebe, D. S., Omelyanchuk, N. A., Levitsky, V. G., & Mironova, V. V. (2016). Meta-analysis of transcriptome data identified TGTCNN motif variants associated with the response to plant hormone auxin in *Arabidopsis thaliana* L. *Journal of Bioinformatics and Computational Biology*, *14*, 1641009.

Zhang, T. Q., Xu, Z. G., Shang, G. D., & Wang, J. W. (2019). A single-cell RNA sequencing profiles the developmental landscape of Arabidopsis root. *Molecular Plant*, *12*, 648–660.

Zhao, Y. (2012). Auxin biosynthesis: A simple two-step pathway converts tryptophan to indole-3-acetic acid in plants. *Molecular Plant*, *5*, 334–338.

Further reading

Brady, S. M., Orlando, D. A., Lee, J. Y., Wang, J. Y., Koch, J., Dinneny, J. R., et al. (2007). A high-resolution root spatiotemporal map reveals dominant expression patterns. *Science*, *318*, 801–806.

CHAPTER FIFTEEN

Small RNAs as plant morphogens

Simon Klesen, Kristine Hill, Marja C.P. Timmermans*
Center for Plant Molecular Biology, University of Tübingen, Tübingen, Germany
*Corresponding author: e-mail address: marja.timmermans@zmbp.uni-tuebingen.de

Contents

Abstract

The coordination of cell fate decisions within complex multicellular structures rests on intercellular communication. To generate ordered patterns, cells need to know their relative positions within the growing structure. This is commonly achieved via the production and perception of mobile signaling molecules. In animal systems, such positional signals often act as morphogens and subdivide a field of cells into domains of discrete cell identities using a threshold-based readout of their mobility gradient. Reflecting the independent origin of multicellularity, plants evolved distinct signaling mechanisms to drive cell fate decisions. Many of the basic principles underlying developmental patterning are, however, shared between animals and plants, including the use of signaling gradients to provide positional information. In plant development, small RNAs can act as mobile instructive signals, and similar to classical morphogens in animals, employ a threshold-based readout of their mobility gradient to generate precisely defined cell fate boundaries. Given the distinctive nature of peptide morphogens and small RNAs,

Current Topics in Developmental Biology, Volume 137
ISSN 0070-2153
https://doi.org/10.1016/bs.ctdb.2019.11.001

how might mechanisms underlying the function of traditionally morphogens be adapted to create morphogen-like behavior using small RNAs? In this review, we highlight the contributions of mobile small RNAs to pattern formation in plants and summarize recent studies that have advanced our understanding regarding the formation, stability, and interpretation of small RNA gradients.

1. Introduction

A fundamental principle of development is the ability to generate complex multicellular structures starting from a single cell. To coordinate the many cell fate decisions that characterize development of multicellular organisms, cells must be able to assess their relative positions within the growing structure. This is achieved through intercellular communication, commonly based on the production and perception of mobile signaling molecules. In animal systems, much of development is dependent upon morphogens, a term first coined by Turing (1952) to describe 'diffusible form-generating substances.' To qualify as a morphogen, mobile signals must fulfill the following two criteria: it must trigger discrete cell fate decisions through a dose-dependent, threshold-based readout of its concentration gradient, and act directly on target cells rather than using a relay of intermediary signals. These principles, first described by Lewis Wolpert in his now 50 years old 'French flag model' (Wolpert, 1969), explain how a field of cells can be subdivided into discrete domains of different fates according to the relative position from the morphogen source.

Plants evolved multicellularity independently of animals, and accordingly utilize distinct mobile signals to drive cell fate decisions. Nevertheless, basic principles underlying developmental patterning are often shared between animals and plants, including the use of signaling gradients to provide positional information. A classic example of a gradient-regulated response is the directional growth of a plant shoot toward a source of light (Darwin & Darwin, 1880; Wiesner, 1878). This response is mediated by the phytohormone auxin, which is actively exported by polarized efflux carriers termed PIN proteins, to form a concentration gradient that is highest on the side furthest from the light (see Fankhauser & Christie, 2015; Finet & Jaillais, 2012). The gradient-readout is mediated by auxin-response factors that promote growth toward the light source. Graded auxin levels are also instructive in other patterning contexts, e.g., the organization of the growing root tip (Blilou et al., 2005; Friml, 2003;

Grieneisen, Scheres, Hogeweg, & Marée, 2012; Grieneisen, Xu, Marée, Hogeweg, & Scheres, 2007), but the status of auxin as a bona fide morphogen has remained controversial, in part because auxin action is not dictated by its absolute levels, but instead appears to sense a differential (Benková, Ivanchenko, Friml, Shishkova, & Dubrovsky, 2009; Finet & Jaillais, 2012; Smith et al., 2006). In fact, the existence of morphogens in plants in general was a topic of much debate (e.g., Benková et al., 2009; Bhalerao & Bennett, 2003; Friml, 2003; Grieneisen et al., 2012). We now know that small RNAs can act as mobile instructive signals in plant development, and similar to morphogens in animal systems, employ a threshold-based readout of their mobility gradient to generate precisely defined cell fate boundaries. Here, we summarize the contributions of mobile small RNAs to plant development and discuss recent studies that have advanced our understanding of how small RNA gradients could be created, maintained, and interpreted.

2. Small RNAs as mobile instructive signals

The idea that small RNAs might act non-cell autonomously was posited over 20 years ago. Grafting experiments showed that transgene-induced gene silencing in tobacco produces a sequence-specific silencing signal that can spread from a silenced rootstock into a non-silenced shoot scion (Palauqui, Elmayan, Pollien, & Vaucheret, 1997). Although small RNAs quickly emerged as candidates for this signal, formal proof to that effect was less easily attained (see Chitwood & Timmermans, 2010). We now know that short interfering RNAs (siRNAs) move from cell to cell via plasmodesmata (microchannels that connect adjacent plant cells), as well as long distance through the vasculature (Buhtz, Pieritz, Springer, & Kehr, 2010; Buhtz, Springer, Chappell, Baulcombe, & Kehr, 2008; Molnar et al., 2010; Pant, Buhtz, Kehr, & Scheible, 2008; Vatén et al., 2011; Yoo et al., 2004). Particularly, the ability of siRNAs to trigger the production of secondary siRNAs, a process called transitivity that relies on the activities of RNA-dependent RNA polymerase 6 (RDR6) and the DICER proteins DCL4 and DCL2, enables siRNAs to propagate the spread of silencing from a single leaf systemically throughout the plant, thereby forming an important component of siRNA-based plant immunity (see Borges & Martienssen, 2015).

In contrast to siRNAs, plant miRNAs were initially reported to behave cell autonomously (Alvarez et al., 2006; Parizotto, Dunoyer, Rahm,

Himber, & Voinnet, 2004). Indeed, with few exceptions, miRNAs do not trigger transitivity (Allen, Xie, Gustafson, & Carrington, 2005; Manavella, Koenig, & Weigel, 2012; Montgomery et al., 2008). However, not only are miRNAs transported through the vascular phloem to coordinate physiological responses between the shoot and root (Buhtz et al., 2010; Lin et al., 2008; Pant et al., 2008; Tsikou et al., 2018), miRNAs can move from cell to cell and act as short-range positional signals in development. For instance, several observations hint to mobile small RNAs as important factors in the reproductive development of *Arabidopsis,* contributing to regulation of genome dosage, epigenetic reprogramming in the male and female germ cells, and to megasporogenesis (Borges et al., 2018; Olmedo-Monfil et al., 2010; Slotkin et al., 2009; Su et al., 2017; Tucker et al., 2012). In addition, the examples described below provide conclusive evidence that small RNAs act as mobile instructive signals in various developmental contexts. It should in this regard be noted that plant miRNAs, unlike their animal counterparts, show high target specificity, typically regulating transcripts derived from closely-related members of just a single gene family to which they show near perfect complementarity (see Willmann & Poethig, 2007; Yu, Jia, & Chen, 2017). The high complementarity between miRNAs and targets in plants also allows for target repression via both transcript cleavage as well as translational repression (Fig. 1A).

2.1 Opposing gradients of mobile small RNAs establish leaf polarity

Flat leaves with distinct cell types on their dorsal/adaxial (top) and ventral/abaxial (bottom) faces are an important innovation in the evolution of land plants that serves to maximize photosynthesis while minimizing water loss to the environment. Development of flat leaf architecture also poses a mechanistically challenging problem; namely, how to create a stable dorsoventral boundary within the plane of a long and wide, but shallow, structure. The acquisition and maintenance of dorsoventral polarity involves an intricate gene regulatory network with several highly conserved transcription factors that promote either dorsal or ventral fate at its core (see Kuhlemeier & Timmermans, 2016). These are expressed in complementary domains delineating the top and bottom side of the developing primordium, respectively (Caggiano et al., 2017; Husbands, Benkovics, Nogueira, Lodha, & Timmermans, 2015). The positional information needed to define these domains is provided in part by small RNAs.

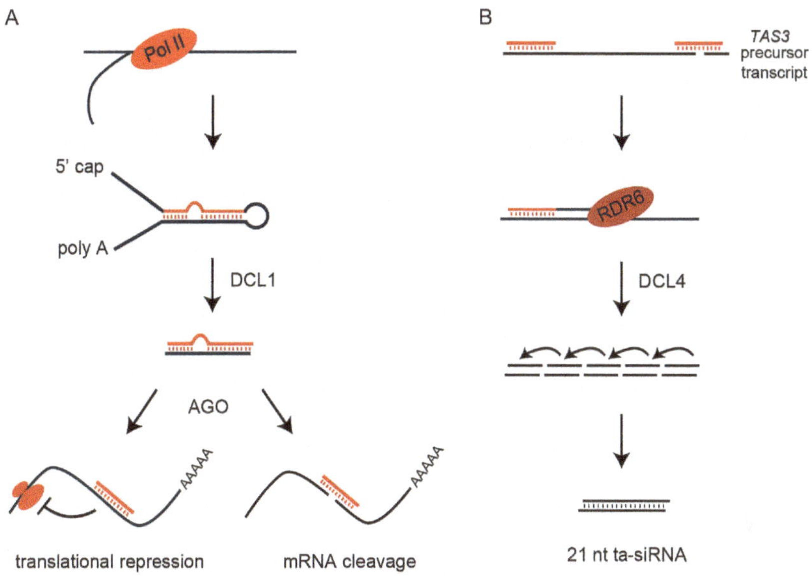

Fig. 1 Pathways for the biogenesis of developmentally-important small RNAs in plants. (A) miRNA precursors transcribed by RNA polymerase II (Pol II) adopt a hairpin-like structure that is processed by DICER-LIKE 1 (DCL1) into a miRNA duplex typically 21 bp in length. The mature miRNA is loaded into an AGO effector complex that in a homology-dependent manner guides the post-transcriptional repression of target mRNAs. Plant miRNAs, unlike their animal counterparts, show near perfect complementarity to target transcripts and direct their cleavage as well as translational repression. (B) A select subset of miRNAs can trigger the production of secondary siRNAs. In the production of tasiARF, *TAS3* precursor transcripts are targeted at two sites by miR390-loaded AGO7, triggering cleavage at the 3′ site. The 5′ cleavage product is subsequently converted into double-stranded RNA by RDR6, and processed into phased 21 nucleotides ta-siRNAs by DCL4. Among the *TAS3*-derived ta-siRNAs, tasiARF represses expression of ARF3/4 in a manner similar to miRNAs. Note: while transitivity is important for tasiARF biogenesis, this amplification mechanism is dispensable for the morphogenic activity of small RNAs.

miR166 contributes to organ polarity by restricting the accumulation of class III HOMEODOMAIN-LEUCIN ZIPPER (HD-ZIPIII) transcription factors, key determinants of dorsal cell fate (Fig. 2A) (Emery et al., 2003; Juarez, Kui, Thomas, Heller, & Timmermans, 2004; Mallory et al., 2004: McConnell et al., 2001). Plants in which this regulatory interaction is perturbed develop a strong radial dorsalized leaf phenotype, reflecting an early role for miR166 in setting up dorsoventral polarity. miR166 is generated specifically in the ventral epidermis of leaf primordia, but was shown to move from the epidermis across the leaf to form a concentration gradient

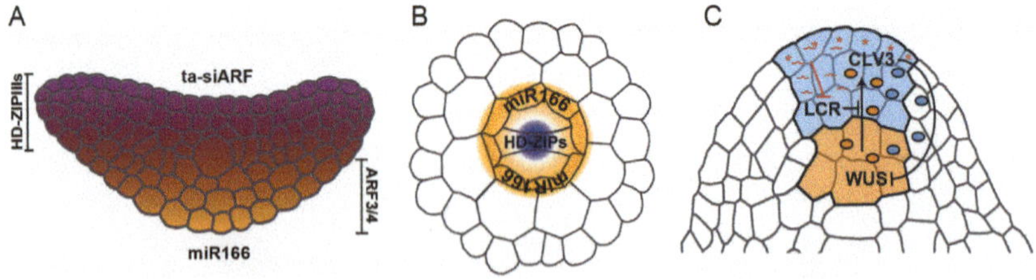

Fig. 2 Mobile small RNA gradients play important roles in plant development. (A) In the developing leaf, tasiARF (purple) and miR166 (orange) form opposing concentration gradients through movement from a source in the top and bottom epidermis, respectively. These gradients generate a morphogen-like, threshold-based readout that limits expression of the respective HD-ZIPIII and ARF3/4 targets (transcribed throughout the developing primordium) to sharply defined domains on the top and bottom side of the developing leaf, respectively. (B) Proper patterning of the root vasculature requires that miR166 (orange) moves from its source in the endodermis to generate a concentration gradient across the central stele that is readout into an inverse gradient of HD-ZIPIII activity (blue) to specify proto- and metaxylem cell fate. (C) Stem cell activity in the shoot apical meristem is maintained via a negative feedback loop in which the WUS transcription factor moves from the organizing center (orange) to induce CLV3 expression in the central zone (blue). This secreted peptide in turn signals a down-regulation of WUS expression, restricting its activity. miR394 (red squiggle) produced in the central zone epidermis (red asterisks) moves into the subjacent two cell layers where it down-regulates its F-box target, LCR, enabling these cells to respond to WUS activity. Polarization of factors that block miRNA mobility to defined cell–cell interfaces creates domains with confined miRNA mobility (black outline), safeguarding cell identities within the stem cell niche.

that dissipates toward the dorsal side (Juarez et al., 2004; Nogueira et al., 2009; Yao et al., 2009). Interestingly this gradient is interpreted into a binary readout and creates a sharply delineated domain of HD-ZIPIII expression that is limited to the two uppermost layers of developing leaf primordia (Skopelitis, Benkovics, Husbands, & Timmermans, 2017).

Maintenance of dorsoventral polarity relies on an additional small RNA gradient formed by tasiARF (Chitwood et al., 2009; Nagasaki et al., 2007; Nogueira, Madi, Chitwood, Juarez, & Timmermans, 2007; Petsch et al., 2015; Yifhar et al., 2012). Biogenesis of this small RNA occurs through the specialized *TAS3* trans-acting siRNA pathway (Fig. 1B), which in leaf primordia is active exclusively on the upper surface (Allen et al., 2005; Chitwood et al., 2009; Montgomery et al., 2008). Similar to miR166, movement of tasiARF from its defined source of biogenesis creates a concentration gradient across the leaf that is read out into a discrete expression domain of its targets, the ventral determinants ARF3 and ARF4, on the bottom side (Fig. 2A) (Chitwood et al., 2009; Skopelitis et al., 2017).

The division of leaf primordia into distinct dorsal and ventral domains thus relies on a novel developmental patterning mechanism in which tasiARF and miR166 form inverse mobility gradients that are read out as the on-off expression of the direct targets, *ARF3/4* and *HD-ZIPIII* genes, respectively.

Given that flat-leaf architecture requires a correctly positioned dorsoventral boundary, the activities of the miR166 and tasiARF gradients must be coordinated and carefully tuned. How the spatiotemporal patterns of miR166 and tasiARF precursor expression are first established remains to be resolved, but once initiated, the larger polarity network directly reinforces this expression (Husbands et al., 2015; Merelo et al., 2016; Nogueira et al., 2007). Integration of the positional information contained within opposing tasiARF and miR166 gradients underlies specification of a stable, uniformly positioned dorsoventral boundary. Thus, the cell-to-cell movement of small RNAs allows the formation of sharply defined target gene expression boundaries, and together, the integrated readouts of opposing small RNA gradients provide a mechanism to specify a robust developmental boundary (Skopelitis et al., 2017).

2.2 A miR166 mobility gradient specifies cell fate within the root

miR166 also serves as a short-range positional signal in patterning of the root, which consists of a central vascular stele surrounded by concentric layers of pericycle, endodermis, cortex, and epidermis (Carlsbecker et al., 2010; Miyashima, Koi, Hashimoto, & Nakajima, 2011). Within the vasculature, the water conducting xylem tissue comprises two cell types, the outer protoxylem and inner metaxylem. These cell fates are specified according to the level of HD-ZIPIII activity determined by the dose-dependent readout of a miR166 mobility gradient that has its source at the endodermis (Fig. 2B) (Carlsbecker et al., 2010; Miyashima et al., 2011). The basis for the endodermal-specific production of miR166 lies in the reciprocal movement of the transcription factor SHORT ROOT (SHR) from the central stele into the adjacent endodermis (Cui et al., 2007; Gallagher, Paquette, Nakajima, & Benfey, 2004; Nakajima, Sena, Nawy, & Benfey, 2001). Here, it captures its interaction partner SCARECROW (SCR), and together activates miR166 expression. Movement of miR166 out of the endodermis also affects positioning of cortex and pericycle (Miyashima et al., 2011). Thus, as for dorsoventral patterning of the leaf, acquisition of discrete cell fates along the radial axis of the root is in part driven by the dose-dependent readout of a small RNA gradient. However, whereas the miR166 and tasiARF gradients in the leaf generate an on-off switch in target gene expression, the miR166 gradient in the root sets up an inverse gradient of its HD-ZIPIII targets, which then drives the acquisition of discrete cell fates.

2.3 miR394 mobility delineates the embryonic shoot stem cell niche

A third example in which cell fate decisions are governed by mobility of a miRNA is found in the regulation of stem cell activity within the embryonic *Arabidopsis* shoot meristem. Plant meristems are specialized niches that orchestrate the balance between stem cell proliferation and organ initiation essential for post-embryonic growth (see Greb & Lohmann, 2016). Stem cells within the shoot apical niche are located within the central zone (CZ) positioned at the meristem tip. This spatial organization is stably maintained, despite ongoing cell divisions. Two opposing signaling centers provide relevant positional cues to maintain stem cell number and position within the growing niche. The organizing center (OC), positioned directly below the stem cells, expresses the homeodomain transcription factor WUSCHEL (WUS), which moves into the CZ where it promotes stem cell identity and activates CLV3 expression (Fig. 2C) (Daum, Medzihradszky, Suzaki, & Lohmann, 2014; Yadav et al., 2011). This secreted peptide in turn signals a downregulation in *WUS* expression, thus establishing a negative feedback loop that maintains WUS levels and thereby stem cell number (see Soyars, James, & Nimchuk, 2016).

In addition to this regulatory loop, classical surgical experiments predicted the need for an epidermal-derived signal in maintaining stem cell activity (see Reinhardt, Frenz, Mandel, & Kuhlemeier, 2005; Steeves & Sussex, 1989). This signal, we now know, involves miR394 (Knauer et al., 2013). miR394 is generated in the surface layer, or protoderm, of the developing embryo, but moves into the subtending two cell layers where it represses expression of the F-box protein, LEAF CURLING RESPONSIVENESS (LCR) (Knauer et al., 2013). As a result, these cells become competent to respond to the stem cell promoting activity of WUS. The limited mobility of miR394 thus defines a zone of stem cell activity and given that the protoderm is propagated by stereotypic anticlinal cell divisions, in addition, stably anchors this zone to the growing shoot tip.

3. Reading out the gradient

3.1 Patterning properties of small RNA gradients are developmental-context dependent

In each of the above examples, the short-range mobility of a small RNA provides positional information essential for the specification of a critical cell fate boundary. The exact manner with which the mobility-derived small RNA gradients pattern their targets, however, appears context dependent. Whereas the movement of miR166 in the root generates an inverse gradient of HD-ZIPIII activity, the opposing tasiARF and miR166 gradients in the leaf create a sharp on-off switch in expression of their respective targets (Carlsbecker et al., 2010; Miyashima et al., 2011; Skopelitis et al., 2017). Why the patterning properties of miR166 in the root and shoot are distinct, even though the targets and range of mobility seem the same, is not currently understood. The readout of a small RNA gradient could be tuned by the gene regulatory network into which it is integrated (see Briscoe & Small, 2015; Cotterell & Sharpe, 2010; Rogers & Schier, 2011). Alternatively, parameters affecting small RNA-target interaction may provide inputs to shape the gradient readout and potentially force it from being inversely graded to being binary or vice versa. For example, AGO10, which specifically binds miR166, is thought to act as a decoy that prevents miR166 from being loaded into a catalytically active AGO1 complex (Zhang & Zhang, 2012; Zhu et al., 2011). As AGO10 shows distinctive tissue specific patterns of expression across the plant (Lynn et al., 1999), it is easy to envision how the readout of the miR166 gradient could be tuned across tissues via pre-patterning at the effector level.

While feedback regulation and pre-patterning are common features of developmental patterning, it was recently shown that these regulatory mechanisms are not essential for the conversion of a small RNA gradient into an on-off boundary of target gene expression. Inversion of the tasiARF and miR166 gradients, by displacing their source to the opposite side of the leaf, inverts their readout (Skopelitis et al., 2017). Irrespective of gradient orientation, a binary switch in target gene expression was observed. The information needed to convert a small RNA gradient into discrete domains of target gene expression must thus be contained within the gradient itself. Moreover, the patterning behaviors of small RNA gradients could be recapitulated in an *rdr6* mutant background, ruling out a contribution of transitivity, as well as in a fluorescence-based synthetic miRNA sensor system (Skopelitis et al., 2017). The latter indicates that the readout of a small RNA mobility gradient relies solely on properties captured in the small RNA-target-AGO interaction. Similar to morphogens in animal systems, plant small RNAs thus have the inherent capacity to generate sharp boundaries of target gene expression through a direct threshold-based readout of their mobility gradients.

3.2 Making the switch

The recognition that small RNAs can function as morphogens raises a number of interesting questions, especially given that their nature is distinct from classical morphogens. Most animal morphogens are extracellular peptide ligands, such that cells can discern their position along the gradient based on the number of activated, ligand-bound receptors at their surface (see Ashe & Briscoe, 2006; Lewis, 2008; Rogers & Schier, 2011). This information is translated via a linear signal transduction pathway into the differential activation of target gene expression, often mediated by the cooperative binding of downstream transcription factors, in a manner depending on whether or not a given signaling intensity threshold is surpassed.

But how might a cell assess where along a small RNA gradient it is positioned to yield the appropriate expression response? Cells on either side of the target boundary can show remarkably subtle (~30%) differences in small RNA levels. Accordingly, the position along the gradient at which the switch is triggered is highly sensitive to the level of small RNA at the source (Skopelitis et al., 2017). Within the context of the developing leaf, a two- to threefold change in epidermal small RNA levels was sufficient to shift the position of the threshold one cell layer. Such sensitivity is also seen for animal morphogens (the *hunchback* gene in flies, for instance, can respond in an all or none fashion to a 10% change in Bicoid concentration), and may be fundamental to generating on-off transitions (Briscoe & Small, 2015; Gregor, Tank, Wieschaus, & Bialek, 2007; Rogers & Schier, 2011). On the other hand, target abundance also affects the readout, as is evident from the sensitivity of developmental programs in the leaf and root to variations in ARF3 or HD-ZIPIII levels (e.g., Carlsbecker et al., 2010; Chitwood et al., 2009; Fahlgren et al., 2006; McConnell et al., 2001; Miyashima et al., 2011). This pinpoints the ratio of small RNA-to-target levels as a means by which the graded information captured within a small RNA gradient is read out. At the high end of the gradient, where the small RNA-to-target ratio exceeds a certain threshold, small RNAs completely eliminate target expression

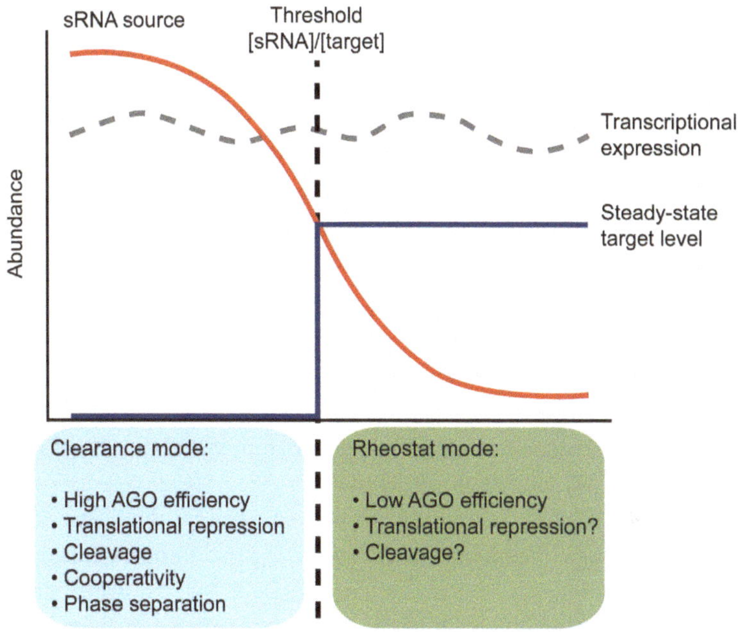

Fig. 3 Small RNA-to-target ratio creates a threshold-based readout of mobility gradients. The cell-to-cell movement of a small RNA from a defined source (left) generates a concentration gradient (red line) across a field of cells. In cells at the high end of the gradient where the small RNA-to-target ratio exceeds a given threshold (black dotted line), small RNAs completely repress target expression (clearance mode). In cells toward the tail end of the gradient where the small RNA-to-target ratio drops below the threshold, small RNAs cause target expression to be reduced and less variable, buffering stochastic fluctuations inherent in gene expression (rheostat mode). Select parameters potentially driving the threshold-based switch in AGO activity are listed. Please see the text for more details.

(Fig. 3). However, once the small RNA-to-target ratio falls below the threshold, the mode of small RNA regulation changes. At the tail end of the gradient, small RNAs no longer clear target expression, but instead cause it to be reduced and considerably less variable (Skopelitis et al., 2017). Two models for the regulation of developmental targets by small RNAs had previously been recognized. In the clearance model, miRNAs clear target transcripts to delineate mutually exclusive domains of accumulation, whereas in the homeostasis model, miRNAs act as rheostats to dampen the noise in target gene expression and refine their domains of activity (Bartel, 2004; Cartolano et al., 2007; Nikovics et al., 2006; Rhoades et al., 2002; Sieber, Wellmer, Gheyselinck, Riechmann, & Meyerowitz,

2007; Vaucheret, Mallory, & Bartel, 2006). The morphogen-like patterning properties of mobile small RNAs thus seem to reflect a highly sensitive switch from a clearance into a rheostat mode of regulation that is dictated by a small RNA-to-target ratio threshold.

A second critical question to be resolved is how the switch in small RNA activity that drives the binary readout and therefore cell fate acquisition might be realized. Switching behavior can follow from strong positive feedback directly coupled to a highly sensitive signaling input. AGO proteins are subject to a number of post-translational modifications that affect the efficacy of AGO complexes via changes in stability, conformation, and composition (see Huberdeau et al., 2017; Jee & Lai, 2014; Lopez-Orozco et al., 2015). For instance, the release of mammalian AGO2 from a miRNA-target mRNA duplex is triggered by its phosphorylation, which in turn is coupled to target loading (Golden et al., 2017). Given that AGO levels are generally limiting in the cell, the phosphorylation and subsequent dephosphorylation of AGO2 provides a timing mechanism to limit the duration of target interaction and thereby tune the overall silencing efficiency. Phosphorylation of AGO2 can also lead to a shift in its mechanism of repression from endonucleolytic cleavage toward translational repression, a point that may be particularly relevant given the properties of AGO proteins mentioned below (Horman et al., 2013). Interestingly, several of these modifications occur in response to specific internal or external cues (Cho, Ryu, Shah, Poulsen, & Yang, 2016; Shen et al., 2013; Von Born, Bernardo-Faura, & Rubio-Somoza, 2018), which could conceivably shape the output of small RNA gradients in a tissue- or state-dependent manner. Knowledge of post-translational regulatory modifications on plant AGO proteins is lacking, but many of the residues modified on animal AGO proteins are conserved also in plants.

Cooperativity provides an alternative, non-mutually exclusive, mechanism via which to create a non-linear, threshold-based response. Like many transcription factors, AGO1, 3 and 4 proteins in animal systems act cooperatively at targets containing multiple small RNA binding sites, allowing for a dose-dependent bimodal silencing response (Broderick, Salomon, Ryder, Aronin, & Zamore, 2011; Denzler et al., 2016; Djuranovic et al., 2010; Klein, Chandradoss, Depken, & Joo, 2017; Mukherji et al., 2011). AGO2, however, does not show this behavior. AGO1, 3, and 4 exert their effects primarily at the translational level, whereas AGO2 also directs the cleavage of perfectly matched small RNA targets. This may hint at a role for translational repression in generating the cooperativity-driven, non-linear behavior of animal small RNAs. Although plant miRNAs and tasiRNAs show extensive complementarity to their targets, most repress translation in addition to guiding the cleavage of target transcripts (Brodersen et al., 2008; Li et al., 2013). Particularly, both tasiARF and miR166 mediate the translational repression of their respective targets (Chitwood et al., 2009; Fahlgren et al., 2006; Li et al., 2013; Pekker, Alvarez, & Eshed, 2005). ARF3 transcripts also bear two tasiARF binding sites and can further be targeted by secondary siRNAs (Allen et al., 2005; Petsch et al., 2015; Yifhar et al., 2012). Assuming conservation in AGO function, cooperativity resulting from AGO occupancy at multiple binding sites may well contribute to the binary readout of this small RNA gradient. However, given that the HD-ZIPIII transcripts carry just a single miR166 target site (Emery et al., 2003; Juarez et al., 2004; Mallory et al., 2004), this form of cooperativity unlike explains the full morphogenic activity of mobile plant small RNAs. In this regard, the observation that AGO proteins condense into phase-separated droplets upon interaction with GW-scaffolding proteins may be particularly intriguing (Sheu-Gruttadauria & MacRae, 2018). Droplet formation was shown to enhance the silencing efficiency of miRNA-target-AGO complexes by over 10-fold, making it quite apparent how phase separation could drive a switching behavior. The recent finding that phosphorylation of RNA-polymerase II underlies a droplet-based switch from transcription initiation to RNA processing may then provide a valuable paradigm on how a signal input might be integrated into such a switch (Guo et al., 2019). Still, the major challenge in understanding the morphogenic activity of small RNAs would be to resolve how a cell assesses a given small RNA-to-target ratio to trigger droplet formation, or otherwise switch miRNA-target-AGO complex activity.

4. Generating the small RNA gradient

To act as morphogens, small RNAs must establish a stable concentration gradient across a field of cells. Exactly how the parameters of the gradient are established and maintained is so far unknown, but this almost certainly reflects a balance between: (1) the production of the small RNA at its source, (2) its rate of movement from source to sink, and (3) the degree of degradation in cells along the field.

4.1 Small RNA turnover

Two distinct degradation pathways have been recognized in plants (see Yu et al., 2017). Small RNA-degrading exonucleases direct the 3′ truncation of small RNAs (Ramachandran & Chen, 2008). In addition, small RNAs can be marked for degradation by 3′ poly-uridylation (Ren, Chen, & Yu, 2012; Zhai et al., 2013; Zhao et al., 2012). The occurrence of overlapping and opposing small RNA gradients predicts that small RNA turnover must be regulated at the level of individual small RNAs. The enzymes mediating these degradation reactions show distinct substrate specificities (Yu et al., 2017; Zhai et al., 2013), supporting the possibility that turnover rate varies between small RNAs. More importantly, small RNA-degrading exonucleases as well as 3′ uridyl-transferases act on small RNAs bound to AGO1 (Yu et al., 2017; Zhai et al., 2013), and this in a manner that appears coupled to target loading. Although the exact mechanism in plants has yet to be investigated, data from other species predicts that AGO proteins undergo a conformational change upon loading of a highly complementary target, such as is the case in plants, which releases the miRNA 3′ end and exposes this for degradation (Ameres et al., 2010; Sheng et al., 2014). Target-induced small RNA decay presents a perfect means by which to differentially regulate the turnover of individual small RNAs. As small RNA turnover would be linked to target levels, it is also easy to envision how such a decay mechanism can be used to tune the effective range and shape of a small RNA gradient across tissues or in response to specific cues. In addition, target-induced small RNA decay is predicted to refine the threshold-based readout of a small RNA gradient (Levine, McHale, & Levine, 2007). The coupling of miRNA activity and turnover may thus serve multiple functions; to shape and stabilize a small RNA gradient and to sharpen the on-off target gene expression boundary this creates.

4.2 Regulation of small RNA mobility

A second criterion critical to the establishment of a gradient is the movement of a small RNA from its source. Despite the central importance of small RNA mobility, also with respect to the coordination of biotic and abiotic stress responses across the plant (Borges & Martienssen, 2015; Buhtz et al., 2010; Lin et al., 2008; Pant et al., 2008; Tsikou et al., 2018), remarkably little is known about how the cell-to-cell movement of small RNAs is mediated, except that plasmodesmata are required (Vatén et al., 2011). The shape of miRNA gradients generated by movement from an epidermal source in the leaf is consistent with the passive diffusion of small RNAs between cells (Chitwood et al., 2009; Skopelitis et al., 2017). Accordingly, the effective range of mobility is correlated to the abundance of the small RNA at the source (Felippes, Ott, & Weigel, 2010; Skopelitis et al., 2017). Specific properties of cells across the mobility field are, however, likely to influence the range and shape of a small RNA gradient. As mentioned, expression of target transcripts in cells between source and sink is predicted to lower small RNA abundance. Likewise, as AGO proteins act cell-autonomously (Zhu et al., 2011), AGO loading is expected to limit the pool of available mobile small RNAs and reduce the length of the gradient. Given that loading into AGO complexes is a selective process (Mi et al., 2008; Montgomery et al., 2008; Zhu et al., 2011), such an effect on mobility may similarly vary between individual miRNAs. This point is nicely illustrated by AGO10, which shows a unique preference for loading of miR166 and specifically prevents its movement out of the central vasculature to maintain stem cell activity in the embryonic shoot meristem above (Liu et al., 2009; Zhu et al., 2011).

Small RNAs move from cell to cell via plasmodesmata (Vatén et al., 2011), providing a further means via which to govern the formation or shape of a small RNA gradient. Plasmodesmata also permit the movement of water, nutrients, hormones, peptides, and small proteins between adjoining cells, and larger proteins, such as transcription factors, may be transported via active processes controlled by unique domains or motifs inherent to the transported protein (see Wu & Gallagher, 2012). Understandably, transport through these channels is precisely regulated, and plasmodesmatal properties, such as density, architecture and aperture, change substantially during development or in instances of stress (see Burch-Smith & Zambryski, 2012; Tilsner, Nicolas, Rosado, & Bayer, 2016; Zavaliev, Sagi, Gera, & Epel, 2009). Interestingly, mechanisms modulating the symplastic diffusion of small proteins do not necessarily impact miRNA mobility. Nonetheless, not all cells are symplastically connected, and in select developmental contexts, mechanisms are in place to specifically limit the movement of small RNAs (Skopelitis et al., 2018). For instance, small RNAs are able to move out but not into phloem cells of the central vasculature. The observed mobility patterns predict the existence of plasmodesmata-associated 'gatekeepers' that polarize at defined cell–cell interfaces to prevent the passage of small RNAs between select cells (Skopelitis et al., 2018).

Gating of small RNA mobility at the central vasculature creates a movement barrier that ensures some small RNA-mediated signaling responses are contained, while others are permitted to propagate systemically. Within meristems, the polarized gating mechanism underpins a domain autonomous behavior. Small RNAs are able to move between stem cells in the niche or within the organizing center but cannot move between these functional domains (Fig. 2C). Likewise, small RNAs are unable to move between stem cells and more determined daughter cells (Skopelitis et al., 2018). The limited movement of miRNAs is thought to help safeguard cell identities within the dynamically growing niche. This is perhaps best illustrated by the action of miR394, which through the repression of LCR promotes stem cell identity (Fig. 2C). The restricted movement of miR394 prevents a shift from organizer to stem cell identity, thus providing a mechanism to maintain stem cell number (Knauer et al., 2013). Likewise, movement of miR166, while essential for the specification of dorsoventral polarity in the incipient primordium, cannot extend into the central zone where its HD-ZIPIII targets are required for stem cell activity (Liu et al., 2009; Zhang, Tucker, Hermann, & Laux, 2017; Zhu et al., 2011).

The molecular components underpinning the regulated movement of small RNAs have proven difficult to identify. Forward genetic screens designed to pinpoint such factors have led to the discovery of numerous small RNA biogenesis components, but failed to uncover genes directly affecting mobility (Brosnan & Voinnet, 2011; Melnyk, Molnar, & Baulcombe, 2011; Taochy et al., 2017). The first insights, which support the idea that the movement of small RNAs is regulated at plasmodesmata, came from a recent biochemical study into the antiviral immune response (Rosas-Diaz et al., 2018). The systemic spread of virus-derived siRNAs is one of the plant's main antiviral defense mechanisms (Burgyán & Havelda, 2011; Melnyk et al., 2011). To combat this defense strategy, viruses evolved various suppressor strategies, one of which targets the plasmodesmata-associated receptor-like kinases, BAM1 and BAM2, to block the movement of siRNA from the vasculature (Rosas-Diaz et al., 2018). It will be interesting to see whether this paradigm holds for miRNA mobility in other developmental contexts, or what other specialized mechanisms exist to regulate small RNA mobility during development. A wide spectrum of proteins, including various receptors-like kinases, is associated with plasmodesmata (Brault et al., 2019; Fernandez-Calvino et al., 2011; Stahl et al., 2013), providing lots of scope to regulate the movement of small RNAs and thereby shape the gradient.

5. Why so complicated?

Why would patterning mechanisms based on the repression of key developmental regulators by miRNA mobility gradients have evolved? Why not simply regulate expression of the respective developmental targets at the transcriptional level? Morphogen gradients, which instruct cells to adopt distinct identities according to their relative position from a fixed source, allow the pattern of cell fate acquisition to be uncoupled from the pattern of cell division. Accordingly, despite variations in the direction or rate of cell division, morphogen gradients enable formation of precise and reliable cell fate boundaries by 'respecifying' the fate of 'mispositioned' cells. The morphogenic behavior of mobile small RNAs thus generates robustness at the tissue or organ level (Skopelitis et al., 2017). In addition, signal gradients are tunable, as illustrated by the fact that relatively subtle changes in miRNA levels at the source can be sufficient to shift the position of a target gene expression boundary (Skopelitis et al., 2017). As a result, morphogen gradients are in principle scalable, such that spatial patterns can be proportionally maintained irrespective of organ size (Ashe & Briscoe, 2006; Briscoe & Small, 2015; Lewis, 2008; Rogers & Schier, 2011). The tunability of a morphogen gradient also allows for plasticity, a point that given their sessile nature may be particularly relevant in plants. Indeed, cell fates within the shoot and root apical meristems are dynamically specified (Gaillochet & Lohmann, 2015). Likewise, the dorsoventral boundary in the leaf, while robustly specified, is flexible in its positioning with the number of dorsal and ventral cell layers dependent on a range of environmental cues (e.g., de Carbonnel et al., 2010; Waites & Hudson, 1995). Subtle changes in small RNA levels at their source permit the flexible positioning of a boundary while maintaining it uniform across the developing organ. Finally, a mobile small RNA gradient can be tuned to create stochasticity (Skopelitis et al., 2017). Stochastic cell fate decisions are often favored in scenarios where variability allows a bet-hedging strategy to better cope with unpredictable environments. An explicit example of this in plants is seen in the regulation of stem cell differentiation in the moss *Physcomitrella patens*. Here, a central source of tasiARF generates a stochastic pattern of ARF expression in stem cells at the plant's edge, creating the variability needed to balance differentiation in response to environmental cues (Plavskin et al., 2016). Thus, the beauty of small RNA gradients, and morphogen gradients in general, is that they provide a means by which to create robust developmental patterns that nonetheless can be flexibly tuned whether as an aspect of environmentally driven phenotypic plasticity or of programmed developmental change.

But why use small RNAs? Small RNAs have properties that make them particularly well suited to drive developmental change. Regulation by small RNAs confers sensitivity and robustness onto gene regulatory networks, in part by dampening intrinsic noise resulting from inherent variability in gene expression (Plavskin et al., 2016; Schmiedel et al., 2015). Both features promote the faithful transfer of information through a signaling network and, consistent with the prevalence of evolutionarily conserved small RNA-target modules in plants as well as animals (Gramzow & Theißen, 2019; Yu et al., 2017), mechanisms underlying these network properties provide a selective advantage during evolution (Frankel et al., 2010; Metzger et al., 2015). Plant small RNAs in addition provide an unprecedented degree of signal specificity, often showing near perfect complementarity to target transcripts, and have a direct mode of action that allows for rapid cell fate transitions (Bartel, 2004; Rhoades et al., 2002). A further conceivable advantage of employing small RNAs as mobile signals in development may be that they represent yet another class of molecules. Patterning processes often occur in close spatial and temporal vicinity, requiring careful coordination between events. Within plant stem cell niches, cells perceive inputs from a multitude of secreted peptides, hormones, mobile transcription factors, as well as mobile small RNAs (see Greb & Lohmann, 2016; Soyars et al., 2016). Thus, perhaps, an additional advantage of employing mobile small RNAs in development is that they broaden the spectrum of available signaling pathways needed to mitigate a 'signaling gridlock.'

6. Concluding remarks

The recent discovery that small RNAs in plants function similar to the classical animal morphogens reveals an intriguing new paradigm in developmental patterning. Morphogenic behavior is an inherent property of small RNAs that relies on their ability to establish a mobility gradient and to trigger a switch in AGO activity dictated by the small RNA-to-target ratio. The elegance of the system lies in the fact that it creates robust on-off cell fate boundaries in a manner that allows for plasticity and that, in principle, is scalable to function across tissues and organs of various sizes. However, the mechanism by which the gradient is established and read out leaves many open questions. How do cells assess a given small RNA-to-target ratio or how is the switch in AGO activity realized? To resolve these questions a deeper understanding of plant AGO complexes is needed. What are their interaction partners and biochemical modifications, and how might these promote cooperative behavior or impact the mode of repression? Similarly, what effects will parameters such as small RNA-target complementarity and sub-cellular compartmentalization have on these processes? Equally important, quantitative insights into the production, turnover, and movement of small RNAs are needed to resolve how stable gradients are formed. It is evident that the movement of small RNAs within the plant stem cell niches is a tightly regulated process, but further experiments are needed to resolve the molecular underpinnings and to unravel the complex interactions that can shape individual small RNA gradients to allow diversity in patterning behaviors. Finally, it remains to be resolved whether animal small RNAs share the capacity to serve as morphogens. The recent observation that layer formation within the neocortex relies on opposing small RNA gradients (Shu et al., 2019), may support this intriguing possibility, especially if we consider that small RNAs in animal systems are sorted into extracellular vesicles and able to move from cell to cell (see Temoche-Diaz et al., 2019). How the distinctive attributes of plant and animal miRNAs will influence the readout of a gradient, only time will tell. For sure, pattern formation by small RNA gradients is a feature common to both kingdoms.

Acknowledgments

The authors thank current and former members of the Timmermans lab for helpful discussions. Work on small RNA regulation and leaf polarity in the Timmermans lab is supported by grants from The Deutsche Forschungsgemeinschaft (SFB 1101 project C06) and an Alexander von Humboldt Professorship. Kristine Hill was supported by a Marie Skodowska-Curie Individual Fellowship (GAP-709293). We apologize to colleagues for work not cited due to space limitations.

References

Allen, E., Xie, Z., Gustafson, A. M., & Carrington, J. C. (2005). microRNA-directed phasing during trans-acting siRNA biogenesis in plants. *Cell*, *121*, 207–221.

Alvarez, J. P., Pekker, I., Goldshmidt, A., Blum, E., Amsellem, Z., & Eshed, Y. (2006). Endogenous and synthetic microRNAs stimulate simultaneous, efficient, and localized regulation of multiple targets in diverse species. *Plant Cell*, *18*, 1134–1151.

Ameres, S. L., Horwich, M. D., Hung, J.-H., Xu, J., Ghildiyal, M., Weng, Z., et al. (2010). Target RNA–directed trimming and tailing of small silencing RNAs. *Science*, *328*, 1534–1539.

Ashe, H. L., & Briscoe, J. (2006). The interpretation of morphogen gradients. *Development*, *133*, 385–394.

Bartel, D. P. (2004). MicroRNAs: Genomics, biogenesis, mechanism, and function. *Cell*, *116*, 281–297.

Benková, E., Ivanchenko, M. G., Friml, J., Shishkova, S., & Dubrovsky, J. G. (2009). A morphogenetic trigger: Is there an emerging concept in plant developmental biology? *Trends in Plant Science*, *14*, 189–193.

Bhalerao, R. P., & Bennett, M. J. (2003). The case for morphogens in plants. *Nature Cell Biology*, *5*, 939.

Blilou, I., Xu, J., Wildwater, M., Willemsen, V., Paponov, I., Friml, J., et al. (2005). The PIN auxin efflux facilitator network controls growth and patterning in Arabidopsis roots. *Nature*, *433*, 39.

Borges, F., & Martienssen, R. A. (2015). The expanding world of small RNAs in plants. *Nature Reviews Molecular Cell Biology*, *16*, 727.

Borges, F., Parent, J.-S., van Ex, F., Wolff, P., Martínez, G., Köhler, C., et al. (2018). Transposon-derived small RNAs triggered by miR845 mediate genome dosage response in Arabidopsis. *Nature Genetics*, *50*, 186.

Brault, M. L., Petit, J. D., Immel, F., Nicolas, W. J., Glavier, M., Brocard, L., et al. (2019). Multiple C2 domains and Transmembrane region proteins (MCTPs) tether membranes at plasmodesmata. *EMBO Reports*, 20.

Briscoe, J., & Small, S. (2015). Morphogen rules: Design principles of gradient-mediated embryo patterning. *Development*, *142*, 3996–4009.

Broderick, J. A., Salomon, W. E., Ryder, S. P., Aronin, N., & Zamore, P. D. (2011). Argonaute protein identity and pairing geometry determine cooperativity in mammalian RNA silencing. *RNA*, *17*, 1858–1869.

Brodersen, P., Sakvarelidze-Achard, L., Bruun-Rasmussen, M., Dunoyer, P., Yamamoto, Y. Y., Sieburth, L., et al. (2008). Widespread translational inhibition by plant miRNAs and siRNAs. *Science*, *320*, 1185–1190.

Brosnan, C. A., & Voinnet, O. (2011). Cell-to-cell and long-distance siRNA movement in plants: Mechanisms and biological implications. *Current Opinion in Plant Biology*, *14*, 580–587.

Buhtz, A., Pieritz, J., Springer, F., & Kehr, J. (2010). Phloem small RNAs, nutrient stress responses, and systemic mobility. *BMC Plant Biology*, *10*, 64.

Buhtz, A., Springer, F., Chappell, L., Baulcombe, D. C., & Kehr, J. (2008). Identification and characterization of small RNAs from the phloem of *Brassica napus*. *The Plant Journal*, *53*, 739–749.

Burch-Smith, T. M., & Zambryski, P. C. (2012). Plasmodesmata paradigm shift: Regulation from without versus within. *Annual Review of Plant Biology*, *63*, 239–260.

Burgyán, J., & Havelda, Z. (2011). Viral suppressors of RNA silencing. *Trends in Plant Science*, *16*, 265–272.

Caggiano, M. P., Yu, X., Bhatia, N., Larsson, A., Ram, H., Ohno, C. K., et al. (2017). Cell type boundaries organize plant development. *eLife*, *6*, e27421.

Carlsbecker, A., Lee, J.-Y., Roberts, C. J., Dettmer, J., Lehesranta, S., Zhou, J., et al. (2010). Cell signalling by microRNA165/6 directs gene dose-dependent root cell fate. *Nature*, *465*, 316–321.

Cartolano, M., Castillo, R., Efremova, N., Kuckenberg, M., Zethof, J., Gerats, T., et al. (2007). A conserved microRNA module exerts homeotic control over *Petunia hybrida* and *Antirrhinum majus* floral organ identity. *Nature Genetics*, *39*, 901–905.

Chitwood, D. H., Nogueira, F. T., Howell, M. D., Montgomery, T. A., Carrington, J. C., & Timmermans, M. C. (2009). Pattern formation via small RNA mobility. *Genes & Development*, *23*, 549–554.

Chitwood, D. H., & Timmermans, M. C. (2010). Small RNAs are on the move. *Nature*, *467*, 415.

Cho, S. K., Ryu, M. Y., Shah, P., Poulsen, C. P., & Yang, S. W. (2016). Post-translational regulation of miRNA pathway components, AGO1 and HYL1, in plants. *Molecules and Cells*, *39*, 581.

Cotterell, J., & Sharpe, J. (2010). An atlas of gene regulatory networks reveals multiple three-gene mechanisms for interpreting morphogen gradients. *Molecular Systems Biology*, *6*.

Cui, H., Levesque, M. P., Vernoux, T., Jung, J. W., Paquette, A. J., Gallagher, K. L., et al. (2007). An evolutionarily conserved mechanism delimiting SHR movement defines a single layer of endodermis in plants. *Science*, *316*, 421–425.

Darwin, C., & Darwin, F. (1880). *The power of movement in plants*. John Murray.

Daum, G., Medzihradszky, A., Suzaki, T., & Lohmann, J. U. (2014). A mechanistic framework for noncell autonomous stem cell induction in Arabidopsis. *Proceedings of the National Academy of Sciences of the United States of America*, *111*, 14619–14624.

de Carbonnel, M., Davis, P., Roelfsema, M. R. G., Inoue, S.-i., Schepens, I., Lariguet, P., et al. (2010). The Arabidopsis PHYTOCHROME KINASE SUBSTRATE2 protein is a phototropin signaling element that regulates leaf flattening and leaf positioning. *Plant Physiology*, *152*, 1391–1405.

Denzler, R., McGeary, S. E., Title, A. C., Agarwal, V., Bartel, D. P., & Stoffel, M. (2016). Impact of microRNA levels, target-site complementarity, and cooperativity on competing endogenous RNA-regulated gene expression. *Molecular Cell*, *64*, 565–579.

Djuranovic, S., Zinchenko, M. K., Hur, J. K., Nahvi, A., Brunelle, J. L., Rogers, E. J., et al. (2010). Allosteric regulation of Argonaute proteins by miRNAs. *Nature Structural & Molecular Biology*, *17*, 144.

Emery, J. F., Floyd, S. K., Alvarez, J., Eshed, Y., Hawker, N. P., Izhaki, A., et al. (2003). Radial patterning of Arabidopsis shoots by class III HD-ZIP and KANADI genes. *Current Biology*, *13*, 1768–1774.

Fahlgren, N., Montgomery, T. A., Howell, M. D., Allen, E., Dvorak, S. K., Alexander, A. L., et al. (2006). Regulation of AUXIN RESPONSE FACTOR3 by TAS3 ta-siRNA affects developmental timing and patterning in Arabidopsis. *Current Biology*, *16*, 939–944.

Fankhauser, C., & Christie, J. M. (2015). Plant phototropic growth. *Current Biology*, *25*, R384–R389.

Felippes, F. F. d., Ott, F., & Weigel, D. (2010). Comparative analysis of non-autonomous effects of tasiRNAs and miRNAs in Arabidopsis thaliana. *Nucleic Acids Research*, *39*, 2880–2889.

Fernandez-Calvino, L., Faulkner, C., Walshaw, J., Saalbach, G., Bayer, E., Benitez-Alfonso, Y., et al. (2011). Arabidopsis plasmodesmal proteome. *PLoS One*, *6*, e18880.

Finet, C., & Jaillais, Y. (2012). Auxology: When auxin meets plant evo-devo. *Developmental Biology*, *369*, 19–31.

Frankel, N., Davis, G. K., Vargas, D., Wang, S., Payre, F., & Stern, D. L. (2010). Phenotypic robustness conferred by apparently redundant transcriptional enhancers. *Nature*, *466*, 490.

Friml, J. (2003). Auxin transport—Shaping the plant. *Current Opininion in Plant Biology*, *6*, 7–12.

Gaillochet, C., & Lohmann, J. U. (2015). The never-ending story: From pluripotency to plant developmental plasticity. *Development*, *142*, 2237–2249.

Gallagher, K. L., Paquette, A. J., Nakajima, K., & Benfey, P. N. (2004). Mechanisms regulating SHORT-ROOT intercellular movement. *Current Biology*, *14*, 1847–1851.

Golden, R. J., Chen, B., Li, T., Braun, J., Manjunath, H., Chen, X., et al. (2017). An Argonaute phosphorylation cycle promotes microRNA-mediated silencing. *Nature*, *542*, 197.

Gramzow, L., & Theißen, G. (2019). Plant miRNA conservation and evolution. In *Plant microRNAs* (pp. 41–50). Springer.

Greb, T., & Lohmann, J. U. (2016). Plant stem cells. *Current Biology*, *26*, R816–R821.

Gregor, T., Tank, D. W., Wieschaus, E. F., & Bialek, W. (2007). Probing the limits to positional information. *Cell*, *130*, 153–164.

Grieneisen, V. A., Scheres, B., Hogeweg, P., & Marée, A. F. (2012). Morphogengineering roots: Comparing mechanisms of morphogen gradient formation. *BMC Systems Biology*, *6*, 37.

Grieneisen, V. A., Xu, J., Marée, A. F., Hogeweg, P., & Scheres, B. (2007). Auxin transport is sufficient to generate a maximum and gradient guiding root growth. *Nature*, *449*, 1008.

Guo, Y. E., Manteiga, J. C., Henninger, J. E., Sabari, B. R., Dall'Agnese, A., Hannett, N. M., et al. (2019). Pol II phosphorylation regulates a switch between transcriptional and splicing condensates. *Nature*, 1–6.

Horman, S. R., Janas, M. M., Litterst, C., Wang, B., MacRae, I. J., Sever, M. J., et al. (2013). Akt-mediated phosphorylation of argonaute 2 downregulates cleavage and upregulates translational repression of MicroRNA targets. *Molecular Cell*, *50*, 356–367.

Huberdeau, M. Q., Zeitler, D. M., Hauptmann, J., Bruckmann, A., Fressigné, L., Danner, J., et al. (2017). Phosphorylation of Argonaute proteins affects mRNA binding and is essential for microRNA-guided gene silencing in vivo. *The EMBO Journal*, *36*, 2088–2106.

Husbands, A. Y., Benkovics, A. H., Nogueira, F. T., Lodha, M., & Timmermans, M. C. (2015). The ASYMMETRIC LEAVES complex employs multiple modes of regulation to affect adaxial-abaxial patterning and leaf complexity. *The Plant Cell*, *27*, 3321–3335.

Jee, D., & Lai, E. C. (2014). Alteration of miRNA activity via context-specific modifications of Argonaute proteins. *Trends in Cell Biology*, *24*, 546–553.

Juarez, M. T., Kui, J. S., Thomas, J., Heller, B. A., & Timmermans, M. C. (2004). microRNA-mediated repression of *rolled leaf1* specifies maize leaf polarity. *Nature*, *428*, 84.

Klein, M., Chandradoss, S. D., Depken, M., & Joo, C. (2017). Why Argonaute is needed to make microRNA target search fast and reliable. In *Seminars in cell & developmental biology* (pp. 20–28): Elsevier.

Knauer, S., Holt, A. L., Rubio-Somoza, I., Tucker, E. J., Hinze, A., Pisch, M., et al. (2013). A protodermal miR394 signal defines a region of stem cell competence in the Arabidopsis shoot meristem. *Developmental Cell*, *24*, 125–132.

Kuhlemeier, C., & Timmermans, M. C. (2016). The Sussex signal: Insights into leaf dorsiventrality. *Development*, *143*, 3230–3237.

Levine, E., McHale, P., & Levine, H. (2007). Small regulatory RNAs may sharpen spatial expression patterns. *PLoS Computational Biology*, *3*, e233.

Lewis, J. (2008). From signals to patterns: Space, time, and mathematics in developmental biology. *Science*, *322*, 399–403.

Li, S., Liu, L., Zhuang, X., Yu, Y., Liu, X., Cui, X., et al. (2013). MicroRNAs inhibit the translation of target mRNAs on the endoplasmic reticulum in Arabidopsis. *Cell*, *153*, 562–574.

Lin, S.-I., Chiang, S.-F., Lin, W.-Y., Chen, J.-W., Tseng, C.-Y., Wu, P.-C., et al. (2008). Regulatory network of microRNA399 and PHO2 by systemic signaling. *Plant Physiology*, *147*, 732–746.

Liu, Q., Yao, X., Pi, L., Wang, H., Cui, X., & Huang, H. (2009). The ARGONAUTE10 gene modulates shoot apical meristem maintenance and establishment of leaf polarity by repressing miR165/166 in Arabidopsis. *The Plant Journal*, *58*, 27–40.

Lopez-Orozco, J., Pare, J. M., Holme, A. L., Chaulk, S. G., Fahlman, R. P., & Hobman, T. C. (2015). Functional analyses of phosphorylation events in human Argonaute 2. *RNA*, *21*, 2030–2038.

Lynn, K., Fernandez, A., Aida, M., Sedbrook, J., Tasaka, M., Masson, P., et al. (1999). The PINHEAD/ZWILLE gene acts pleiotropically in Arabidopsis development and has overlapping functions with the ARGONAUTE1 gene. *Development*, *126*, 469–481.

Mallory, A. C., Reinhart, B. J., Jones-Rhoades, M. W., Tang, G., Zamore, P. D., Barton, M. K., et al. (2004). MicroRNA control of PHABULOSA in leaf development: Importance of pairing to the microRNA 5′ region. *The EMBO Journal, 23*, 3356–3364.

Manavella, P. A., Koenig, D., & Weigel, D. (2012). Plant secondary siRNA production determined by microRNA-duplex structure. *Proceedings of the National Academy of Sciences of the United States of America, 109*, 2461–2466.

McConnell, J. R., Emery, J., Eshed, Y., Bao, N., Bowman, J., & Barton, M. K. (2001). Role of PHABULOSA and PHAVOLUTA in determining radial patterning in shoots. *Nature, 411*, 709–713.

Melnyk, C. W., Molnar, A., & Baulcombe, D. C. (2011). Intercellular and systemic movement of RNA silencing signals. *The EMBO Journal, 30*, 3553–3563.

Metzger, B. P., Yuan, D. C., Gruber, J. D., Duveau, F., & Wittkopp, P. J. (2015). Selection on noise constrains variation in a eukaryotic promoter. *Nature, 521*, 344.

Merelo, P., Ram, H., Caggiano, M. P., Ohno, C., Ott, F., Straub, D., et al. (2016). Regulation of MIR165/166 by class II and class III homeodomain leucine zipper proteins establishes leaf polarity. *Proceedings of the National Academy of Sciences of the United States of America, 113*, 11973–11978.

Mi, S., Cai, T., Hu, Y., Chen, Y., Hodges, E., Ni, F., et al. (2008). Sorting of small RNAs into Arabidopsis argonaute complexes is directed by the 5′ terminal nucleotide. *Cell, 133*, 116–127.

Miyashima, S., Koi, S., Hashimoto, T., & Nakajima, K. (2011). Non-cell-autonomous microRNA165 acts in a dose-dependent manner to regulate multiple differentiation status in the Arabidopsis root. *Development, 138*, 2303–2313.

Molnar, A., Melnyk, C. W., Bassett, A., Hardcastle, T. J., Dunn, R., & Baulcombe, D. C. (2010). Small silencing RNAs in plants are mobile and direct epigenetic modification in recipient cells. *Science, 328*, 872–875.

Montgomery, T. A., Howell, M. D., Cuperus, J. T., Li, D., Hansen, J. E., Alexander, A. L., et al. (2008). Specificity of ARGONAUTE7-miR390 interaction and dual functionality in *TAS3* trans-acting siRNA formation. *Cell, 133*, 128–141.

Mukherji, S., Ebert, M. S., Zheng, G. X., Tsang, J. S., Sharp, P. A., & van Oudenaarden, A. (2011). MicroRNAs can generate thresholds in target gene expression. *Nature Genetics, 43*, 854.

Nagasaki, H., Itoh, J.-i., Hayashi, K., Hibara, K.-i., Satoh-Nagasawa, N., Nosaka, M., et al. (2007). The small interfering RNA production pathway is required for shoot meristem initiation in rice. *Proceedings of the National Academy of Sciences of the United States of America, 104*, 14867–14871.

Nakajima, K., Sena, G., Nawy, T., & Benfey, P. N. (2001). Intercellular movement of the putative transcription factor SHR in root patterning. *Nature, 413*, 307–311.

Nikovics, K., Blein, T., Peaucelle, A., Ishida, T., Morin, H., Aida, M., et al. (2006). The balance between the *MIR164A* and *CUC2* genes controls leaf margin serration in Arabidopsis. *The Plant Cell, 18*, 2929–2945.

Nogueira, F. T., Chitwood, D. H., Madi, S., Ohtsu, K., Schnable, P. S., Scanlon, M. J., et al. (2009). Regulation of small RNA accumulation in the maize shoot apex. *PLoS Genetics, 5*, e1000320.

Nogueira, F. T., Madi, S., Chitwood, D. H., Juarez, M. T., & Timmermans, M. C. (2007). Two small regulatory RNAs establish opposing fates of a developmental axis. *Genes & Development, 21*, 750–755.

Olmedo-Monfil, V., Durán-Figueroa, N., Arteaga-Vázquez, M., Demesa-Arévalo, E., Autran, D., Grimanelli, D., et al. (2010). Control of female gamete formation by a small RNA pathway in Arabidopsis. *Nature, 464*, 628.

Palauqui, J. C., Elmayan, T., Pollien, J. M., & Vaucheret, H. (1997). Systemic acquired silencing: Transgene-specific post-transcriptional silencing is transmitted by grafting from silenced stocks to non-silenced scions. *The EMBO Journal*, *16*, 4738–4745.

Pant, B. D., Buhtz, A., Kehr, J., & Scheible, W. R. (2008). MicroRNA399 is a long-distance signal for the regulation of plant phosphate homeostasis. *The Plant Journal*, *53*, 731–738.

Parizotto, E. A., Dunoyer, P., Rahm, N., Himber, C., & Voinnet, O. (2004). In vivo investigation of the transcription, processing, endonucleolytic activity, and functional relevance of the spatial distribution of a plant miRNA. *Genes & Development*, *18*, 2237–2242.

Pekker, I., Alvarez, J. P., & Eshed, Y. (2005). Auxin response factors mediate Arabidopsis organ asymmetry via modulation of KANADI activity. *The Plant Cell*, *17*, 2899–2910.

Petsch, K., Manzotti, P. S., Tam, O. H., Meeley, R., Hammell, M., Consonni, G., et al. (2015). Novel DICER-LIKE1 siRNAs bypass the requirement for DICER-LIKE4 in maize development. *The Plant Cell*, *27*, 2163–2177.

Plavskin, Y., Nagashima, A., Perroud, P.-F., Hasebe, M., Quatrano, R. S., Atwal, G. S., et al. (2016). Ancient trans-acting siRNAs confer robustness and sensitivity onto the auxin response. *Developmental Cell*, *36*, 276–289.

Ramachandran, V., & Chen, X. (2008). Degradation of microRNAs by a family of exoribonucleases in Arabidopsis. *Science*, *321*, 1490–1492.

Reinhardt, D., Frenz, M., Mandel, T., & Kuhlemeier, C. (2005). Microsurgical and laser ablation analysis of leaf positioning and dorsoventral patterning in tomato. *Development*, *132*, 15–26.

Ren, G., Chen, X., & Yu, B. (2012). Uridylation of miRNAs by *hen1 suppressor1* in Arabidopsis. *Current Biology*, *22*, 695–700.

Rhoades, M. W., Reinhart, B. J., Lim, L. P., Burge, C. B., Bartel, B., & Bartel, D. P. (2002). Prediction of plant microRNA targets. *Cell*, *110*, 513–520.

Rogers, K. W., & Schier, A. F. (2011). Morphogen gradients: From generation to interpretation. *Annual Review of Cell and Developmental Biology*, *27*, 377–407.

Rosas-Diaz, T., Zhang, D., Fan, P., Wang, L., Ding, X., Jiang, Y., et al. (2018). A virus-targeted plant receptor-like kinase promotes cell-to-cell spread of RNAi. *Proceedings of the National Academy of Sciences of the United States of America*, *115*, 1388–1393.

Schmiedel, J. M., Klemm, S. L., Zheng, Y., Sahay, A., Blüthgen, N., Marks, D. S., et al. (2015). MicroRNA control of protein expression noise. *Science*, *348*, 128–132.

Shen, J., Xia, W., Khotskaya, Y. B., Huo, L., Nakanishi, K., Lim, S.-O., et al. (2013). EGFR modulates microRNA maturation in response to hypoxia through phosphorylation of AGO2. *Nature*, *497*, 383.

Sheng, G., Zhao, H., Wang, J., Rao, Y., Tian, W., Swarts, D. C., et al. (2014). Structure-based cleavage mechanism of *Thermus thermophilus* Argonaute DNA guide strand-mediated DNA target cleavage. *Proceedings of the National Academy of Sciences of the United States of America*, *111*, 652–657.

Sheu-Gruttadauria, J., & MacRae, I. J. (2018). Phase transitions in the assembly and function of human miRISC. *Cell*, *173*, 946–957, e916.

Shu, P., Wu, C., Ruan, X., Liu, W., Hou, L., Fu, H., et al. (2019). Opposing gradients of microRNA expression temporally pattern layer formation in the developing neocortex. *Developmental Cell*, *49*, 764–785, e764.

Sieber, P., Wellmer, F., Gheyselinck, J., Riechmann, J. L., & Meyerowitz, E. M. (2007). Redundancy and specialization among plant microRNAs: Role of the *MIR164* family in developmental robustness. *Development*, *134*, 1051–1060.

Skopelitis, D. S., Benkovics, A. H., Husbands, A. Y., & Timmermans, M. C. (2017). Boundary formation through a direct threshold-based readout of mobile small RNA gradients. *Developmental Cell*, *43*, 265–273, e266.

Skopelitis, D. S., Hill, K., Klesen, S., Marco, C. F., von Born, P., Chitwood, D. H., et al. (2018). Gating of miRNA movement at defined cell-cell interfaces governs their impact as positional signals. *Nature Communications*, *9*, 3107.

Slotkin, R. K., Vaughn, M., Borges, F., Tanurdžić, M., Becker, J. D., Feijó, J. A., et al. (2009). Epigenetic reprogramming and small RNA silencing of transposable elements in pollen. *Cell*, *136*, 461–472.

Smith, R. S., Guyomarc'h, S., Mandel, T., Reinhardt, D., Kuhlemeier, C., & Prusinkiewicz, P. (2006). A plausible model of phyllotaxis. *Proceedings of the National Academy of Sciences of the United States of America*, *103*, 1301–1306.

Soyars, C. L., James, S. R., & Nimchuk, Z. L. (2016). Ready, aim, shoot: Stem cell regulation of the shoot apical meristem. *Current Opinion in Plant Biology*, *29*, 163–168.

Stahl, Y., Grabowski, S., Bleckmann, A., Kühnemuth, R., Weidtkamp-Peters, S., Pinto, K. G., et al. (2013). Moderation of Arabidopsis root stemness by CLAVATA1 and ARABIDOPSIS CRINKLY4 receptor kinase complexes. *Current Biology*, *23*, 362–371.

Steeves, T. A., & Sussex, I. M. (1989). *Patterns in plant development*. Cambridge University Press.

Su, Z., Zhao, L., Zhao, Y., Li, S., Won, S., Cai, H., et al. (2017). The THO complex non-cell-autonomously represses female germline specification through the TAS3-ARF3 module. *Current Biology*, *27*, 1597–1609, e1592.

Taochy, C., Gursanscky, N. R., Cao, J., Fletcher, S. J., Dressel, U., Mitter, N., et al. (2017). A genetic screen for impaired systemic RNAi highlights the crucial role of DICER-LIKE 2. *Plant Physiology*, *175*, 1424–1437.

Temoche-Diaz, M. M., Shurtleff, M. J., Nottingham, R. M., Yao, J., Fadadu, R. P., Lambowitz, A. M., et al. (2019). Distinct mechanisms of microRNA sorting into cancer cell-derived extracellular vesicle subtypes. *eLife*, *8*, e47544.

Tilsner, J., Nicolas, W., Rosado, A., & Bayer, E. M. (2016). Staying tight: Plasmodesmal membrane contact sites and the control of cell-to-cell connectivity in plants. *Annual Review of Plant Biology*, *67*, 337–364.

Tsikou, D., Yan, Z., Holt, D. B., Abel, N. B., Reid, D. E., Madsen, L. H., et al. (2018). Systemic control of legume susceptibility to rhizobial infection by a mobile microRNA. *Science*, *362*, 233–236.

Tucker, M. R., Okada, T., Hu, Y., Scholefield, A., Taylor, J. M., & Koltunow, A. M. (2012). Somatic small RNA pathways promote the mitotic events of megagametogenesis during female reproductive development in Arabidopsis. *Development*, *139*, 1399–1404.

Turing, A. M. (1952). The chemical basis of morphogenesis. *Philosophical Transactions of the Royal Society London B*, *237*, 37–72.

Vatén, A., Dettmer, J., Wu, S., Stierhof, Y.-D., Miyashima, S., Yadav, S. R., et al. (2011). Callose biosynthesis regulates symplastic trafficking during root development. *Developmental Cell*, *21*, 1144–1155.

Vaucheret, H., Mallory, A. C., & Bartel, D. P. (2006). AGO1 homeostasis entails coexpression of *MIR168* and *AGO1* and preferential stabilization of miR168 by AGO1. *Molecular Cell*, *22*, 129–136.

Von Born, P., Bernardo-Faura, M., & Rubio-Somoza, I. (2018). An artificial miRNA system reveals that relative contribution of translational inhibition to miRNA-mediated regulation depends on environmental and developmental factors in Arabidopsis thaliana. *PLoS One*, *13*, e0192984.

Waites, R., & Hudson, A. (1995). *phantastica*: A gene required for dorsoventrality of leaves in *Antirrhinum majus*. *Development*, *121*, 2143–2154.

Wiesner, J. (1878). *Die heliotropischen Erscheinungen im Pflanzenreich*. Vienna, Austria: Kaiserlich-Koniglichen Hof-und Staatsdruckerei.

Willmann, M. R., & Poethig, R. S. (2007). Conservation and evolution of miRNA regulatory programs in plant development. *Current Opinion in Plant Biology, 10*, 503–511.

Wolpert, L. (1969). Positional information and the spatial pattern of cellular differentiation. *Journal of Theoretical Biology, 25*, 1–47.

Wu, S., & Gallagher, K. L. (2012). Transcription factors on the move. *Current Opinion in Plant Biology, 15*, 645–651.

Yadav, R. K., Perales, M., Gruel, J., Girke, T., Jönsson, H., & Reddy, G. V. (2011). WUSCHEL protein movement mediates stem cell homeostasis in the Arabidopsis shoot apex. *Genes & Development, 25*, 2025–2030.

Yao, X., Wang, H., Li, H., Yuan, Z., Li, F., Yang, L., et al. (2009). Two types of cis-acting elements control the abaxial epidermis-specific transcription of the *MIR165a* and *MIR166a* genes. *FEBS Letters, 583*, 3711–3717.

Yifhar, T., Pekker, I., Peled, D., Friedlander, G., Pistunov, A., Sabban, M., et al. (2012). Failure of the tomato trans-acting short interfering RNA program to regulate AUXIN RESPONSE FACTOR3 and ARF4 underlies the wiry leaf syndrome. *Plant Cell, 24*, 3575–3589.

Yoo, B.-C., Kragler, F., Varkonyi-Gasic, E., Haywood, V., Archer-Evans, S., Lee, Y. M., et al. (2004). A systemic small RNA signaling system in plants. *Plant Cell, 16*, 1979–2000.

Yu, Y., Jia, T., & Chen, X. (2017). The 'how'and 'where'of plant micro RNAs. *New Phytologist, 216*, 1002–1017.

Zavaliev, R., Sagi, G., Gera, A., & Epel, B. L. (2009). The constitutive expression of Arabidopsis plasmodesmal-associated class 1 reversibly glycosylated polypeptide impairs plant development and virus spread. *Journal of Experimental Botany, 61*, 131–142.

Zhai, J., Zhao, Y., Simon, S. A., Huang, S., Petsch, K., Arikit, S., et al. (2013). Plant microRNAs display differential 3′ truncation and tailing modifications that are ARGONAUTE1 dependent and conserved across species. *Plant Cell, 25*, 2417–2428.

Zhang, Z., Tucker, E., Hermann, M., & Laux, T. (2017). A molecular framework for the embryonic initiation of shoot meristem stem cells. *Developmental Cell, 40*, 264–277, e264.

Zhang, Z., & Zhang, X. (2012). Argonautes compete for miR165/166 to regulate shoot apical meristem development. *Current Opinion in Plant Biology, 15*, 652–658.

Zhao, Y., Yu, Y., Zhai, J., Ramachandran, V., Dinh, T. T., Meyers, B. C., et al. (2012). The Arabidopsis nucleotidyl transferase HESO1 uridylates unmethylated small RNAs to trigger their degradation. *Current Biology, 22*, 689–694.

Zhu, H., Hu, F., Wang, R., Zhou, X., Sze, S.-H., Liou, L. W., et al. (2011). Arabidopsis Argonaute10 specifically sequesters miR166/165 to regulate shoot apical meristem development. *Cell, 145*, 242–256.

Further reading

Husbands, A. Y., Chitwood, D. H., Plavskin, Y., & Timmermans, M. C. (2009). Signals and prepatterns: New insights into organ polarity in plants. *Genes & Development, 23*, 1986–1997.

Liu, Q., Wang, F., & Axtell, M. J. (2014). Analysis of complementarity requirements for plant microRNA targeting using a Nicotiana benthamiana quantitative transient assay. *Plant Cell, 26*, 741–753.

Martin, H. C., Wani, S., Steptoe, A. L., Krishnan, K., Nones, K., Nourbakhsh, E., et al. (2014). Imperfect centered miRNA binding sites are common and can mediate repression of target mRNAs. *Genome Biology, 15*, R51.

Kolibri 360 GmbH
Ferdinand-Jühlke-Straße 7
99095 Erfurt, Deutschland
produktsicherheit@kolibri360.de